D1552822

LEWIS ACIDS and SELECTIVITY in ORGANIC SYNTHESIS

NEW DIRECTIONS in ORGANIC and BIOLOGICAL CHEMISTRY

Series Editor: C.W. Rees, CBE, FRS
Imperial College of Science, Technology and Medicine, London, UK

Published and Forthcoming Titles

Chirality and the Biological Activity of Drugs
Roger J. Crossley

Enzyme-Assisted Organic Synthesis
Manfred Schneider and Stefano Servi

***C*-Glycoside Synthesis**
Maarten Postema

Organozinc Reagents in Organic Synthesis
Ender Erdik

Activated Metals in Organic Synthesis
Pedro Cintas

Capillary Electrophoresis: Theory and Practice
Patrick Camilleri

Cyclization Reactions
C. Thebtaranonth and Y. Thebtaranonth

Mannich Bases: Chemistry and Uses
Maurilio Tramontini and Luigi Angiolini

Vicarious Nucleophilic Substitution and Related Processes in Organic Synthesis
Mieczyslaw Makosza

Aromatic Fluorination
James H. Clark and Tony W. Bastock

Lewis Acids and Selectivity in Organic Synthesis
M. Santelli and J.-M. Pons

Dianion Chemistry in Organic Synthesis
Charles M. Thompson

Asymmetric Synthetic Methodology
David J. Ager and Michael B. East

Synthesis Using Vilsmeier Reagents
C. M. Marson and P. R. Giles

The Anomeric Effect
Eusebio Juaristi

Chiral Sulfur Reagents
M. Mikołajczyk, J. Drabowicz, and P. Kiełbasiński

Chemical Approaches to the Systhesis of Peptides and Proteins
Paul Lloyd-Williams, Fernanado Albericio, and Ernest Giralt

Concerted Organic Mechanisms
Andrew Williams

LEWIS ACIDS and SELECTIVITY in ORGANIC SYNTHESIS

Maurice Santelli
Jean-Marc Pons

CRC Press
Boca Raton New York London Tokyo

Library of Congress Cataloging-in-Publication Data

Santelli, Maurice
Lewis acids and selectivity in organic synthesis / Maurice Santelli, Jean-Marc Pons.
p. cm. -- (New directions in organic and biological chemistry)
Includes bibliographical references and index.
ISBN 0-8493-7866-4 (alk. paper)
1. Organic compounds--Synthesis. 2. Lewis acids. I. Pons, Jean-Marc. II. Title.
III. Series.
QD262.S244 1995
547.2--dc20

95-35887
CIP

International Standard Book Number 0-8493-7866-4
Library of Congress Card Number 95-35887
Printed in the United States of America 1 2 3 4 5 6 7 8 9 0
Printed on acid-free paper

à Françoise et Florence,

Però ch'amore no si pò vedere
E no si tratta corporalemente
Manti ne son di sì folle sapere
Che credono ch'amor sïa nïente

Pier dela Vigna (1180–1249)

M. S.

à Françoise, Thomas et Vincent

J.-M. P.

Foreword

In his landmark treatise published in 1923, G.N. Lewis laid the foundation for our modern understanding of the valence bond. In formulating the concept of bonding by the sharing of electron pairs, Lewis provided the following definition of donors and acceptors: "...with complete generality we may say that a basic substance is one which has a lone pair of electrons which may be used to complete the stable group of another atom, and that an acidic substance is one which can employ a lone pair from another molecule in completing the stable group of one of its own atoms." (G.N. Lewis, *Valence and the Structure of Atoms and Molecules*; the Chemical Catalog Co., Inc.; New York, 1923) Although this definition in principle encompasses all phenomena, including those usually identified with Bronsted and Lowry, the appellation of electron pair donors and acceptors as *Lewis* bases and *Lewis* acids is firmly and fittingly ingrained in the language of chemistry.

Within the microcosm of organic chemistry, the process of complexation is most commonly associated with the interaction of (electron rich) heteroatomic functional groups and main group, transition metal or lanthanide-based Lewis acids. The role of Lewis acids in the Friedel Crafts alkylation and acylation reactions is one of the earliest manifestations. However, because of its central role in synthetic organic chemistry the subset of carbonyl containing functional groups has received the greatest attention. Indeed, it is in this arena, namely the reactions of carbonyl compounds (or their equivalents) in the presence of Lewis acids, that some of the most spectacular advances in organic synthesis methodology have taken place over the past 30 years.

The primary consequence of complexation of a carbonyl group is the dramatic enhancement in electrophilic reactivity of the carbonyl carbon. This phenomenon has allowed the development of mild nucleophilic reagents (e.g., organosilanes, enoxysilanes and the related stannanes) for carbon-carbon bond formation. Further, complexation of α,β-unsaturated carbonyl compounds by Lewis acids leads to enormous acceleration in Diels-Alder, ene and Michael (Sakurai) reactions at the pendant double bond. Finally, Lewis acid complexation of acetals endangers unique reactivity towards nucleophiles which these otherwise protective groups are designed to mask.

The recognition of the pervasive role of Lewis acids in promoting and controlling reactivity of carbonyl compounds has also stimulated a wide range of fundamental structural studies. Originally the realm only of X-ray crystallography and calorimetry, modern spectroscopic methods — IR, multinulear variable temperature NMR — and computational methods have been brought to bear on elucidating critical issues of bonding, conformation and reactivity of Lewis acid carbonyl complexes.

Finally, one of the most exciting and continuously developing aspects of the chemistry of Lewis acids is their potential as chiral catalysts for asymmetric carbon-carbon bond forming reactions. Since the first reports on the influence of chirally modified Lewis acids on the stereochemical course of the Diels-Alder reaction, a major enterprise of chiral Lewis acid catalysis has risen to prominence as the ultimate goal of many fundamental catalyzed processes. Indeed, the advances already on record in Diels Alder, ene, directed Aldol and alkylation reactions are nothing short of spectacular. In combination with powerful spectroscopic and computational methods, the design of new, efficient and selective catalysis from all corners of the periodic table continues to provide intellectual and practical challenges.

Any serious attempt to comprehensively survey the edifice of information on Lewis acid promoted reactions is a daunting task and Professors Santelli and Pons have met this challenge admirably. While the sheer volume of information has necessitated the judicious selection of coverage, the most important reactions; ene, allylation of aldehydes and acetals, Sakurai and Diels Alder reactions are all covered. Each chapter follows a logical and through progression from historical development through current mechanistic thinking, then on to synthetic applications organized from the most simple to the most complex incarnations with asymmetric catalysis serving as the crowning section.

This first-of-its-kind treatment will long be noted as the definitive resource in an important subdomain of organic synthesis and will serve advanced graduate students and active researchers for many years to come.

Scott E. Denmark
Professor of Chemistry

University of Illinois, Urbana–Champaign
Urbana, Illinois, 1995

Preface

The quest for selectivity in synthesis, and particularly stereoselectivity, is a common concern to most organic chemists all over the world. This is illustrated by an ever increasing number of publications and recently, several books on the topic.[1]

We thought however that Lewis acids deserved a specific treatment, with respect to the "sea-change" their use in organic synthesis has suffered over the last decade. So much ground has indeed been covered since the 31st of December 1808 when Gay-Lussac reported the discovery of fluoroboric gas[2] and the preparation of the first Lewis acid-Lewis base complex. He indeed established that "1 volume of fluoroboric gas combines with 1 volume of ammoniacal gas to yield a neutral salt" (just in the same way that "1 volume of oxygene and 2 volumes of hydrogene lead to the formation of water").[3] Later the discovery of aluminum chloride properties by Friedel and Crafts,[4] the rationalization of their properties by Lewis, and the demonstration of their dramatic effect in Diels-Alder reaction by Yates and Eaton set the foundations for the spectacular asymmetric induction promoted in a number of reactions by real catalytic (1–10 mol%) amounts of these reagents. Although we originally intended to provide the reader with the most possibly complete coverage of Lewis acid promoted C–C bond formation methods, the amount of information and our desire to go into as much detail as possible, particularly by using a fairly large number of schemes (over a thousand) has compelled us to concentrate on only six chapters, leaving aside important reactions such as the aldol condensation or, the [2+2] and [2+3] cycloadditions: (1) Lewis acid-carbonyl complexes, (2) Ene reaction, (3) Allylsilanes and allylstannanes addition to aldehydes, (4) Nucleophilic substitution on acetals, (5) Sakurai reaction, and (6) Diels-Alder reaction. In every chapter, after giving a short historical background and recent information on the mechanism of the reaction, we concentrated on highly stereoselective examples with specific highlight on enantioselectivity induced by chiral Lewis acids; the 1220 references include some up to the end of 1993.

---

1. *Catalytic Asymmetric Synthesis*, Ojima, I., Ed., VCH Publishers, Weinheim, **1993**. Nogradi, M., *Stereoselective Synthesis*, VCH Publishers, Weinheim, **1995**. Seyden-Penne, J., *Chiral Auxilliaries and Ligands in Asymmetric Synthesis*, Wiley, New York, **1995**. Kagan, H., *Asymmetric Synthesis*, Thieme, Stuttgart, **1995**.
2. Gay-Lussac, L.J.; Thénard, L.J. *Ann. Chim. Phys.* **1809**, *69*, 204.
3. Gay-Lussac, L.J. *Mém. Phys. Chim. Soc. Arcueil* **1809**, *2*, 207–234.
4. Friedel, C.; Crafts, J.-M. *C. R. Acad. Sc. (Paris)* **1884**, 449–532.

We are indebted to Mrs. Colette Nouguier for typing the manuscript, Mr. Philippe Escoffier for his assistance in collecting references, and CRC Press people, particularly Mr. Navin Sullivan, for their kindness and patience in waiting for the typescript. We would also like to thank Mr. Neil Carpenter and Mr. Jean Pons for helping us to provide a readable English manuscript and, lastly, acknowledge Prof. Scott E. Denmark, a leading authority in the field, for contributing a foreword.

Maurice Santelli
Jean-Marc Pons

Marseille, May 1995

About the Authors

Maurice Santelli was born in Marseille in 1939. He received his PhD in chemistry working with Prof. M. Bertrand (homoallenylic participation, non-classical ion). He had a postdoctoral position at the University of Cambridge (U.K.) in 1973 (Prof. R.A. Raphael). After an appointment at the University of Oran (Algeria) (1975–77), he is presently Professor of Chemistry at the University of Aix-Marseille. His main research areas are physical organic chemistry, low valent transition metal chemistry, electrophilic activation, and the synthesis of bioactive products.

Jean-Marc Pons was born in Aix-en-Provence in 1958. He studied chemistry in Marseille where he completed his PhD on the reactivity of low valent titanium reagents towards enones with Prof. M. Santelli. He entered the CNRS in 1982 and worked in that area until 1988. At that time, he joined Prof. P.J. Kocienski's group (Southampton, U.K.), as a postdoctoral fellow, where he got involved in natural products synthesis. He returned to Marseille in 1989 and is interested in β-lactone and silylketene chemistry, Lewis acid-promoted [2+2] cycloaddition, low valent transition metal chemistry and the semi-empirical modelization of cycloaddition reactions.

Contents

Chapter 1. Lewis Acids and Their Complexes With Lewis Bases

I. Introduction and History 1
II. Physical Properties of Lewis Acids 1
III. Lewis Acid-Carbonyl Compound Complexes 2
III.1 Theoretical studies 4
III.1.a Carbonyl compound-cation complexes 4
III.1.b Carbonyl compound-neutral Lewis acid complexes 5
III.1.c α,β-Unsaturated carbonyl compound-Lewis acid complexes 7
III.2 NMR studies 8
III.2.a Stereochemistry 9
III.2.b Conformational studies 9
III.2.c Chelation 12
III.3 X-ray studies 14
IV. Lewis Acid Complexes With Other Lewis Bases 15
V. Conclusion 16
References 16

Chapter 2. Lewis Acid-Promoted Ene Reaction

I. Introduction and History 21
I.1 The thermal reaction 21
I.2 The Lewis acid-promoted reaction 22
II. Mechanism 23
II.1 The thermal reaction 23
II.2 The Lewis acid-promoted reaction 26
III. Intermolecular Reaction 31
III.1 Reaction involving achiral aldehydes as enophiles 31
III.1.a Chloral 31
III.1.b Formaldehyde 37
III.1.c Glyoxylate esters 40
III.1.d Other achiral aldehydes 46
III.2 Reaction involving chiral aldehydes as enophiles 50
III.3 Reaction with various enophiles 56
IV. Intramolecular Reaction 60
IV.1 Introduction 60
IV.2 Cyclization of Type I 60
IV.2.a Cyclization of hex-4-enal derivatives 60
IV.2.b Cyclization of hept-5-enal derivatives 62
IV.2.c Cyclization of oct-6-enal derivatives and related compounds 63

IV.2.d Cyclization of other derivatives.................................. 67
IV.3 Cyclization of Type II.. 68
IV.4 Cyclization of Type III.. 72
V. Asymmetric Reaction Catalyzed by Chiral Catalyst........................ 76
V.1 Intramolecular reaction..76
V.2 Intermolecular reaction..79
VI. Conclusion.. 83
References.. 84

Chapter 3. Lewis Acid-Promoted Allylsilanes and Allystannanes Addition to Aldehydes and Ketones

I. Introduction and History.. 91
II. Intermolecular Addition to Achiral Aldehydes................................95
II.1 Mechanism..95
II.2 Stereoselective addition..105
II.2.a Reaction between achiral aldehydes and unfunctionalized allylsilanes or allylstannanes............... 105
II.2.b Reaction between achiral aldehydes and functionalized allylsilanes or allylstannanes.................................... 112
II.2.c Reaction between achiral aldehydes and chiral and homochiral allylsilanes or allylstannanes..................... 118
II.2.d Catalytic asymmetric allylation of achiral aldehydes........ 128
III. Intermolecular Addition to Chiral Aldehydes................................131
III.1 Reaction with unfunctionalized chiral aldehydes and ketones... 131
III.1.a Unfunctionalized α-chiral aldehydes............................ 131
III.1.b Unfunctionalized chiral ketones.................................. 133
III.2 Reaction with functionalized chiral aldehydes and ketones........ 134
III.2.a Addition of allylsilanes or allylstannanes...................... 134
III.2.b Addition of crotylsilanes or crotylstannanes.................. 147
III.2.c Addition of allenylsilanes or allenylstannanes................ 153
III.2.d Addition of functionalized allylstannanes or allylsilanes...156
IV. Intramolecular Addition of Allylstannane or Allylsilane Moieties to Aldehydes and Ketones...163
IV.1 Intramolecular addition without cyclization..........................163
IV.2 Cyclization by intramolecular addition of allylsilane or allylstannane moieties... 164
IV.2.a Preparation of 3-cycloalkenols (Type I)........................165
IV.2.b Preparation of 2-vinylcycloalkanols (Type II)................167
IV.2.c Preparation of 3-methylenecycloalkanols (Type III)......... 173
V. Conclusion..176
References...177

Chapter 4. Lewis Acid-Promoted Acetal Substitution Reactions

I. Introduction and History........ 185
II. Mechanism........ 187
III. Intermolecular Reaction........ 196
III.1 Reaction of allylsilanes or allylstannanes with achiral acetals........ 196
III.2 Reaction of allylsilanes or allylstannanes with chiral acetals........ 199
III.3 Reaction of enoxysilanes with acetals........ 202
III.4 Reaction of other silyl nucleophiles with acetals........ 204
IV. Intermolecular Reaction with Acetal Templates........ 205
IV.1 Acetals derived from homochiral pentan-2,4-diols........ 205
IV.1.a Reaction with allylsilanes and allylstannanes........ 206
IV.1.b Reaction with other silyl nucleophiles........ 209
IV.2 Acetals derived from homochiral butan-1,3-diols........ 210
IV.3 Acetals derived from homochiral butan-2,3-diols........ 211
IV.4 Acetals derived from other homochiral alcohols........ 213
V. Anomeric Allylation of Carbohydrate Derivatives........ 214
V.1 Allylation of glycopyranosides and glycofuranosides........ 214
V.2 Allylation of glycals........ 217
VI. Intramolecular Reaction........ 218
VI.1 Reaction of silyl nucleophiles with acetals........ 218
VI.2 Biomimetic cationic polyolefin cyclization........ 221
VII. Conclusion........ 225
References........ 225

Chapter 5. Sakurai Reaction.
Lewis Acid-Promoted Allylsilanes and Allylstannanes 1,4-Addition to Unsaturated Ketones

I. Introduction and History........ 231
II. Mechanism........ 234
III. Reaction Stereochemistry........ 237
III.1 Diastereofacial selectivity........ 237
III.2 1,2-Diastereoselectivity in intermolecular reactions........ 238
IV. [3+2] Annulation Reaction........ 244
V. Reaction Involving Allenyl- and Propargylsilanes........ 249
V.1 Addition of allenylsilanes........ 249
V.2 Addition of propargylsilanes........ 250
VI. Intramolecular Reaction........ 251
VI.1 Addition of allylsilane or allylstannane moieties........ 251
VI.2 Addition of propargylsilane moieties........ 262
VII. Conclusion........ 262
References........ 263

Chapter 6. Lewis Acid-Promoted Diels-Alder Reaction

I. Introduction and History... 267
 I.1 The thermal reaction... 267
 I.2 The Lewis acid-promoted reaction.. 268
 I.3 Stereochemical issues.. 269
II. Intermolecular Reaction. Regio- and Diastereofacial Selectivities.. 270
 II.1 Cycloaddition of buta-1,3-diene and methylbuta-1,3-dienes with various dienophiles...................... 270
  II.1.a Buta-1,3-diene... 270
  II.1.b (*E*)-Piperylene.. 271
  II.1.c Isoprene.. 274
 II.2 Cycloaddition of substituted 1,3-dienes with various dienophiles.. 276
 II.3 Cycloaddition of cyclopentadiene with various dienophiles.. 280
 II.4 Cycloaddition of allenic dienophiles with various dienes......... 283
III. Intramolecular Reaction.. 284
 III.1 Cyclization of Type I.. 285
  III.1.a Cyclization of nona-1,3,8-triene derivatives................. 285
  III.1.b Cyclization of deca-1,3,9-triene derivatives................. 286
  III.1.c Cyclization of 1,3,ω-triene derivatives........................ 287
 III.2. Cyclization of Type II.. 288
IV. Asymmetric Reaction.. 289
 IV.1 Chiral dienophiles... 290
  IV.1.a Chiral acrylate and alkylacrylate derivatives.................. 290
  IV.1.b Functionalized vinyl derivatives................................. 295
 IV.2 Chiral dienes... 299
 IV.3 Chiral Lewis acids... 300
V. Hetero Diels-Alder Reaction.. 307
 V.1 Lewis acid-mediated cyclocondensation of activated dienes with aldehydes... 307
  V.1.a Cyclocondensation with achiral aldehydes......................... 309
  V.1.b Cyclocondensation with chiral aldehydes.......................... 312
 V.2 Reaction with other heterodienophiles................................. 315
 V.3 Reaction involving heterodienes.. 316
VI. Conclusion.. 318
References... 318

Index... 329

1. Lewis Acids and Their Complexes With Lewis Bases

I. Introduction and History

Why are BF_3, $AlCl_3$, $TiCl_4$, $SnCl_4$ among many others called after the name of Lewis and qualified as acids? Part of the answer can be found in the introduction of the small biographical sketch of G.N. Lewis published by the *Encyclopaedia Britannica.*[1]

"Gilbert Newton Lewis (1875–1946): U.S. chemist whose theory of the electron pair fostered understanding of the covalent bond and extended the concept of acids and bases."

Indeed, Lewis, who is also known for his book on *Thermodynamics and the Free Energy of Chemical Substances*[2] and was the first to isolate deuterium and to prepare a pure sample of heavy water in 1933, is famous for his 1923 book *Valence and the Structure of Atoms and Molecules.*[3]

It is in this book that he developed his conception of the shared pair bond and proposed the famous octet rule, two ideas that laid the foundations for a better understanding of chemistry and particularly of reaction mechanisms. In the same book, Lewis proposed his general definition of acids and bases which states that acids are molecules capable of receiving an electron pair into the shell of one of its atoms while bases are molecules with an electron pair capable of entering such a shell to create an electron pair bond. Actually, as reported by Seaborg,[4] although Lewis proposed that definition of acids and bases in the early twenties, he only returned to this issue to provide experimental support in the late thirties. Interestingly enough, Lewis' experimental procedures involved BCl_3, $SnCl_4$, SO_2 and $AgCl_4$ as acids; the second one being nowadays one of the very standard Lewis acids.

A wealth of details on Lewis' achievement in chemistry, which makes him one of the top chemists of the century, can be found in the first three 1984 issues of the *Journal of Chemical Education*[5] which reports on the symposium held in Las Vegas in 1982 "*Gilbert Newton Lewis: 1875–1946*"; (see also ref.6).

II. Physical Properties of Lewis Acids

Although it would be very interesting and useful, on account of the variations (as a function of the nature of the Lewis acid) observed in many reactions, to include a detailed description and/or a discussion on the physical properties of Lewis acids, this has been considered to lie far beyond the scope of this book.

Carlson, in an approach towards a selection strategy for screening experiments in Lewis acid-catalyzed reactions, has provided twenty physical descriptors for over a hundred Lewis acids.[7] Table 1 is a selection of seven descriptors of the five more frequently encountered Lewis acids; the other descriptors available in Carlson's paper are: coordinate bond energy, negative of standard Gibbs energy of formation, standard entropy, heat capacity, density, standard enthalpy of formation of M^{n+} species, lattice energy, ionic bond energy, partial charge on ligand atom, ionization potential, magnetic susceptibility and atomic electronegativity of central atom in different oxidation states.

	1	2	3	4	5	6	7
BF_3	272	1.29	bp −99.9	36.9	71.5	0.504	0
$AlCl_3$	168.5	2.06	mp 190	24.3	39.5	0.576	1.97
$ZnBr_2$	78.6	2.24	mp 394	15.8	35.7	0.266	/
$TiCl_4$	192.4	2.19	bp 136.4	24.9	/	/	0
$SnCl_4$	122.3	2.31	bp 114.1	18	41.4	0.352	0

Table 1: Physical properties of some Lewis acids (ref. 7). 1) negative of standard enthalpy of formation ($kcal.mol^{-1}$). 2) mean bond length (Å). 3) melting or boiling point (°C). 4) mean bond energy ($kcal.mol^{-1}$). 5) covalent bond energy ($kcal.mol^{-1}$). 6) partial charge on central atom (e). 7) dipole moment (gas phase)(D).

III. Lewis Acid-Carbonyl Compound Complexes

Among the Lewis acid-catalyzed reactions, those involving the formation of complexes between carbonyl compounds and Lewis acids are particularly important since these complexes play a fundamental role in organic and bioorganic chemistry. Thus, most reactions studied in this book involve the formation of such carbonyl compound-Lewis acid complexes: the carbonyl ene reaction (Chapter 2), the allylsilanes or allylstannanes addition to aldehydes (Chapter 3) and to conjugated enones (Sakurai reaction, Chapter 5), and the Diels-Alder reaction (Chapter 6). Moreover, for all these reactions, much of the discussion about their mechanism and their stereochemical preferences is based on a minimum knowledge on the carbonyl compound-Lewis acid complexes involved. Indeed, this knowledge is a necessary prerequisite when trying to shed some light on these reactions and particularly when trying to determine the origin of their stereochemical outcome.

Among the issues to be addressed, the main ones are:

- which type of coordination, σ or π?
- which conformation will the complex adopt?
- are the complexes formed 1:1 or 2:1 (carbonyl:acid) complexes?
- what will be the influence of the complexation on the reactivity of the carbonyl compounds and on the rate of the reaction?

A carbonyl compound can interact with a Lewis acid either through one of the oxygen atom lone pairs (σ-bonding: complex **A**) or through the π-system of the carbonyl moiety (π-bonding: complex **B**).

A B

In the first case, a σ bond is formed between the oxygen atom and the metal center of the Lewis acid, the latter being roughly in the same plane as the carbonyl moiety and its substituents. No facial differentiation is available but the Lewis acid will be *syn* or *anti* to particular substituents (provided that the α angle is different from 180°).

In the second case, π-bonding, the outcome of the complexation is a η^2-metallooxirane complex. The Lewis acid is out of the carbonyl moiety plane and blocks a particular face, directing the attack of the nucleophilic reagent on the opposite face, but the two substituents of the carbonyl moiety are no longer differentiated.

Along with the *syn* and *anti* conformations possible with σ-complexes, another conformational issue is of particular importance: will σ-complexes between α,β-unsaturated carbonyl compounds and Lewis acids adopt *s-cis* and *s-trans* conformation? The answer to that question will result in the exposure of one or the other face of the unsaturated system to a given nucleophile and will obviously be of great importance when dealing with enantioselective reactions.

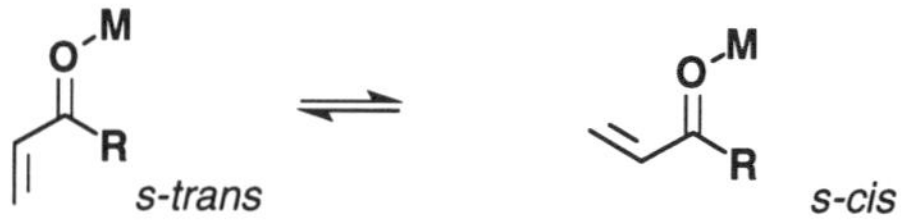

s-trans *s-cis*

Another interesting problem is the ability of α- or β-alkoxy carbonyl compounds, or more generally, carbonyl compounds bearing a Lewis base such as an oxygen or a nitrogen atom in a reasonable proximity, to undergo double chelation with suitable Lewis acids.

Information about these complexes can be collected from three main sources:

- Theoretical studies (Section III.1)
- NMR spectroscopy (Section III.2)
- X-ray crystallography (Section III.3)

Such an approach was used recently by Shambayati and Schreiber on one side,[8] and Reetz on the other,[9] in two excellent reviews on that subject. However, other spectroscopic techniques, such as IR,[10–18] UV[19–22] and Raman,[23] or calorimetric studies (which give values of the enthalpies of complexation)[24–26] have also been used successfully.

III.1 Theoretical studies

Theoretical studies of the role and effect of Lewis acids on carbonyl compounds and particularly on Diels-Alder reactions started in the early seventies. Hence, it took a little over a decade after the famous result from Yates and Eaton on the marked effect of $AlCl_3$ on the Diels-Alder reaction,[27] for the first theoretical papers on the Lewis acid effect in Diels-Alder reactions to appear.[28–30] These studies however do not focus on the carbonyl group-Lewis acid interaction and deal more with the frontier–orbital interaction. At the same period, the first studies on the interaction between formaldehyde and Li^+ cation were published.[31–33]

III.1.a Carbonyl compound-cation complexes

In 1979, ab initio SCF calculations, conducted at the STO-3G basis set level, were carried out to determine the relative Li^+ affinities with various R_2CO bases and the structure of the $R_2CO–Li^+$ complexes.[34] This study established that the predictable order of affinities of the lithium ion for the bases R_2CO, with respect to R is $NH_2 > CH_3 > OH > H > F$. All complexes, except $(OH)_2CO–Li^+$ which has C_s symmetry, have C_{2v} symmetry with a Li–O–C angle of 180° and the R_{OLi} distance varying from 1.54 Å (STO-3G) or 1.706 Å (STO 6.31 G) for $(NH_2)_2CO–Li^+$ to 1.637Å (STO-3G) or 1.767 Å (STO 6.31 G) for $H_2CO–Li^+$. Then, Del Bene studied the interaction between formaldehyde and H^+ (C_S complex with R_{OH} = 0.967 Å and H–O–C = 124.5°),[35] and compared the electronic features of both $H_2CO–Li^+$ and $H_2CO–H^+$complexes.[36] She concluded that the Li–O interaction is primarily electrostatic (ion-dipole attraction), with little covalent character while the H–O interaction is primarily covalent.

The interaction between acetone and a sodium cation was investigated by various semi–empirical and ab initio methods. The calculation revealed an energy minimum for a C_{2v} complex ($(CH_3)_3CO–Na^+$) with C–O–Na = 180° and R_{ONa} = 2.2 Å.[37] This result and the preceding ones proved to be coherent with those obtained from an extensive ab initio study (conducted at the STO 6.31G* // 3.21G level) on the proton, lithium and sodium affinities with first and second row bases including formaldehyde and acetaldehyde.[38]

The complexation of formaldehyde with first- and second-row cations (H^+, Li^+, BeH^+, BH_2^+, CH_3^+, Na^+, MgH^+, AlH_2^+, SiH_3^+) was also studied.[39] The authors concluded that both bent and linear coordination are possible, depending on the nature of the Lewis acid: a bent geometry is preferred when the main interaction is a charge transfer from the carbonyl oxygen atom to a σ type acceptor orbital of the Lewis acid ($H_2CO–H^+$, $H_2CO–CH_3^+$, $H_2CO–SiH_3^+$, $H_2CO–BH_2^+$); a linear geometry is favored if a π type acceptor orbital is available on the Lewis acid that can interact with the lone-pair orbital of the carbonyl oxygen atom and if electrostatic interactions predominate ($H_2CO–Li^+$, $H_2CO–BeH^+$, $H_2CO–Na^+$, $H_2CO–MgH^+$).

Finally, in the course of his work on the ene reaction (see Chapter 2, Sections II.2 and V.2), Mikami studied the reactivity of trihaloacetaldehydes in terms of the balance of LUMO energy level vs. electron density on the carbonyl carbon atom.[40]

III.1.b Carbonyl compound-neutral Lewis acid complexes

In 1986, Reetz published one of the first X-ray structures and NMR studies on an aldehyde-Lewis acid complex.[41] Thus, he described in his paper the benzaldehyde-BF_3 adduct (see Section III.3) in which BF_3 is complexed *anti* to the phenyl group. In the same paper, he also reported MNDO calculations on the acetaldehyde-BF_3 complex. It appeared that the *anti* adduct (R_{OB} = 1.629Å; C–O–B = 133.6°; C–C–O–B = 180°) is the most stable conformer. However, the *syn* complex (R_{OB} = 1.635Å; C–O–B = 141.5°; C–C–O–B = 0°) lies only 1.8 kcal.mol^{-1} higher in energy. In contrast with results reported on carbonyl complexation to metal cations, the linear form does not correspond to a minimum of the energy surface but is rather the lowest transition state between the *syn* and *anti* conformers.

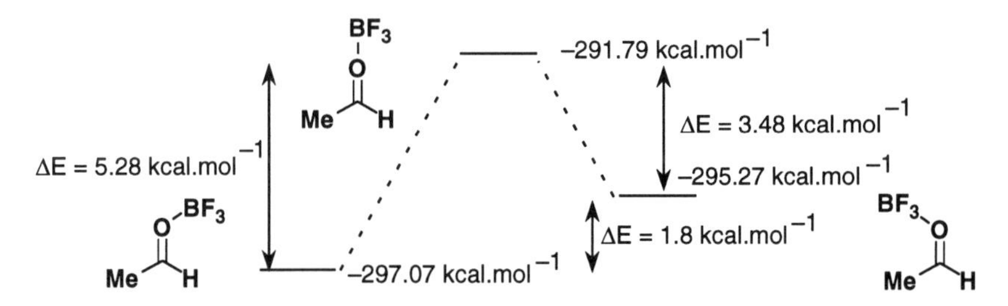

The same year, Nelson performed also MNDO calculations on complexes between *trans*-2,3-dimethylcyclopropanone and various charged and uncharged Lewis acids. It appeared that neutral boron or aluminum compound led to bent forms which are more stable than linear ones by 1.2 to 8.8 kcal.mol^{-1}. However, linear forms are favored with cations having fewer electronegative elements (Table 2).[42]

	C=O–X bent		C=O–X linear	
X	angle	ΔH_f	angle	ΔH_f
BeH_2	149.8	−27.04	179.9	−26.63
BF_2^+	133.5	−11.27	180.0	−6.85
BF_3	133.4	−270.8	180.0	−265.6
AlF_2^+	157.2	−61.8	179.6	−61.7
AlF_3	143.5	−322.0	180.0	−320.8

Table 2: Calculated heat of formation (kcal.mol^{-1}) for bent and linear **C=O–X** conformation of dimethylcyclopropanone-**X** complexes.

Several ab initio studies dealing with carbonyl compound-neutral Lewis acid complexes were also published. Gung, who experimentally observed a dramatic reversal of diastereofacial selectivity, in BF_3-OEt_2 promoted addition of chiral allyl stannanes to aldehydes, when aliphatic aldehydes were changed to aromatic aldehydes[43] (see Chapter 3, Section II.2.c), studied the difference in strength of BF_3-complexes of aromatic and aliphatic aldehydes.[44,45] From this study, conducted at the MP2/6-31G* level of theory, it appeared that the most stable rotamer of the *syn* CH_3CHO-BF_3 complex is only 1.22 kcal.mol^{-1} less stable than the most stable *anti* CH_3CHO-BF_3 complex. However, on the other hand, the corresponding difference between the *syn* and *anti* PhCHO-BF_3 complexes is of 5.31 kcal.mol^{-1}. That difference was seen by Gung as an indication of an only weak complexation between aliphatic aldehydes and BF_3, that could account for the reversal of diastereofacial selectivity experimentally observed[43] and is consistent with some NMR investigation reported by Denmark (see Section III.2).

Wiberg studied rotational barriers in aldehydes (formaldehyde, propanal) and ketones (acetone) coordinated to neutral Lewis acids (BH_3, BF_3, AlH_3, and $AlCl_3$).[46] In agreement with previous works, all complexes were found to prefer bent geometries. For formaldehyde complexes linear structures are 6.10 kcal.mol^{-1} higher in energy and out of plane structure (π-complexes) even higher. Complexes between formaldehyde and BCl_3 and BBr_3 were also studied by Oliva,[47] as well as R^1R^2CO–BH_2F complexes by Goodman.[48]

Theoretical studies on transition metal complexes of formaldehyde exist, but, in most of them electron rich metals were chosen. The paper from Branchadell and Oliva, on the complexes between formaldehyde and $TiCl_4$, is therefore of particular interest.[49] Indeed, $TiCl_4$ is widely used as a Lewis acid and, unlike boron or aluminum Lewis acids, can form 2:1 complexes through coordination of two carbonyl compounds, form chelate with bidentate carbonyl compounds (keto ester, alkoxy ketone or aldehyde) or 1:1 dimeric complexes. In their study, carried out using the 3-21G and MIDI-3 bases sets, they considered the H_2CO–$TiCl_4$, the $(H_2CO)_2TiCl_4$ and the $(H_2CO$–$TiCl_4)_2$ complexes. For the 1:1 complex, it appeared that out of seven optimized structures, corresponding to stationary points of the energy surface of the system and being distorted trigonal bipyramids, only two (**A** and **B**) were energy minima. In both structures the formaldehyde is placed in the axis of the bipyramid and presents a bent mode of coordination. In the more stable structure **A**, formaldehyde is *syn* with respect to one of the in–plane chlorine atoms whereas it is *anti* in structure **B**.

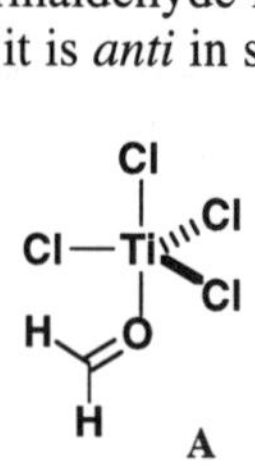

A 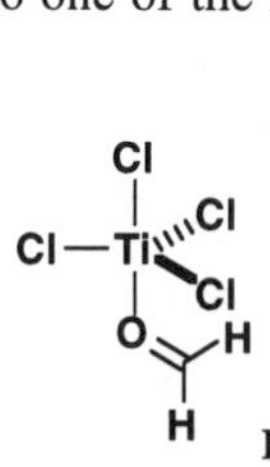 **B**

Three different structures of the 2:1 complex were considered, each being an energy minimum; in these octahedral structures the two molecules of H_2CO are placed in *cis* (**C**) or *trans* (**D** and **E**). Calculations showed *cis* isomer **C** to be more stable than *trans* isomers **D** and **E** and the formation of the *cis* 2:1 complex more exothermic than the formation of the 1:1 complex (Table 3).

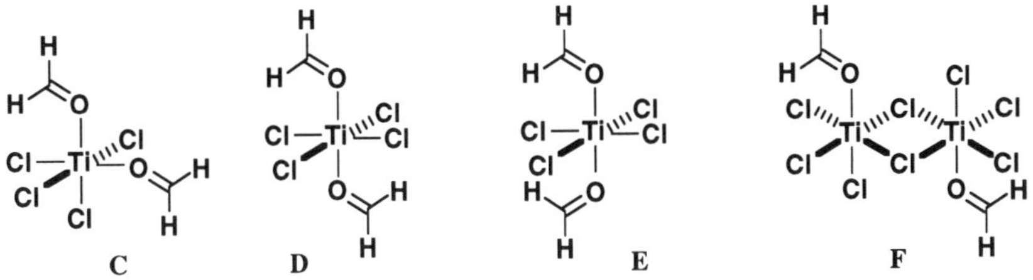

C D E F

	C	D	E
Formation energies	–32.5	–29.6	–30.8

Table 3: Formation (kcal.mol^{-1}) of complexes **C**, **D** and **E** relative to complex **A** and **H_2CO**.

It was shown that in the $(H_2CO–TiCl_4)_2$ complex **F**, obtained through dimerization of the 1:1 complex, the interaction between $TiCl_4$ and formaldehyde is stronger. The results are in good agreement with known X-ray structures that include complexes of dimeric $TiCl_4$ (from monodentate carbonyl compounds) and 1:1 octahedral complexes from chelating esters or ketones (with the two carbonyl group *cis* coordinated to the titanium atom) (see Section III.3).

III.1.c α,β-Unsaturated carbonyl compound-Lewis acid complexes

As already mentioned, the effect of carbonyl complexation on α,β-unsaturated carbonyl compounds was studied, from the early seventies on,[28–30] mainly in connection with the increased regio- and stereoselectivity of Lewis acid-catalyzed Diels-Alder reactions. Thus, the influence on frontier molecular orbitals of acrolein with respect to its protonation or to its complexation by lithium or sodium cation was studied by CNDO/2[29,30,50] or STO-3G[51] calculations. Under complexation, the most noticeable changes in acrolein electronic structure are:

- a lowering of the LUMO energy level
- an inversion of the atomic coefficients in the LUMO between carbon atoms 2 and 4.

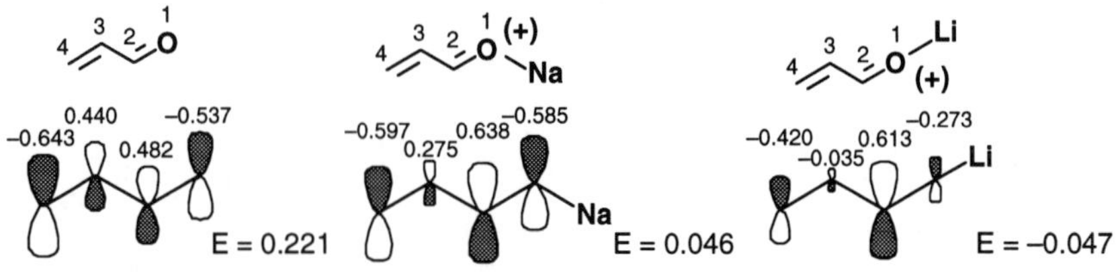

LUMO energies (eV) and coefficients for acrolein and its complexes with Na^+ and Li^+ (STO-3G)(ref.51).

The geometry and frontier orbital energies and coefficients of the acrolein-BF_3 complex were studied by Ottenbrite, by using ab initio methods (Table 4).[52]

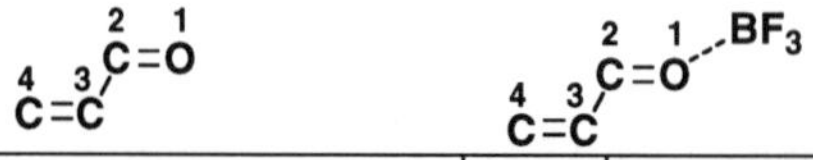

STO 3-G/4-31G	HOMO	HOMO coeff.				LUMO	LUMO coeff.			
		O_1	C_2	C_3	C_4		O_1	C_2	C_3	C_4
acrolein	−9.05	0.436	0.236	−0.53	−0.58	5.96	0.561	−0.471	−0.481	0.663
acrolein - BF_3	−10.53	−0.30	−0.11	0.328	0.329	3.33	0.543	−0.612	−0.327	0.629

Table 4 : FMO energies (eV) and coefficients of acrolein and acrolein-BF_3 complex (ref.52)

These results were analyzed with respect to the role of Lewis acids in Diels-Alder reactions (see Chapter 6). Also in connection with Diels-Alder reaction, Houk addressed the question of the conformations of acrolein, acrylic acid, methylacrylate and their complexes with H^+, Li^+ and BH_3[53]: *s-cis* vs. *s-trans* conformation. The outcome is of great significance in reactions dealing with chiral starting materials. He showed that although acrylic acid and methylacrylate prefer *s-cis* conformation, acrolein and all the Lewis acid complexes prefer *s-trans* conformation.

The same kind of result was obtained by Laszlo when considering the crotonaldehyde-BF_3 complex, with ab initio (STO-3G) and semi-empirical (MNDO) methods.[54,55] His study which is a theoretical justification of the determination of the acidity of Lewis acid experimentally (NMR) proposed by Childs, will be discussed in Section III.2.

Finally, theoretical studies are now directed towards the modelization of transition states involved in Lewis acid-promoted reactions such as [2+2] cycloaddition reactions[56,57] or Diels-Alder reactions.[58,59]

III.2 NMR studies

The discovery in the late fifties that ^1H NMR signals of Lewis bases and particularly carbonyl compounds were shifted upon complexation with a Lewis acid has laid the foundation for extensive studies in the field. Particularly interesting, although beyond the scope of this section is the use of lanthanide shift reagents in structure determination and conformational analysis. NMR gives information on the Lewis acid-carbonyl compound complex [stoichiometry, conformation (*syn/anti* and *s-cis/s-trans* for α,β-unsaturated compounds), formation of a chelate with another Lewis base in the molecule] which are of direct importance with respect to the stereooutcome of the reactions involving these complexes.

III.2.a Stereochemistry

Detailed structural informations on Lewis acid-carbonyl compound complexes obtained by means of NMR studies go back to the late sixties and the seventies.[60–66]

In most of these studies boron trihalides and particularly BF_3, were used as Lewis acids and therefore conclusions rest upon ^1H, ^{11}B, ^{13}C and ^{19}F NMR results. It was shown that BX_3 and carbonyl compounds form 1:1 complexes.[67] Moreover Forsen established in 1970 that the $Et_2CO–BF_3$ complex shows below –120° two separate signals for the protons of the two methyl groups;[68] this observation was attributed to a *syn/anti* isomerism and therefore established that the Lewis acid coordinate to the carbonyl in a non-linear fashion. The activation energy of this equilibrium was estimated to be *ca* 8 $kcal.mol^{-1}$ from the value of the coalescence temperature (–110°C).

^{13}C NMR proved to be a better tool to study this conformational exchange since the average shielding changes are much larger than in ^1H NMR.[69] Thus, a number of studies devoted to the Lewis acid induced shifts in ^{13}C NMR spectra of various ketones,[70,71] including terpenes[72] and steroids,[73,74] can be found in the literature. These studies showed that $TiCl_4$ and $SnCl_4$ complexes were generally 2:1 complexes (carbonyl compound:Lewis acid).

Brownstein described the complex formed between acetyl chloride and $TiCl_4$,[75] and more generally complexes of interest in Friedel-Crafts reactions.[76,77] ^{27}Al NMR was also used to propose a donor-acceptor structure (rather that an ion-pair form) for the acetyl chloride-$AlCl_3$ complex in CH_2Cl_2.[78]

Rabinovitz studied aromatic aldehyde-BF_3 complexes by ^1H, ^{13}C, ^{19}F NMR and IR and showed that the $CHO–BF_3$ group behaves as a pseudo substituent with strong electron withdrawing character.[79–81]

III.2.b Conformational studies

Relatively few investigations deal with that problem. An extensive study due to Childs however shed some light on that issue;[82] indeed the systematic study of the complexation of crotonaldehyde, tiglic aldehyde, pent-3-enone and methyl crotonate with over ten different Lewis acids indicated an *anti s-trans* conformation for the complexes in solution (for a calorimetric study on the same complexes, see ref.83). Figures 1 and 2 indicate the variation of the proton and carbon chemical shifts of crotonaldehyde when complexed to various Lewis acids.

Childs then proposed a classification of Lewis acids with respect to their acidity. As mentioned in Section III.1, Laszlo brought some theoretical support to that experimental work (Figure 3).[54,55] He studied the rates of the ene reaction between β-pinene and methyl acrylate as a function of the Lewis acid involved and showed that a linear relationship exists between the Gibbs free energy change (ΔG*) and the calculated π* molecular orbital energy change. Thus, the following classification, with respect to the Lewis acidity, was proposed: $BCl_3 > AlBr_3 > AlCl_3 > EtAlCl_2 > Me_2AlCl > Me_3Al$.

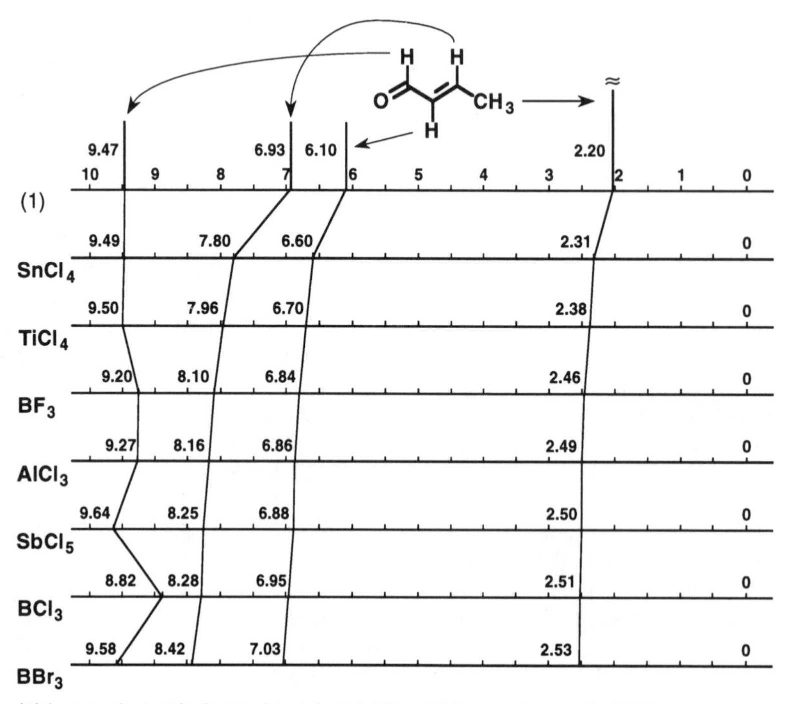

(1) In ppm (± 0.01). Approximately 0.3 M in dichloromethane at –20°C.

Figure 1: ^{1}H NMR chemical shifts of crotonaldehyde-Lewis acid complexes (ref.82).

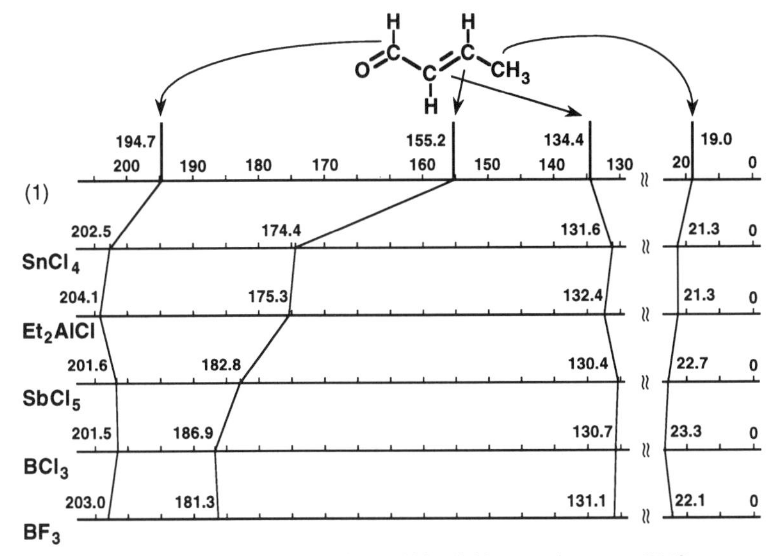

(1) In ppm (±0.01). Approximately 0.3 M in dichloromethane at –20°C.

Figure 2: ^{13}C NMR chemical shifts of crotonaldehyde-Lexis acid complexes (ref.82).

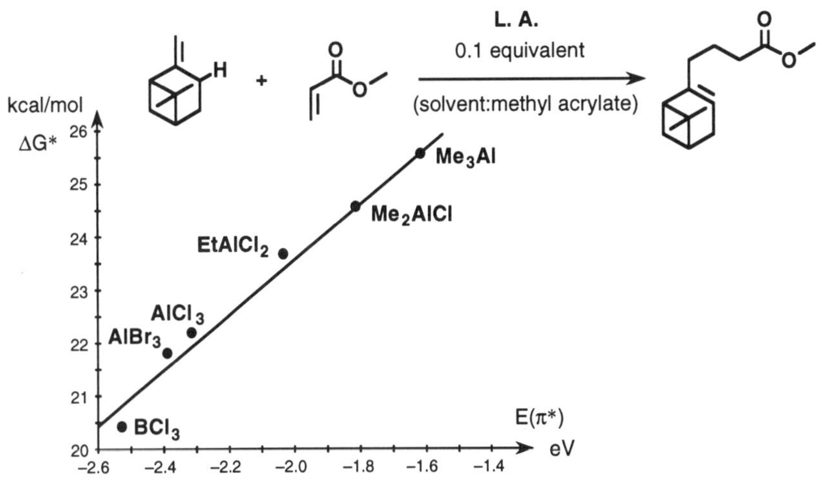

Figure 3: E(π*) (MNDO calculated π* level for the complex between crotonaldehyde and the Lewis acid) vs. ΔG* (experimental Gibbs free energy for the ene reaction)(ref.55).

The preference of complexes for an *s-trans* conformation was recently confirmed by Corey,[84] and Denmark on the occasion of a detailed study directed towards the structure and conformation of aldehyde-Lewis acid complexes.[85] Indeed, Denmark showed that (*E*)-hept-2-enal, complexed either to BF_3 (1:1 complex) or $SnCl_4$ (2:1 complex) adopts an *s-trans* conformation with the Lewis acid complexed *anti* to the aldehyde residue.

III.2.c Chelation

As will be shown in Chapters 2–6, the formation of chelates is of crucial importance with respect to the stereooutcome of Lewis acid-catalyzed reactions; indeed, an amazing number of examples exploit the chelation-controlled (or anti-Cram) phenomena. However the first experimental evidences for the formation of such chelates are fairly recent. Thus, Keck reported ^{1}H NMR evidence for the formation of complexes **A** and **B** from the corresponding β-benzyloxy aldehydes and $TiCl_4$ (similar results were obtained with $SnCl_4$ and $MgBr_2$).[86] The formation of **A** and **B** provides a coherent explanation for the reactivity of the two aldehydes towards allylsilane and allylstannane nucleophiles (see Chapter 3, Section III.2.a).

A B

Nearly simultaneously, Reetz provided the first evidence for a Cram-chelate involving a chiral α-benzyloxy ketone.[87] Later, he was able to monitor the reaction of aldehyde **A** with CH_3TiCl_3 at –45°C by ^{13}C NMR and to observe the formation of two discrete species which were attributed to structures **B** and **C**[88] (see also ref.89 for other examples of NMR evidence of chelates).

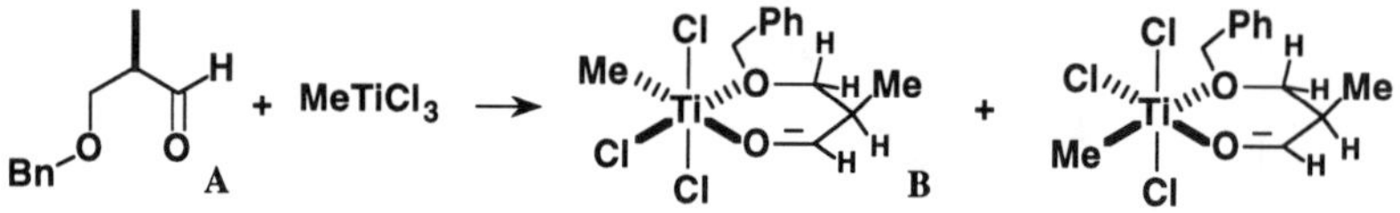

A B C

However chelation does not always occur; Keck provided direct evidence for the absence of chelation between β-silyloxy aldehydes and Lewis acids.[90] ^{1}H NMR established that the 1:1 complex **B** was formed from aldehyde **A** and $SnCl_4$ while the 2:1 complex **D** was formed from **C** and $SnCl_4$.

This feature, in agreement with calculations,[91] accounts for the observed stereooutcome of a number of reactions involving β-silyloxy aldehydes (Chapter 3, Section III.2). However, the interpretation based on the lower basicity of silyl ether compared to alkyl ethers and the widening of the C–O–Si angle,[8,92] which is commonly cited to account for such results, was recently disputed by Eliel and Frye who proposed a simple steric one.[93]

Other NMR evidence of the formation of chelates exists in the literature, including *N*-acyloxazolidinone-Et_2AlCl[94] and carbomethoxy substituted dioxolanes-Et_2AlCl complexes.[95]

The role of Lewis acids based on NMR evidence was reported to support the discussion on the mechanism of a number of reactions, the Mukaiyama aldol reaction,[96] the allylstannane addition to aldehydes[97–101] (see also Chapter 3, Section II.1), the 1,6- and 1,8-addition of organocuprates to Michael acceptors,[102] and the addition of allylsilanes to aldehydes and α,β-unsaturated ketones (see also Chapter 5, Section II).[103] In that last case, in addition to the formation of a complex, Denmark established that titanium enolate **A** was formed as the intermediate of the reaction after the addition of allyltrimethylsilane to the enone-$TiCl_4$ complex.

Only fairly recently, Corcoran, in a very nicely set-up work, addressed specifically the issue of the mode of complexation of Lewis acids with carbonyl compounds, π- or σ-complexation.[104] His results led him to exclude the possibility that the reaction under consideration could proceed even through a small concentration of a highly reactive ketone-$TiCl_4$ π-complexed intermediate.

The complexation of less commonly used Lewis acids such as $LiClO_4$,[105] $NaClO_4$, $Mg(ClO_4)_2$, $Ba(ClO_4)_2$,[106] $TiCl_3(OR)$,[107] $SbCl_5$, $AgSbF_6$,[108] monodentate and bidentate $VOCl_2(OR)$[109] and Lewis acids derived from Mo and W nitrosyles was also reported.[110]

III.3 X-ray studies

The study of carbonyl compound-Lewis acid complexes by X-ray analysis is well precedented in the literature. Schreiber, in an important review[111] published in 1990 (see also ref.8), reported the result of a systematic survey of the Cambridge Structural Database on crystal structures of Lewis acid-carbonyl compound complexes (including cationic alkali-metal, main group, early and late transition metal Lewis acids).

The average values of bond length (d M–O), valence angle (α C–O–M) and torsion angle (θ X–C–O–M) for boron, aluminum, titanium and tin complexes resulting from that study are reported in Table 5.

M	d M–O	α C–M–O	θ X–C–M–O
B	1.58 ± 0.02 Å	115 ± 4 °	0–11 °
Al	1.88 ± 0.09 Å	136 ± 4 °	0–18 °
Ti	2.14 ± 0.07 Å	125 ± 12 °	0–3 °
Sn	2.3 ± 0.1 Å	127 ± 10 °	0–3 °

Table 5: Average bond lengths, valence and torsion angles for **B**, **Al**, **Ti** and **Sn** complexes (ref.111).

In addition to these average values, it appeared that BF_3 and $AlCl_3$ prefer to form 1:1 complexes while Sn derived, like Ti derived Lewis acids prefer to form 2:1 complexes, 1:1 chelated adducts, or even 1:1 dimeric complexes for Ti (*vide infra*); these complexes all satisfy the desire of Ti and Sn for hexacoordination and in the case of dimeric structures, bridging chlorine atoms are involved.

Although, as indicated in Table 5, the Lewis acid generally lies within a few degrees of the plane of the carbonyl group (σ-complexes), extreme cases of out-of-plane bonding are known.

With aldehydes, the Lewis acid generally lies *anti* to the aldehyde residue while with esters, the steric bulk is of crucial importance.

The following crystal structures are selected examples. PhCHO–BF_3 reported by Reetz, was the first bimolecular crystal structure involving a carbonyl compound and a boronic Lewis acid.[41] Methacrolein-BF_3 is the only other example of such a complex we are aware of. It indicates that the α,β-enal is in *s-trans* conformation and that the Lewis acid is *anti* to the aldehyde residue.[84]

O BF3 H   O BF3 H

$TiCl_4$ generally forms 1:1 complexes with carbonyl compounds either through a dimeric structure involving bridging chlorine atoms[112] such as **A**,[113] or through a chelate complex[114–116] such as **B**[117] and **C**[118] (**C** being the first crystal structure of a Lewis acid-chiral dienophile complex, see Chapter 6, Section IV.1.a). The structure of a 1:2 complex between a bidentate titanium trichloroalkoxide and pinacolone was also described by Wuest[119] (for a related example, see refs.120 and 121 for a review on multidentate Lewis acids).

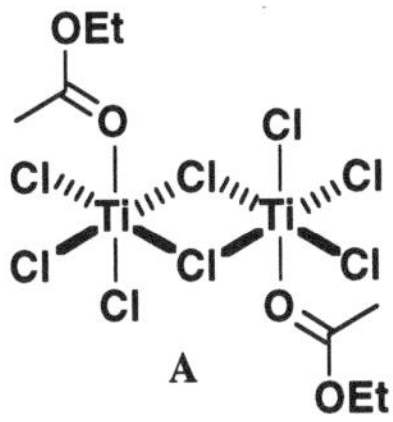

In agreement with the results observed in solution by NMR (Section III.2), $SnCl_4$ prefers to form 2:1 complexes[122,123] such as **D**[97] and 1:1 chelated aducts such as **E**.[124] However, 1:1 complexes between dichlorodimethyltin and tetramethylurea[125] and salicylaldehyde[126] were also described.

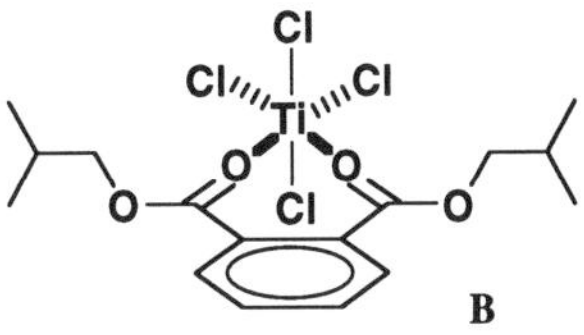

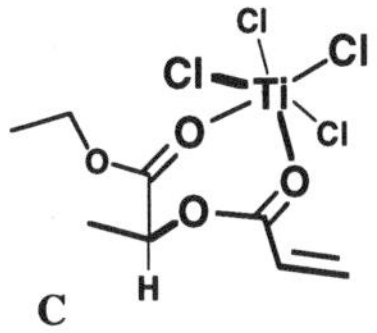

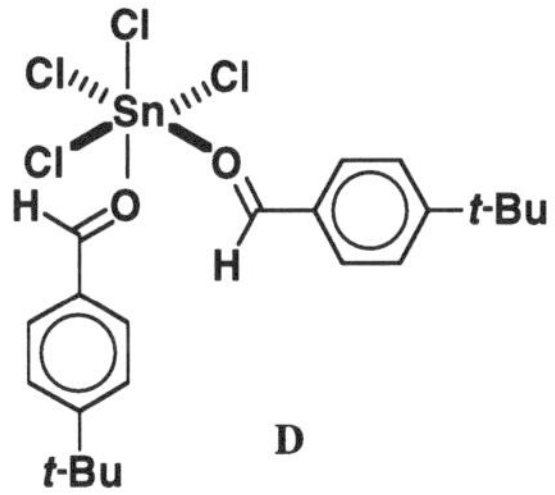

Other complexes involving less common Lewis acids are also known. Thus, in a combined X-ray and NMR study on the complexation of α-bromo ketones with hard and soft Lewis acids, Laube showed that $SbCl_5$ (hard) binds to the oxygen atom whereas Ag^+ (soft) coordinates to both the bromine and oxygen atoms and thus forms a chelate.[127]

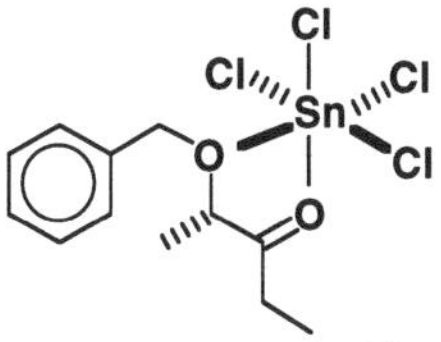

IV. Lewis Acid Complexes With Other Lewis Bases

Complexes resulting from the interaction between Lewis acids and Lewis bases different from carbonyl compounds have also been studied. Thus, ether-, amino-, cyano-, isocyanate- and phosphate-Lewis acid complexes were studied from a theoretical point of view[128–130] and/or by spectroscopic means.[131–135] Crystal structures have also been reported; thus among others, the synthesis and structure of $TiCl_3OC_2H_5.2C_2H_5OH.(18\text{-crown-6})_2$,[136] (18-crown-6).$MCl_4$ (M = Ti, Sn)[137] and of $MgBr_2(THF)_2$ and $MgBr_2(THF)_4(H_2O)_2$[138] are known. The X-ray structure of complexes resulting from the interaction between *tert*-butyl isocyanide and $TiCl_4$ or vanadium (V) complexes[139] and between nitromethane and $TiCl_4$ (1:1 dimeric complex)[140] are also known.

Of more significance with respect to the contents of our book is the complexation of acetals (see Chapter 4). Denmark, in a work directed towards the study of the mechanism of allyl metal-acetal addition,[141] was able to establish by means of low temperature ^{13}C NMR the formation of Lewis acid-acetal complexes. With BF_3-OEt_2 as Lewis acid, the 1:1 complex **A** (one methoxy group being complexed, the other one remaining free) was formed. The introduction of 0.5 equivalent of $SnCl_4$ induced the formation of the 2:1 complex **B** in which a single methoxy group of each acetal is complexed to the tin atom. However upon addition of 1 equivalent of $SnCl_4$ the formation of a new complex was observed in which both methoxy groups of the acetal are complexed; most probably this new complex, formed to the exclusion of any other one and particularly **B**, is the 1:1 complex **C**.

Me O BF3 O Me A

Me O SnCl4 O 2 Me B

Me O SnCl4 O Me C

V. Conclusion

Many complexes formed upon the interaction between Lewis acids and Lewis bases (particularly carbonyl compounds) have been studied from a theoretical point of view or observed by different spectroscopic techniques. Although not necessarily always true it is possible to propose, as Schreiber and Shambayati did,[8] "complexation rules" for Lewis acid-carbonyl complexes:

- σ complexation (for the most commonly used Lewis acids)
- complexation *syn* to the small substituent
- *s-trans* preference for α,β-unsaturated compounds

As shown in the following chapters, part of the mechanistic understanding of the Lewis acid induced reactions comes from this better knowledge of Lewis acid-carbonyl compound complexes.

References

1. *The New Encyclopaedia Britannica*, Encyclopaedia Britannica, Inc., 15th edition, Chicago, **1988**, vol. 7, p. 313.
2. Lewis, G.N.; Randall, M. *Thermodynamics and the Free Energy of Chemical Substances,* McGraw–Hill Book Co., New York, **1923**.
3. Lewis, G.N. *Valence and the Structure of Atoms and Molecules,* Chemical Catalog Company, Inc., New York, **1923**.
4. Seaborg, G.T. *J. Chem. Educ.* **1984**, *61*, 93–100.
5. *J. Chem. Educ.*, **1984**, *61*, January, February and March issues.
6. Borman, S. *Chem. Eng. News* **1993**, *71*, (44), 25–27.
7. Carlson, R; Lundstedt, T; Nordahl, Å; Prochazka, M. *Acta Chem. Scand.* **1986**, *B40*, 522–533.
8. Shambayati, S.; Schreiber, S.L. in *Comprehensive Organic Chemistry,* Trost, B.M.; Fleming, I., Eds. Pergamon, Oxford, **1991**, vol. 1, chap. 1.10, pp. 283–324.
9. Reetz, M.T. *Acc. Chem. Res.* **1993**, *26*, 462–468.

10. Weber, R.; Susz, B.–P. *Helv. Chim. Acta* **1967**, *50*, 2226–2232.
11. Susz, B.–P.; Weber, R. *Helv. Chim. Acta* **1970**, *53*, 2085–2097.
12. Filippini, F.; Susz, B.–P. *Helv. Chim. Acta* **1971**, *54*, 835–845.
13. Dabrowski, J.; Katcka, M. *J. Mol. Struct.* **1972**, *12*, 179–183.
14. Chevrier, B.; Weiss, R. *Angew. Chem., Int. Ed. Engl.* **1974**, 13, 1–10.
15. Langseth, W.; Tangbol, K. *Inorg. Chim. Acta* **1983**, *72*, 43–45.
16. Lysenko, Y.A.; Troshina, E.A. *Zh. Obsh. Khim.* **1983**, *53*, 895–901.
17. Lysenko, Y.A.; Shewchenko, A.Y. *Zh. Obsh. Khim.* **1985**, *55*, 668–672.
18. Braca, G.; Raspolli Galletti, A.M.; Sbrana, G.; Zanni, F. *J. Mol. Cat.* **1986**, *34*, 183–194.
19. Klein, D.P.; Gladysz, J.A. *J. Am. Chem. Soc.* **1992**, *114*, 8710–8711.
20. Lewis, F.D.; Oxman, J.D.; Gibson, L.L.; Hampsch, H.L.; Quillen, S.L. *J. Am. Chem. Soc.* **1986**, *108*, 3005–3015.
21. Lewis, F.D.; Howard, D.K.; Oxman, J.D.; Upthagrove, A.L.; Quillen, S.L. *J. Am. Chem. Soc.* **1986**, *108*, 5964–5968.
22. Lewis, F.D.; Reddy, G.D.; Elbert, J.E.; Tillberg, B.E.; Meltzer, J.A. *J. Org. Chem.* **1991**, *56*, 5311–5318.
23. Goetz, G.J.; Leroy, M.J.F. *Z. Anorg. Allg. Chem.* **1976**, *424*, 59–67.
24. Maria, P.–C.; Gal, J.–F. *J. Phys. Chem.* **1985**, *89*, 1296–1304.
25. Maria, P.–C.; Gal, J.–F.; de Franceschi, J.; Farjin, E. *J. Am. Chem. Soc.* **1987**, *109*, 483–492.
26. Gal, J.–F.; Morris, D.G.; Rouillard, M. *J. Chem. Soc., Perkin Trans. 2* **1992**, 1287–1293.
27. Yates, P.; Eaton, P. *J. Am. Chem. Soc.* **1960**, *82*, 4436.
28. Nguyen Trong Anh ; Seyden–Penne, J. *Tetrahedron* **1973**, *29*, 3259–3265.
29. Houk, K.N.; Strozier, R.W. *J. Am. Chem. Soc.* **1973**, *95*, 4094–4096.
30. Imammura, A.; Hirano, T. *J. Am. Chem. Soc.* **1975**, *97*, 4192–4198.
31. Russegger, P.; Schuster, P. *Chem. Phys. Lett.* **1973**, *19*, 254–259.
32. Bernardi, F.; Pedulli, G.–F. *J. Chem. Soc., Perkin Trans.2* **1975**, 194–197.
33. Tae–Kyu Ha ; Wild, U.P.; Kühne, R.O.; Loesch, C.; Schaffhenser, T.; Stachel, J.; Wokaun, A. *Helv. Chim. Acta* **1978**, *61*, 1193–1199.
34. Del Bene, J.E. *Chem. Phys.* **1979**, *40*, 329–335.
35. Del Bene, J.E. *J. Am. Chem. Soc.* **1978**, *100*, 1673.
36. Del Bene, J.E. *Chem. Phys. Lett.* **1979**, *64*, 227–229.
37. Weller, T.; Lochmann, R.; Meiler, W.; Köhler, H.–J. *J. Mol. Struct. (Theochem)* **1982**, *90*, 81–87.
38. Smith, S.F.; Chandrasekhar, J.; Jorgensen, W.L. *J. Phys. Chem.* **1982**, *86*, 3308–3318.
39. Raber, D.J.; Raber, N.K.; Chandrasekhar, J.; von R. Schleyer, P. *Inorg. Chem.* **1984**, *23*, 4076–4080.
40. Mikami, K.; Yajima, T.; Terada, M.; Uchimaru, T. *Tetrahedron Lett.* **1993**, *34*, 7591–7594.
41. Reetz, M.T.; Hullmann, M.; Massa, W.; Berger, S.; Rademacher, P.; Heymanns, P. *J. Am. Chem. Soc.* **1986**, *108,* 2405–2408.
42. Nelson, D.J. *J. Org. Chem.* **1986**, *51*, 3185–3186.
43. Gung, B.W.; Peat, A.J.; Snook, B.M.; Smith, D.T. *Tetrahedron Lett.* **1991**, *32*, 453.
44. Gung, B.W. *Tetrahedron Lett.* **1991**, *32*, 2867–2870.
45. Gung, B.W.; Wolf, M.A. *J. Org. Chem.* **1992**, *57*, 1370–1375.
46. Le Page, T.J.; Wiberg, K.B. *J. Am. Chem. Soc.* **1988**, *110,* 6642–6650.
47. Branchadell, V.; Oliva, A. *J. Am. Chem. Soc.* **1991**, *113*, 4132–4136.
48. Goodman, J.M. *Tetrahedron Lett.* **1992**, *33*, 7219–7222.
49. Branchadell, V.; Oliva, A. *J. Am. Chem. Soc.* **1992**, *114,* 4357–4364.

50. Dargelos, A.; Liotard, D.; Chaillet, M. *Tetrahedron* **1972**, *28*, 5595.
51. Lefour, J.–M.; Loupy, A. *Tetrahedron* **1978**, *34*, 2597–2605.
52. Guner, O.F.; Ottenbrite, R.M.; Shillady, D.D.; Alston, P.V. *J. Org. Chem.*, **1987**, *52*, 391–394.
53. Loncharich, R.J.; Schwartz, T.R.; Houk, K.N. *J. Am. Chem. Soc.* **1987**, *109*, 14–23.
54. Laszlo, P.; Teston, M. *J. Am. Chem. Soc.* **1990**, *112*, 8750–8754.
55. Laszlo, P.; Teston–Henry, M. *Tetrahedron Lett.* **1991**, *32*, 3837–3838.
56. Yamazaki, S.; Fujitsuka, H.; Yamabe, S.; Tamura, H. *J. Org. Chem.* **1992**, *57*, 5610–5619.
57. Lecea, B.; Arrieta, A.; Roa, G.; Ugalde, J.M.; Cossio, F.P. *J. Am. Chem. Soc.* **1994**, *116*, 9613–9619.
58. Suarez, D.; Sordo, T.L.; Sordo, J.A. *J. Am. Chem. Soc.* **1994**, *116*, 763–764.
59. Suarez, D.; Gonzalez, J.; Sordo, T.L.; Sordo, J.A. *J. Org. Chem.* **1994**, *59*, 8058–8064.
60. Fratiello, A.; Onak, T.P.; Schuster, R.E. *J. Am. Chem. Soc.* **1968**, *90,* 1194–1198.
61. Hinckley, C.C. *J. Am. Chem. Soc.* **1969**, *91,* 5160–5162.
62. Torri, G.; Rosset, J.–P.; Azzaro, M. *Bull. Soc. Chim. Fr.* **1970**, 2167–2168.
63. Filippini, F.; Susz, B.–P. *Helv. Chim. Acta* **1971**, *54,* 1175–1178.
64. Rosset, J.–P.; Torri, G.; Pagliardini, A.; Azzaro, M. *Tetrahedron Lett.* **1971**, 1319–1320.
65. Stilbs, P.; Forsen, S. *Tetrahedron Lett.* **1974**, 3185–3186.
66. Bose, A.K.; Srinivasan, P.R.; Trainor, G. *J. Am. Chem. Soc.* **1974**, *96*, 3670–3671.
67. Gillespie, R.J.; Hartman, J.S. *Can. J. Chem.* **1968**, *46*, 2147–2157.
68. Henriksson, U.; Forsen, S. *J. Chem. Soc., Chem. Comm.* **1970**, 1229–1230.
69. Hartman, J.S.; Stilbs, P.; Forsen, S. *Tetrahedron Lett.* **1975**, 3497–3500.
70. Fratiello, A.; Kubo, R.; Chow, S. *J. Chem. Soc., Perkin Trans. 2* **1976**, 1205–1209.
71. Torri, J.; Azzaro, M. *Bull. Soc. Chim. Fr.* **1978**, part.2, 283–291.
72. Bose, A.K.; Srinivasan, P.R. *Tetrahedron Lett.* **1975**, 1571–1574.
73. Fratiello, A.; Stover, C.S. *J. Org. Chem.* **1975**, *40*, 1244–1248.
74. Schuster, R.E.; Bennet, R.D. *J. Org. Chem.* **1973**, *38*, 2904–2908.
75. Tan, L.K.; Brownstein, S. *Inorg. Chem.* **1984**, *23*, 1353–1355.
76. Tan, L.K.; Brownstein, S. *J. Org. Chem.* **1983**, *48*, 3389–3393.
77. Glavincevski, B.; Brownstein, S. *J. Org. Chem.* **1982**, *47*, 1005–1007.
78. Wilinski, J.; Kurland, R.J. *J. Am. Chem. Soc.* **1978**, *100*, 2233–2234.
79. Rabinovitz, M.; Grinvald, A. *Tetrahedron Lett.* **1971**, 641–644.
80. Rabinovitz, M.; Grinvald, A. *J. Am. Chem. Soc.* **1972**, *94*, 2724–2729.
81. Grinvald, A.; Rabinovitz, M. *J. Chem. Soc., Perkin Trans. 2* **1974**, 94–98.
82. Childs, R.F.; Mulholland, D.L.; Nixon, A. *Can. J. Chem.* **1982**, *60*, 801–808.
83. Childs, R.F.; Mulholland, D.L.; Nixon, A. *Can. J. Chem.* **1982**, *60*, 809–812.
84. Corey, E.J.; Loh, T.–P.; Sarshar, S.; Azimioara, M. *Tetrahedron Lett.* **1992**, *33*, 6945–6948.

85. Denmark, S.E.; Alinstead, N.G. *J. Am. Chem. Soc.* **1993**, *115*, 3133-3139.
86. Keck, G.E.; Castellino, S. *J. Am. Chem. Soc.* **1986**, *108*, 3847-3849.
87. Reetz, M.T.; Hüllmann, M.; Seitz, T. *Angew. Chem., Int. Ed. Eng.* **1987**, *26*, 477-479.
88. Reetz, M.T.; Raguse, B.; Seitz, T. *Tetrahedron* **1993**, *49*, 8561-8568.
89. Reetz, M.T. in *Selectivities in Lewis Acid Promoted Reactions*, Schinzer, D., Ed.; Kluwer, Dordrecht, Holland, **1989**, pp. 107-125.
90. Keck, G.E.; Castellino, S. *Tetrahedron Lett.* **1987**, *28*, 281-284.
91. Kahn, S.D.; Keck, G.E.; Hehre, W.J. *Tetrahedron Lett.* **1987**, *28*, 279-280.
92. Shambayati, S.; Blake, J.F.; Wierschke, S.G.; Jorgensen, W.L.; Schreiber, S.L. *J. Am. Chem. Soc.* **1990**, *112*, 697-703.
93. Chen, X.; Hortelano, E.R.; Eliel, E.L.; Frye, S.V. *J. Am. Chem. Soc.* **1992**, *114*, 1778-1784.
94. Castellino, S.; Dwight, W.J. *J. Am. Chem. Soc.* **1993**, *115*, 2986-2987.
95. Castellino, S.; Volk, D.E. *Tetrahedron Lett.* **1993**, *34*, 967-970.
96. Reetz, M.T.; Raguse, B.; Marth, C.F.; Hügel, H.M.; Bach, T.; Fox, D.N.A. *Tetrahedron* **1992**, *48*, 5731-5742.
97. Denmark, S.E.; Henke, B.R.; Weber, E. *J. Am. Chem. Soc.* **1987**, *109*, 2512-2514.
98. Denmark, S.E.; Wilson, T.; Willson, T.M. *J. Am. Chem. Soc.* **1988**, *110*, 984-986.
99. Keck, G.E.; Andrus, M.B.; Castellino, S. *J. Am. Chem. Soc.* **1989**, *111*, 8136-8141.
100. Denmark, S.E.; Weber, E.J.; Wilson, T.M.; Willson, T.M. *Tetrahedron* **1989**, *45*, 1053.
101. Keck, G.E.; Castellino, S.; Andrus, M.B. in *Selectivities in Lewis Acid Promoted Reactions,* Schinzer, D., Ed.; Kluwer , Dordrecht, Holland, **1989**, pp. 73-105.
102. Krause, N. *J. Org. Chem.* **1992**, *57*, 3509-3512.
103. Denmark, S.E.; Almstead, N.G. *Tetrahedron* **1992**, *48*, 5565-5578.
104. Corcoran, R.C.; Ma, J. *J. Am. Chem. Soc.* **1992**, *114*, 4536-4542.
105. Pagni, R.M.; Kabalka, G.W.; Bains, S.; Plesco, M.; Wilson, J.; Bartmess, J. *J. Org. Chem.*, **1993**, *58*, 3130-3133.
106. Casaschi, A.; Desimoni, G.; Faita, G.; Invernizzi, A.G.; Lanati, S.; Righetti, P. *J. Am. Chem. Soc.*, **1993**, *115*, 8002-8007.
107. Bachaud, B.; Wuest, J.D. *Organometallics* **1991**, *10*, 2015-2025.
108. Laube, T.; Weidenhaupt, A.; Hunziker, R. *J. Am. Chem. Soc.* **1991**, *113*, 2561-2567.
109. Viet, M.T.P.; Sharma, V.; Wuest, J.D. *Inorg. Chem.* **1991**, *30*, 3026-3032.
110. Faller, J.W.; Ma, Y. *J. Am. Chem. Soc.* **1991**, *113*, 1579-1586.
111. Shambayati, S.; Crowe, W.E.; Schreiber, S.L. *Angew. Chem., Int. Ed. Engl.* **1990**, *29*, 256-272.
112. Brun, L. *Acta Cryst.* **1966**, *20*, 739-749.
113. Bassi, I.W.; Calcaterra, M.; Intrito, R. *J. Organomet. Chem.* **1977**, *127*, 305-313.
114. Utko, J.; Sobota, P.; Lis, T. *J. Organomet. Chem.* **1987**, *334*, 341-345.
115. Maier, G.; Seipp, U.; Boese, R. *Tetrahedron Lett.* **1987**, *28*, 4515-4516.
116. Viard, B.; Poulain, M.; Grandjean, D.; Amandrut, J. *J. Chem. Res. (S)* **1983**, 84-85.
117. Sobota, P.; Utko, J.; Lis, T. *J. Organomet. Chem.* **1990**, *393*, 349-358.

118. Poll, T.; Metter, J. O.; Helmchen, G. *Angew. Chem., Int. Ed. Engl.* **1985**, *24*, 112–114.
119. Bachand, B.; Bélanger–Gariepy, F.; Wuest, J.D. *Organometallics* **1990**, *9*, 2860–2862.
120. Simard, M.; Vaugeois, J.; Wuest, J.D. *J. Am. Chem. Soc.* **1990**, *115*, 370–372.
121. Kuivila, H.G. *Heteroat. Chem.* **1990**, 245–265. *Chem. Abst.* **1991**, *115*, 29412q.
122. Lewis, F.D.; Oxman, J.D.; Huffman, J.C. *J. Am. Chem. Soc.* **1984**, *106*, 466–468.
123. Calogero, S.; Valle, G.; Russo, U. *Organometallics* **1984**, *3*, 1205–1210.
124. Reetz, M.T.; Harms, K.; Reif, W. *Tetrahedron Lett.* **1988**, *29*, 5881.
125. Valle, G.; Calogero, S.; Russo, U. *J. Organomet. Chem.* **1982**, *228*, C79–C82.
126. Cunningham, D.; Donek, I.; Frazer, M.J.; McPartlin, M.; Matthews, J.D. *J. Organomet. Chem.* **1975**, *90*, C23–C24.
127. Laube, T.; Weidenhaupt, A.; Hunziker, R. *J. Am. Chem. Soc.* **1991**, *113*, 2561–2567.
128. Derouault, J.; Forel, M.J. *Inorg. Chem* **1977**, *16*, 3207–3213.
129. Jonas, V.; Frenking, G.; Reetz, M.T. *J. Am. Chem. Soc.* **1994**, *116*, 8741–8753.
130. Rauk, A.; Hunt, I.R.; Keay, B.A. *J. Org. Chem.* **1994**, *59*, 6808–6816.
131. Lopez–Garriga, J.J.; Babcock, G.T.; Harrison, J.F. *J. Am. Chem. Soc.* **1986**, *108*, 7131–7133.
132. Derouault, J.; Granger, P.; Forel, M.J. *Inorg. Chem* **1977**, *16*, 3214–3218.
133. Owens, C.; Woods, J.M.; Filo, A.K.; Pytlewski, L.L.; Chasau, D.E.; Korayannis, N.M. *Inorg. Chim. Acta* **1979**, *37*, 89–94.
134. Cowley, A.H.; Cushner, M.C.; Riley, P.E. *J. Am. Chem. Soc.*, **1980**, *102*, 624–628.
135. Macias, A.; Rodriguez, A.; Bastida, R.; DeBlas, A.; Sousa, A.; Rodriguez, D. *Anales Quim.* **1988**, *84*, 279–286.
136. Strel'tsova, N.R.; Ivakina, L.V.; Bel'skii, V.K.; Storozhenko, P.A.; Bulychev, B.M.; Gorbunov, A.I. *Zh. Obsh. Khim.* **1988**, *58*, 861–865.
137. Bott, S.G.; Prinz, H.; Alvanipour, A.; Atwood, J.L. *J. Coord. Chem.* **1987**, *16*, 303–309.
138. Sarma, R.; Ramirez, F.; McKeever, B.; Chaw, Y.F.; Marecek, J.F.; Nierman, D.; McCaffrey, T.M. *J. Am. Chem. Soc.* **1977**, *99*, 5289–5295.
139. Carofiglio, T.; Floriani, C.; Chiesi–Villa, A.; Guastini, C. *Inorg. Chem.* **1989**, *28*, 4417–4419.
140. Boyer, M.; Jeannin, Y.; Rocchiccioli–Deltcheff, C.; Thouvenot, R. *J. Coord. Chem.* **1978**, *7*, 219–226.
141. Denmark, S.E.; Willson, T.M. in *Selectivities in Lewis Acid Promoted Reactions,* Schinzer, D., Ed.; Kluwer Academic Publishers, Dordrecht, Holland, **1989**, pp. 247–263.

2. Lewis Acid-Promoted Ene Reaction

I. Introduction and History

I.1 The thermal reaction

In 1943, Alder found that propene reacts with maleic anhydride, in benzene at high temperature and pressure, to give allylic succinic anhydride as the only product.[1]

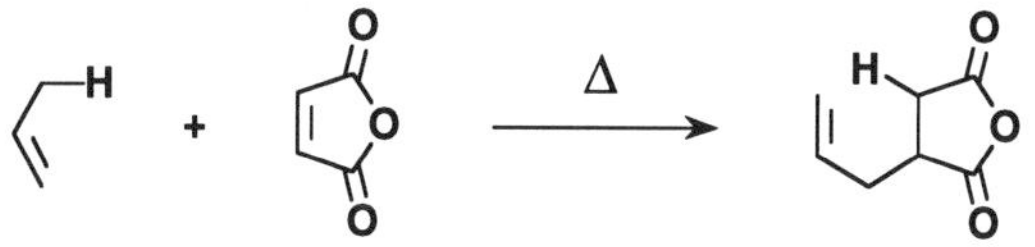

He described this reaction, initially called "Alder-ene reaction", as the "indirect substitutive addition" of a compound bearing a double bond (enophile) to an olefin possessing an allylic hydrogen atom (ene). It involves an allylic shift of the double bond, a transfer of the allylic hydrogen atom of the ene moiety to the enophile and the creation of a bond between the two unsaturated termini. Over the years, the ene reaction has attracted much attention; hence several reviews deal with this important subject.[2–12]

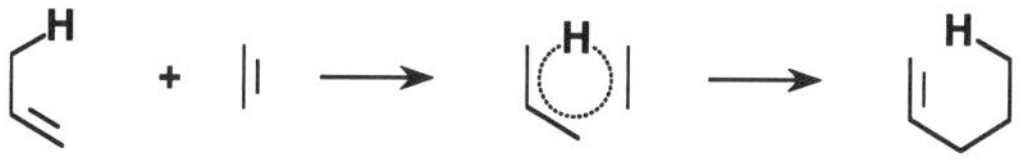

The first example of an intramolecular ene reaction was described even earlier by Schmidt in 1927; he reported the thermal cyclization of citronellal into isopulegol[13] (the acid-induced cyclization was observed as long ago as 1896[14]). The reversibility of the reaction was first reported three years later, in 1930, by Grignard (vapor phase heating of isopulegol at 500°C) who demonstrated the migration of the C–C double bond.[15]

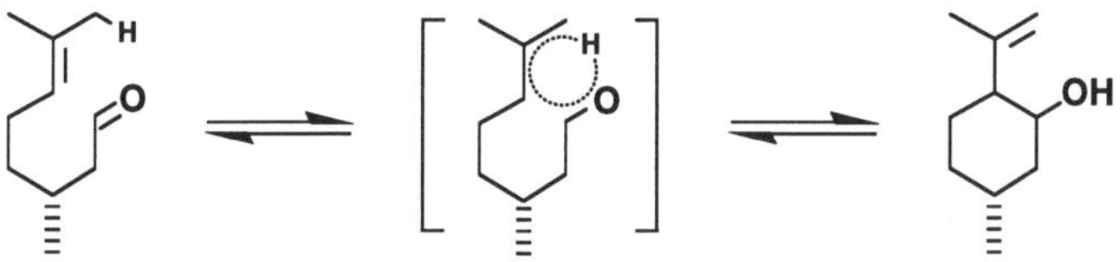

The ene reaction, which is closely related to the Diels-Alder reaction, can occur with simple unactivated hydrocarbons, although it is well known that the best results are obtained between alkenes and electron-deficient enophiles such as alkenes, ketones, or azo compounds substituted by electron-withdrawing groups.[16]

Nevertheless, the activation energy of an ene reaction is higher than for the corresponding Diels-Alder reaction (dimethyl mesoxalate and alkenes, $\Delta H^* =$ 75 to 96 $kJ.mol^{-1}$, $\Delta S^* = -120$ to -170 $J.mol^{-1}K^{-1}$).[16] Therefore ene reactions typically occur at higher temperatures, a fact that has limited the synthetic use of this reaction and can explain why it was overshadowed for many years by the corresponding Diels-Alder reaction. As for Diels-Alder reactions, activation volume is negative; thus, Gladysz reported that 40 kbar pressures must be used to perform ene reactions with β-pinene at room temperature.[18]

I.2 The Lewis acid-promoted reaction

Since enophiles, like dienophiles in Diels-Alder reactions, should be electron deficient, complexation of Lewis acids with enophiles bearing donor atoms should increase the ene reaction rate. Indeed, Lewis acid catalysis allows these reactions to be performed under milder, and therefore more synthetically attractive conditions. Thus, although the thermal reaction of (+)-limonene with formaldehyde is unsatisfactory even at 180–200°C, Blomquist established that, in the presence of BF_3-dihydrate, in CH_2Cl_2-acetic anhydride, the expected acetate is formed in 69% yield after 70 minutes at room temperature;[19] he also observed the same phenomenon with camphene.[20]

+ $CH_2=O$ → (BF_3, H_2O; Ac_2O / CH_2Cl_2; 69% yield) OAc

A few years later, Snider showed that naturally-occurring (–)-β-pinene reacts in the presence of $AlCl_3$ at room temperature with several enophiles such as acroleine, methyl acrylate, methyl vinyl ketone in better yields that under thermal conditions.[21]

(–)-β-pinene ≡ H + CO_2Me → ($AlCl_3$ (0.1 equiv.); benzene / 25°C; 70% yield) CO_2Me H

However, to the best of our knowledge, the first time a Lewis acid ene reaction was observed was in 1954. Thus, Colonge reported that chloral (1 equivalent) and isobutene (2 equivalents), in the presence of $AlCl_3$ (10% mol) in petroleum ether, led to an unsaturated alcohol, he thought was allylic alcohol **A**.[22] Two years later, Normant showed that alcohol **A** was not the only product of the reaction and that substantial amount of homoallylic alcohol **B** was also present; they accounted for its formation through an ene-mechanism.[23]

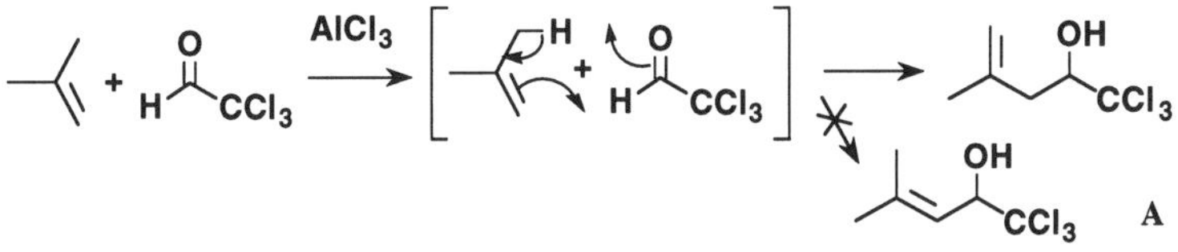

Over the years, various Lewis acids were found to promote the ene reaction at or below room temperature with excellent stereo- and regiochemical control.

II. Mechanism

II.1 The thermal reaction

The mechanism of the ene reaction has been the subject of controversial discussions, the main questions being:

- Does the reaction proceed through a concerted pericyclic mechanism (symmetry-allowed $[2\pi_s+2\pi_s+2\sigma_s]$ process) or a stepwise mechanism involving a zwitterionic or diradical intermediate?
- Is the formation of such an intermediate either fast and reversible, followed by slow hydrogen atom transfer, or slow and rate-determining, followed by a fast hydrogen atom transfer?
- Will the endo/exo selectivity be important and predictable?

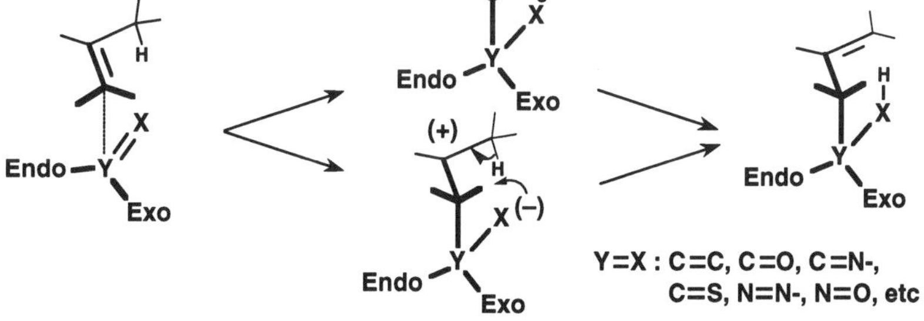

Y=X : C=C, C=O, C=N-, C=S, N=N-, N=O, etc

Both experimental data and theoretical calculations are available to try to answer these questions. Most recent experimental studies are based on kinetic isotope effects (K.I.E.; for a general reference, see[24]). However, Ohloff established thirty years ago that the thermal ene cleavage of (–)-6-hydroxymethyl-1-*p*-menthene smoothly forms (+)-1-*p*-menthene with 100% conservation of chirality. If the retro-ene reaction were to proceed via a methylcyclohexenyl radical or an allylic carbocationic intermediate, much racemic menthene would have to be formed. Microscopic reversibility demands that the reverse reaction be also completely stereospecific. The reaction of (+)-1-*p*-menthene with formaldehyde in acidic medium, led to (–)-6-hydroxymethyl-1-*p*-menthene with more than 98% retention of configuration at the C-4 carbon atom.[25] This is indeed clearly in favor of a concerted mechanism.

However, such a concerted mechanism was challenged in favor of a stepwise mechanism. In the same study, it was also established that *cis*-but-2-ene reacts with maleic anhydride to give a *threo-erythro* mixture in a 85:15 ratio whereas *trans*-but-2-ene affords the same mixture in a ca 43:57 ratio; hence, here, like in the corresponding Diels-Alder reaction, the endo approach is the rule.[26]

R^1 = H, R^2 = Me, *cis*-butene; R^1 = Me, R^2 = H, *trans*-butene

R^1 = H, R^2 = Me, *threo*; R^1 = Me, R^2 = H, *erythro*

More recently, Achmatowicz studied the ene reaction between dimethyl mesoxalate and various alkenes. Thermodynamic parameters indicated a concerted mechanism involving a "late" (product-like) transition state.[17] These conclusions were then confirmed by a kinetic isotope effects study: primary isotope effect $k_{H'}/k_{D'}$ = 2.16; secondary isotope effect $k_{H''}/k_{D''}$ = 1.05.[27] Similarly, the temperature independent isotope effects led Kwart to propose a pseudopericyclic transition state involving a nonlinear hydrogen atom transfer in the ene reaction between alkenes and TsN=S=O (k_H/k_D = 2.86)[28] or diethyl mesoxalate (k_H/k_D = 2.56).[29] However, in two studies devoted to the study of the mechanism of the triazolinedione-tetramethylethylene[30] and pentafluoro-nitrosobenzene-tetramethylethylene[31] ene reactions, intra- and intermolecular isotope effects led Greene to propose a two-step process involving the formation of an intermediate, via the rate determining step, in which no or little breaking of the C–H bond had occurred. This was corroborated by a recent work by Kresze who established, again by means of a K.I.E. study, that nonconcerted two-step mechanisms occur with a series of heteroenophiles.[32]

As for theoretical studies, orbital symmetry considerations are consistent with a concerted pathway involving a *supra-supra* facial interaction, corresponding to a pericyclic reaction, symbolized by a $[\pi2_s+\pi2_s+\sigma2_s]$

process.[33] This implies that essentially the double bond of the olefin is a donor while the C–H bond and the enophile are acceptors.[34]

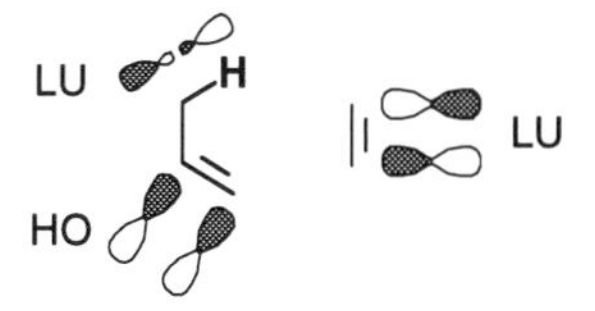

The first ab initio study is fairly recent. Thus, in 1987, Houk calculated the geometry of the transition state of the parent ene reactions between propene and ethene or formaldehyde.[35] In both reactions, a single transition state was located, indicating a concerted mechanism; moreover, the two transition states were similar and described as having, as cyclopentane, an envelope-like conformation. Both transition states display two long C–C bonds and the transferring hydrogen atom is placed between the two carbon atoms of one of these two bonds (for a review, see "Transition Structures of Hydrocarbon Pericyclic Reactions"[36]).

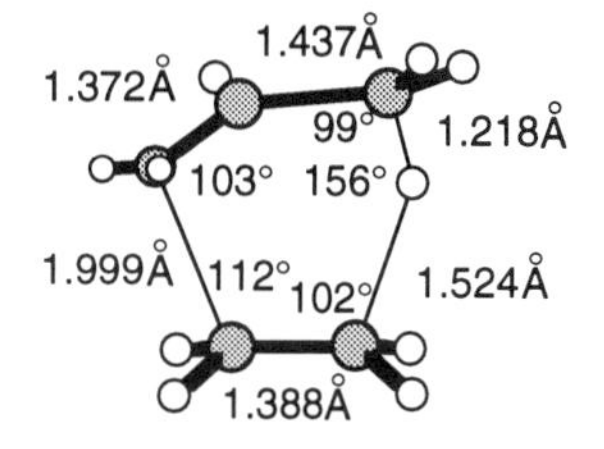

RHF/3-21G transition structure of the ene reaction between propene and ethene.[35]

3-21G transition structure of the propene-formaldehyde ene reaction.[35]

Calculations were also performed on propene and methyl acrylate. As compared with the parent ene reaction, there is less asynchronous character in the bond reorganization (more C–C bond formation) both in the exo and the endo transition structure.[37,38] The relatively small stability difference, found on that occasion as on others in favor of the exo transition state,[39,40] between exo and endo transition structures, reflects subtle exo/endo selectivity in the ene reaction.

As already mentioned, the ene reaction usually requires high temperatures, however tetracyclo[5.4.$1^{2,6}$.$1^{8,11}$]trideca-3,9-diene undergoes thermal ene reaction at an exceptionally moderate temperature (a half-life of 6 ±2 h could be estimated at 45°C). As a matter of fact, in this birdcage molecule, the allylic moiety and the C=C double bond constitute a system whose geometry is similar to that of an exo transition state.[41]

45 °C

DMSO

II.2 The Lewis acid-promoted reaction

Lewis acid catalysis not only causes dramatic rate enhancements, but can also induce mechanistic changes and hence influence the regio- and stereo outcome of the reaction. A 6-membered chair-like transition-state model can sometimes account for the stereochemistry of the adducts.[42]

H H O–L.A. R H

However, this is not always the case and the Lewis acid promoted ene reaction mechanism has been studied by various experimental means, particularly once again through kinetic isotope effects.

Thus, inter- and intramolecular kinetic isotope effect of the $SnCl_4$ catalyzed and thermal reactions of diethyl mesoxalate **A** with allylbenzene have been compared: the introduction of the Lewis acid results in a substantial reduction in the magnitude of the isotope effect (thermal: $k_H/k_D = 3.3$; $SnCl_4$ catalyzed: $k_H/k_D = 1.1$).[43] The catalyzed reaction led to two alcohols obtained in similar proportions.

C_6H_5 + EtO_2C–CO–CO_2Et (**A**) → [$SnCl_4$ (0.2 equiv.), 2–3 min.; room temp. $CDCl_3$] → 51–54% + 49–46%

Moreover, both inter- and intramolecular effects are affected to a very similar degree.

{C_6H_5 (D, D) / C_6H_5 (H, H)} + A → {51% D / H = 1.03–1.18} + {49% D / H = 1.03–1.18}

C_6H_5 (H, D) + A → 54% D / H = 1.08–1.17 (D(H)) + 46%

Similar small intermolecular kinetic isotope effects are observed during the reaction of 2,3-dimethylbut-2-ene with formaldehyde-Me_2AlCl ($k_H/k_D = 1.4 \pm 0.15$) or methyl propiolate-$EtAlCl_2$ ($k_H/k_D = 1.1 \pm 0.15$).

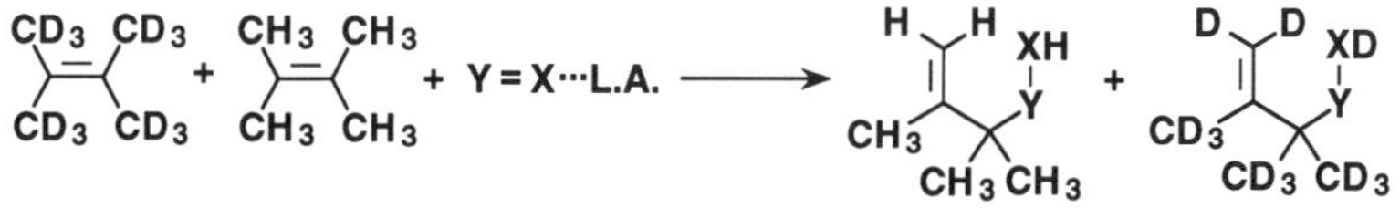

In contrast, a substantial intramolecular isotope effect is observed in the Lewis acid catalyzed ene reactions of 2,3-dimethylbut-2-ene-d_6 and formaldehyde-Me_2AlCl (k_H/k_D = 3.3 ±0.15), methyl propiolate-$EtAlCl_2$ (k_H/k_D = 2.5 ±0.15), diethyl mesoxalate-$SnCl_4$ ((k_H/k_D = 2.1 ±0.15) and acetyl chloride-$SbCl_5$ in the presence of NEt(i-Pr)$_2$ (k_H/k_D = 1.9 ±0.2).

If the reaction is concerted, the intermolecular and intramolecular isotope effects should be of similar magnitude. The large intramolecular isotope effects observed with methyl propiolate and formaldehyde are most significant. They are not consistent with a simple stepwise mechanism, proceeding through a zwitterionic intermediate, which cannot have a primary isotope effect and should have secondary isotope effects of the order of 1.1–1.4.[44]

Beak reported quite similar results in the $SnCl_4$ catalyzed ene reaction of methylenecyclohexane with diethyl mesoxalate. An intermolecular isotope effect of 1.1 ±0.2 and an intramolecular effect of 1.2 ±0.1 were observed. These small values for both effects suggest that this reaction does not proceed via a fully developed freely rotating zwitterion intermediate. He proposed a stepwise mechanism that allows analysis of the isotope effects in terms of the partitioning of a geometrically defined intermediate.[45,46]

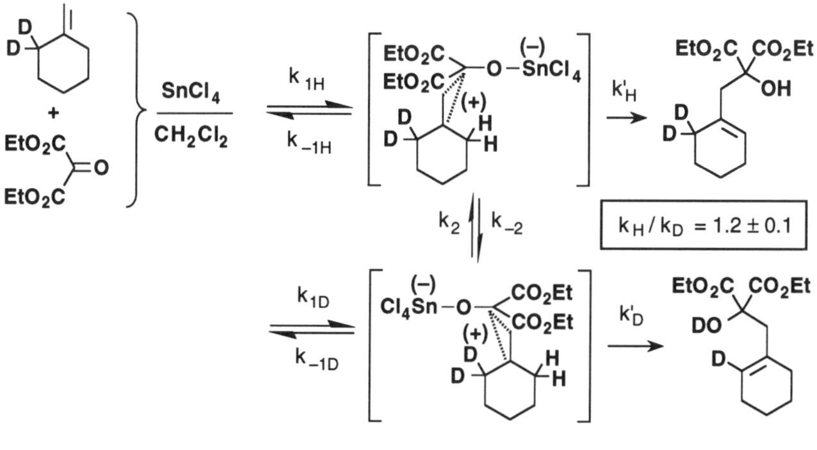

$k_H / k_D = 1.2 \pm 0.1$

These results clearly indicate that the C–H bond breaking is not an important component of the rate-determining step. They are consistent with either a stepwise reaction in which the first step is irreversible or a concerted reaction with a very unsymmetrical transition state, i.e., an asynchronous process.

Such a mechanism involving the formation of a three-membered complex as a transition state or intermediate had previously been proposed by Kwart.[47] Indeed, it was invoked to account for the formation of an oxetane ring rather than an allylic alcohol when allylbenzene and diethyl mesoxalate were reacted under $SnCl_4$ catalysis.

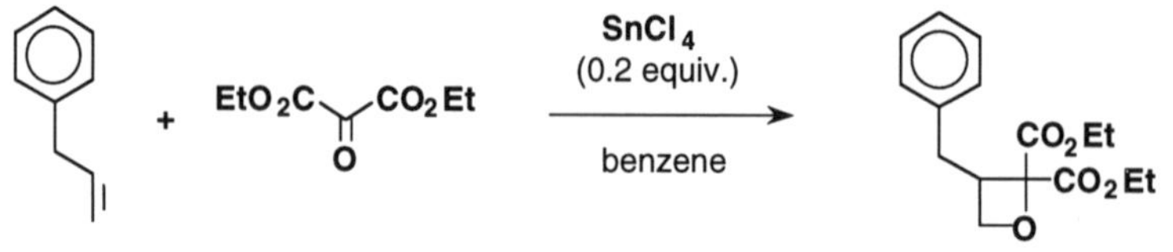

$SnCl_4$ (0.2 equiv.), benzene

It was also shown that in the presence of Lewis acids, control of selectivity is more electronically determined and less dependent on steric effects. Thus, the relative reactivities of the mono- and trisubstituted C=C bonds of 2-methylhepta-2,6-diene are 1.00 to 0.09.[48]

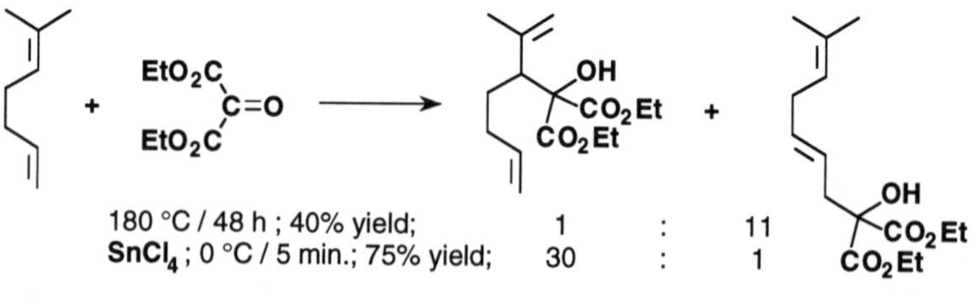

180 °C / 48 h ; 40% yield;	1	:	11
$SnCl_4$; 0 °C / 5 min.; 75% yield;	30	:	1

Moreover, a large accelerating effect of electron-donating substituents was observed for $SnCl_4$-catalyzed ene reactions of diethyl mesoxalate with a series of 1-arylcyclopentenes. The ρ value (ρ = –3.9 ±0.3) indicated a cationic charge for the benzylic carbon atom C–1. This result is reminiscent of those observed for solvolytic reactions in which benzylic cation intermediates are considered as important intermediates, e.g., hydrolysis of cumyl chloride (ρ = –4.48 ±0.12).[49] In contrast, a small accelerating effect of electron-donating substituents is detected for the corresponding thermal reactions (ρ = –1.2 ±0.2).

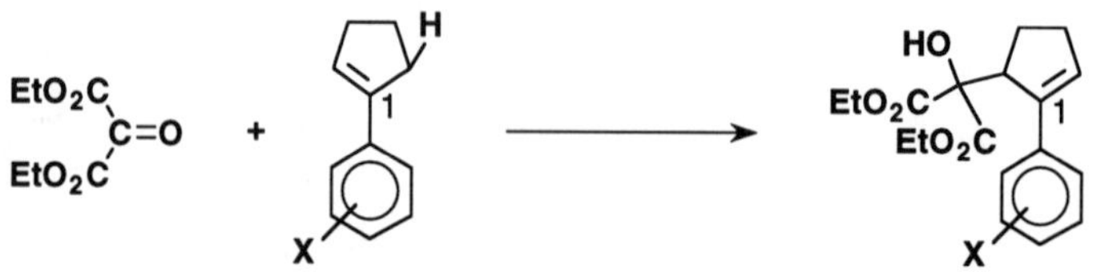

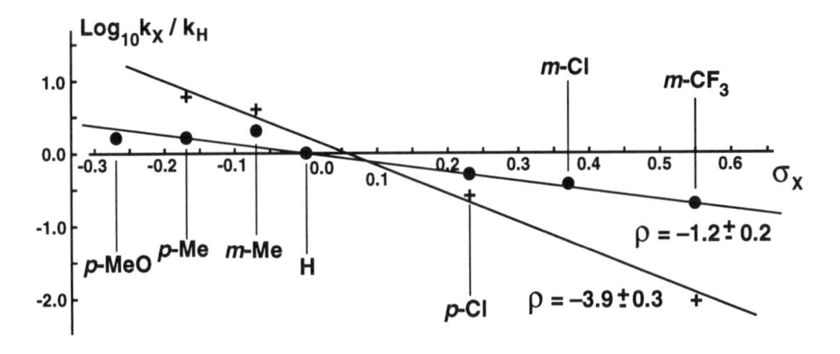

Substituent effects on the thermal and $SnCl_4$-catalyzed ene reaction of 1-arylcyclopentenes and diethyl mesoxalate (thermal, $\rho = -1.2 \pm 0.2$, catalyzed, $\rho = -3.9 \pm 0.3$).

However, it is unlikely to consider the formation of a zwitterion with a carbocation as component. Indeed, β-pinene, which is a powerful enophile, leads to a substituted α-pinene derivative.[48]

H
(−)-β-pinene
+
EtO_2C
C=O
EtO_2C
$Hg(OTf)_2$
(0.2 equiv.)
benzene
90% yield
H
H
O--$Hg(OTf)_2$
EtO_2C
CO_2Et
H
OH
EtO_2C
CO_2Et
≡
OH
CO_2Et
CO_2Et

If the reaction were to proceed via a carbonium ion such as **A**, much product would arise from the rearranged carbonium ion **B**.

H
H
O--L.A.
EtO_2C
CO_2Et
H
(−)
H
(+)
O−L.A.
EtO_2C
CO_2Et
A
H
(−)
H
(+)
O−L.A.
EtO_2C
CO_2Et
B

In fact, it was found that the isolated products retain the pinene carbon skeleton. Another significant example is the ene reaction between vinyl cyclopropane **A** and a series of aliphatic and aromatic aldehydes; no products resulting from a carbocationic process were obtained.[50] Similarly, the reaction of methylenecyclobutane with pivalaldehyde in the presence of Me_2AlCl lead only to the corresponding cyclobutenylmethylcarbinol (43% yield).[51]

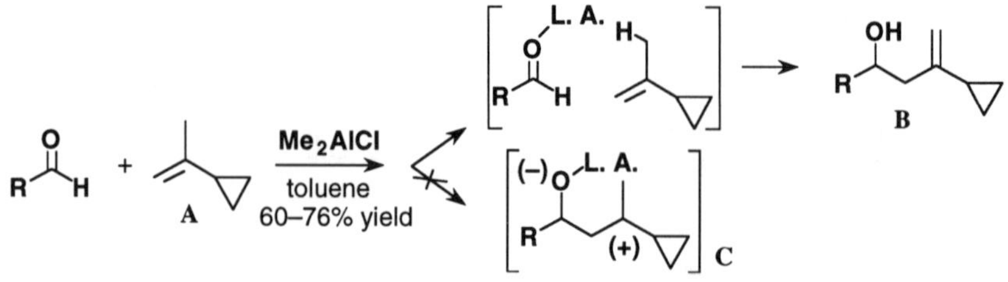

Dimethylaluminum chloride, which is a mild Lewis acid and a proton scavenger, catalyzes the ene reaction of aldehydes with alkenes bearing a disubstituted vinylic carbon atom. Proton-initiated rearrangements do not occur, since the alcohol-Lewis acid complex formed in the ene reaction reacts rapidly to give methane and a nonacidic aluminum alkoxide.

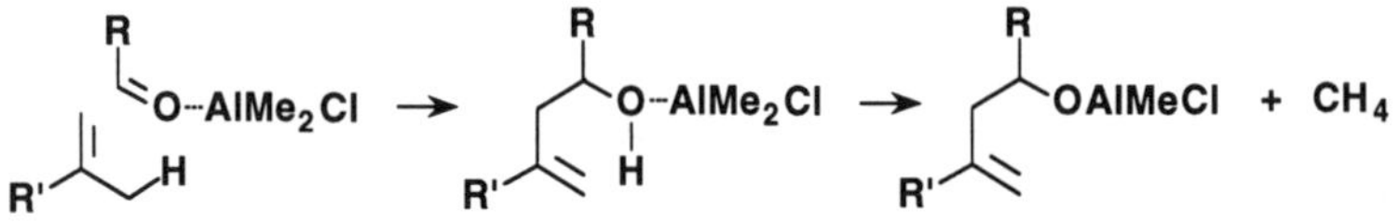

The dichotomy between concerted and cationic mechanisms of Lewis acid mediated ene reactions was also studied by means of $^{13}CH_3$-labeling experiments on 1,6-dienoate. In a concerted process, the hydrogen atom should be transferred selectively from the *trans*-positioned allylic methyl group and therefore the ^{13}C atom present in the product must be the terminal olefinic one. Both the thermal and catalyzed reactions are in agreement with that statement.

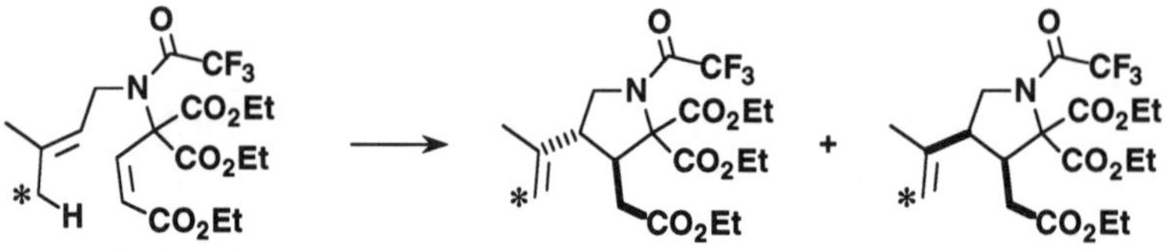

180°C, 5 min. or 70 °C, 8 h	3	:	1
Et_2AlCl, 20 equiv., –78°C, 5 min. or **Et_2AlCl** , 3 equiv., –78°C, 8 h	1	:	0

In contrast, if the C–C bond closure and the H-transfer occur in two independent kinetic steps through a carbocationic intermediate, the ^{13}C-label would be scrambled between the methyl and the methylene in the cyclization product. This is clearly not the case.[52]

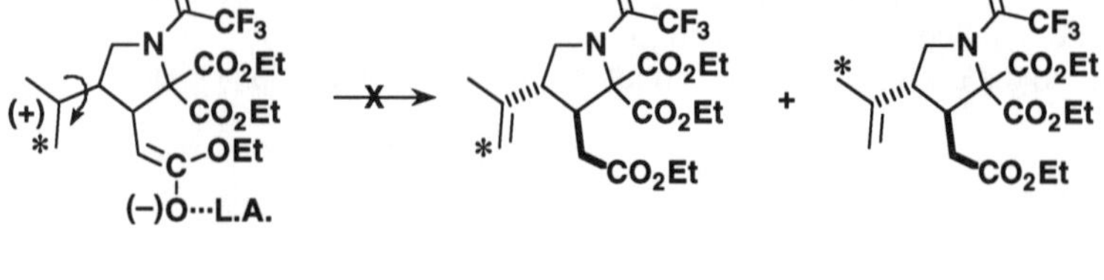

Labeling experiments have shown that the stereoelectronic "*cis*-effect" (allylic hydrogens on the disubstituted side of the alkene linkage are abstracted more readily than those on the monosubstituted side)[53] provides useful explanation to the stereooutcome of the reaction. Indeed, it appears from the following experiment that the abstracted atom of the ene comes from the *cis* methyl group (for a detailed discussion on the "*cis*-effect", see Section III.1.a).

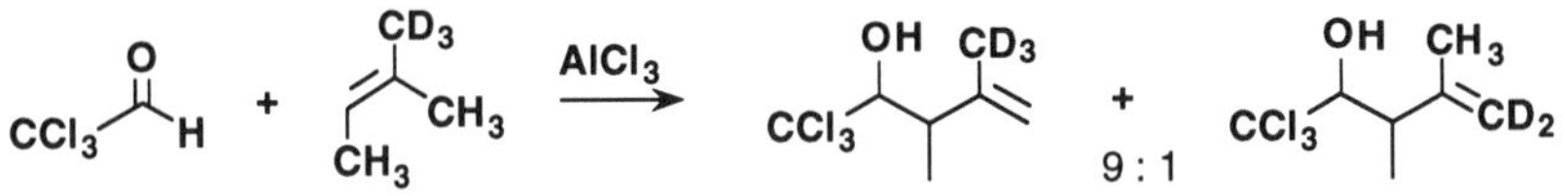

Finally, a semi-empirical and ab initio study of trihaloacetaldehyde-H^+ complexes was reported.[54] Particularly, LUMO level vs. electron density on the carbonyl carbon was studied and correlated to experimental observations (see Section V-2).

In conclusion, one can say that:

- the thermal reaction proceeds through a concerted mechanism involving a late transition state, except with hetero enophiles.
- the Lewis acid induced reaction is a stepwise process but the intermediate involved does not generally display a real carbocationic center.

III. Intermolecular Reaction

III.1 Reaction involving achiral aldehydes as enophiles

III.1.a Chloral

Chloral is among the most reactive enophile. It is used both under thermal[55] and Lewis acid catalysis conditions. Thus, unsaturated cyclic and bicyclic hydrocarbons react at room temperature with chloral in the presence of $AlCl_3$ to yield oxatricyclic compounds.[56] The addition of norbornene to chloral gives rise to cyclic ethers **A** which display the oxabrexane skeleton.

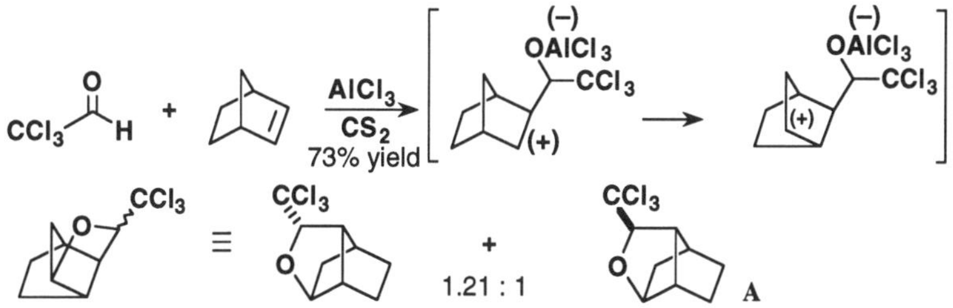

Treatment of ethers **A** with potassium hydroxide affords cyclic enol ether **B**, a potent synthetic intermediate.

Also very interesting is the addition of cycloocta-1,5-diene to the chloral-$AlCl_3$ complex; it leads to an oxatriquinane derivative.[56]

Because of the typical carbocationic rearrangements observed, these reactions are likely to proceed through a stepwise mechanism involving zwitterionic intermediates.

Gill proposed to explain his results on the Lewis acid catalyzed chloral ene reaction, performed at room temperature, by invoking an envelope model transition state rather than a six-membered ring one; the Lewis acid being complexed *anti* or *syn* to the exo substituent. Thus, the following topology (and their enantiomers) accounts for most of his results.[57]

This topology has been used throughout the chapter, even by using the enantiomers of the original reactants.

Alk-1-enes were found to be less reactive than most alkenes, and ketonic compounds were formed as side products in almost all reactions.

The formation of dichloromethylketones could result from a chloride anion transfer followed by hydride shift.

For example, the Lewis acid catalyzed reaction between hex-1-ene and chloral led to two adducts, alcohol **A** and ketone **B**; if performed under thermal conditions, the same reaction proved to be very sluggish and led only to tarry products. Alcohol **A** is obtained as a mixture of *trans:cis* isomers (95:5) which indicates the high stereoselectivity in the transfer of the allylic hydrogen atom during the reaction (*trans* selectivities have been observed in ene reactions with alk-1-enes). The following rate-determining pseudopericyclic transition state accounts for this result.

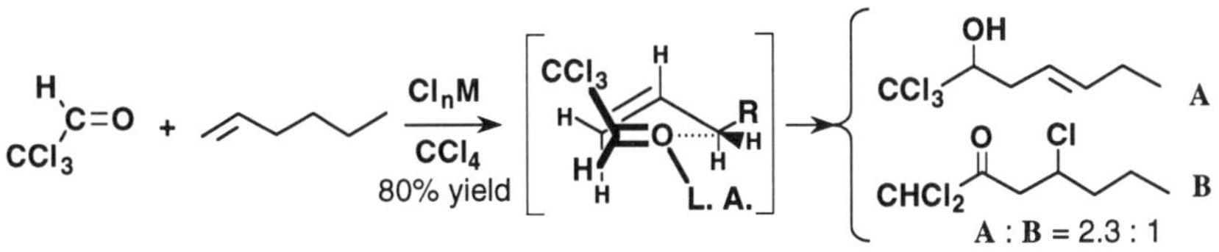

Cl_nM, CCl_4, 80% yield; L. A.; **A** : **B** = 2.3 : 1

The fact that Lewis acids generally complex *anti* to the R group of the aldehyde is illustrated by the endo/exo selectivity observed for the ene reaction between chloral and (+)-β-pinene. Indeed, depending on the conditions, Lewis acid catalyzed or thermal, the reaction leads predominantly either to endo adduct **B** or to exo adduct **A**.

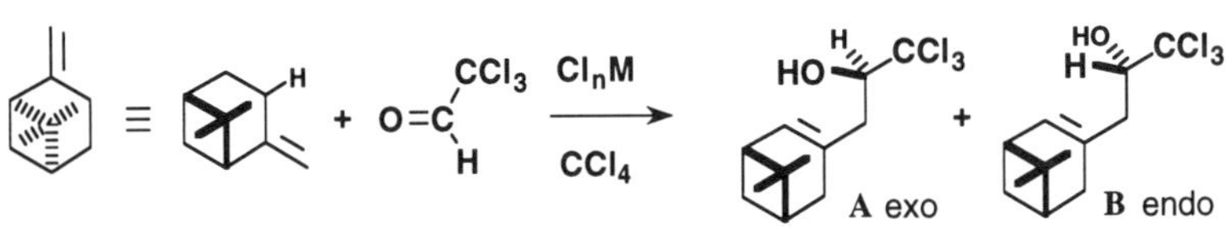

Cl_nM, CCl_4; **A** exo + **B** endo

In the thermal reaction, the exo/endo selectivity (83:17) appears to be determined by steric interactions. Nonbonded repulsions should be minimized if chloral approaches the β-pinene from the methylene bridge side with the trichloromethyl group orientated away from the hydrocarbon skeleton. Hence, the major diastereoisomer **A** results from a transition state with an exo trichloromethyl group.

exo → **A**

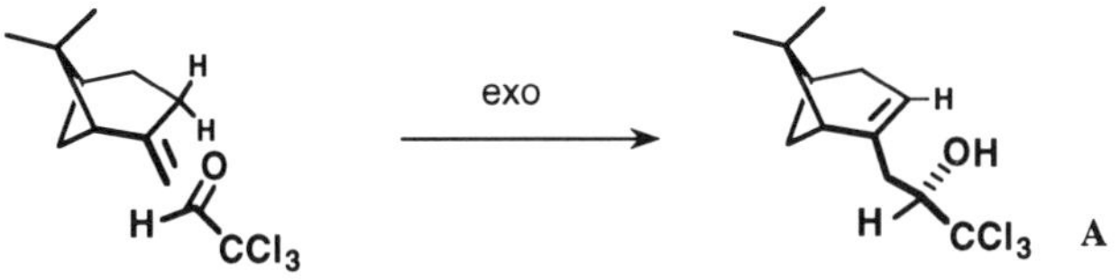

Reverse stereoselectivity is observed for the Lewis acid catalyzed reaction. The most likely mode of coordination between chloral and the Lewis acid is the one that places the Lewis acid *anti* to the bulky trichloromethyl group (see Chapter 1, Section III), thereby changing the preferred approach between the ene and enophile. Reaction then occurs selectively through a transition state in which the trichloromethyl group takes an endo position and the Lewis acid an exo one. Such a configuration minimizes steric interactions between the bulky Lewis acid and the hydrocarbon skeleton. Adduct **B** is produced with 100%

stereoselectivity in the $TiCl_4$-catalyzed reaction (2 mol%). It is therefore assumed that $TiCl_4$ is more sterically demanding than the aldehyde trichloromethyl group.[58]

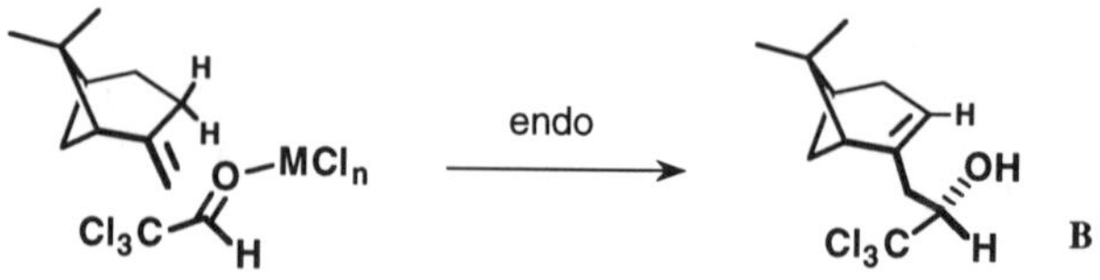

endo

B

The diastereoselectivity of the ene addition between chloral **A** and 2-methylbut-2-ene **B** depends on the reaction conditions. The thermal reaction affords only 15% of the adducts with *anti* alcohol **D** being the major component. In contrast, the $AlCl_3$ catalyzed reaction proceeds with *syn*-selectivity leading predominantly to homoallylic alcohol **C** (X-ray structure analysis of 3,5-dinitrobenzoate ester).[59,60]

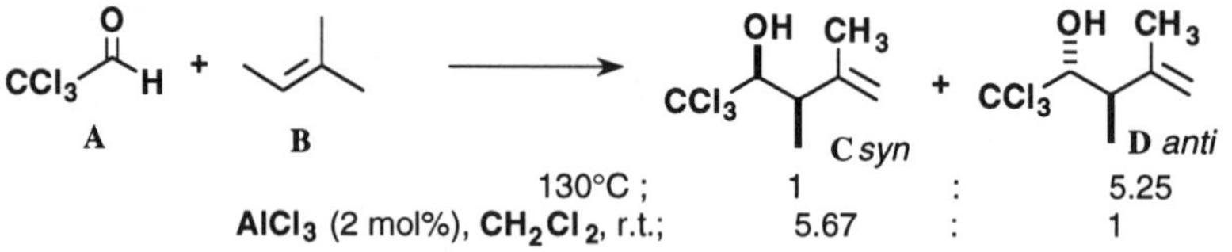

Conditions	C *syn*		D *anti*
130°C ;	1	:	5.25
$AlCl_3$ (2 mol%), CH_2Cl_2, r.t.;	5.67	:	1

These results can be accounted for in the following way. For the thermal reaction, in the four possible rate-determining pseudopericyclic transition states **E–H**, the major steric interaction occurs between the R (=CCl_3) group and the terminal methyl group. This interaction is minimized in transition states **E** and **H** which lead to the formation of *anti*-isomer **D**.

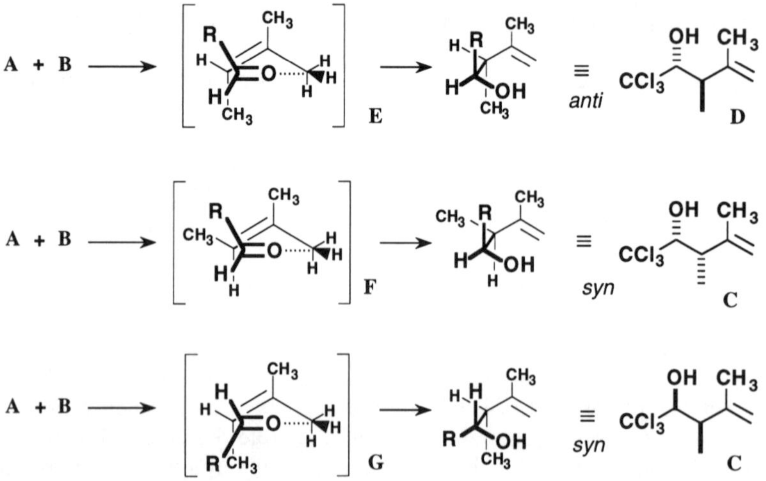

A + B → E → ≡ *anti* D

A + B → F → ≡ *syn* C

A + B → G → ≡ *syn* C

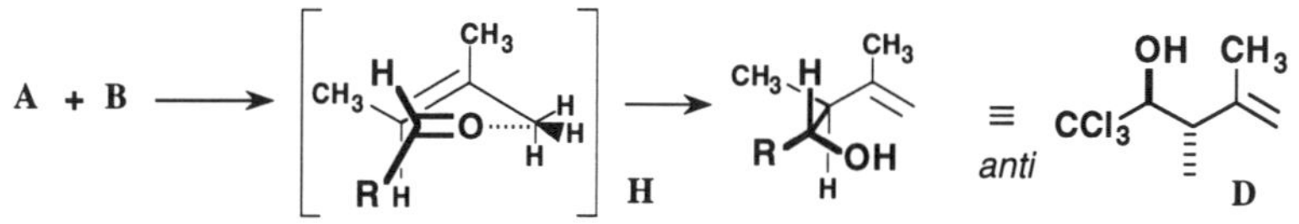

Obtainment of the *syn*-isomer **C** under $AlCl_3$ catalysis can be accounted for by the intervention of the pseudopericyclic transition state **I** in which the major steric interaction occurs between the terminal methyl and R groups. The alternative pseudopericyclic transition state **J** which leads to the minor *anti*-isomer **D** minimizes that interaction but implies that the Lewis acid and the terminal methyl group are on the same side and that the R groups are in an endo position.

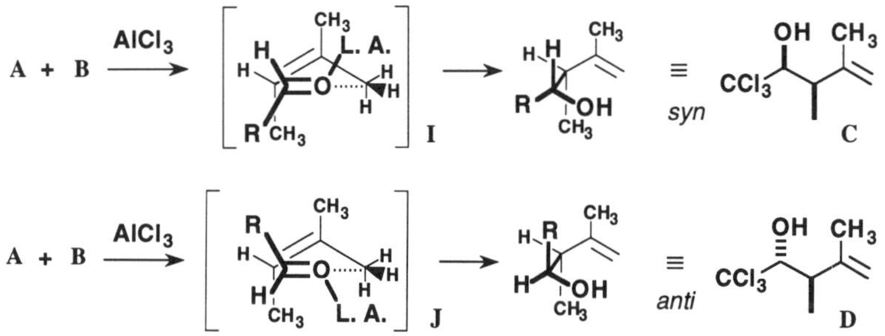

Both transition states **I** and **J** involve the attack of the hydrogen atom of the *cis*-methyl group; this stereoelectronic phenomenon is known as the "*cis* effect" (see Section II.2), it was established by deuterium labeling experiments. Indeed, distinction between the two methyl groups is possible by specific deuterium substitution. Thus, when the reaction is performed between chloral **A** and deuterated 2-methylbut-2-ene ***d*-B**, *syn*-isomers ***d*-C** and ***d*-C'** constituted 87% of ene adducts with an allylic H vs. D abstraction ratio of about 9:1. The same ratio is observed for *anti*-isomers ***d*-D** and ***d*-D'** (13% of ene adducts).

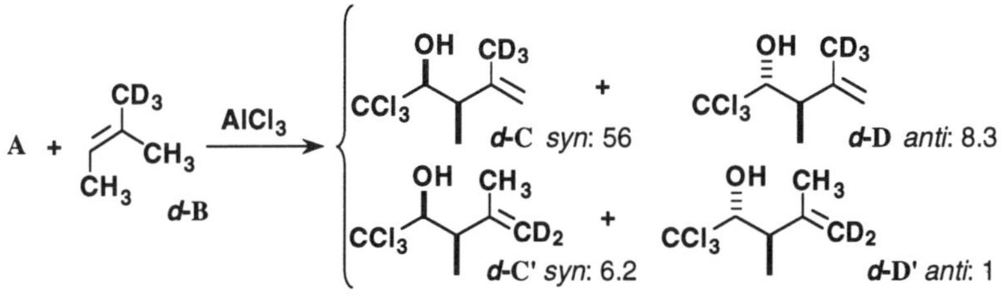

The formation of ***d*-C** as the major product is in agreement with the "*cis* effect" which involves the pseudopericyclic transition state ***d*-H** with the R group in an exo position.

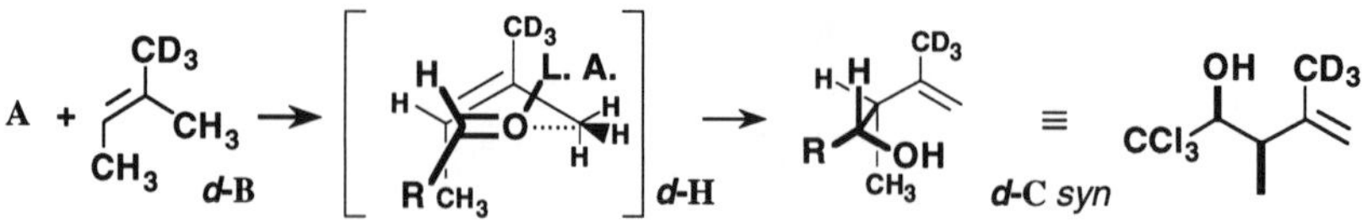

Minor *syn*-isotopomer ***d*-C'** results from a reaction path involving the transfer, through transition state ***d*-J**, of a deuterium atom from the *trans* methyl group.

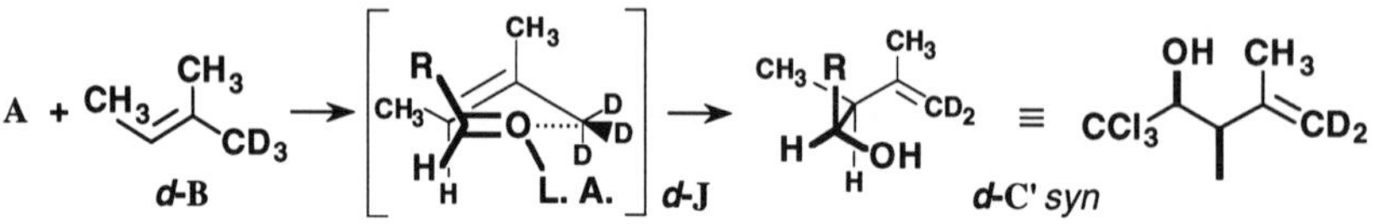

The ene reaction of chloral with both (*E*)- and (*Z*)-3-methylpent-2-ene shows a preference for the "*cis* effect". However, the presence of an ethyl group *syn* to the vinylic hydrogen in the (*E*)-isomer has an appreciable perturbing influence.

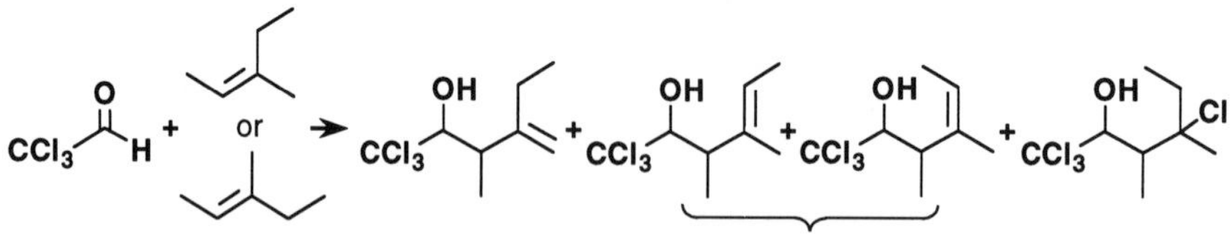

$AlCl_3$	(*E*)-isomer: 80% yield;	3.75	:	3.6	:	1
(3 mol%)	(*Z*)-isomer: 78% yield;	1	:	6.7	:	0

Simple cycloalkenes such as cyclohexene also react smoothly with chloral to give ene adducts with good to excellent diastereoselectivity.[61]

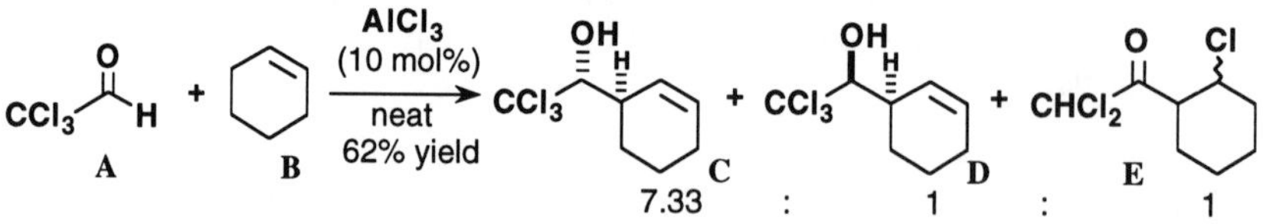

Nevertheless, in this case, major diastereomer **C** was found (by X-ray crystallographic analysis) to be *anti* due to greater steric repulsions between the R (=CCl_3) group and the cycle.

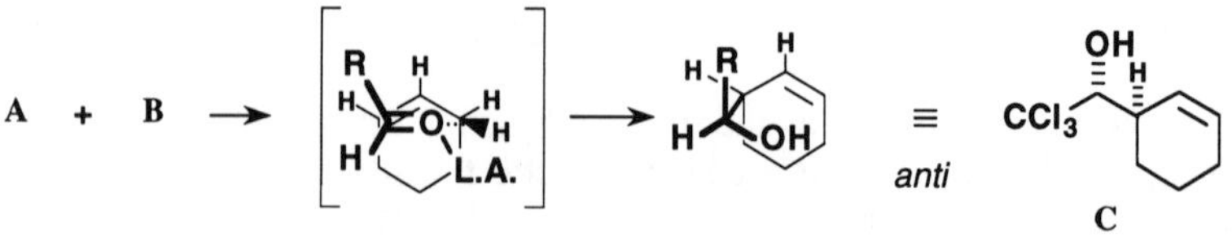

Better overall yield and diastereoselectivity was observed with cyclooctene.[59]

$AlCl_3$ (6 mol%)
CH_2Cl_2
70% yield; 16 : 1
5% yield

Olefin reactivity towards chloral-$AlCl_3$ was determined by competitive method and showed a ca 900-fold variation in rate.

Olefin	k_{rel}	Olefin	k_{rel}
β-pinene	27.3	hex-1-ene	1.7
(+)-limonene	9.8	oct-1-ene	1.0
α-methylstyrene	9.2	cyclohexene	0.16
1-methylcyclohexene	2.6	allylbenzene	0.09
		1-chlorocyclohexene	0.03

Table 1. Relative reactivities of alkenes with chloral-$AlCl_3$.[60]

Finally, chloral can also react with terminal olefin under $FeCl_3$ catalysis as shown by Sekiya in his synthesis of deoxyobtusilactone A.[62]

III.1.b Formaldehyde

Formaldehyde is also a reactive enophile that often is used for the preparation of primary homoallylic alcohols. The well-known Prins reaction is an acid catalyzed stepwise addition of formaldehyde to an alkene.[63]

Lewis acid catalyzed reactions of formaldehyde with alkenes can occur without selectivity.[64,65] In contrast, Snider prepared antibiotic (±)-pseudomonic acid A through a sequence involving two ene reactions. He first prepared, from formaldehyde and hexa-1,5-diene, alcohol **A** which was then protected into acetate **B**. In the presence of an excess of formaldehyde and $EtAlCl_2$, **B** led to **E**, a crucial intermediate en route to pseudomonic acid A. The formation of **E** results from an ene reaction followed by a Diels-Alder cyclization of complex **D**.[66,67]

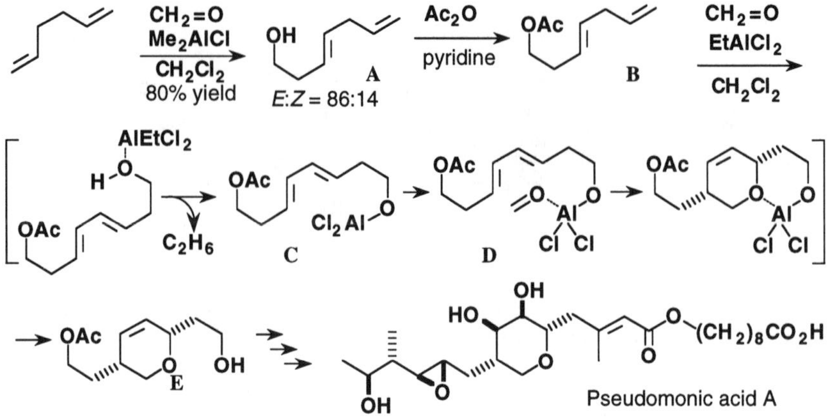

The Me_2AlCl promoted addition of formaldehyde to (*E*)- and (*Z*)-3-methylpent-2-ene leads to a mixture of regioisomers.[68]

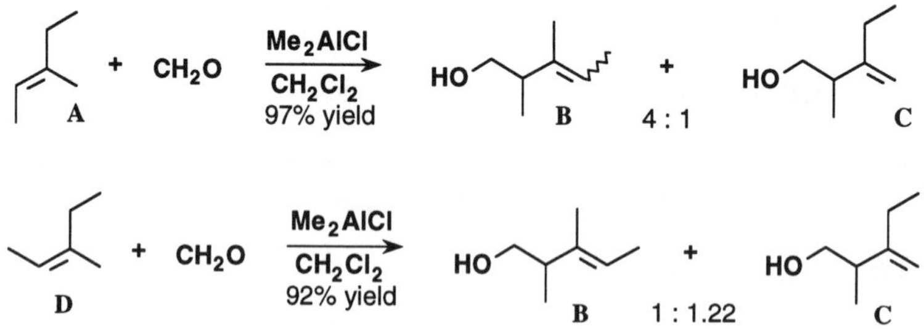

The preferential abstraction of a hydrogen atom from the alkyl group *syn* to the vinylic hydrogen atom may be due to steric interactions between the Lewis acid and the substituent on the less substituted end of the double bond. In that case, steric interactions seem to counterbalance the stereoelectronic "*cis*-effect".

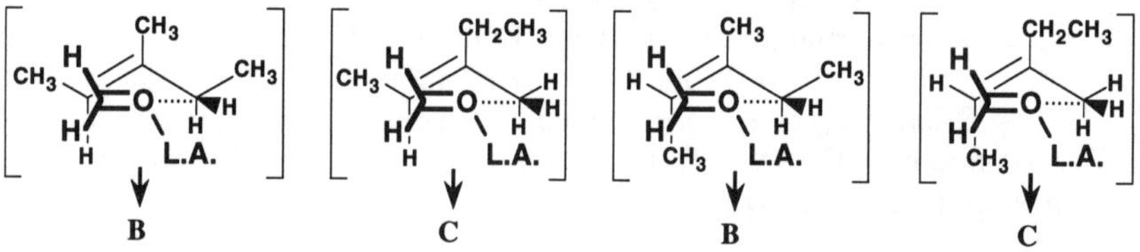

Yamamoto showed that methylaluminum bis(2,6-diphenylphenoxide) (MAPH) is capable of forming a 1:1 complex, stable at 0°C for 5 h, with formaldehyde (from trioxane). This complex induces eminent chemo- and regioselectivity in ene reactions.[69] With 1-methylcyclohexene, the hindered catalyst MAPH gives rise only to alcohol **E** which results from transition state **C** whereas when Me_2AlCl is used, a mixture of **D** and **E** is obtained. Presumably, with MAPH, transition states **A** and **B** are defavored because of steric interactions between the bulky Lewis acid and the cycle.

CH_2Cl_2 → A or B → D; C → E

L.A. = Me_2AlCl (92% yield); **D** : **E** = 1.44 : 1

L.A. = (MAPH) (80% yield); **D** : **E** = 0 : 1

Wovkulich proposed a very convenient synthesis of Prelog-Djerassi lactone from formaldehyde and 1-ethylidene-2-methylcyclopentane.[70] The regioselectivity of the ene reaction is at least 96% from (*Z*)-isomer **A** and 88% from (*E*)-isomer **B**.

	Conditions	C		D		E
A	paraformaldehyde, $BF_3 \cdot OEt_2$ (4 mol%), 60–70% yield	21.5	:	2.5	:	1
B	paraformaldehyde, $BF_3 \cdot OEt_2$ (4 mol %), 65% yield	1	:	1.4	:	17.6

The proton abstraction occurs preferentially on the ring allylic carbon atom *anti* to the olefinic methyl group. With (*Z*)-isomer **A**, formaldehyde adds predominantly to the *Si* face of the double bond to yield alcohol **C**. Addition to the more hindered *Re* face of the double bond yields isomeric alcohol **D**. The abstraction of the methine hydrogen atom is a thermodynamically favored process and leads to alcohol **E**. However, transition state leading to **E** is particularly hindered; this accounts for only 4% of **E**. With (*E*)-isomer **B**, formaldehyde adds to the *Si* face of the double bond to yield alcohol **E**. The transition state involved is kinetically and thermodynamically favored.

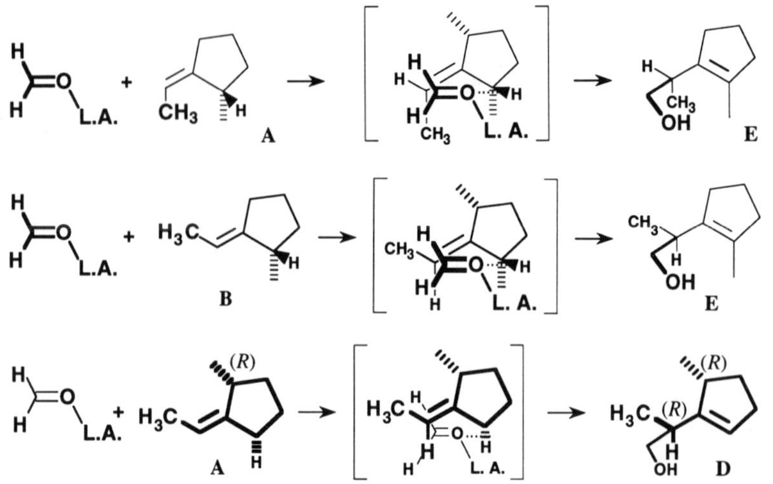

Lewis acid catalyzed ene reactions sometimes can lead predominantly to products resulting from carbocationic intermediates. For example, the Me_2AlCl catalyzed ene reaction between formaldehyde and terminal alkynes gives a 2:3 mixture of the ene adduct and the (*Z*)-chloroalcohol resulting from a zwitterionic intermediate.[71]

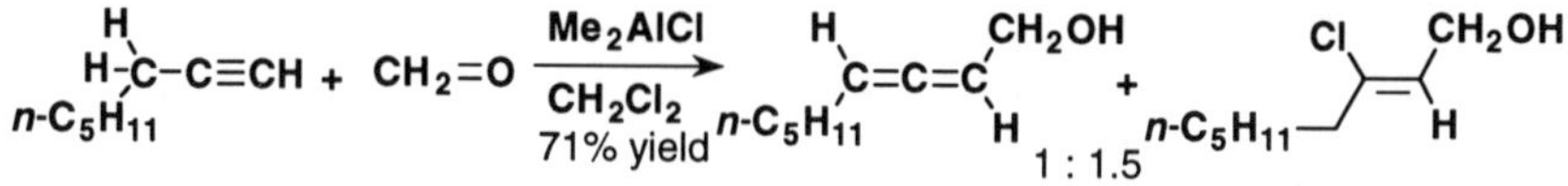

The reactivity of terminal alkynes was taken advantage of to synthesize methyl 8-hydroxyocta-5,6-dienoate, an antifungal agent isolated from *Sapium japonicum*, from methyl hept-6-ynoate. In this case, a slightly stronger Lewis acid, methylaluminum sesquichloride (prepared from 2 equivalents $AlCl_3$ and 2 equiv. $AlMe_3$) and 4 equivalents CH_2O were used to yield 25% of the desired allenic adduct.

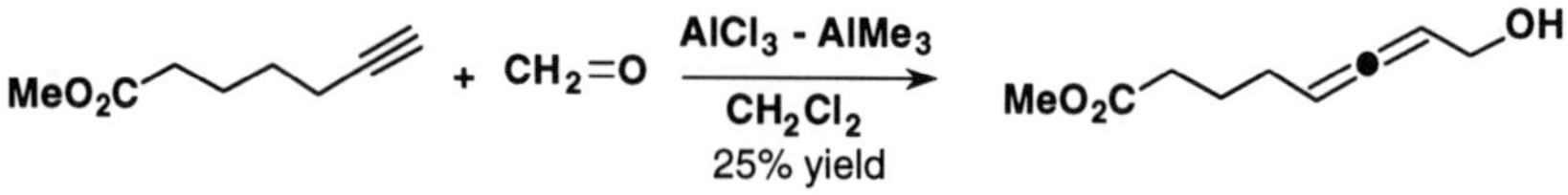

III.1.c Glyoxylate esters

Glyoxylate-ene reaction is synthetically very useful.[72] Particularly, Mikami developed a bulky silyl ether-directed carbonyl-ene reaction in which allylic silyl ethers react with methyl glyoxylate with both high regio- and stereocontrol. Thus, *anti* (*E*)-methyl ester **B** is obtained as a single isomer, irrespectively of the geometry of ene **A**.[73]

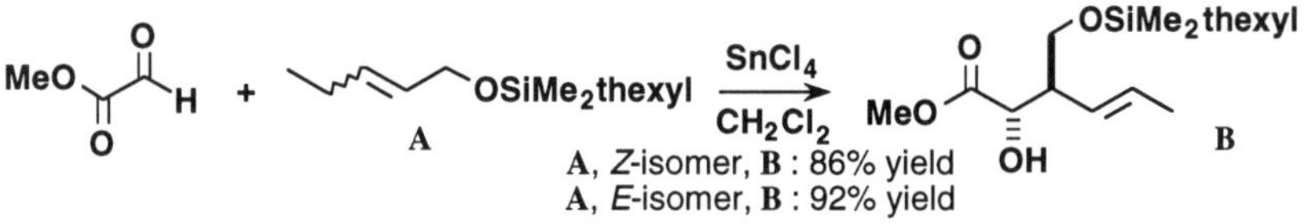

A, *Z*-isomer, **B** : 86% yield
A, *E*-isomer, **B** : 92% yield

The exclusive formation of *anti* (*E*)-methylester **B** is explained as resulting from the intervention of exo transition state **C** (from (*E*)-isomer of **A**) and endo transition state **D** (from (*Z*)-isomer of **A**) which both minimize steric repulsions between alkoxy groups.

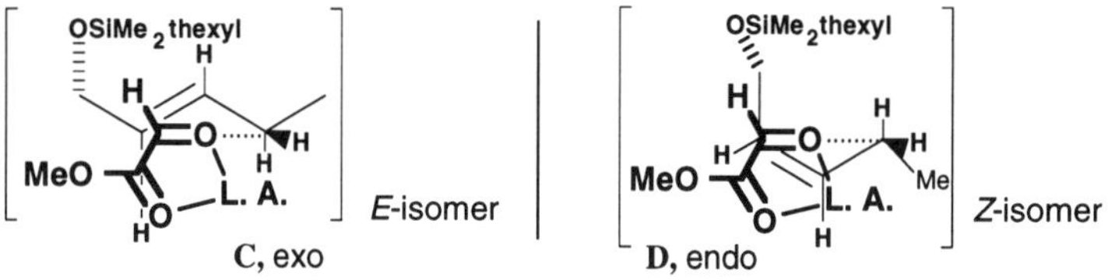

C, exo — *E*-isomer | **D**, endo — *Z*-isomer

Bishomoallylic silyl ethers react with methyl glyoxylate to yield stereoselectively (>91%) two products, ene adduct **A** and substituted tetrahydrofuran **B**. The formation of **B** results from a Prins reaction and constitutes a remarkably stereocontrolled route to the tetrahydrofuran unit of polyether antibiotics such as tetronasin or pamamycin-607.[74]

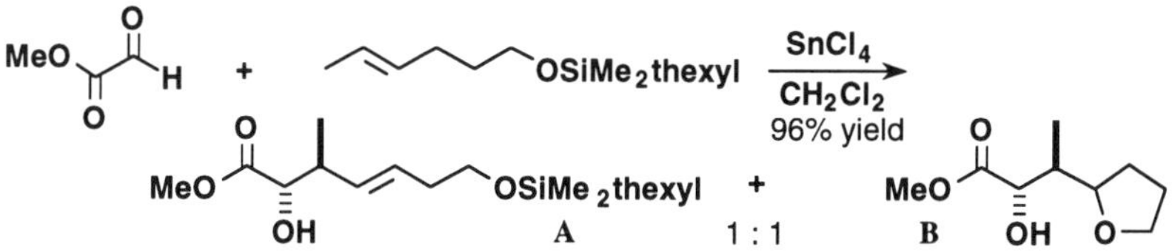

$SnCl_4$, CH_2Cl_2, 96% yield; **A** : **B** = 1 : 1

Mikami also developed an efficient method for 1,4- and 1,5-remote stereocontrol.[75]

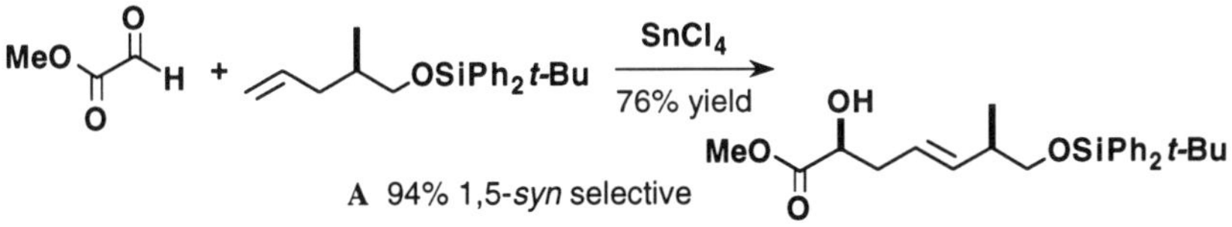

$SnCl_4$, 76% yield; **A** 94% 1,5-*syn* selective

The following exo transition state, with a conformation of the chain that minimizes steric repulsions, explains the 1,5-relationship observed.

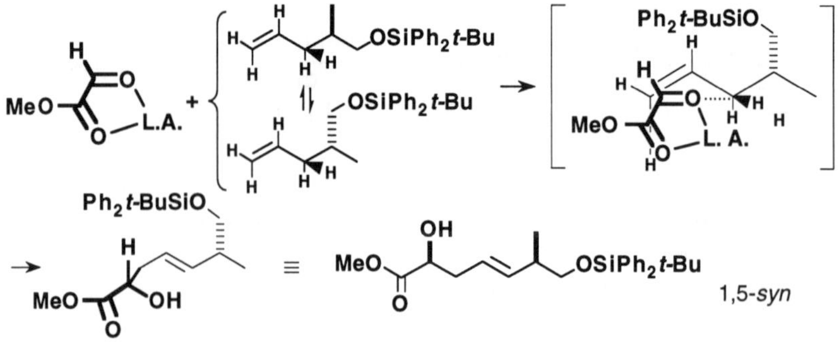

The high diastereoselectivity of the glyoxylate ene reaction was used by Nakai to synthesize a (22*R*)-hydroxy-23-carboxylate steroid side chain from a (*Z*)-$\Delta^{17(20)}$-steroidal olefin, methyl glyoxylate and Me_2AlCl.[42]

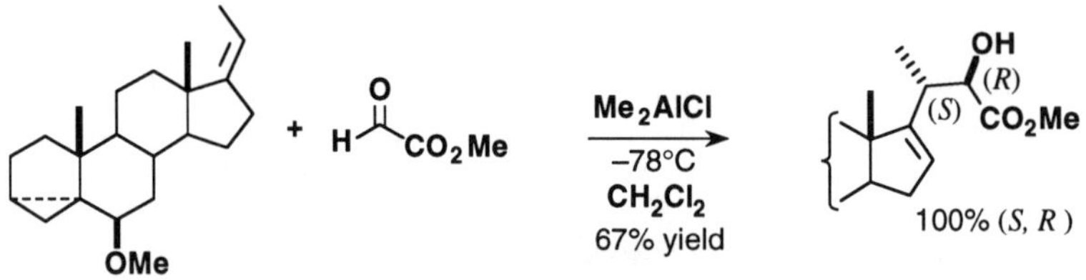

The stereoselectivity observed with Me_2AlCl was rationalized by postulating that the catalyst being complexed to the glyoxylate in an *anti*-fashion (i.e., not forming a chelate), the endo transition state, leading to the product, is favored.

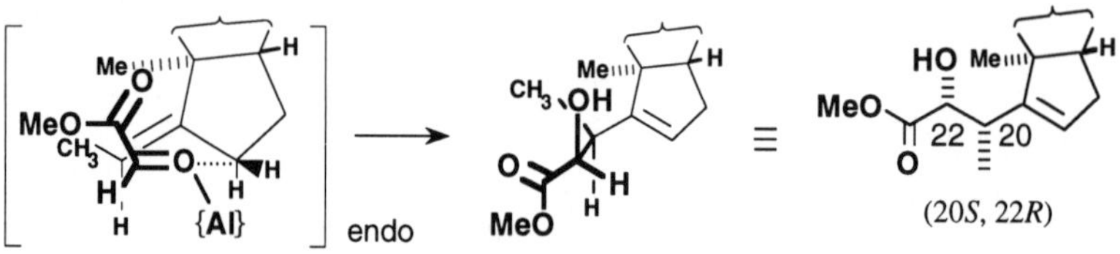

The use of vinylsilanes as enes was found to alter the regiochemical outcome of the glyoxylate-ene reaction. Thus, in sharp contrast with results obtained with (*E*)-pent-2-ene (formation of a mixture of regioisomers **A** and **B**) the use of (*Z*)-2-trimethylsilylpent-2-ene under the same reaction conditions led exclusively to vinylsilane **A**.[76]

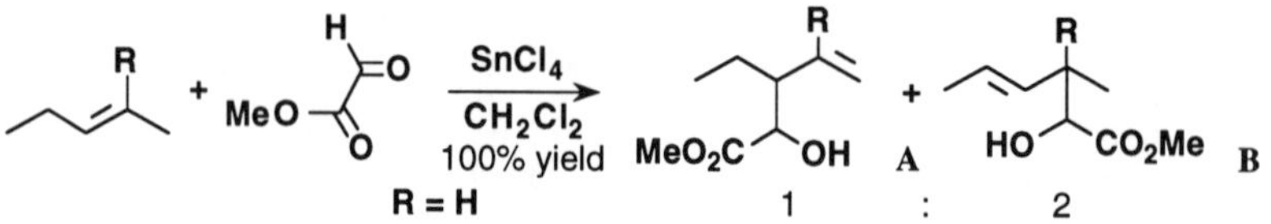

	A	:	B
R = H	1	:	2
R = SiMe3	1	:	0

The regioselectivity observed with vinylsilanes might provide a probe for the mechanistic study of the reaction. Thus, a cationic process should provide product **B** via the favorable β-silyl cation **D** and not, as is the case, product **A** which, under such a process, arises from unfavorable α-silyl cation **C**.

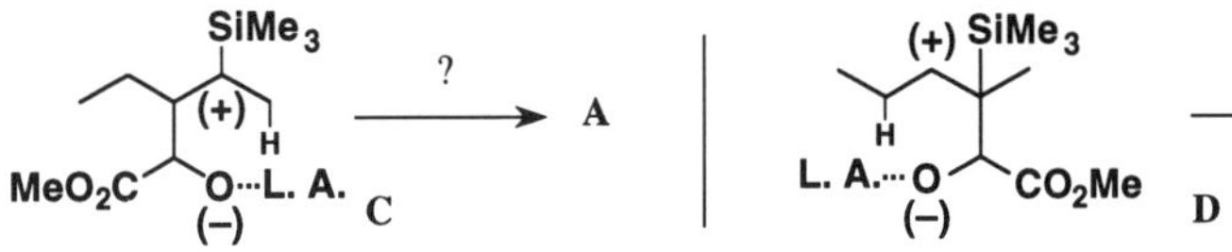

The introduction of a trimethylsilyl group also increases the diastereoselectivity of the reaction. With (*Z*)-vinylsilane **A**, the reaction leads almost exclusively to *anti* alcohol **B** whereas with (*E*)-but-2-ene an *anti:syn* (4.1:1) mixture was obtained.

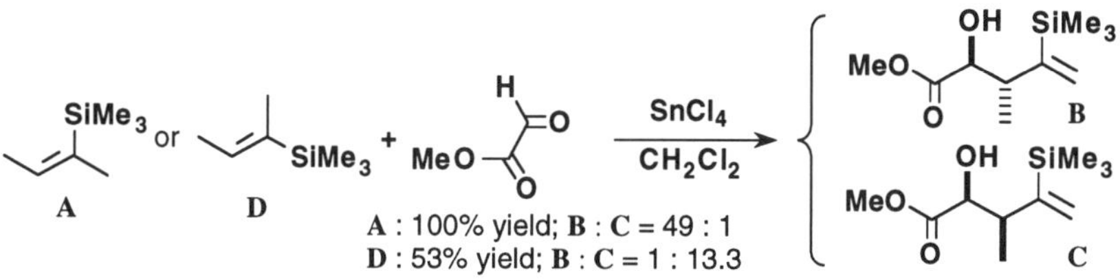

A : 100% yield; **B** : **C** = 49 : 1
D : 53% yield; **B** : **C** = 1 : 13.3

An exo transition state accounts for the preferential formation of hydroxy ester **B**.

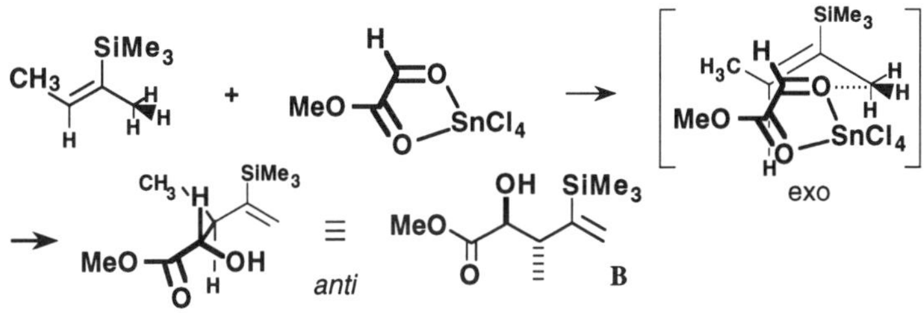

When the same reaction is performed from (*E*)-isomer **D**, it leads predominantly to *syn* alcohol **C** (86% d.e.). Again, an exo transition state can account for that result.

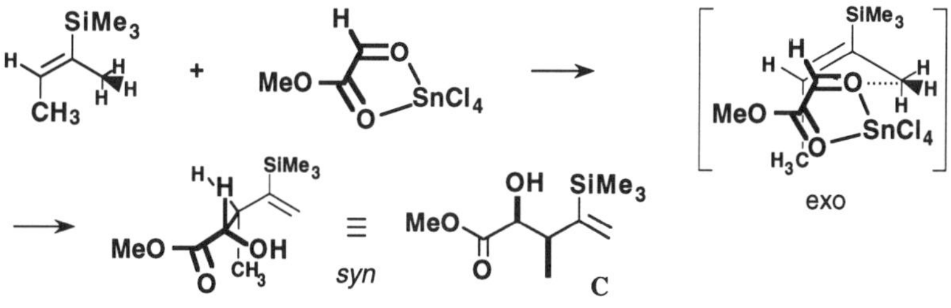

Anti alcohol **B** would in that case result from an endo transition state.

In that case, the introduction of the trimethylsilyl group altogether reversed the stereooutcome of the reaction since from (*Z*)-but-2-ene the major product was the *anti* isomer.

In contrast, when promoted by $TiCl_4$ instead of $SnCl_4$, the reaction did not provide the expected ene product but gave the substitution product as a single (*E*)-isomer.[77]

The glyoxylate ene reaction of but-2-ene can be achieved with a high level of either *erythro-* or *threo*-selectivity depending on the choice of the Lewis acid employed.[76]

Ene	R	L. A.	Yield	anti	:	syn
Z-2-butene	R = Me	$SnCl_4$	100% yield	2.6	:	1
E-2-butene	R = Me	$SnCl_4$	100% yield	4.6	:	1
Z-2-butene	R = *i*-Pr	$SnCl_4$	100% yield	2.4	:	1
E-2-butene	R = *i*-Pr	$SnCl_4$	100% yield	11.5	:	1
Z-2-butene	R = Me	Me_2AlOTf	65% yield	1	:	11.5
E-2-butene	R = Me	$MeAl(OTf)_2$	41% yield	1	:	1.9
Z-2-butene	R = *i*-Pr	MeAlCl(OTf)	44% yield	1	:	4.3

It appears that the $SnCl_4$-promoted reaction exhibits *anti*-selectivity, irrespective of the ene geometry. This result can be rationalized in terms of the reasonable postulate that $SnCl_4$ forms a chelate with the glyoxylate. With (*Z*)-but-2-ene, endo transition state **A**, the most favored, gives rise to the *anti* diastereomer.

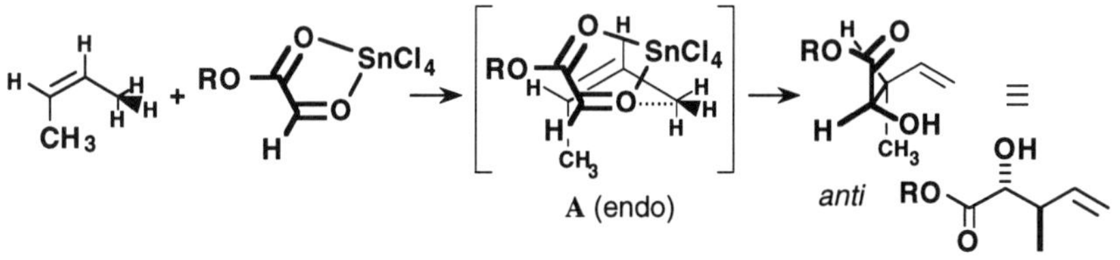

H H CH_3 H H H + RO O $SnCl_4$ O H → [RO O H $SnCl_4$ H O H H H CH_3] → RO H O H OH CH_3 ≡ OH RO O

A (endo) *anti*

Indeed, exo transition state **B**, precursor of the *syn* diastereomer, suffers from steric repulsions between the carboalkoxy group and the *cis*-methyl group.

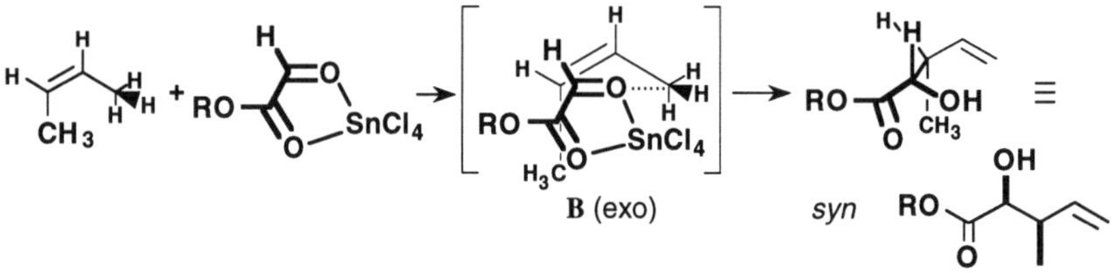

H H CH_3 H H H + RO H O O $SnCl_4$ → [H H H O H H RO O $SnCl_4$ H_3C] → RO O H H OH CH_3 ≡ OH RO O

B (exo) *syn*

From (*E*)-but-2-ene, exo transition state **C**, which minimizes steric repulsion, gives rise to major *anti* diastereomer. On the other hand, endo transition state **D** again presents a severe interaction between the alkoxycarbonyl and *trans*-methyl groups.

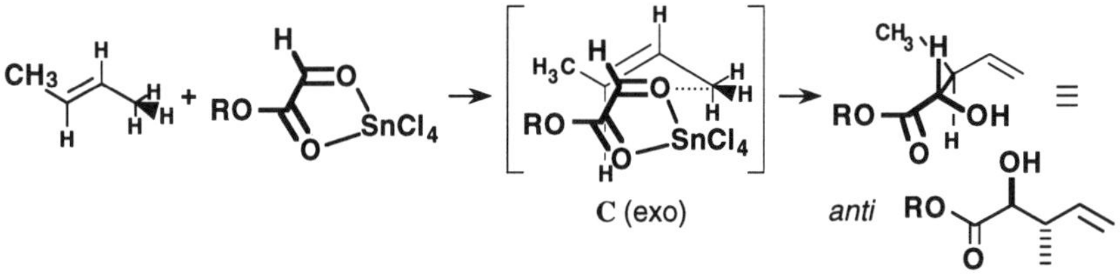

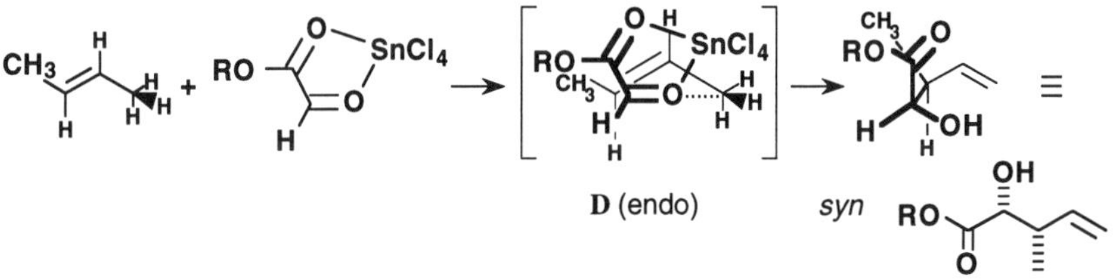

CH_3 H H H H H + RO H O O $SnCl_4$ → [H_3C H H O H H RO O $SnCl_4$ H] → RO O CH_3 H OH H ≡ OH RO O

C (exo) *anti*

CH_3 H H H H H + RO O $SnCl_4$ O H → [RO O H $SnCl_4$ CH_3 O H H H H] → RO CH_3 O H OH H ≡ OH RO O

D (endo) *syn*

The *syn*-selectivity observed with aluminum derivatives as catalysts could result from an *anti* complexation of the glyoxylate (see Section III.1.a). With (*Z*)-but-2-ene, the selectivity is low, because exo and endo transition states are congested. The *cis*-methyl group interacts either with the alkoxycarbonyl group in exo transition state **E** or with the bulky Lewis acid in endo transition state **F**.

E (exo) *syn*

F (endo) *anti*

Phenylglyoxal undergoes $SnCl_4$-induced ene reaction with olefins to give γ,δ-unsaturated-α-hydroxyketones in good yields.[78] $FeCl_3$-promoted ene reaction between ethylglyoxylate and allylic amino acid derivatives gave highly functionalized unsaturated pimelic acids.[79]

III.1.d Other achiral aldehydes

The ene reaction between aldehydes and alkenes provides a potentially valuable route to homoallylic alcohol.[80] The reaction of isovaleraldehyde with limonene occurs exclusively at the less substituted double bond.[81]

Me_2AlCl, CH_2Cl_2, 79% yield

The Me_2AlCl-catalyzed ene reactions of aliphatic aldehydes with (*Z*)- and (*E*)-3-methylpent-2-ene **A** and **B** led to complex mixtures of *erythro* and *threo* adducts and double bond position isomers.[68]

A + aldehyde — Me_2AlCl, CH_2Cl_2, 62% yield → **C** : **D** : **E** : **F** : **G** = 4 : 12.8 : 4 : 2.2 : 1

B + aldehyde — Me_2AlCl, CH_2Cl_2, 67% yield → **C** : **D** : **E** : **F** : **G** = 2 : 7 : 0 : 12.3 : 1

C **D** **E** **F** **G**

Exo transition states can be invoked to explain the formation of **D**, **E** or **F** while only an endo transition state can account for the formation of **D**.

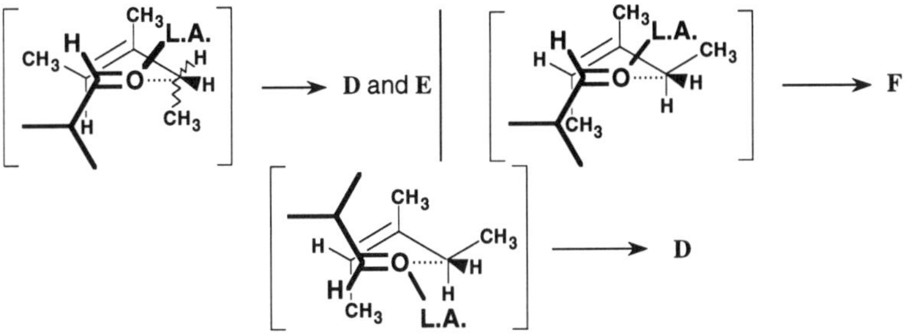

Ciufolini observed that catalytic amounts (0.5 mol%) of $Yb(fod)_3$ catalyze a bimolecular ene reaction between aldehydes and enol ethers, in which the oxygen functionality is located at the central carbon atom of the allylic system. Traces of acetic acid or addition of silica gel to the reaction mixture enhance the catalytic activity.[82]

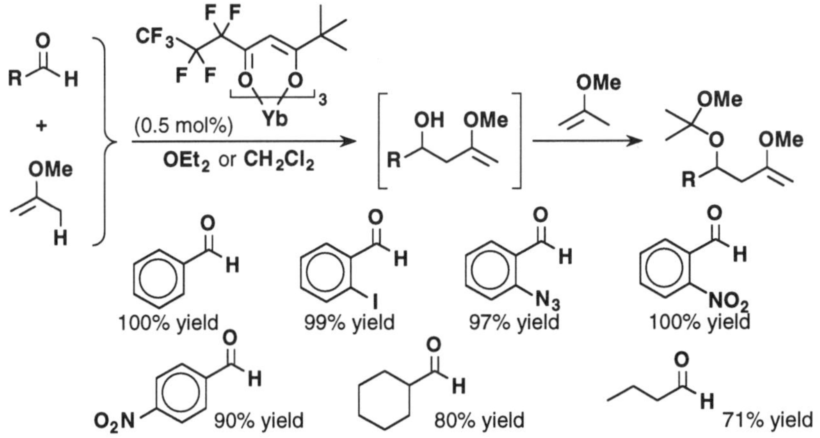

Moreover, the $Yb(fod)_3$ catalyzed addition of 4-nitrobenzaldehyde to 1-methoxycyclohexene proved to be stereoselective.

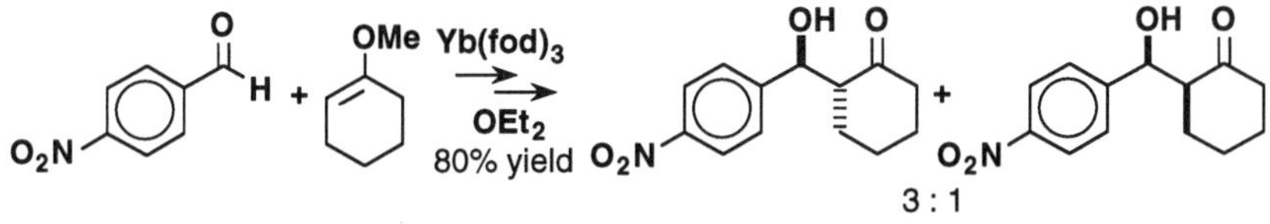

Similar results were obtained from enol ethers by Kuwajima by using Me_2AlCl or Et_2AlI as Lewis acids.[83]

The ene reaction of a (*Z*)-$\Delta^{17(20)}$-steroidal olefin with acetylenic aldehydes in the presence of Me_2AlCl produces acetylenic steroid side chains with high diastereoselectivity and diastereofacial selectivity.[84]

R = SiMe3, Me, *i*-Pr

The reaction proceeds on the α-face of the steroid via an endo transition state. In that way, the large steric repulsion between the catalyst and the cyclopentene ring is avoided.

Koreeda also studied the stereochemistry of the Me_2AlCl-mediated ene reaction of aldehydes with a (*Z*)-$\Delta^{(17,20)}$-steroidal olefin.[85] Results indicate an unexpected dependence on the size of the aldehyde employed. Aliphatic aldehydes led preferentially to alcohols **A** while aromatic aldehydes, relatively more congested, favored the generation of alcohols **B**.

R = **isoamyl**, 91% yield, **A** : **B** = 12 : 1
R = **phenyl**, 92% yield, **A** : **B** = 1 : 9
R = **phenyl** (room temp.), 80% yield, **A** : **B** = 1.33 : 1
R = ***o*-tolyl**, 60% yield, **A** : **B** = 1 : >25
R = ***o*-nitrophenyl**, >53% yield, **A** : **B** = 1 : 1.5
R = **2-furyl**, 56% yield, **A** : **B** = 1 : 19
R = **cyclohexyl**, 60% yield, **A** : **B** = 14 : 1
R = ***tert*-butyl**, no ene adduct

The formation of alcohol **A** from aliphatic aldehydes results from the preferential formation of endo transition state **C** while the alternative exo transition state **D** leads to the formation of alcohol **B**. **B** could also result from transition state **E** in which the Lewis acid is complexed *syn* to the aldehyde alkyl group.

A highly enantioselective reaction was developed by Kuwajima on the bases of chirality transfer. The reaction of various aldehydes with (*R*)-3(*tert*-butyldimethylsiloxy)-2-(ethylthio)but-1-ene **A** affords alcohols **C** with high enantiomeric excess. Endo transition state **B** can account for the stereochemistry obtained.[86]

Me_2AlCl, solvent

R = **n-hexyl**; solvent : toluene; 82% yield; 86% e.e.
R = **n-hexyl**; solvent : hexane; 78% yield; 94% e.e.
R = **phenyl**; solvent : CH_2Cl_2 ; 96% yield; 96% e.e.
R = **phenyl**; solvent : toluene; 91% yield; 96% e.e.
R = **o-toluyl**; solvent : toluene; 93% yield; 97% e.e.
R =**o-TBSO-phenyl**; solvent : CH_2Cl_2; 96% yield; 99% e.e.
R = **2-furyl**; solvent : toluene; 76% yield; 86% e.e.
R = **styryl**; solvent : toluene; 70% yield; 90% e.e.

Aromatic aldehydes add to β-pinene under Lewis acid catalysis to yield the corresponding homoallylic alcohols.[87]

Mikami observed that γ-lactols can be used as enophiles. He investigated the diastereofacial selection and diastereoselectivity of the lactol-ene reaction with a (Z)-$\Delta^{17(20)}$-steroidal olefin and found that best results were obtained with Me_2AlCl.[88]

Me_2AlCl, CH_2Cl_2, 47% yield; 4 / 1

Ene reactions between lactol-derived oxonium ion intermediates and methylene cyclohexane exhibit a high level of diastereofacial selectivity to afford predominantly 2,6-*trans*-tetrahydropyrans.[89]

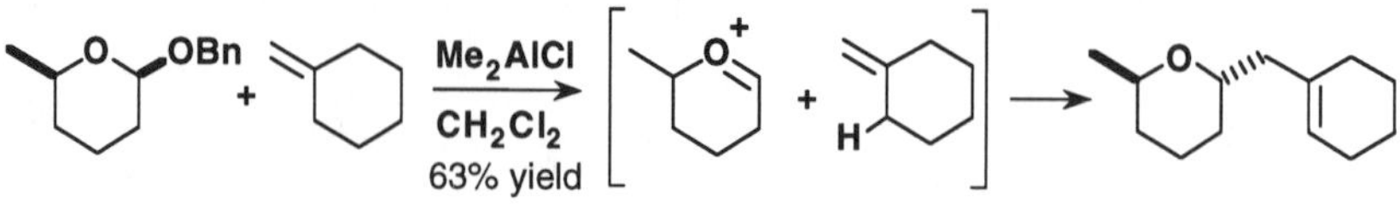

Me_2AlCl, CH_2Cl_2, 63% yield

III.2 Reaction involving chiral aldehydes as enophiles

The ene reaction takes place with efficient 1,3-asymmetric induction when starting 3-formyl-Δ^2-isoxazolines.[90,91]

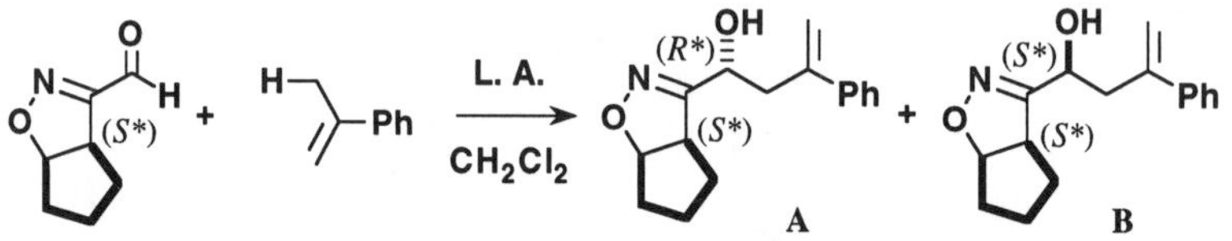

L. A. : $SnCl_4$, 70% yield, **A** : **B** = 199 : 1 L. A. : Et_2AlCl, 59% yield, **A** : **B** = 0 : 1
L. A. : $TiCl_2(Oi\text{-}Pr)_2$, 74% yield, **A** : **B** = 19 : 1

The changing stereoselectivity of the reaction is attributed to chelation and nonchelation control. With bidentate Lewis acids such as $SnCl_4$ or $TiCl_2(Oi\text{-}Pr)_2$, the olefin preferentially attacks on the *Re* face of the aldehyde while with a monodentate acid like Et_2AlCl, the *Si* face is more accessible since the isoxazoline adopts an *s-trans* conformation.

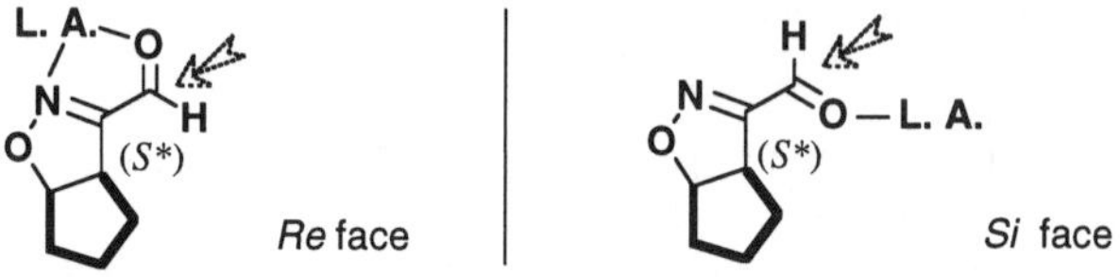

Re face *Si* face

High yields and complete stereoselectivity were also observed for the formation of alcohol **C** from the same isoxazoline and methylenecyclohexane.

$SnCl_4$, CH_2Cl_2, 93% yield

Aminoaldehyde-ene reaction involving enophiles with a bulky amino protective group were also developed and high level of *syn* diastereofacial control was obtained.[92]

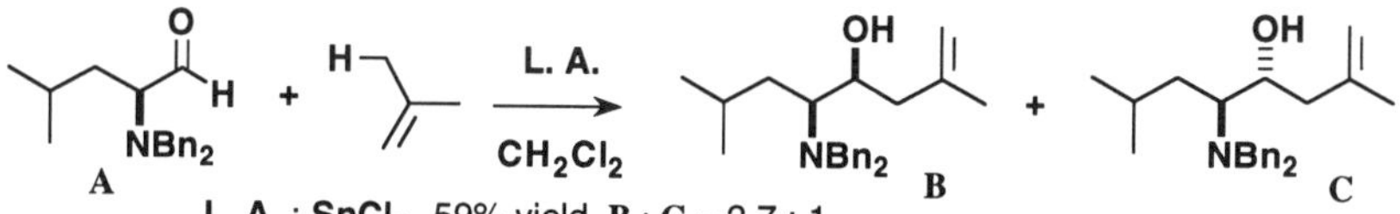
L. A.
CH2Cl2
NBn2
OH
A
B
C

L. A. : $SnCl_4$, 59% yield, **B** : **C** = 2.7 : 1
L. A. : $EtAlCl_2$, 71% yield, **B** : **C** = 99 : 1

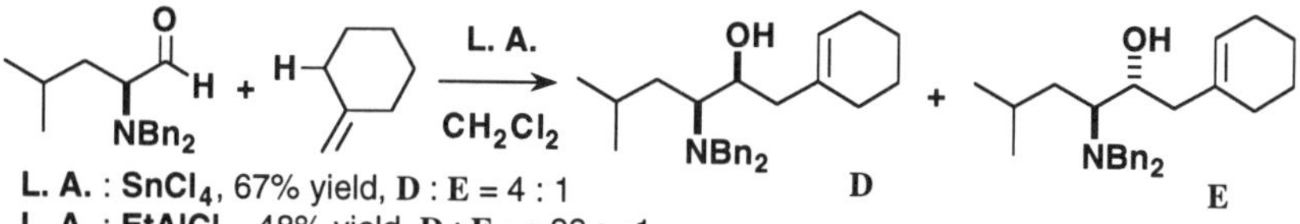
L. A.
CH2Cl2
NBn2
OH
D
E

L. A. : $SnCl_4$, 67% yield, **D** : **E** = 4 : 1
L. A. : $EtAlCl_2$, 48% yield, **D** : **E** = >99 : <1

The reaction of chiral α- and β-benzyloxyaldehydes with 2-methylbut-2-ene led to high levels of both diastereofacial selection and diastereoselectivity.[93]

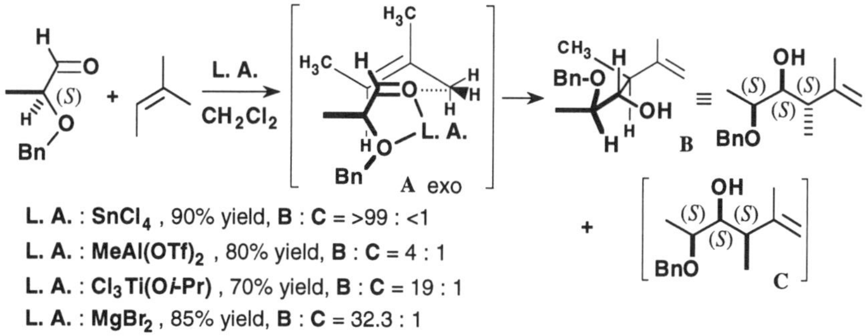
L. A.
CH2Cl2
A exo
B
C
BnO
(S)
L. A. : SnCl4 , 90% yield, B : C = >99 : <1
L. A. : MeAl(OTf)2 , 80% yield, B : C = 4 : 1
L. A. : Cl3Ti(Oi-Pr) , 70% yield, B : C = 19 : 1
L. A. : MgBr2 , 85% yield, B : C = 32.3 : 1

The reaction was applied to the synthesis of a steroid side chain. One single stereoisomer was obtained in quantitative yield.

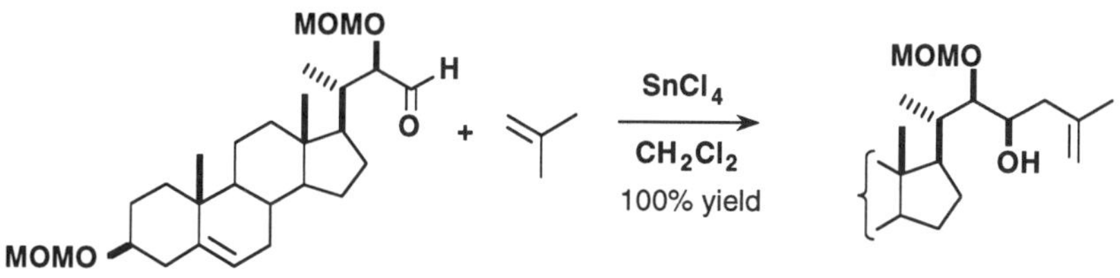
MOMO
SnCl4
CH2Cl2
100% yield
OH

α-Haloaldehydes react as enophiles, with reasonable *anti*-selectivity, under effective nonchelation control.[94]

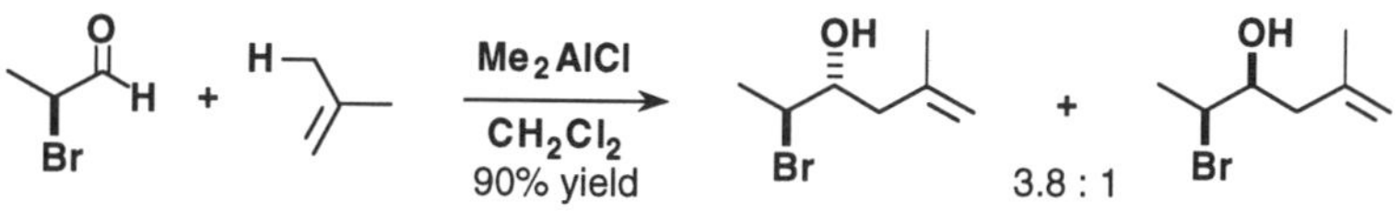
Me2AlCl
CH2Cl2
90% yield
Br
OH
3.8 : 1

Partial asymmetric induction in the Lewis acid-promoted ene reaction of (–)-menthylglyoxylate with pent-1-ene was observed.[95]

Whitesell observed that the $SnCl_4$-promoted reaction of 8-phenylmenthyl glyoxylate with hex-1-ene and (*E*)-but-2-ene provides ene adducts with asymmetric induction levels consistently above 93%.[96–98] Similar results were also observed with 8-phenylmenthyl pyruvate.[99]

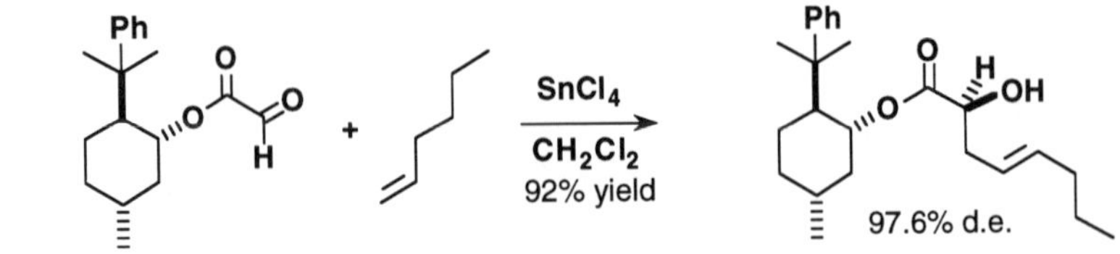

Such an induction implies that the addition selectively occurs on the *Si*-face of the aldehyde through a chelated exo transition state in which a favorable intramolecular interaction between the phenyl group and the α-carbonyl ester moiety exists. Photophysical probes of intramolecular interactions between the phenyl group and the α-carbonyl ester moiety of 8-phenylmenthyl glyoxylate or pyruvate are given by Whitesell.[100]

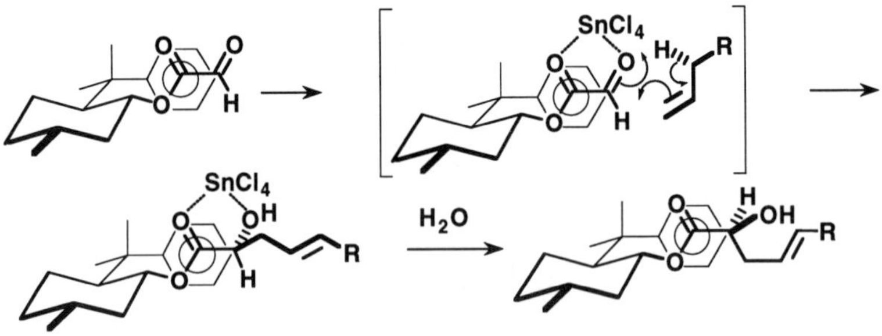

The facial selectivity inherent to 8-phenylmenthyl glyoxylate can be combined with the facial selectivity present in a chiral alkene such as **A** or **A'**. Thus, a kinetic resolution occurred and only enantiomer **A** reacts with the glyoxylate to afford α-hydroxy ester **C** through transition state **B**.[101–103]

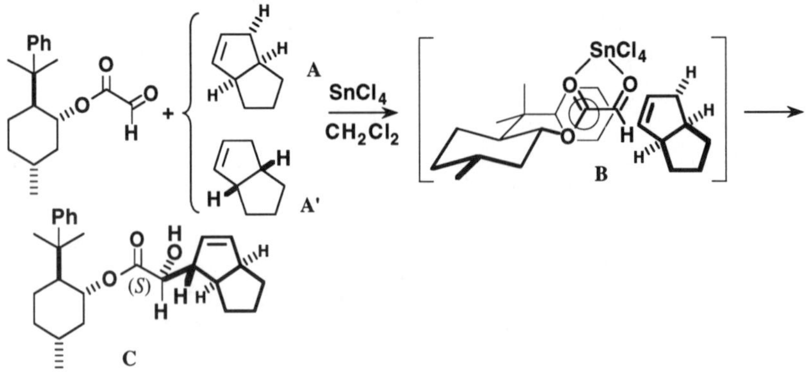

Whitesell even showed that *trans*-2-phenylcyclohexanol can be used as a substitute for 8-phenylmenthol. Thus, the stereochemical outcome of the reaction with hex-1-ene is similar to the one observed when 8-phenylmenthol was used as chiral auxiliary.[104]

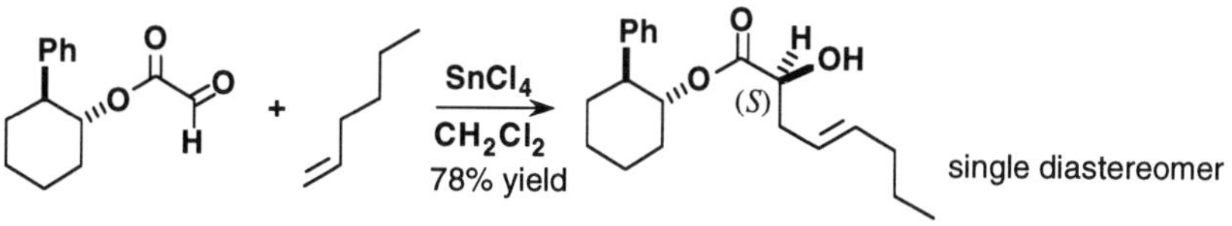

Attack on the *Si*-face of the aldehyde through an exo transition state accounts for the stereochemistry of the adduct.

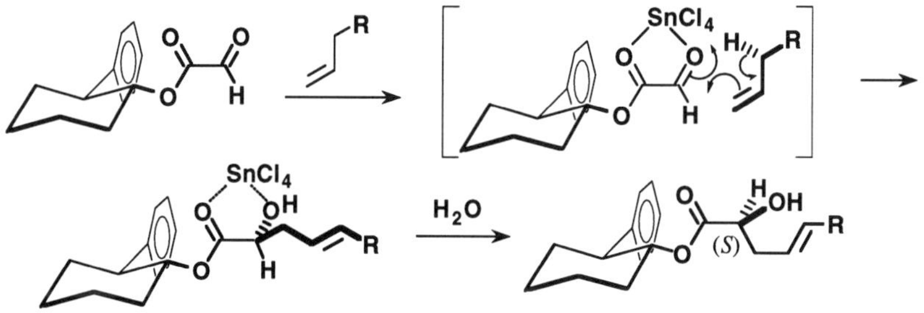

Kuwajima used 2-(alkylthio)allyl ethers as ene compounds. Under chelation conditions, these enes react with α-benzyloxypropanal to afford *syn* monoprotected diols exclusively.[105]

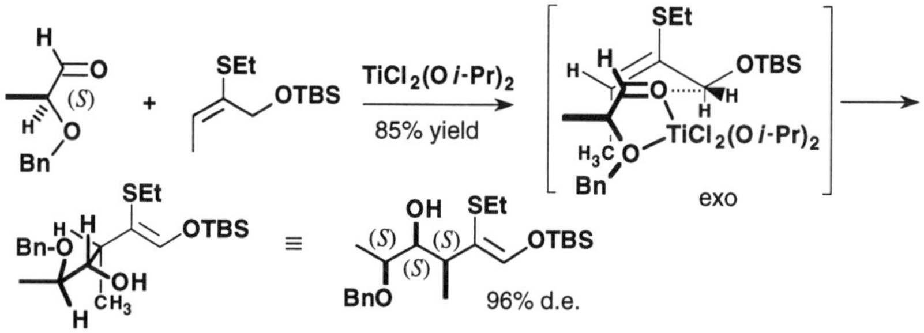

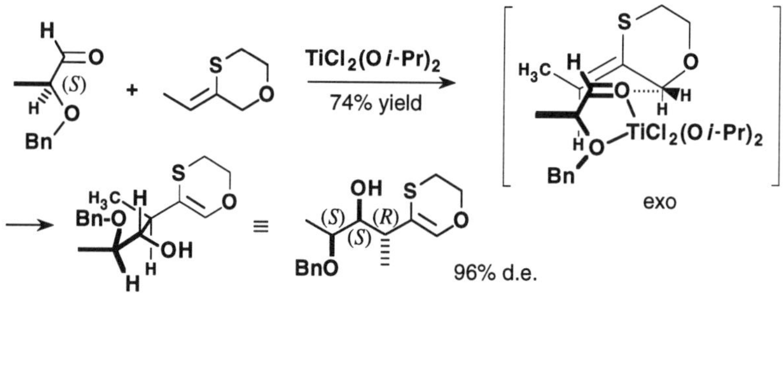

This made a good contrast with the *anti* selectivity observed in the ene reaction of the same allyl silyl ether with achiral aldehydes.[107,107]

This methodology, which enables the construction of three contiguous diastereomeric centers, was applied to a stereoselective synthesis of brassinolide side chain.[105]

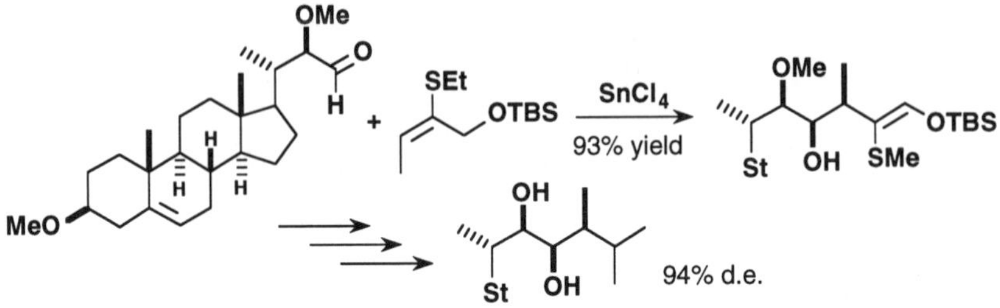

Vinylic sulfides can also act as ene components in carbonyl-ene reaction with α-alkoxy aldehydes. The degree and sense of 2,3-simple diastereoselection depends critically not only on the geometry but also on the steric bulk of the vinylic sulfides.[108] Thus, *anti* diastereomers are mainly obtained from (*Z*)-vinyl sulfides while *syn* diastereomers are formed from (*E*)-vinyl sulfides. In both cases, exo transition states with chelation control were postulated to account for the stereochemistry observed. However, in the second case, an interaction occurs between the methyl group of the sulfide and the benzyloxy group.

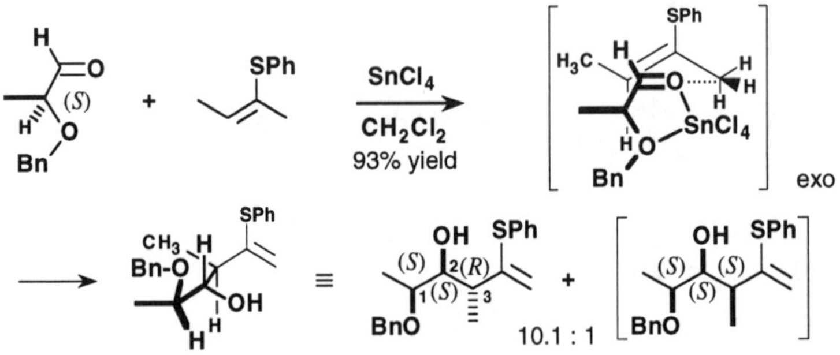

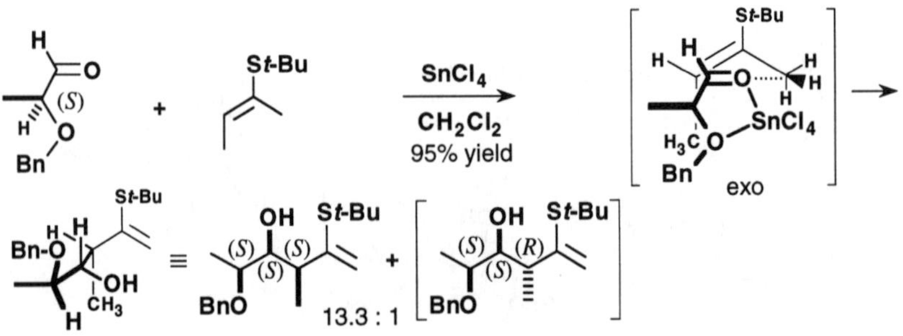

Hanessian proposed a remarkable application of the ene reaction to the synthesis of alicyclic and cyclic compounds with alternate and remote C-methyl substitution patterns.[109] Thus, reaction of (*aR*)-(4-methylcyclohexane) ethylidene **B** with 2,3-di-*O*-benzyl L-glyceraldehyde **A** in the presence of $SnCl_4$ led to alcohol **C**, while treatment of **A** with enantiomeric (*aS*)-(4-methylcyclohexane) ethylidene **D** led predominantly to diastereomeric alcohol **E** (schemes are drawn with the enantiomers of the original reactants). Oxidative cleavage of **C** and **E** led to acyclic compounds with *anti* or *syn* 1,5-relationship between the two methyl groups.

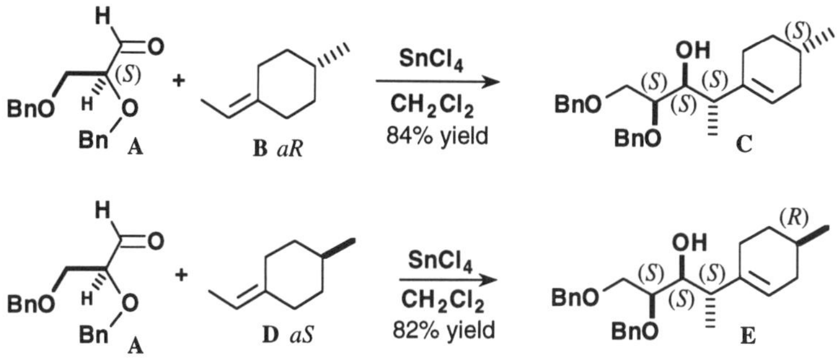

This result can be rationalized in the following way: In the first case, alcohol **C** results from the attack of the complexed aldehyde on the *pro-S* face of the more favored conformation (Me equatorial) of ene **B**; the reaction then proceeds through transition state **F** in which interaction between the complexed aldehyde **A** and the methyl group bonded to the double bond of **B** is minimized. In the second case, alcohol **E** is obtained despite the fact that its formation results from the attack of the complexed aldehyde on the *pro-S* face of the least favored conformation (Me axial) of ene **D.** As a matter of fact, the reaction then proceeds through transition state **G** in which the above-mentioned interaction is minimized.

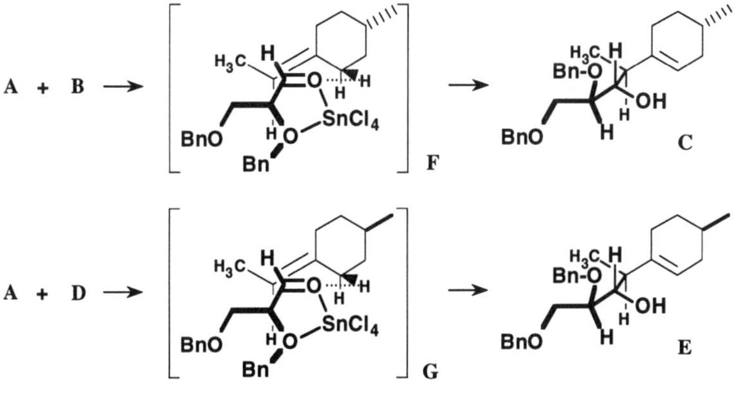

The stereochemistry of alcohols **C** and **E** therefore results in both cases from an addition on the same face of the ene.

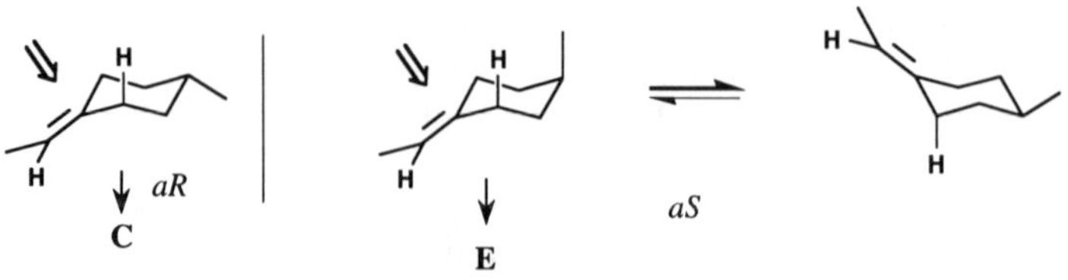

III.3 Reaction with various enophiles

As already mentioned, diethyl mesoxalate is a good enophile. Indeed, vinylcyclohexene and diethyl mesoxalate lead in the presence of $SnCl_4$ to furan **B** which presumably results from the cyclization of ene adduct **A**. **B** can then be oxidized into spiro lactone **C**.[110]

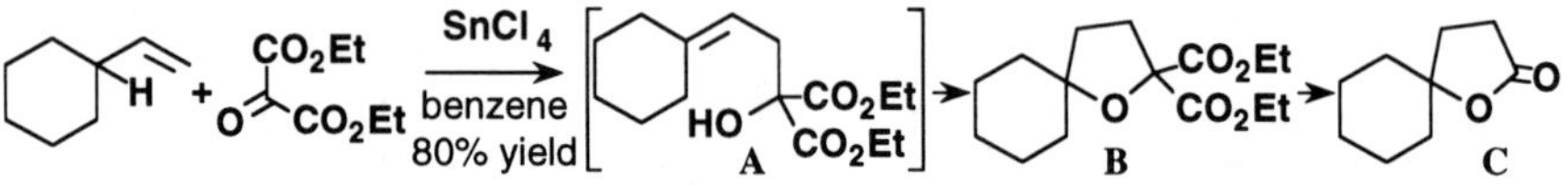

Cyclobutenes are often obtained through [2+2] cycloaddition from dimethyl acetylene dicarboxylate and alkenes,[111] but ene adducts can also arise. $EtAlCl_2$ can promote the ene reaction of methyl propiolate with various alkenes.[112]

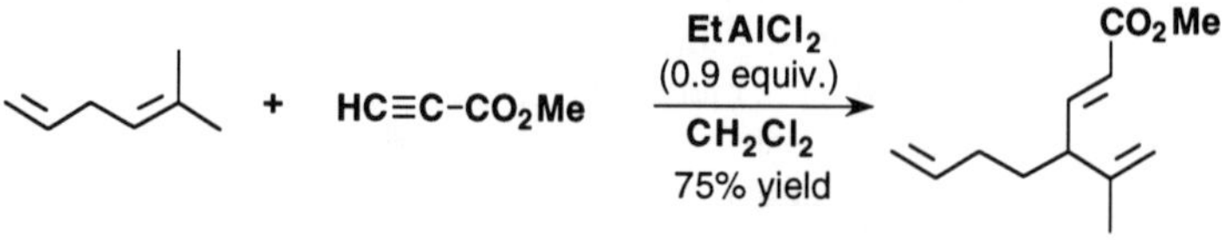

Snider found that methyl propiolate undergoes Lewis acid catalyzed reactions with alkenes in high yield. Mono- and 1,2-disubstituted alkenes give mainly cyclobutenes.[113] The formation of ene adducts occurs exclusively from alkenes containing at least one disubstituted carbon. $EtAlCl_2$ is the best catalyst, because of its ability to act as a proton scavenger as well as as a Lewis acid.[114] Thus, the $AlCl_3$-promoted reaction of methylenecyclohexane with methyl propiolate led to a mixture of **A** and **B** because part of the alkene is isomerized before the addition.

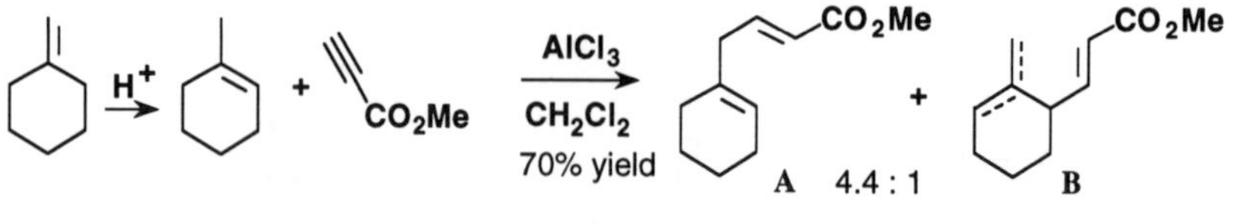

In contrast, the use of $EtAlCl_2$ only promotes the predicted ene adduct **A** with an excellent yield.

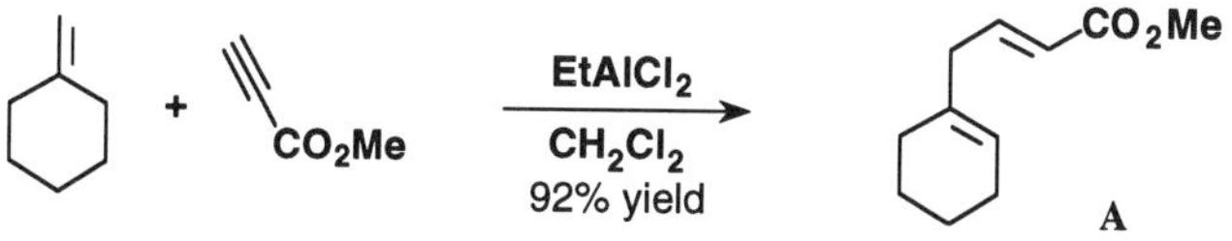

The usefulness of $EtAlCl_2$ is emphasized by Dauben in the C-20 stereospecific introduction of a steroid side chain.[115,116] Similarly, Batcho and Snider, while usual Lewis acids failed to induce the reaction, obtained the desired product with 89% yield when using $EtAlCl_2$.[117]

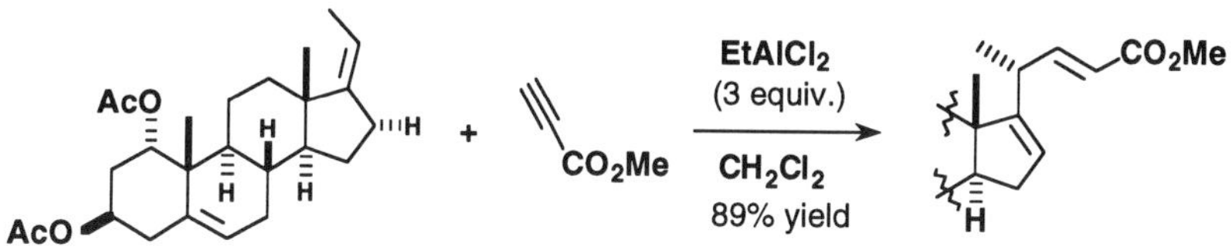

In the course of a synthesis of γ-ionone, Monti observed a ZnI_2-promoted ene reaction between a substituted allylsilane and but-3-yn-2-one.[118]

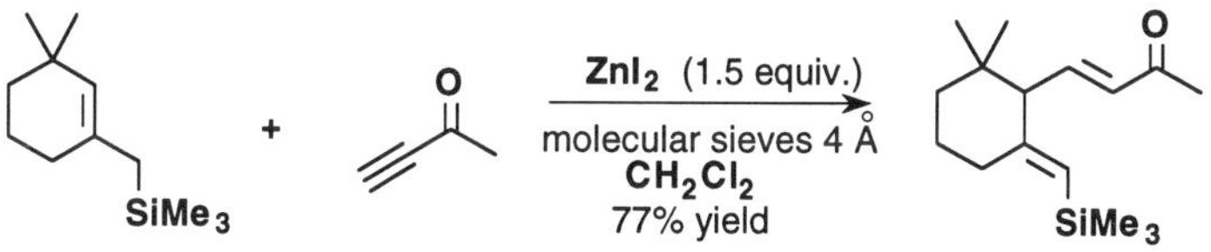

Alkylidenecycloalkanes react with α,β-unsaturated ketones in the presence of Me_2AlCl to give bicyclic alcohols resulting from two sequential ene reactions.[119,120] The first ene reaction leads to **A** which then undergoes intramolecular ene reaction (see Section IV.2).

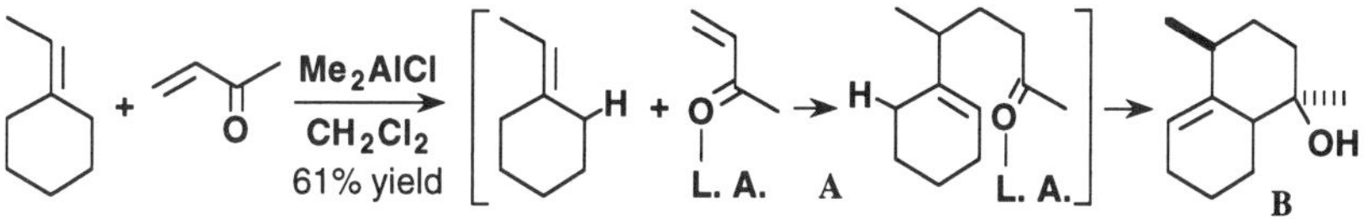

In contrast, the reaction between isopropyl vinyl ketone and (*Z*)-$\Delta^{17(20)}$-steroidal olefin **A** ends after one ene reaction to give 24-oxocholesteryl acetate.

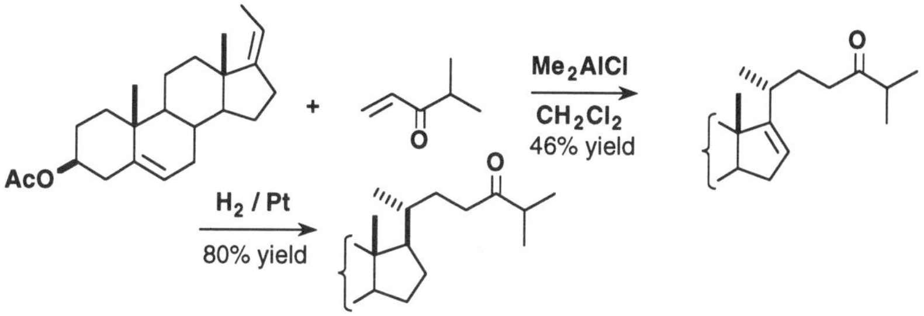

Methyl acrylates[121] and particularly α-haloacrylates also act as efficient enophiles.[122–124]

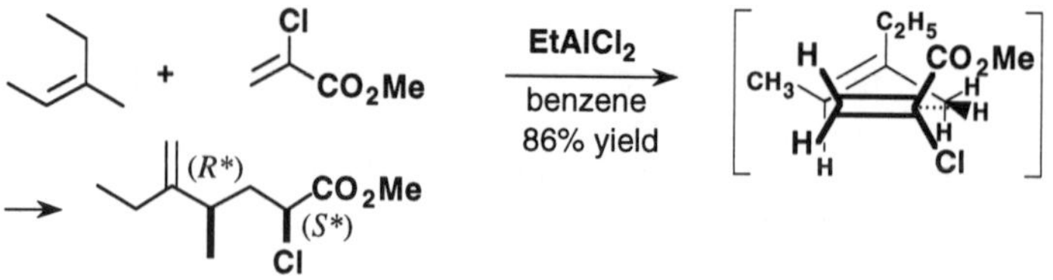

Remarkably, these reactions are highly stereoselective and regioselective and yield exclusively the ene adduct resulting from the endo addition of the ester group and the transfer of a hydrogen atom from the alkyl group *syn* to the vinylic hydrogen atom.

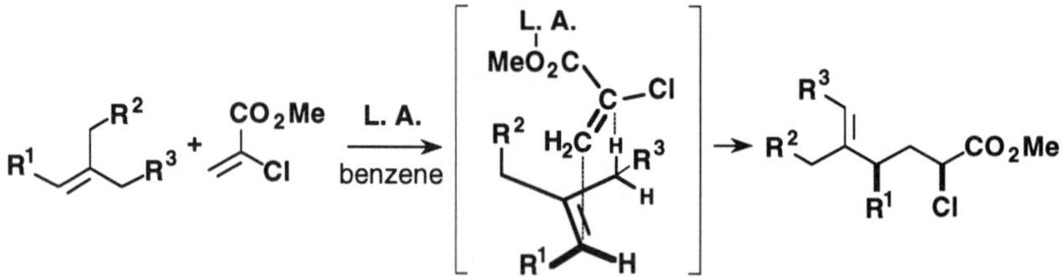

Methyl α-chloroacrylate was thus used in the first step of a synthesis of (±)-nitramine, an interesting alkaloid which possesses a 2-azaspiro[5.5] undecane skeleton.[125]

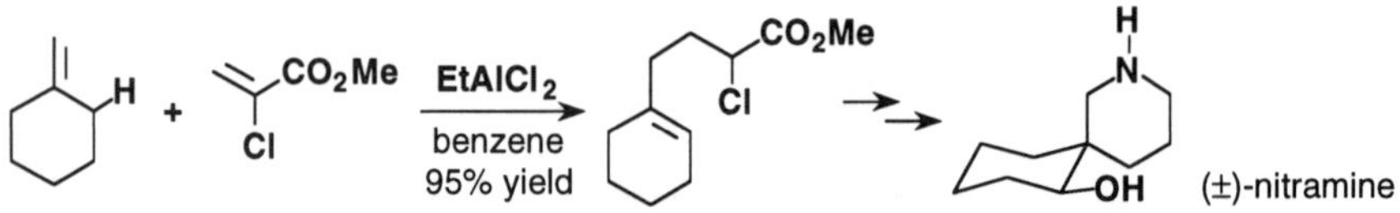

2-Phosphonoacrylates were also used as enophiles. Under $EtAlCl_2$ catalysis, they lead, in the presence of most alkenes, to adducts which are useful reagents for the phosphonate modification of the Wittig reaction.[126]

As observed by Taguchi, 2-trifluoromethylpropenoic esters are also good enophiles.[127] Although trifluoroethyl esters react under $EtAlCl_2$ catalysis without any stereoselectivity to give 1:1 mixtures of esters, the $TiCl_4$-catalyzed ene reaction of D-pantolactone ester **A** occurs with high enantioselectivity. The alkene approaches the *Re* face of the chelated $TiCl_4$-enophile complex to yield ene product **B** and hydrogen chloride adduct **C** with good d.e.

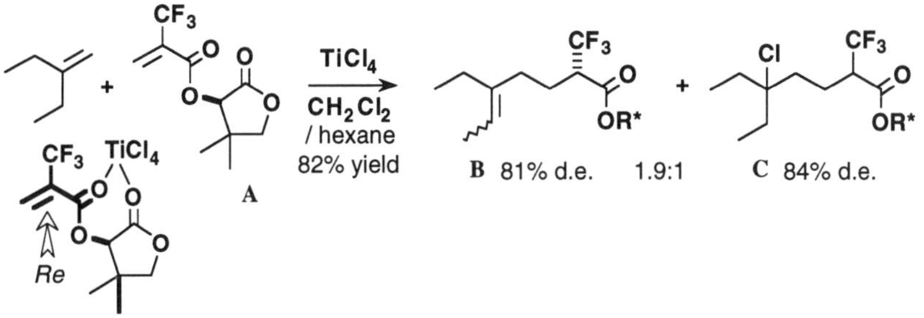

In contrast, methyl α-cyanoacrylate reacts with alkenes to give low to moderate yields of a complex mixture of 1:1 and 2:1 adducts.[128]

Reaction of alkenes with *N*-sulfinylcarbamate of *trans*-2-phenylcyclohexanol **A** provided allylic sulfinamides **B** with absolute stereochemical control at both sulfur and carbon stereocenters.[129,130] These adducts could then be transformed into allylic alcohols **C** through a 2,3-rearrangement of derived sulfoxides.

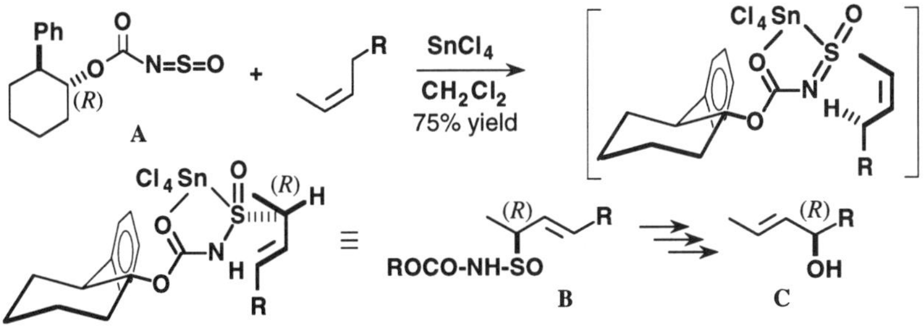

Partial asymmetric induction of the Lewis acid catalyzed ene reaction of (–)-methyl *N*-(*p*-toluenesulfonyl)iminoacetate with isobutene was observed.[131]

The asymmetric ene reactions of α-imino esters, prepared fom 8-phenylmenthyl glyoxylate provide an efficient entry to asymmetric synthesis of α-amino acids.[132]

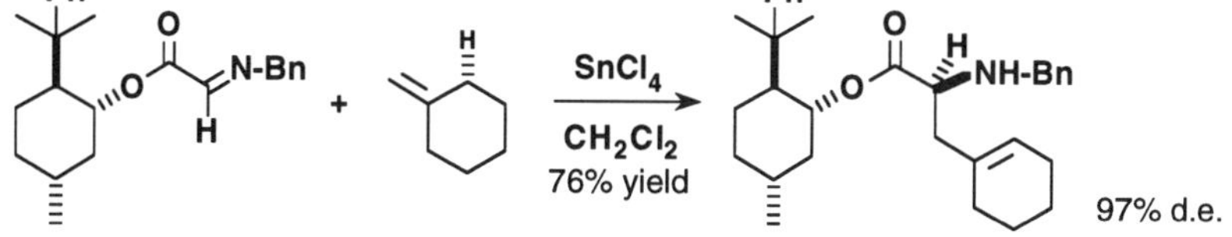

Finally, ethyl chloro(phenylthio)acetate and benzene sulfinyl chloride were also involved in ene reactions. Thus, ethyl chlorophenylthioacetate reacts with alk-1-enes in the presence of $SnCl_4$ to yield ene adduct **A**.[133]

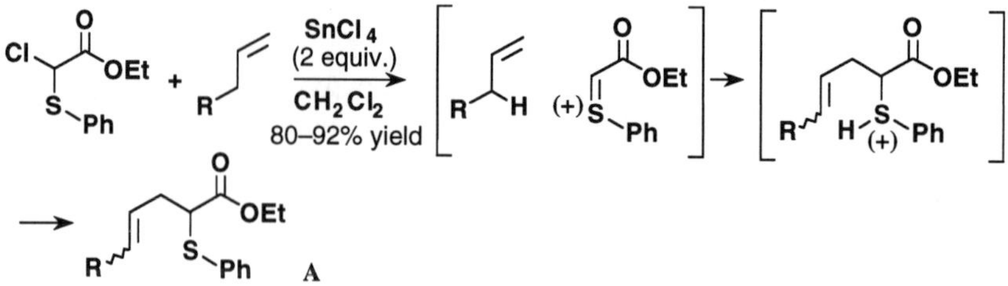

The $ZnCl_2$ promoted ene reaction of benzene sufinyl chloride with linear isoprenoids proceeds chemoselectively (exclusive attack at the terminal trisubstituted C=C bond) to yield allylic sulfoxides.[134–136]

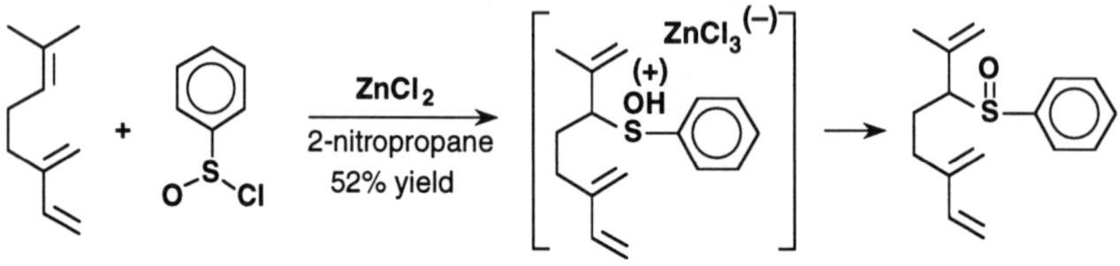

IV. Intramolecular Reaction

IV.1 Introduction

The possible modes of carbocyclic ring formation from olefinic aldehydes was classified as follows by Oppolzer[5] and Andersen.[137] They depend on the way the enophile moiety is linked to the ene part of the molecule:

- Type I: the enophile is linked to the terminal olefinic atom.
- Type II: the enophile is linked to the central atom.
- Type III: the enophile is linked to the allylic terminal atom.

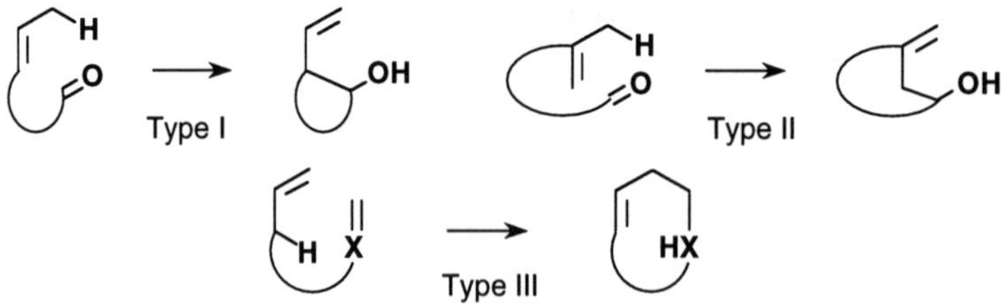

IV.2 Cyclization of Type I

IV.2.a Cyclization of hex-4-enal derivatives

Hex-4-enal derivatives do not usually lead, under Lewis acid catalysis, to ene products. Thus, Lewis acid-mediated cyclization of (*E*)-hex-4-enal derivative **A** leads to *trans*-cyclopentanone **E**.[138] This cyclization involves a cationic process and therefore the formation of carbocation **B**, which then undergoes two successive 1,2-hydride shifts to yield ketone **E** through **C** and **D**.

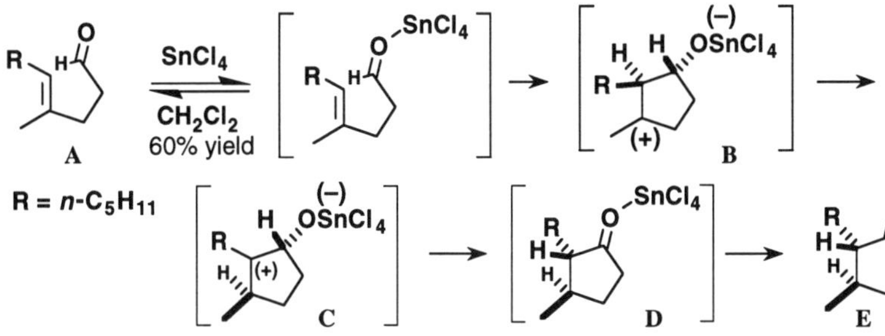
SnCl4
CH2Cl2
60% yield
A
B
C
D
E
R = n-C5H11
(–)
(+)
OSnCl4
SnCl4

This mechanism was established by Cookson[138] on the basis of deuterium labeling experiments.

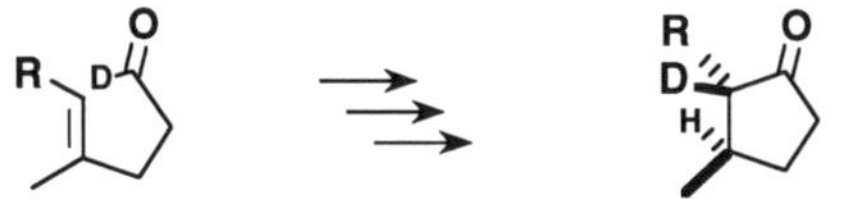

Isomeric (*Z*)-aldehyde **F** undergoes cyclization into the corresponding *cis*-cyclopentanone.

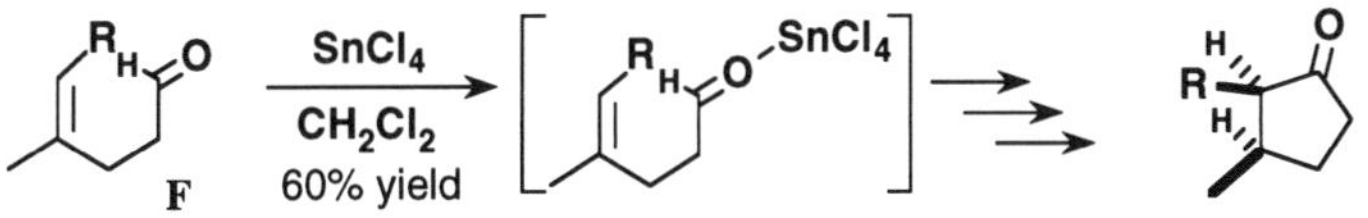
SnCl4
CH2Cl2
60% yield
F
SnCl4

Similarly, the $MeAlCl_2$-initiated cyclization of enone or dienone **A** provides functionalized *trans*-fused hydrindanone or hydrindenone **D**.[139] Cyclization of **A** can occur via zwitterionic intermediate **B** which then undergoes hydride and methyl shifts.

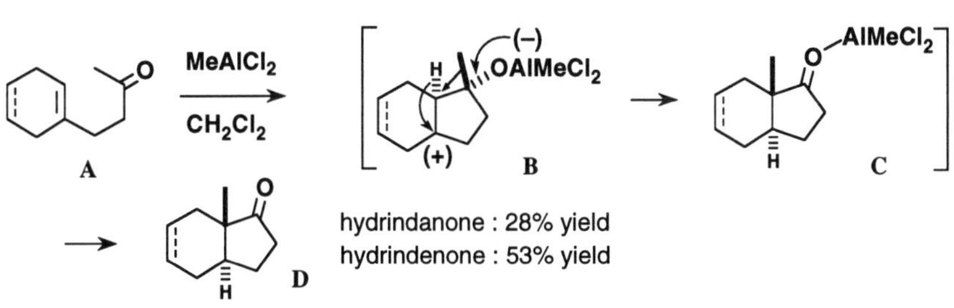
MeAlCl2
CH2Cl2
A
B
C
D
(–)
(+)
OAlMeCl2
AlMeCl2
hydrindanone : 28% yield
hydrindenone : 53% yield

Hydrindenone **D** can then be converted into enedione **E**, a key intermediate en route to 11-oxo steroids.

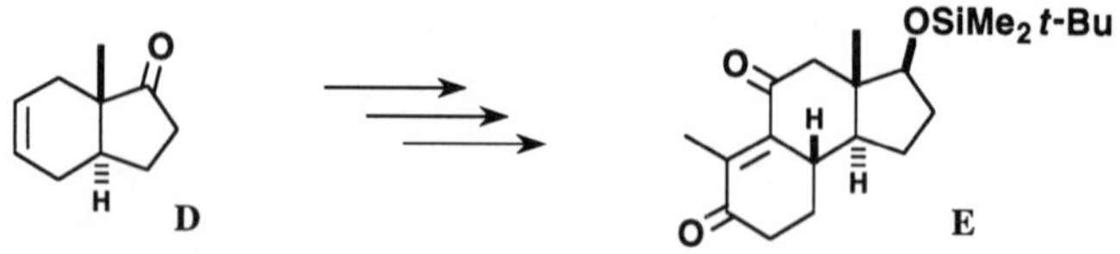

IV.2.b Cyclization of hept-5-enal derivatives

Hept-5-enal derivatives can lead, depending on the Lewis acid involved, either to ene products or to cyclopentanone derivatives. Thus, Snider observed that 2,6-dimethylhept-5-enal **A** leads predominantly to cyclopentanols **B** and **C** when using Me_2AlCl as catalyst. Their formation occurs most probably via a concerted mechanism that involves a loss of CH_4 by the carbonyl-Me_2AlCl complex, thereby preventing reversal of the ene reaction.[140]

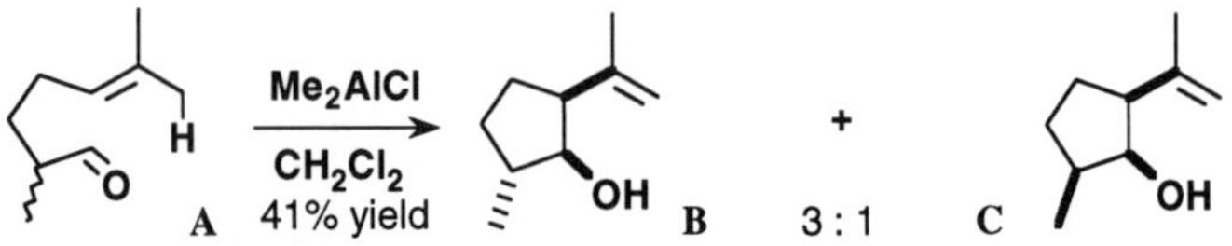

In contrast, with $MeAlCl_2$, a more acidic Lewis acid, cyclopentanones **D** and **E** are the major products of the reaction.

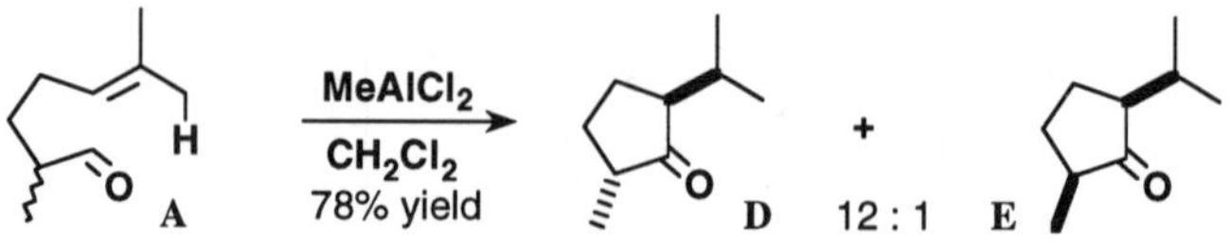

The formation of cyclopentanone **D** could result from a cationic process involving two 1,2-hydride shifts. Thus, homoallylic alcohol **B** could give cations **F** and then **G**, which then leads to ketone **D**.

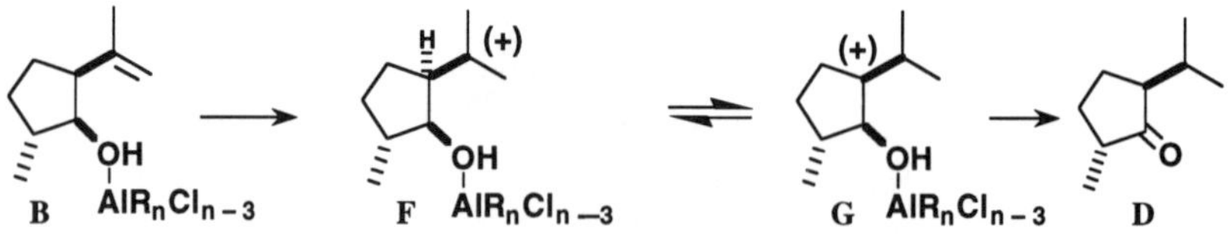

From 3,7-dimethyloct-6-en-2-one, the stepwise process, involving a methyl shift, occurs and the corresponding cyclopentanone is obtained.

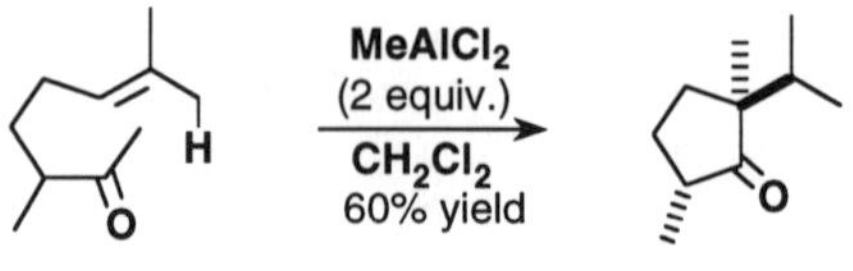

Cyclopentanones **D** and **E** had previously been obtained upon treatment of **A** with BF_3-OEt_2.[141]

Similarly, oct-5-enal derivatives undergo ene reactions with Me_2AlCl and cation-olefin cyclizations with 2 equivalents Me_2AlCl, $MeAlCl_2$ or $EtAlCl_2$ to give a zwitterion which reacts to give several products including ene adducts.[142]

IV.2.c Cyclization of oct-6-enal derivatives and related compounds

3,7-Dimethyloct-6-enal and non-6-enal undergo only ene reaction with all catalysts.[142] Snider examined the cyclization of (*E*)- and (*Z*)-non-6-enal to explore the effect of the double bond stereochemistry on the stereochemical outcome of the reaction. Thus, while (*Z*)-isomer **A** gives exclusively *cis*-substituted ene adduct **C** through exo transition state **B**, (*E*)-isomer **D** leads mainly to *trans*-isomeric alcohols **E** and **F**.

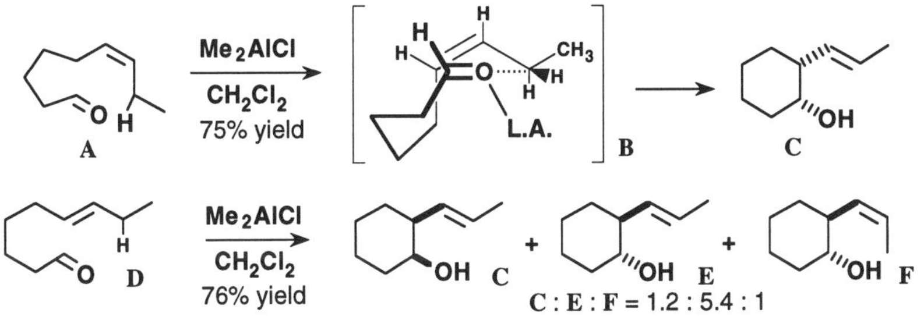

A —(Me_2AlCl, CH_2Cl_2, 75% yield)→ [B] → C

D —(Me_2AlCl, CH_2Cl_2, 76% yield)→ C + E + F

C : E : F = 1.2 : 5.4 : 1

When starting from **D**, alcohol **C** results from the formation of endo transition state **G** whereas the more favored exo transition state **H** can account for the formation of the major ene adduct, *trans-anti* alcohol **E**; minor alcohol **F** results from the formation of the less favorable exo transition state **I**.

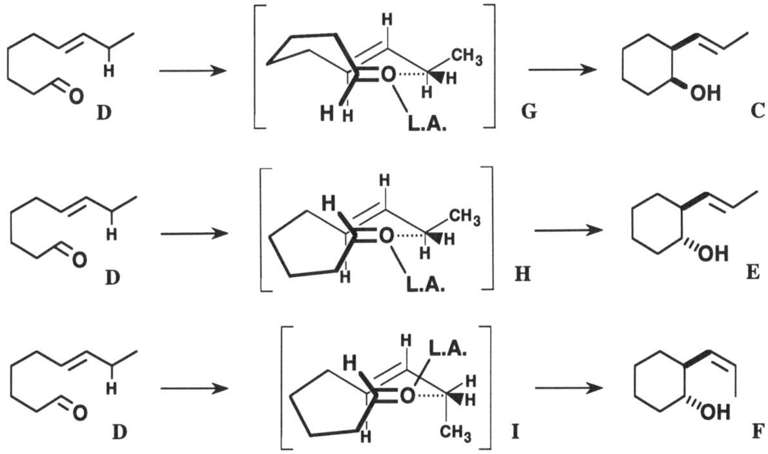

D → [G] → C

D → [H] → E

D → [I] → F

Cyclization of (+)-citronellal **A** into isopulegol is a very well-known process for historical reasons (see Introduction, Section I.1) and because (–)-

isopulegol **B** is an important intermediate in the manufacture of (–)-menthol. Nakatani proposed a highly stereoselective ene reaction using $ZnBr_2$ as catalyst, where the asymmetric induction is controlled by the chiral center on the chain.[143]

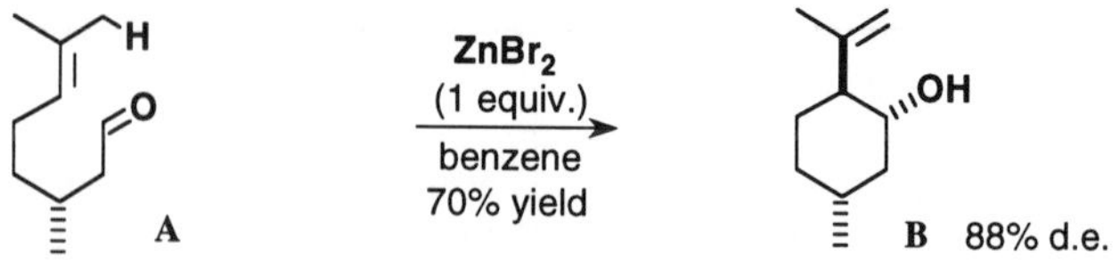

A — $ZnBr_2$ (1 equiv.), benzene, 70% yield → B 88% d.e.

Another very good example of asymmetric induction is reported by Sakai. Thus, the Rh(I) catalyzed cyclization of chiral oct-6-enal derivative **A** leads overwhelmingly to cyclohexanol **B**.[144] This 99:1 ratio is thought to result from steric repulsions between one of the methyl group of the chiral 1,3-dioxane and C-5 in the transition state leading to isomeric cyclohexanol **C**. Subsequent hydrolysis of **B** led to hydroxy cyclohexanone **D**.

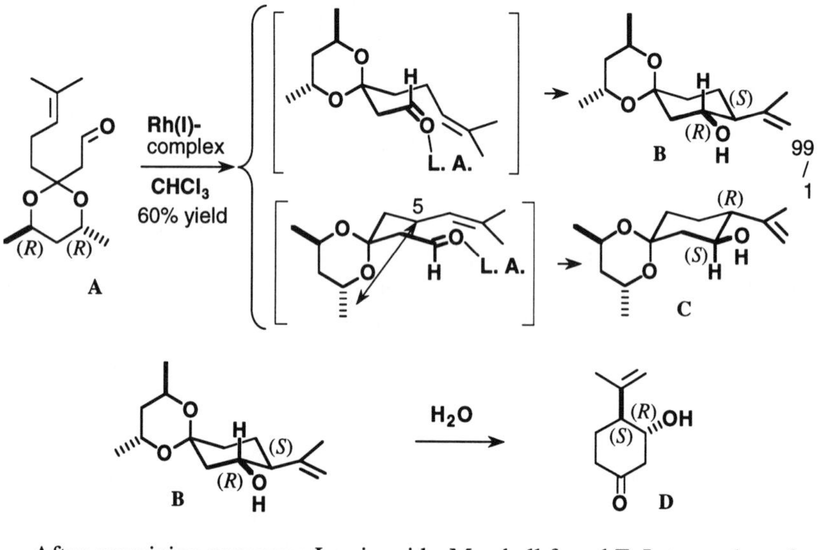

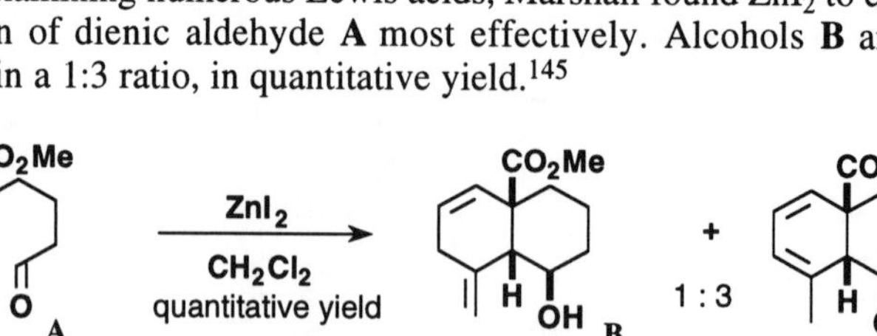

A — Rh(I)-complex, $CHCl_3$, 60% yield → B : C = 99 / 1

B — H_2O → D

After examining numerous Lewis acids, Marshall found ZnI_2 to catalyze the cyclization of dienic aldehyde **A** most effectively. Alcohols **B** and **C** were obtained, in a 1:3 ratio, in quantitative yield.[145]

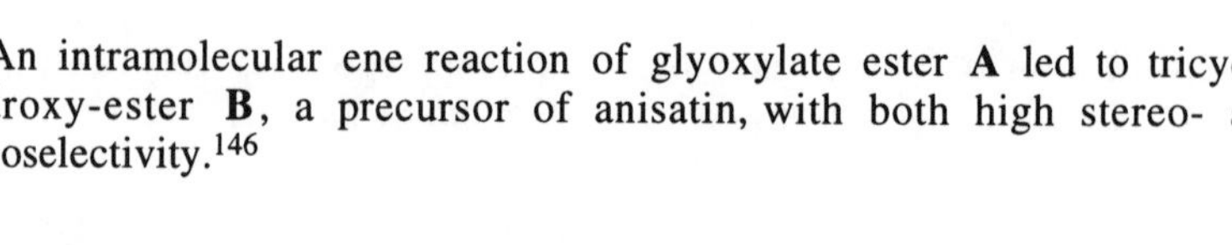

A — ZnI_2, CH_2Cl_2, quantitative yield → B + C (1 : 3)

An intramolecular ene reaction of glyoxylate ester **A** led to tricyclic hydroxy-ester **B**, a precursor of anisatin, with both high stereo- and regioselectivity.[146]

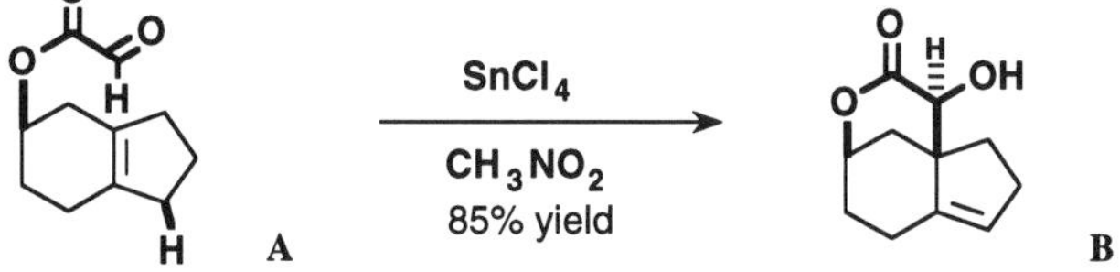

Ene reaction can also occur from unsaturated imines, dienic nitriles, esters or ketones or unsaturated epoxides. Thus, Solladié studied the cyclization of an unsaturated imine prepared from benzylamine and (+)-citronellal. He obtained isomeric benzylmenthylamines as the major products of the reaction.[147–149]

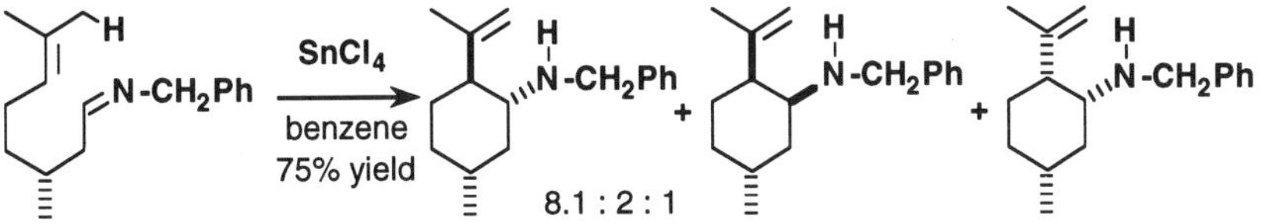

Chiral α-cyanovinylic sulfoxide **A** also served as an efficient chiral enophile in an Et_2AlCl-mediated intramolecular ene reaction.[150]

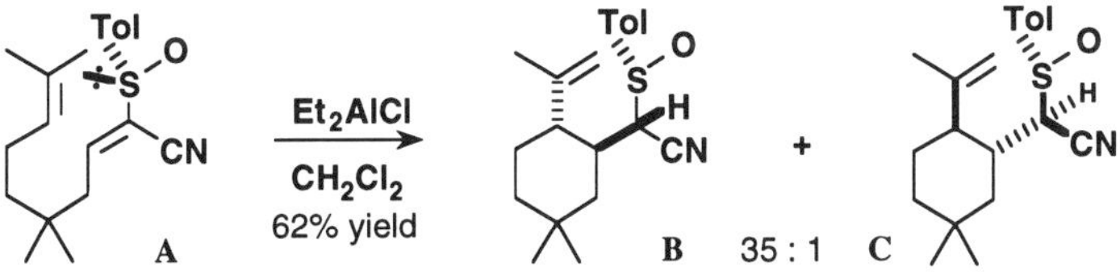

The reaction takes place preferentially on the *Re*-face of the enophile as the *Si*-face is more sterically hindered (presence of the tolyl group).

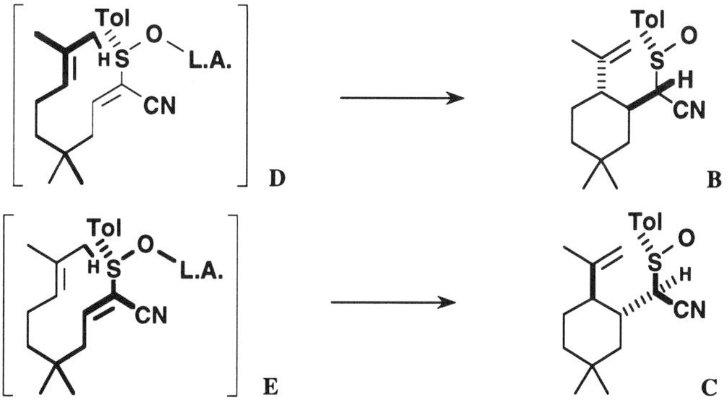

Intramolecular ene reaction on 1,7-dienes bearing an activated double bond occurs also with high diastereoselectivity to give *trans*-1,2-disubstituted cyclohexanes.[151–155] The presence of a chiral center at C-4 can even induce good asymmetric induction.[156] Particularly, chiral 1,7-diene **A** gave predominantly ene adduct **B** which, through a sequence involving another (intermolecular) ene reaction, led to veticadinol.[157]

CO_2Me CO_2Me **A** — $FeCl_3$ / Al_2O_3, CH_2Cl_2, 92% yield → CO_2Me CO_2Me **B** 82.3 : 1 + CO_2Me CO_2Me **C**

CO_2Me CO_2Me **B** → CO_2Me — paraformaldehyde, Me_2AlCl, CH_2Cl_2, 81% yield → OH CO_2Me → → → OH H H veticadinol

The cyclization of dienone **A** (used as a 1:1 mixture of isomers) depends on the Lewis acid used and the experimental conditions. Thus, although treatment of **A** with 2 equivalents $MeAlCl_2$ gives a conjugated enone resulting from hydrid shifts, better yields in ene adduct **D** were observed with BF_3-OEt_2 or $SnCl_4$.[158]

H O **A** — BF_3-OEt_2 (1.5 equiv), CH_2Cl_2, 79% yield → [H H O–L. A.] **B** → [H H O–L. A.] **C** → H O H **D**

High asymmetric induction was observed by Oppolzer who treated 8-phenylmenthol dienic ester **A** with Et_2AlCl. He obtained ene adduct **B** with 90% d.e. and then transformed it into (+)-α-allokainic acid **C**, a monocyclic amino diacid.[159,160]

O CF_3 N CO_2Et CO_2Et Ph H CO_2 **A** — Et_2AlCl, CH_2Cl_2, 60% yield → O CF_3 N CO_2Et CO_2Et Ph CO_2 **B** 90% d.e. → → → 73% yield NH (S) (S) (R) CO_2H CO_2H **C** α-allokainic acid

Finally, treatment of olefinic epoxy ketone **A** with $SnCl_4$ led to bicyclic ketol **E**. Such a transformation can be accounted for by an intramolecular ene reaction leading to ene adduct **C** which then undergoes the well-known epoxide-to-ketone transformation to yield ketol **D**, a precursor of **E**. **D** is in fact the kinetic product of the reaction and can undergo isomerization into **E** either under the reaction conditions or upon treatment with NaOH.[161]

O O H A $SnCl_4$ CH_2Cl_2 94% yield O L. A. O H B HO O – L. A. C

HO O D $SnCl_4$ or NaOH O OH E

Sharpless observed an epoxy alcohol rearrangement leading to diol **C** when treating **A** with $VO(OEt)_3$. This reaction can be regarded as an intramolecular ene reaction with the epoxide moiety being the enophile; however, it suffers from low total mass recovery (<40%).[162]

H O OH A $VO(OEt)_3$ CH_2Cl_2 <40% yield H O L. A. O H B OH OH C

The use of $Ti(O\textit{i}\text{-}Pr)_4$ did not lead to better results. However, this Lewis acid promoted a stereoselective ene reaction from epoxy alkyne **A**; thus, allenic diol **C** was obtained, through **B**, as a single isomer. This constitutes a rare case of an efficient cyclization process leading to an allene derivative from a simple acetylene terminator.

O HO A $Ti(O\text{-}i\text{-}Pr)_4$ CH_2Cl_2 50% yield L. A. H O HO B HO HO C

IV.2.d Cyclization of other derivatives

The scope of the reaction was expanded by Marshall who prepared 12-, 14- and 16-membered rings.[163] Thus, dienic ynal **A** undergoes efficient cyclization into cyclic dienynol **B**.

O
H
TIPSO A
$EtAlCl_2$
CH_2Cl_2
66% yield
OH
TIPSO B

In the same way, trienic ynal **C** was cyclized into the 14-membered propargyl alcohol **D**.

O
H
C
$EtAlCl_2$
CH_2Cl_2
79% yield
OH
D
1 : 1 mixture of diastereomers
5 : 1 mixture of diastereomers with BF_3-OEt_2

IV.3 Cyclization of Type II

Works on the cyclization of ω-olefinic trifluoromethyl ketones demonstrated that olefin regiochemistry and stereoselectivity are better controlled than in related nonfluorinated series[164] (the reactivity of trifluoromethyl ketones in intermolecular reactions had been previously studied by Kumadaki[165–167]). Thus, Bégué observed a very clean $TiCl_4$-mediated cyclization of trifluoromethyl ketone **A** into cyclopentenol **B**.

Ph
EtO_2C
CF_3 O A
$TiCl_4$
CH_2Cl_2
90% yield
Ph
EtO_2C
CF_3 OH B

With γ,δ-ethylenic ketones, cyclization occurs through a stepwise mechanism involving a zwitterionic intermediate such as **E**.[168]

EtO_2C CH_3
O
F_3C
C
$TiCl_4$
CH_2Cl_2
92% yield
CH_3 CO_2Et
L.A.–O
CF_3
D
CH_3 CO_2Et
(–)
L.A.-O
CF_3 H (+)
E
CH_3 CO_2Et
HO
CF_3
F

In contrast, a Lewis acid mediated ene reaction occurs with δ,ε-ethylenic ketones.

CH_3
EtO_2C
CF_3 O
G
Et_2AlCl
CH_2Cl_2
65% yield
CH_3
EtO_2C
H
O
H
H
CF_3
L.A.
H
CH_3
EtO_2C
CF_3
HO H
I

Lewis acid induced cyclization of δ,ε-olefinic trifluoromethyl ketone **A** provides (trifluoromethyl)decalols in high yield. The ene reaction occurs when $EtAlCl_2$, Me_2AlCl or Me_3Al are used as catalysts while $TiCl_4$ induces a stepwise process.

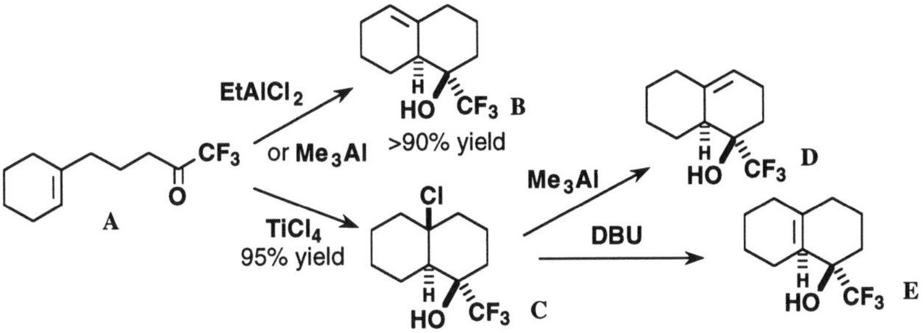

In order to synthesize a precursor of trichodermol, Andersen studied the cyclization of olefinic aldehydes by intramolecular ene reaction and particularly the reactivity of aldehyde **A** towards a variety of Lewis acids, $SnCl_4$, $TiCl_4$, $SnBr_4$, ZnI_2, BF_3-OEt_2, $EtAlCl_2$ and Et_2AlCl. $SnCl_4$ proved to be the best behaved system. Tricyclic alcohol **C** can be viewed as an ene adduct, while chlorhydrin **D** comes from an ionic pathway.[169]

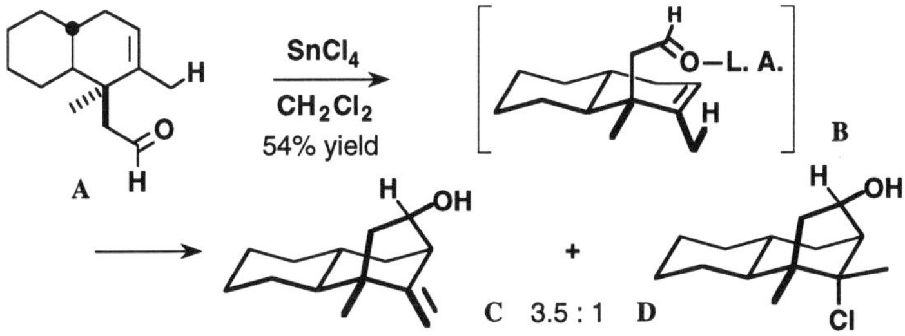

In contrast, aldehyde **A** undergoes cyclization in quantitative yield to give the axial isomer of the *trans* hydrindanol **B** as the exclusive product.

Cyclization of aldehyde **A** gave mainly alkylidene cyclohexanol **C** through transition state **B** in which the methyl group selectively adopted the synperiplanar conformation.[170]

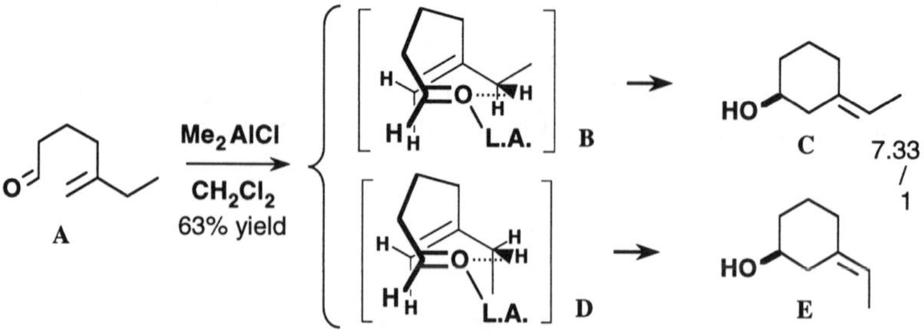

Marshall demonstrated that the ene cyclization of unsaturated aldehyde **A** occurs by internal proton or deuton transfer from the vinylic CHD group to the aldehyde carbonyl group.[171] Transition states involve an α-methyl group in the opposite side of the allylic moiety. In the major process, the allylic methyl group is synperiplanar to the chain.

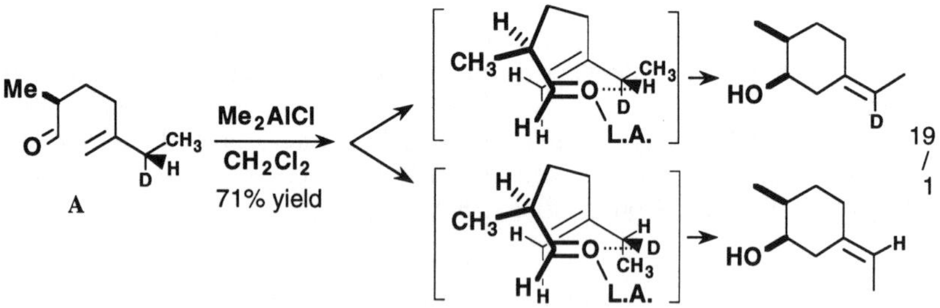

A highly stereocontrolled intramolecular ene reaction of δ,ε-unsaturated aldehydes was proposed by Yamamoto who used an exceptionally bulky Lewis acid, methylaluminum bis(4-bromo-2,6-di-*tert*-butylphenoxide) (MABR).[172] Thus, while the Me_2AlCl-mediated cyclization of aldehyde **A** is known to afford *cis*-alcohol **B**,[170] in marked contrast the use of MABR gave rise to isomeric *trans*-alcohol **C**.

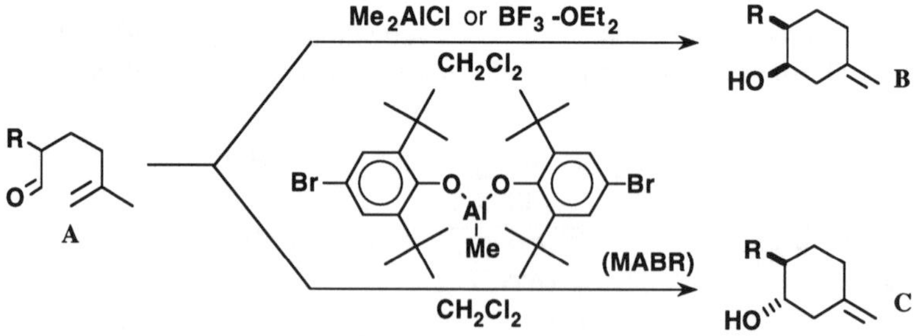

This striking result can be explained in the following way:
In transition state **E**, which can account for the formation of alcohol **F**, from aldehyde **D** and BF_3-OEt_2, the α-methyl group is on the opposite side with respect to the ethylenic double bond.

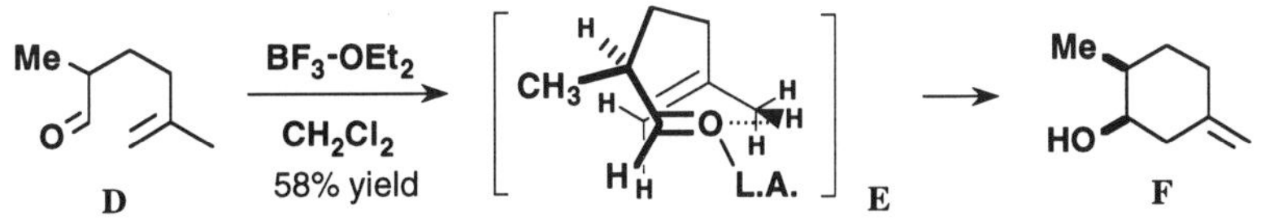

However, under the influence of the bulky MABR, the α-methyl group is, in transition state **G**, on the same side of the ethylenic double bond leading therefore to *trans*-alcohol **H**.

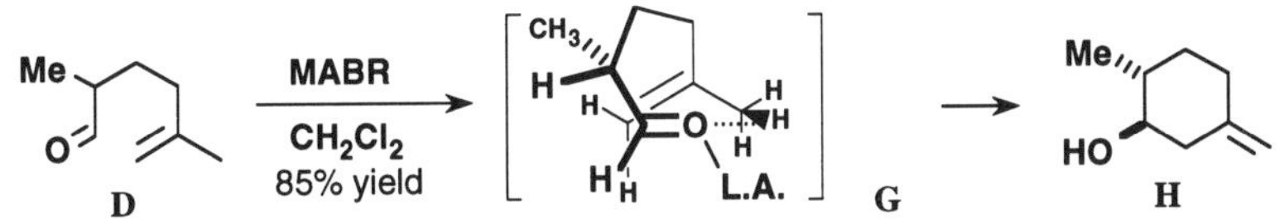

With aldehydes **I** and **K** possessing a trisubstituted double bond, MABR also promotes a highly stereoselective cyclization leading to the formation of alcohols **J** and **L**, respectively.

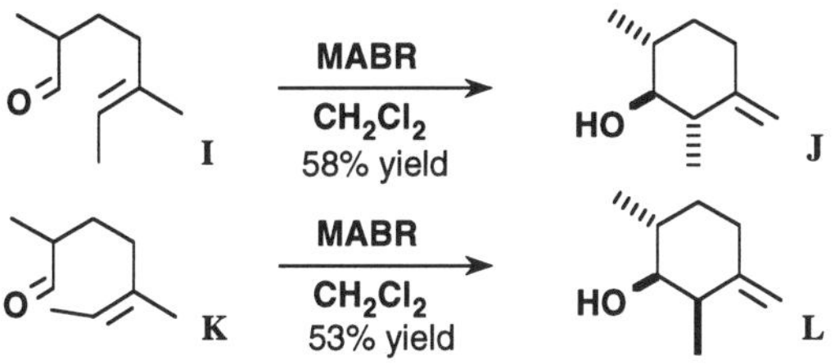

With the aim of preparing the hexahydronaphthalene unit of pravastatin, Barrish prepared aldehyde **A** which underwent, under Lewis acid catalysis, a smooth ene reaction affording bicyclic alcohol **B.**[173]

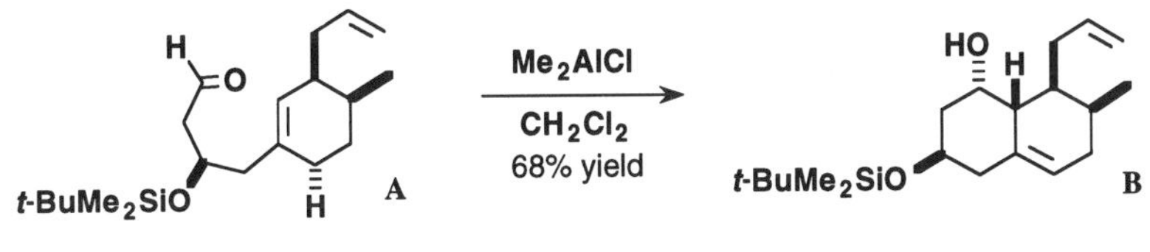

The $AlCl_3$ catalyzed cyclization of unsaturated ketone **A** leads to a 3:1 mixture of alkylidene isomers **B** and **C**; **B** being a potential intermediate for the synthesis of pumiliotoxin derivatives.[174]

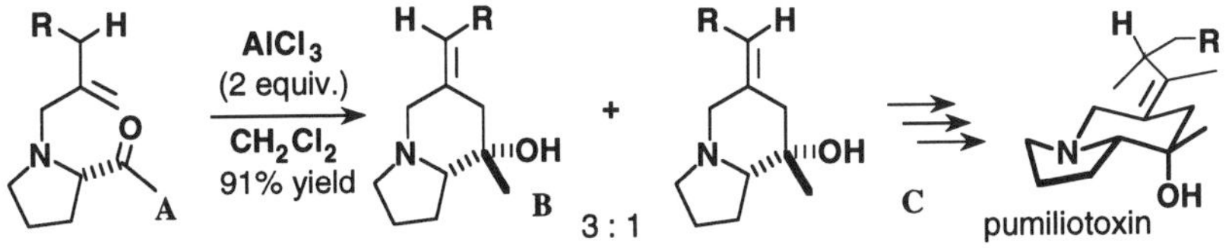

One-pot sequential ene reactions (inter- + intramolecular) can be obtained.[119,120] Particularly, octenol **C** was directly prepared through unsaturated aldehyde **B** from acrolein and methylene cyclohexane **A**.[175]

When oxirane **A** (1:1 isomeric mixture) is treated with BF_3-OEt_2, alcohol **C** is obtained as a 1:1 mixture of diastereomers, presumably via the initial formation of aldehyde **B** then followed by intramolecular ene cyclization.[137]

Hydroazulenols can be cleanly synthesized by Lewis acid-promoted intramolecular ene reaction.[176–179] Thus, Marshall and Anderson observed the quantitative and highly stereoselective cyclization of aldehyde **A**. Endo transition state **D** can account for the stereochemistry of major isomer **B**.

IV.4 Cyclization of Type III

The cyclization of (hexenylimino)malonate **A**, in the presence of TMS-OTf, led to substituted piperidine **B**, as the major product, and annulated piperidine lactone **C**.[180,181]

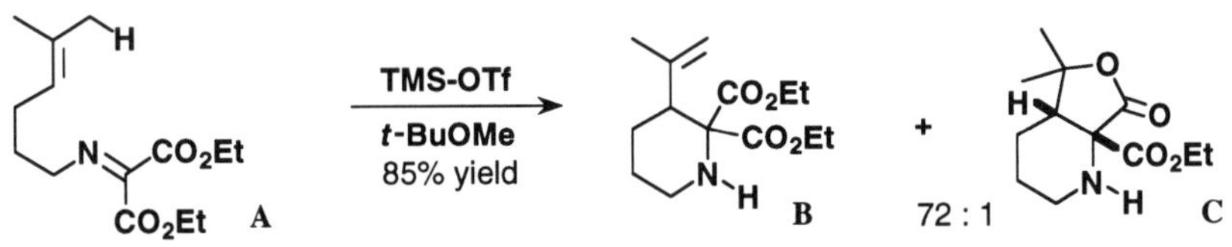

However the cyclization of the corresponding (heptenylimino)malonate **D** afforded, under TMS-OTf catalysis, only fused cycloamino lactone **E** in good yield and high diastereoisomeric excess.[182]

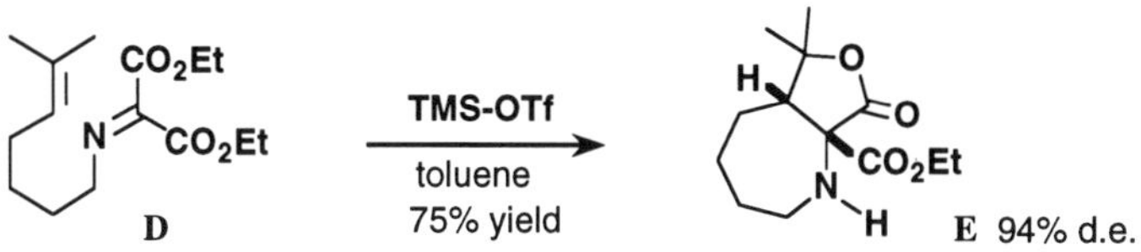

The preparation of Δ^4-oxocenes through Lewis acid promoted cyclization of hex-5-enyl acetals occurs with both high regio- and stereoselectivity.[183–185] The regiochemical outcome of the reaction can be accounted for if the transformation occurs in a concerted way.

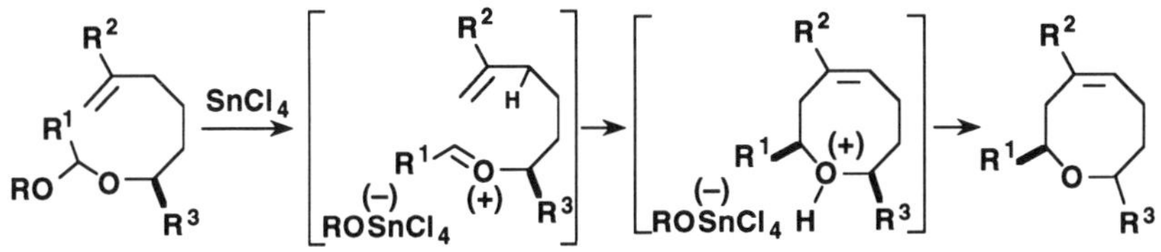

Vinyl sulfides, used as cyclization terminators, proved to provide the highest yields of Δ^4-oxocene products. Thus, vinyl sulfide acetal **A** undergoes clean cyclization with BF_3-OEt_2 in *t*-BuOMe to yield **B** which was then desulfurized to give Δ^4-oxocene **C**.

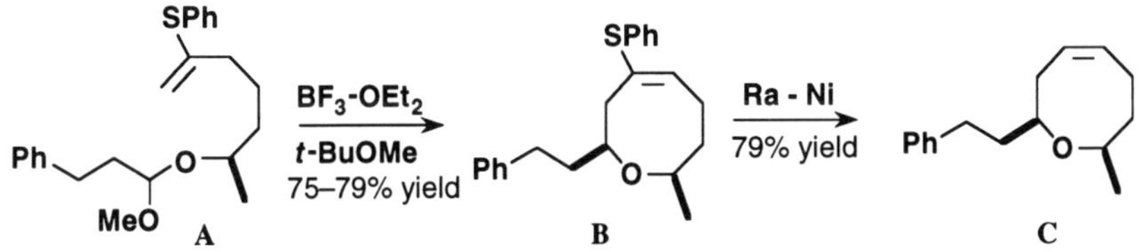

The intramolecular and intermolecular hydrogen-deuterium isotope effects of the following reaction were found to be identical, $(k_H/k_D)_{inter} = (k_H/k_D)_{intra} = 1.65$. The equivalence of these isotope effects provides further evidence that the formation of the Δ^4-oxocenes occurs through a concerted process in which the C–H bond cleavage accompanies the C–C bond formation.

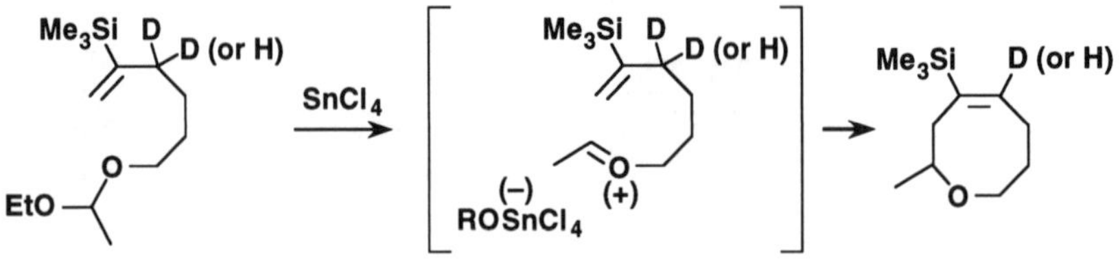

An exo transition state can account for the formation of the (*E*)-double bond through a concerted process.

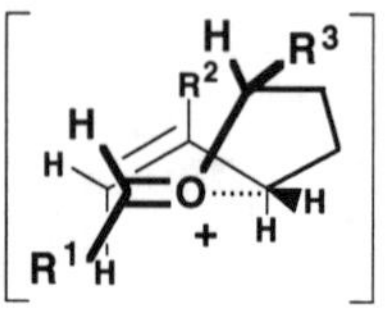

Postulated exo transition state leading to (*E*)-double bond.

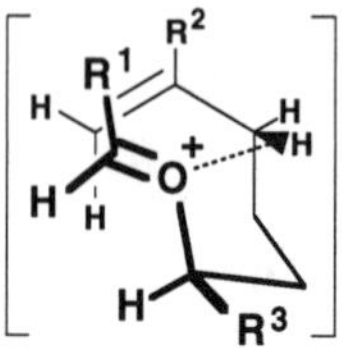

Endo transition state for the concerted cyclization.

Trisubstituted double bond stereoisomeric acetals **A** and **B** underwent cyclization to give mixtures of Δ^4-oxocenes and 3-alkylideneoxepanes. Thus, (*Z*)-vinylsilane acetal **A** afforded oxocene **C**, as the major product (2:1), along with (*Z*)-pentylidene oxepane **D** while (*E*)-vinylsilane **B** afforded (*E*)-pentylideneoxepane **E** with high selectivity (**E**:**C** = 17:1).

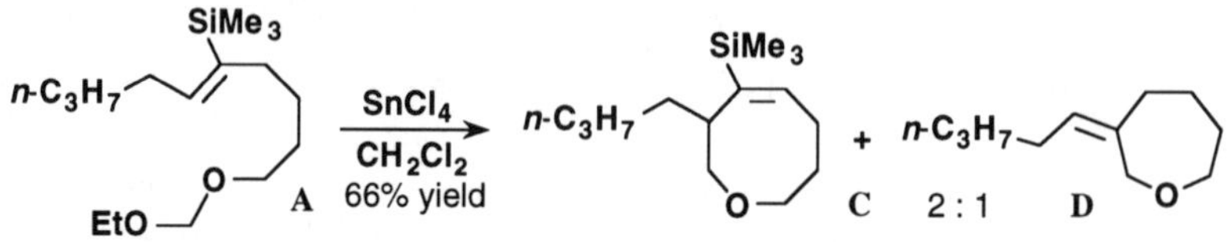

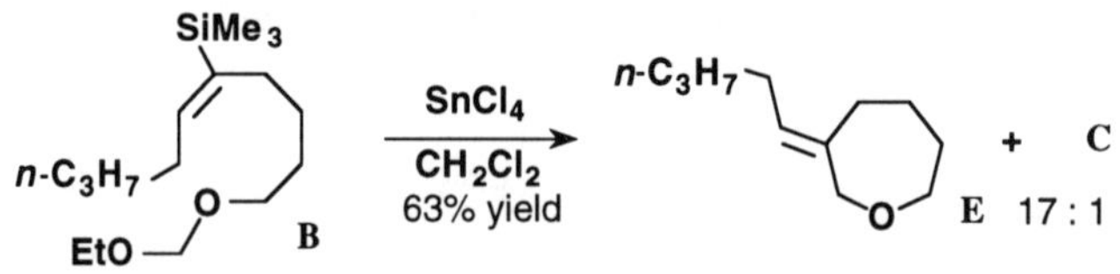

The predominant formation of **C** from (*Z*)-isomer **A** can be explained by the intervention of an exo transition state.

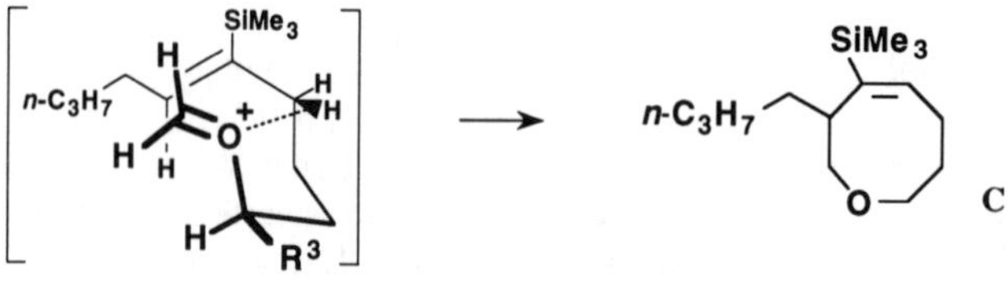

Similarly, the almost exclusive formation of **E** from (*E*)-isomer **B** can also be accounted for by invoking an exo transition state.

SiMe3 H H H n-C3H7 + O H H → Me3Si n-C3H7 O+ H → O n-C3H7 **E**

These data must however be interpreted cautiously since minor changes in Lewis acid, ene, or enophile can lead to changes in mechanism.

In contrast, ketal **A**, a key intermediate in the synthesis of (–)-magellanine and (+)-magellaninone, undergoes cyclization through a carbocationic process which involves the rearrangement of carbocation **B**.[186]

Et3SiO OMe OMe **A** — $SnCl_4$ / CH_2Cl_2, 60% yield → (+) OMe H Et3SiO → (+) OMe Et3SiO **B**

→ O H OMe ≡ O H OMe

Finally, it is worth noting that the formation of Δ^4-oxocenes through Type III intramolecular ene reaction is to some extent compatible with the presence of strong leaving groups on the molecule.[187]

Me3Si Cl OTs O OEt t-BuPh2SiO — $SnCl_4$ / CH_2Cl_2, 37% yield → Me3Si Cl OTs O t-BuPh2SiO

Sutherland observed that some epoxy cyclohexanones can undergo cyclization into decalone derivatives.[188] Some products of the reaction can be considered as resulting from an intramolecular ene cyclization. Use of weak Lewis acids and/or nonpolar solvents leads to ring contraction.

O O **A** → [O–L. A. O] **B** → O OH **C** + OH O OH **D**

→ [O O–L. A.] **E** → O O **F**

L. A. = $TiCl_4$: 55% yield; **C** : **D** = 1.2 : 1

L. A. = $SnCl_4$: **C** : **F** = 1 : 1.7

The cyclization of similar compounds but bearing a disubstituted double bond occurs in better yields.

BF_3-OEt_2 or $SnCl_4$ or BCl_3, CH_2Cl_2, >60% yield

Epoxy-acetylenic compounds can also undergo cyclization into decalin or hydrindane derivatives through a cationic process.[189]

$TiCl_4$, CH_2Cl_2, 90% yield

$SnCl_4$, CH_2Cl_2, 94% yield

Finally, this section ends with a particular intramolecular ene reaction that does not belong to any of the above-mentioned classification.[190]

$EtAlCl_2$, CH_2Cl_2, 82% yield

V. Asymmetric Reaction Catalyzed by Chiral Catalyst

Chiral catalysts, used in reaction between achiral ene and enophile components or with unsaturated ene, can promote enantiofacial control. Indeed chiral ene adducts are often obtained with remarkable high levels of enantiomeric excess.[191,192]

V.1 Intramolecular reaction

The first successful example of asymmetric cyclization of a prochiral unsaturated aldehyde promoted by a chiral Lewis acid is due to Yamamoto who used chiral catalyst **A** which was prepared *in situ* from dimethylzinc and optically pure (*R*)-(+)-BINOL.[193,194] Cyclization of 3-methylcitronellal **B**, with 3 mol. equivalents of **A**, afforded methylisopulegol **C** as a single *trans* isomer with 90% e.e.

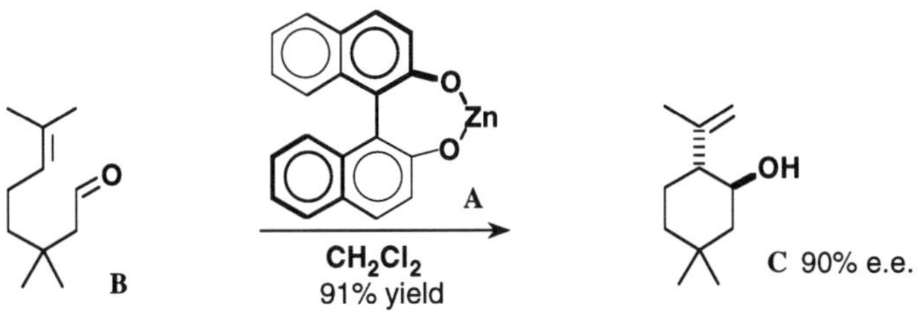

However, in sharp contrast, exposure of aldehyde **D** to the same chiral catalyst **A** afforded *trans* alcohol **E** as a racemic mixture.

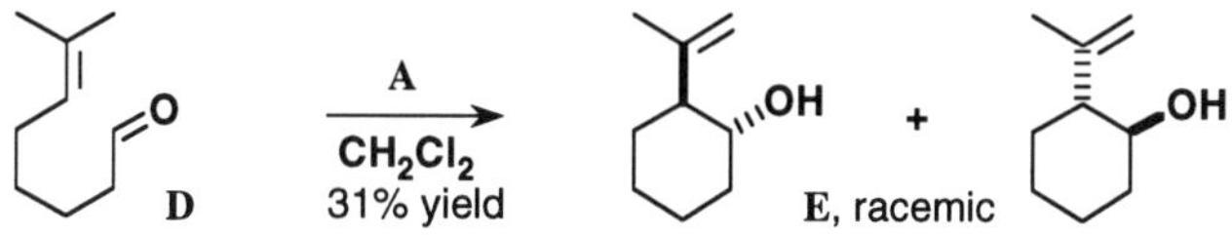

Surprisingly, a separate treatment of (*R*)- and (*S*)-citronellal **F** and **H** with chiral catalyst **A** gives rise to (*R*)- and (*S*)-isopulegol **G** and **I**, respectively. This clearly establishes that the asymmetric induction is totally controlled by the C-3 chiral center of the aldehyde and is independent of the chirality of the catalyst.

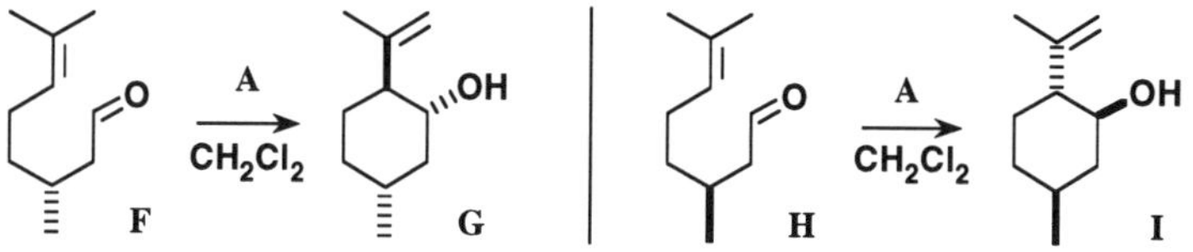

Cyclization of (*Z*)-methylfarnesal **J** gives rise to alcohol **K** with high enantiomeric excess. In contrast, (*E*)-methylfarnesal **L** leads mainly to alcohol **M** but with low optical purity.

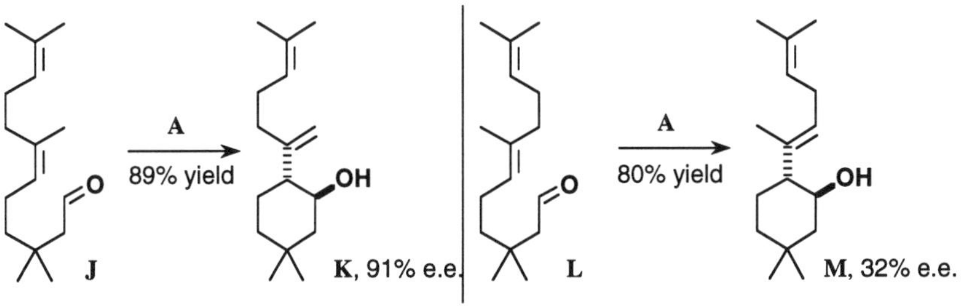

A highly diastereo- and enantioselective intramolecular ene reaction was developed by Narasaka who employed a chiral titanium reagent **A** generated *in situ* from $TiCl_2(Oi\text{-}Pr)_2$ and a tartrate-derived chiral 1,4-diol in the presence of molecular sieves.[195]

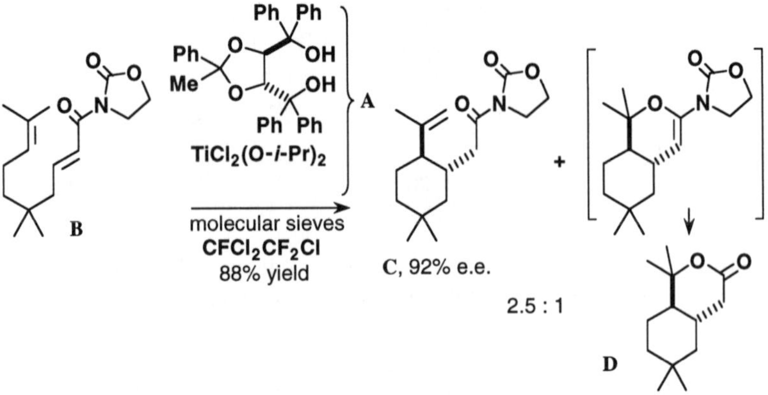

B

Ph Ph

Ph

Me

OH

OH

Ph Ph

$TiCl_2(O\text{-}i\text{-}Pr)_2$

A

molecular sieves

$CFCl_2CF_2Cl$

88% yield

C, 92% e.e.

2.5 : 1

D

The same catalyst was used to perform an efficient synthesis of optically pure (–)-ε-cadinene from dithioketals **E** and **F**.

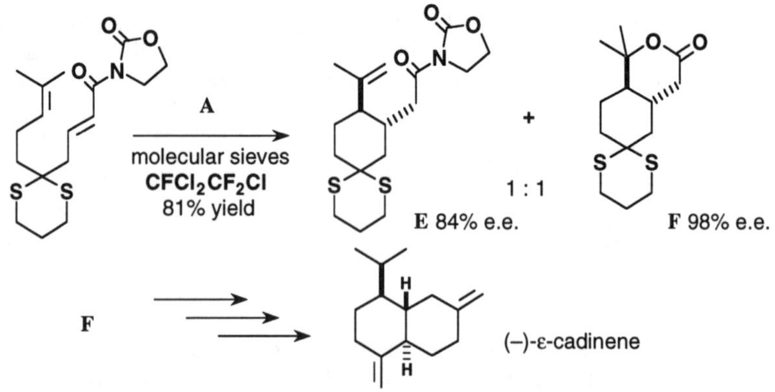

A

molecular sieves

$CFCl_2CF_2Cl$

81% yield

E 84% e.e.

1 : 1

F 98% e.e.

F

(–)-ε-cadinene

The chiral BINOL-derived titanium perchlorate was shown to serve efficiently as an asymmetric catalyst for the ene cyclization of unsaturated aldehydes such as **A**.[196]

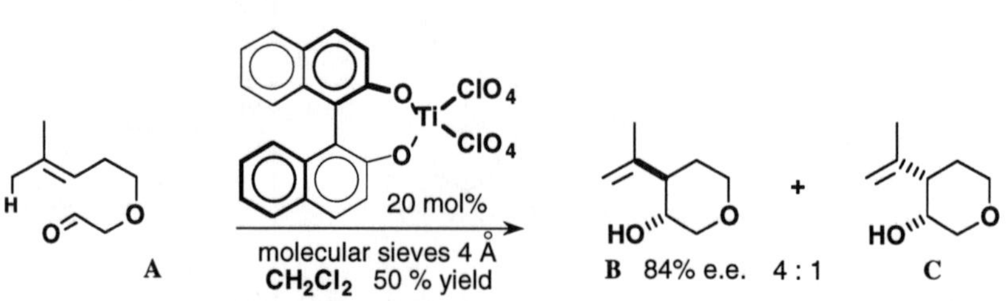

ClO_4

Ti

ClO_4

20 mol%

A

molecular sieves 4 Å

CH_2Cl_2 50 % yield

B 84% e.e. 4 : 1

C

An exo transition state involving the addition of the ene moiety on the *Re* face of the carbonyl group accounts for the stereochemistry of major adduct **B**.

The intramolecular carbonyl ene reaction of achiral malondialdehyde **A** was conducted in the presence of another BINOL titanium complex catalyst. Cyclized adduct **B** was obtained with modest enantioselectivity. **B** is an intermediate in the synthesis of the highly functionalized carbon ring skeleton of trichothecene anguidine.[197] Transition state **D**, with the methyl cyclopentadienyl group in pseudo-equatorial position, should prevail and accounts for the formation of **B**.

B 38% e.e. 4.5 : 1 C

V.2 Intermolecular reaction

The first example of a chiral Lewis acid-promoted asymmetric ene reaction between a prochiral aldehyde and an alkene is also due to Yamamoto who used catalyst **A**.[198] Moreover, the presence of 4 Å molecular sieves (activated powder) allows the use of only 20 mol% of **A**.

A 20 mol%, molecular sieves 4 Å, CH_2Cl_2 56% yield

B 84% e.e.

Efficient asymmetric catalysis conditions in glyoxylate-ene reaction were discovered by Nakai. An extensive screening of various chiral catalysts derived from optically active diol was made. These results clearly indicate the effectiveness of those which exhibit a C_2 symmetry and among those, BINOL-derived chiral catalysts prepared in situ, were found to afford the best enhanced levels of enantiocontrol. Moreover, the role of molecular sieves, which are essential for obtaining high enantioselectivity, was clarified; they make the alkoxy-ligand exchange reaction easier but are not involved in the ene reaction step.[199–201]

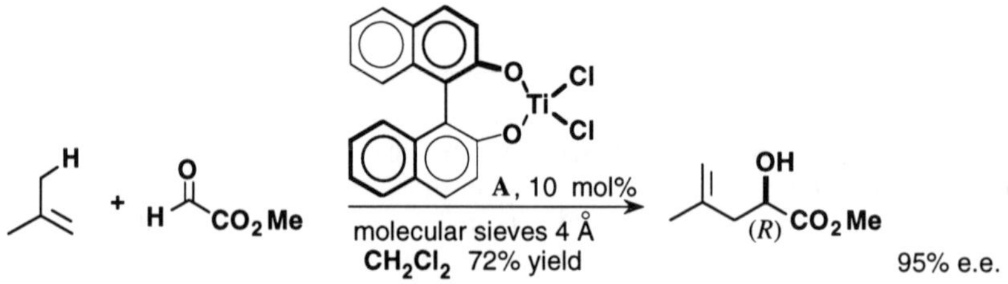

Reaction between isoprene and glyoxylate, catalyzed by a chiral titanium complex in low concentration, provides the carbonyl-ene adduct and the hetero Diels-Alder product both obtained with extremely high enantioselectivity.[202]

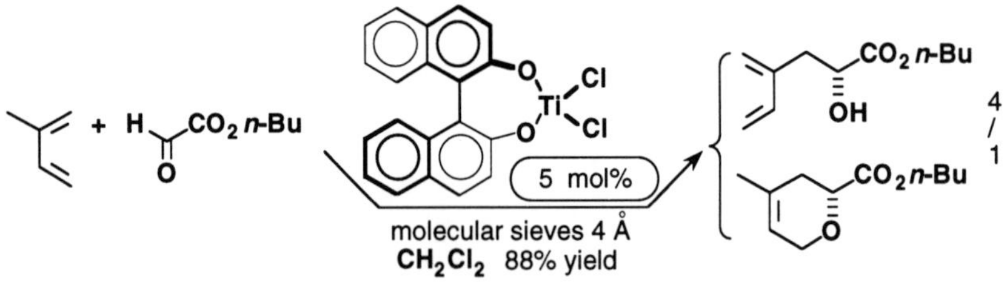

The ene adduct was then converted to optically pure (*R*)-(–)-ipsdienol, an aggregation pheromone of bark beetles.

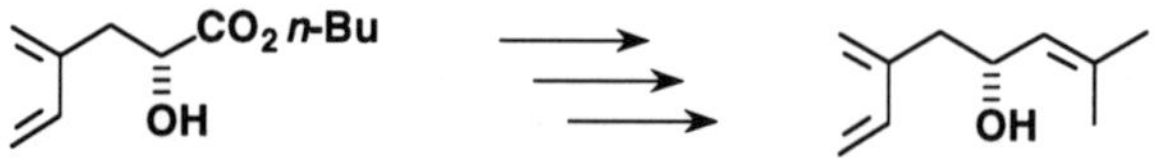

When a ligand with C_2 symmetry is bound to a metal, reactants experience the same chiral environment regardless of the side from which they approach.

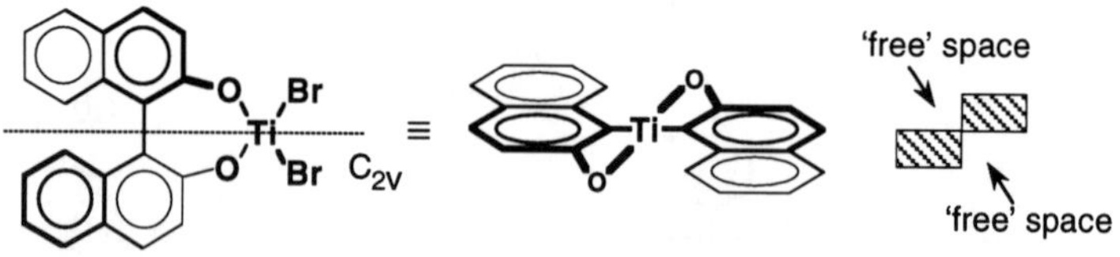

The following transition states, involving chelation control, can be tentatively proposed to account for the enantioselectivity of the glyoxylate ene reaction.

Moreover, Mikami and Nakai observed that the e.e. of the products significantly exceeded the enantiomeric purity of the chiral auxiliaries involved. This remarkable "asymmetric amplification" is known as the positive nonlinear effect (NLE).[203,204]

The following graph shows the variation of e.e. of the carbonyl ene adduct **A** as a function of the e.e. of BINOL ligand; it indicates that the use of BINOL of 35–40% e.e. is good enough to provide the same level of e.e. for ene adduct **A** than when prepared from enantiomerically pure BINOL.

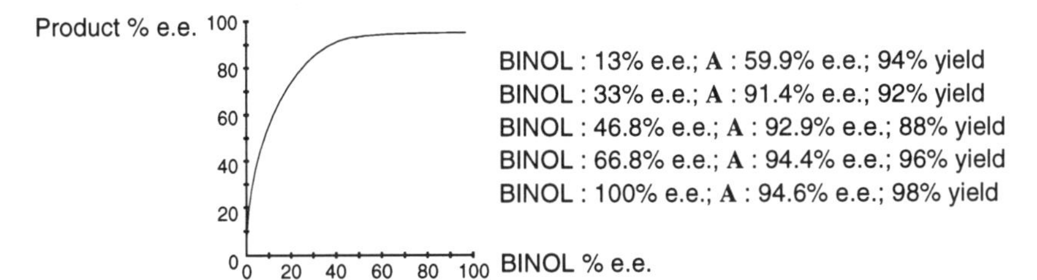

Some experimental results led Mikami and Nakai to consider the role of the dimeric structure of the BINOL-derived complex. It appeared likely that the remarkable NLE results from a marked difference in the stability of the diastereomeric dimers (*R*)(*R*) and (*R*)(*S*). The meso dimer (*R*)(*S*) is the most stable and hence less reactive; this is underlined by the fact that the chiral titanium complex derived from 100% e.e. of BINOL reacts 35 times faster than the complex derived from racemic BINOL. As a result, even if appreciable amounts of (*S*)-BINOL exist, they are associated with an equivalent amount of

(*R*)-BINOL to form a rather stable complex. Thus, the remaining monomeric (*R*)-BINOL can induce high e.e.

$$(R)(R) \rightleftharpoons 2\,(R)$$
$$(R)(S) \rightleftharpoons (R) + (S)$$

Asymmetric catalysis with BINOL-titanium complex of the glyoxylate-ene reaction of vinylic sulfides affords mainly *anti* enantiomerically pure diastereomers.[205]

PhS, H, +, O, H, CO_2Me, O, Ti, Cl, Cl, 10 mol%, molecular sieves 4 Å, CH_2Cl_2 94% yield, PhS OH CO_2Me, *anti*, >99% e.e., 19 : 1, PhS OH CO_2Me, *syn*, >90% e.e.

Applied on a simpler case, this reaction allowed the preparation of (*R*)-(–)-ipsdienol with remarkable optical purity.

PhS, H, +, O, H, CO_2Me, O, Ti, Cl, Cl, 0.5 mol%, molecular sieves 4 Å, CH_2Cl_2 94% yield, PhS OH CO_2Me, 99% e.e., OH, (*R*)-(–)-ipsdienol

Mikami discovered that under BINOL-titanium complex **A** catalysis, ketone silyl enols such as **B** and glyoxylate esters do not lead to a Mukaiyama aldol adduct but to the ene adduct **C**. Moreover, **C** is formed with complete control of absolute and relative stereochemistry.[206]

Me_3SiO, **B**, +, O, H, CO_2Me, O, Ti, Cl, Cl, **A**, 0.5 mol%, molecular sieves 4 Å, CH_2Cl_2 58% yield, Me_3SiO OH CO_2Me, **C** 99% e.e.

Optically pure benzocycloalkenes can be synthesized from methyl-glyoxylate ester and benzomethylenecycloalkene.[207]

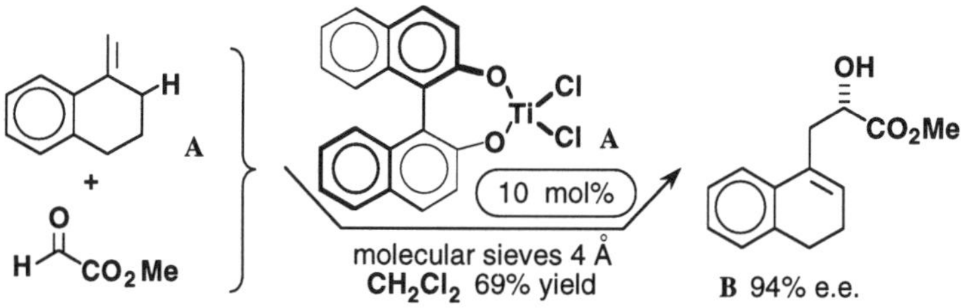

The glyoxylate-ene reaction, catalyzed by a chiral BINOL-titanium complex of symmetrical bis-allylic silyl ethers such as **A**, occurs with asymmetric desymmetrization. *Syn* adduct **B** was obtained with high levels of diastereo- and enantioselectivity, irrespective of the aldehyde stoichiometry.[208]

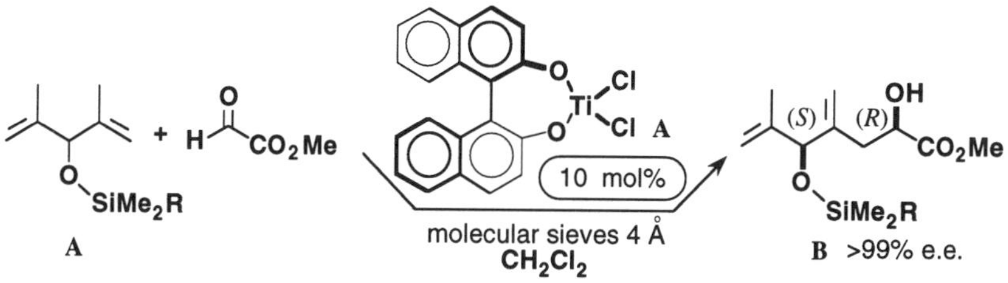

Mikami showed that the use of fluoral provides an efficient means for the asymmetic synthesis of CF_3-containing compounds with high levels of enantiomeric excess.[54]

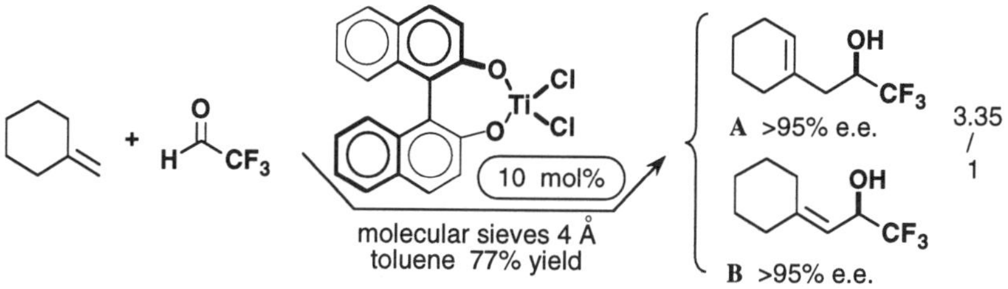

Modest enantiomeric excess (50–60%) were observed in the catalytic ene reaction of 2-methylpropene with 2,3-dichloropropanal and chloral in the presence of (–)-menthyloxyaluminum dichloride.[209]

VI. Conclusion

The Lewis acid catalyzed ene reaction involving carbonyl enophile or related compounds is a versatile synthetic method. Particularly, it provides a useful means for the carbofunctionalization of unactivated alkenes and also a very easy way of introducing an allylic moiety, from what can be seen as the simplest allylic nucleophile.

Of the reactions discussed in this book, this reaction is probably the one that is the most dependent on the introduction of Lewis acids. Thus, when the thermal reaction requires high temperatures and has a very limited scope, Lewis acids allow it to be carried at low temperatures, lead to ene adducts with high yields and also provide very impressive regio- and stereoselectivity.

Finally, the Lewis acid catalyzed ene reaction is one of the very first reactions in which asymmetric induction was obtained, with remarkable enantiomeric excess, from optically active Lewis acids used in catalytic amounts (down to 1 mol%). This reaction also provides one of the most dramatic nonlinear effects in asymmetric induction. Indeed, ene adduct with e.e. greater than 90% could be obtained by using Lewis acids with only 30–40% e.e. in catalytic amounts.

References

1. Alder, K.; Pascher, F.; Schmitz, A. *Chem. Ber.* **1943**, *76*, 27.
2. Hoffmann, H.M.R. *Angew. Chem., Int. Ed. Engl.* **1969**, *8*, 556–577
3. Keung, E.C.; Alper, H. *J. Chem. Educ.* **1972**, *49*, 97–100.
4. Conia, J.M.; Le Perchec, P. *Synthesis*, **1975**, 1–19.
5. Oppolzer, W.; Snieckus, V. *Angew. Chem., Int. Ed. Engl.* **1978**, *17*, 476–486.
6. Taber, D.F. *Intramolecular Diels–Alder and Ene Reactions*, Springer Verlag, Berlin, **1984**.
7. Snider, B.B. *Acc. Chem. Res.* **1980**, *13*, 426–432.
8. Oppolzer, W. *Pure & Appl. Chem.* **1981**, *53*, 1181–1201.
9. Mikami, K.; Terada, M.; Shimizu, M.; Nakai, T. *J. Synth. Org. Chem. Jpn.* **1990**, *48*, 292–303.
10. Oppolzer, W. *Pure Appl. Chem.* **1990**, *62*, 1941–1948.
11. Snider, B.B. in *Comprehensive Organic Chemistry* , Trost, B.M.; Fleming, I., Eds., Pergamon Press, Oxford, **1991**, Vol. 2, Chap. 2.1, p. 527–561.
12. Snider, B.B. in *Comprehensive Organic Chemistry* Trost, B.M.; Fleming, I., Eds., Pergamon Press, Oxford, **1991**, Vol. 5, Chap. 1.1, p. 1–27.
13. Treibs, W.; Schmidt, H. *Ber. Dtsch. Chem. Ges.* **1927**, *60*, 2335–2341.
14. Tiemann, F.; Schmidt, R. *Ber. Dtsch. Chem. Ges.* **1896**, *29*, 903–926.
15. Grignard, V.; Dœuvre, J. *C. R. Acad. Sci.* **1930**, *190*, 1164–1167.
16. *Nobel Lectures – Chemistry, 1942–1962*, Elsevier Publishing Co.: Amsterdam–London–New York, **1964**, pp 253–305.
17. Achmatowicz Jr., O.; Szymoniak, J. *J. Org. Chem.* **1980**, *45*, 1228–1232.
18. Gladysz, J.A.; Yu, Y.S. *J. Chem. Soc., Chem. Commun.* **1978**, 599–600.
19. Blomquist, A.T.; Himics, R.J. *J. Org. Chem.* **1968**, *33*, 1156–1159.
20. Blomquist, A.T.; Himics, R.J.; Meador, J.D. *J. Org. Chem.* **1968**, *33*, 2462–2465.
21. Snider, B.B. *J. Org. Chem.* **1974**, *39*, 255–256.
22. Colonge, J.; Perrot, A. *C. R. Acad. Sci.* **1954**, *239*, 541–543.
23. Normant, H.; Ficini, J. *Bull. Soc. Chim. Fr.* **1956**, 1441–1443.
24. Shiner , V.J., Jr. *Isotope Effects in Chemical Reactions*, Cillins, C.J.; Bowman, N.S., Eds., Van Nostrand Reinhold Co., New York, **1971**.
25. Ohloff, G. *Chem. Ber.* **1960**, *93*, 2673–2681.
26. Berson, J.A.; Wall, R.G.; Perlmutter, H.D. *J. Am. Chem. Soc.* **1966**, *88*, 187–188.
27. Achmatowicz, O.; Szymoniak, J. *J. Org. Chem.* **1980**, *45*, 4774–4776.

28. Münsterer, H.; Kresze, G.; Brechbiel, M.; Kwart, H. *J. Org. Chem.* **1982**, *47*, 2677–2679.
29. Kwart, H.; Brechbiel, M.W. *J. Org. Chem.* **1982**, *47*, 3353–3355.
30. Seymour, C.A.; Greene, F.D. *J. Am. Chem. Soc.* **1980**, *102*, 6384–6385.
31. Seymour, C.A.; Greene, F.D. *J. Org. Chem.* **1982**, *47*, 5226–5227.
32. Starflinger, W.; Kresze, G.; Huss, K. *J. Org. Chem.* **1986**, *51*, 37–40.
33. Woodward, R.B.; Hoffmann, R. *Angew. Chem. Int. Ed. Engl.*, **1969**, *8*, 781–932.
34. Fukui, K. *Theory of Orientation and Stereoselection,* Springer–Verlag: New York, **1975**, p.88.
35. Loncharich, R.J.; Houk, K.N. *J. Am. Chem. Soc.* **1987**, *109*, 6947–6952.
36. Houk, K.N.; Li, Y.; Evanseck, J.D. *Angew. Chem., Int. Ed. Engl.*, **1992**, *31*, 682–708.
37. Uchimaru, T.; Tsuzuki, S.; Tanabe, K.; Hayashi, Y. *J. Chem. Soc., Chem. Commun.* **1989**, 1861–1862.
38. Uchimaru, T.; Tsuzuki, S.; Tanabe, K.; Hayashi, Y. *Bull. Chem. Soc. Jpn.* **1990**, *63*, 2246–2251.
39. Thomas IV, B.E.; Loncharich, R.J.; Houk, K.N. *J. Org. Chem.* **1992**, *57*, 1354–1362.
40. Thomas IV, B.E.; Houk, K.N. *J. Am. Chem. Soc.* **1993**, *115*, 790–792.
41. Brown, J.M. *J. Chem. Soc. (B)* **1969**, 868–872.
42. Mikani, K.; Loh, T.P.; Nakai, T. *Tetrahedron Lett.* **1988**, *29*, 6305–6308.
43. Stephenson, L.M.; Orfanopoulos, M. *J. Org. Chem.* **1981**, *46*, 2201–2202.
44. Snider, B.B.; Ron, E. *J. Am. Chem. Soc.* **1985**, *107*, 8160–8164.
45. Song, Z.; Chrisope, D.R.; Beak, P. *J. Org. Chem.* **1987**, *52*, 3938–3940.
46. Song, Z.; Beak, P. *J. Am. Chem. Soc.* **1990**, *112*, 8126–8134.
47. Kwart, H.; Brechbiel, M. *J. Org. Chem.* **1982**, *47*, 5409–5411.
48. Salomon, M.F.; Pardo, S.N.; Salomon, R.G. *J. Org. Chem.* **1984**, *49*, 2446–2454.
49. Wells, P.R. *Linear Free Energy Relationships*, Academic Press, London, New York, 1968, p. 13.
50. Nagasawa, T.; Suzuki, K. *Synlett* **1993**, 29–31.
51. Snider, B.B.; Rodini, D.J.; Kirk, T.C.; Cordova, R. *J. Am. Chem. Soc.* **1982**, *104*, 555–563.
52. Oppolzer, W.; Mirza, S. *Helv. Chim. Acta* **1984**, *67*, 730–738.
53. Houk, K.N.; Williams, Jr., J.C.; Mitchell, P.A.; Yamaguchi, K. *J. Am. Chem. Soc.* **1981**, *103*, 949–951.
54. Mikami, K.; Yajima, T.; Terada, M.; Uchimaru, T. *Tetrahedron Lett.* **1993**, *34*, 7591–7594.
55. Vilkas, M.; Dupont, G.; Dulou, R. *Bull. Soc. Chim. Fr.* **1955**, 799–805.
56. Fritz, H.; Weis, C.D.; Winkler, T. *Helv. Chim. Acta* **1975**, *58*, 1345–1357.
57. Gill, G.B.; Wallace, B. *J. Chem. Soc., Chem. Commun.* **1977**, 380–382
58. Gill, G.B.; Wallace, B. *J. Chem. Soc., Chem. Commun.* **1977**, 382–383.
59. Benner, J.P.; Gill, G.B.; Parrott, S.J.; Wallace, B. *J. Chem. Soc., Perkin Trans. 1* **1984**, 291–313.
60. Benner, J.P.; Gill, G.B.; Parrott, S.J.; Wallace, B.; Begley, M.J. *J. Chem. Soc., Perkin Trans. 1* **1984**, 315–329.
61. Begley, M.J.; Benner, J.P.; Gill, G.B. *J. Chem. Soc., Perkin Trans. 1* **1981**, 1112–1118.
62. Morito, T.; Sekiya, M. *Chem. Pharm. Bull.* **1982**, *30*, 3513–3516.
63. Adams, D.R.; Bhatnagar, S.P. *Synthesis* **1977**, 661–672.
64. Snider, B.B.; Cordova, R.; Price, R.T. *J. Org. Chem.* **1982**, *47*, 3643–3646.

65. Snider, B.B.; Phillips, G.B. *J. Org. Chem.* **1983**, *48*, 464–469.
66. Snider, B.B.; Phillips, G.B. *J. Am. Chem. Soc.* **1982**, *104*, 1113–1114.
67. Snider, B.B.; Phillips, G.B.; Cordova, R. *J. Org. Chem.* **1983**, *48*, 3003–3010.
68. Cartaya–Marin, C.P.; Jackson, A.C.; Snider, B.B. *J. Org. Chem.* **1984**, *49*, 2443–2446.
69. Maruoka, K.; Concepcion, A.B.; Hirayama, N.; Yamamoto, H. *J. Am. Chem. Soc.* **1990**, *112*, 7422–7423.
70. Wovkulich, P.M.; Uskokovic, M.R. *J. Org. Chem.* **1982**, *47*, 1600–1602.
71. Rodini, D.J.; Snider, B.B. *Tetrahedron Lett.* **1980**, *21*, 3857–3860.
72. Snider, B.B.; van Straten, J.W. *J. Org. Chem.* **1979**, *44*, 3567–3571.
73. Mikami, K.; Shimizu, M.; Nakai, T. *J. Org. Chem.* **1991**, *56*, 2952–2953.
74. Mikami, K.; Shimizu, M. *Tetrahedron Lett.* **1992**, *33*, 6315–6318.
75. Mikami, K.; Shimizu, M. *Yuki Gosei Kagakù Kyokaishi* **1993**, *51*, 3–13; *Chem. Abstr.* **1993**, *118*, 254084r. See also ref.192.
76. Mikami, K.; Loh, T.P.; Nakai, T. *J. Am. Chem. Soc.* **1990**, *112*, 6737–6738.
77. Mikami, K.; Wakabayashi, H.; Nakai, T. *J. Org. Chem.* **1991**, *56*, 4337–4339.
78. Achmatowicz, O.; Bialecka–Florjanczyk, E.; Golinski, J.; Rozwadowski, J. *Synthesis* **1987**, 413–415.
79. Agouridas, K.; Girodeau, J.M.; Pineau, R. *Tetrahedron Lett.* **1985**, *26*, 3115–3118.
80. Snider, B.B. in *Selectivities in Lewis Acid Promoted Reactions*, Schinzer, D.; Ed., Kluwer Academic Publ., Dordrecht, Holland, 1989, p. 147–167.
81. Snider, B.B.; Rodini, D.J. *Tetrahedron Lett.* **1980**, *21*, 1815–1818.
82. Deaton, M.V.; Ciufolini, M.A. *Tetrahedron Lett.* **1993**, *34*, 2409–2412.
83. Shoda, H.; Nakamura, T.; Tanino, K.; Kuwajima, I. *Tetrahedron Lett.* **1993**, *34*, 6281–6284.
84. Mikami, K.; Loh, T.–P.; Nakai, T. *J. Chem. Soc., Chem. Commun.* **1988**, 1430–1431.
85. Houston, T.A.; Tanaka, Y.; Koreeda, M. *J. Org. Chem.* **1993**, *58*, 4287–4292.
86. Tanino, K.; Shoda, H.; Nakamura, T.; Kuwajima, I. *Tetrahedron Lett.* **1992**, *33*, 1337–1340.
87. Majewski, M.; Bantle, G.W. *Synth. Commun.* **1990**, *20*, 2549–2558.
88. Mikami, K.; Kishino, H.; Matsueda, H.; Loh, T.–P. *Synlett* **1993**, 497–@ 498.
89. Mikami, K.; Kishino, H. *J. Chem. Soc., Chem. Commun.* **1993**, 1843–1844.
90. Kamimura, A.; Yamamoto, A. *Chem. Lett.* **1990**, 1991–1994.
91. Kamimura, A.; Yoshihara, K.; Marumo, S.; Yamamoto, A.; Nishiguchi, T.; Kakehi, A.; Hori, K. *J. Org. Chem.* **1992**, *57*, 5403–5413.
92. Mikami, K.; Kaneko, M.; Loh, T.–P.; Terada, M.; Nakai, T. *Tetrahedron Lett.* **1990**, *31*, 3909–3912.
93. Mikami, K.; Loh, T.–P.; Nakai, T. *Tetrahedron: Asymetry* **1990**, *1*, 13–16.
94. Mikami, K.; Loh, T.–P.; Nakai, T. *J. Chem. Soc., Chem. Commun.* **1991**, 77–78.
95. Achmatowicz, O.,Jr.; Szechner, B. *J. Org. Chem.* **1972**, *37*, 964–967.
96. Whitesell, J.K.; Bhattacharya, A.; Aguilar, D.A.; Henke, K. *J. Chem. Soc., Chem. Commun.* **1982**, 989–990.
97. Whitesell, J.K.; Bhattacharya, A.; Buchanan, C.M.; Chen, H.–H.; Deyo, D.; James, D.; Liu, C.–L.; Minton, M.A. *Tetrahedron* **1986**, *42*, 2993–3001.

98. Whitesell, J.K.; Lawrence, R.M.; Chen, H.–H. *J. Org. Chem.* **1986**, *51*, 4779–4784.
99. Whitesell, J.K.; Deyo, D.; Bhattacharya, A. *J. Chem. Soc., Chem. Commun.* **1983**, 802–803.
100. Whitesell, J.K.; Younathan, J.N.; Hurst, J.R.; Fox, M.A. *J. Org. Chem.* **1985**, *50*, 5499–5503.
101. Whitesell, J.K.; Allen, D.E. *J. Org. Chem.* **1985**, *50*, 3025–3026.
102. Whitesell, J.K. *Acc. Chem. Res.* **1985**, *18*, 280–284.
103. Whitesell, J.K. *Chem. Rev.* **1992**, *92*, 953–964.
104. Whitesell, J.K.; Chen, H.–H; Lawrence, R.M. *J. Org. Chem.* **1985**, *50*, 4663–4664.
105. Nakamura, T.; Tanino, K.; Kuwajima, I. *Tetrahedron Lett.* **1993**, *34*, 477–480.
106. Nakamura, T.; Tanino, K.; Kuwajima, I. *Chem. Lett.* **1992**, 1425–1428.
107. Tanino, K.; Nakamura, T.; Kuwajima, I. *Tetrahedron Lett.* **1990**, *31*, 2165–2168.
108. Mikami, K.; Sakuda, S.–i. *J. Chem. Soc., Chem. Commun.* **1993**, 710–712.
109. Hanessian, S.; Beaudoin, S. *Tetrahedron Lett.* **1992**, *33*, 7659–7662.
110. Salomon, R.G.; Roy, S.; Salomon, M.F. *Tetrahedron Lett.* **1988**, *29*, 769–772.
111. McCulloch, A.W.; McInnes, A.G. *Tetrahedron Lett.* **1979**, 1963–1966.
112. Snider, B.B.; Roush, D.M.; Rodini, D.J.; Gonzalez, D.; Spindell, D. *J. Org. Chem.* **1980**, *45*, 2773–2785
113. Snider, B.B.; Rodini, D.J.; Conn, R.S.E.; Sealfon, S. *J. Am. Chem. Soc.* **1979**, *101*, 5283–5293.
114. Snider, B.B.; Rodini, D.J.; Karras, M.; Kirk, T.C.; Deutsch, E.A.; Cordova, R.; Price, R.T. *Tetrahedron* **1981**, *37*, 3927–3934.
115. Dauben, W.G.; Brookhart, T. *J. Am. Chem. Soc.* **1981**, *103*, 237–238.
116. Dauben, W.G.; Brookhart, T. *J. Org. Chem.* **1982**, *47*, 3921–3923.
117. Batcho, A.D.; Berger, D.E.; Uskokovic, M.R.; Snider, B.B. *J. Am. Chem. Soc.* **1981**, *103*, 1293–1295.
118. Audran, G.; Monti, H.; Léandri, G.; Monti, J.–P. *Tetrahedron Lett.* **1993**, *34*, 3417–3418.
119. Snider, B.B.; Deutsch, E.A. *J. Org. Chem.* **1982**, *47*, 745–747.
120. Snider, B.B.; Deutsch, E.A. *J. Org. Chem.* **1983**, *48*, 1822–1829.
121. Greuter, H.; Bellus, D. *Synth. Commun.* **1976**, *6*, 409–415.
122. Snider, B.B.; Duncia, J.V. *J. Am. Chem. Soc.* **1980**, *102*, 5926–5928.
123. Snider, B.B.; Duncia, J.V. *J. Org. Chem.* **1981**, *46*, 3223–3226
124. Duncia, J.V.; Lansbury , P.T., Jr.; Miller, T.; Snider, B.B. *J. Am. Chem. Soc.* **1982**, *104*, 1930–1936.
125. Snider, B.B.; Cartaya–Marin, C.P. *J. Org. Chem.* **1984**, *49*, 1688–1691.
126. Snider, B.B.; Phillips, G.B. *J. Org. Chem.* **1983**, *48*, 3685–3689.
127. Hanzawa, Y.; Suzuki, M.; Sekine, T.; Murayama, T.; Taguchi, T. *J. Chem. Soc., Chem. Commun.* **1991**, 721–722.
128. Snider, B.B.; Phillips, G.B. *J. Org. Chem.* **1981**, *46*, 2563–2566.
129. Whitesell, J.K.; Carpenter, J.F. *J. Am. Chem. Soc.* **1987**, *109*, 2839–2840.
130. Whitesell, J.K.; Yaser, H.K. *J. Am. Chem. Soc.* **1991**, *113*, 3526–3529.
131. Achmatowicz, O., Jr.; Pietraszkiewicz, M. *Pol. J. Chem.* **1982**, *56*, 511–519.
132. Mikami, K.; Kaneko, M.; Yajima, T. *Tetrahedron Lett.* **1993**, *34*, 4841–4842.

133. Akiba, K.–y.; Takasu, Y.; Wada, M. *Tetrahedron Lett.* **1985**, *26*, 2463–2466.
134. Moiseenkov, A.M.; Dragan, V.A.; Shavnya, A.V.; Veselovsky, V.V. *Izv. Akad. Nauk. SSSR, Ser. Khim.* **1986**, 1692.
135. Dragan, V.A.; Moiseenkov, A.M. *Mendeleev Commun.* **1992**, 150.
136. Moiseenkov, A.M.; Dragan, V.A.; Koptenkova, V.A.; Veselovsky, V.V. *Synthesis*, **1987**, 814–815.
137. Andersen, N.H.; Ladner, D.W. *Synth. Commun.* **1978**, 449–461.
138. Cookson, R.C.; Smith, S.A. *J. Chem. Soc., Chem. Commun.* **1979**, 145–146.
139. Snider, B.B.; Kirk, T.C. *J. Am. Chem. Soc.* **1983**, *105*, 2364–2368.
140. Karras, M.; Snider, B.B. *J. Am. Chem. Soc.* **1980**, *102*, 7951–7953.
141. Kulkarni, B.S.; Rao, A.S. *Org. Prep. Proced. Inc.* **1978**, *10*, 73–77.
142. Snider, B.B.; Karras, M.; Price, R.T.; Rodini, D.J. *J. Org. Chem.* **1982**, *47*, 4538–4545.
143. Nakatani, Y.; Kawashima, K. *Synthesis* **1978**, 147–148.
144. Funakoshi, K.; Togo, N.; Sakai, K. *Tetrahedron Lett.* **1989**, *30*, 1095–1098.
145. Marshall, J.A.; Wuts, P.G.M. *J. Org. Chem.* **1977**, *42*, 1794–1798.
146. Lindner, D.L.; Doherty, J.B.; Shoham, G.; Woodward, R.B. *Tetrahedron Lett.* **1982**, *23*, 5111–5114.
147. Demailly, G.; Solladié, G. *Tetrahedron Lett.* **1977**, 1885–1888.
148. Demailly, G.; Solladié, G. *Tetrahedron Lett.* **1980**, *21*, 3355–3358.
149. Demailly, G.; Solladié, G. *J. Org. Chem.* **1981**, *46*, 3102–3108.
150. Hiroi, K.; Umemura, M. *Tetrahedron* **1993**, *49*, 1831–1840.
151. Tietze, L.F.; Beifuss, U. *Liebigs Ann. Chem.* **1988**, 321–329.
152. Tietze, L.F.; Beifuss, U. *Angew. Chem., Int. Ed. Engl.* **1985**, *24*, 1042–1043.
153. Tietze, L.F.; Beifuss, U. *Tetrahedron Lett.* **1986**, *27*, 1767–1770.
154. Tietze, L.F.; Beifuss, U.; Ruther, M.; Rühlmann, A.; Antel, J.; Sheldrick, G.M. *Angew. Chem., Int. Ed. Engl.* **1988**, *27*, 1186–1187.
155. Tietze, L.F.; Beifuss, U.; Ruther, M. *J. Org. Chem.* **1989**, *54*, 3120–3129.
156. Tietze, L.F.; Beifuss, U. *Synthesis*, **1988**, 359–362.
157. Tietze, L.F.; Beifuss, U.; Antel, J.; Sheldrick, G.M. *Angew. Chem., Int. Ed. Engl.* **1988**, *27*, 703–705.
158. Snider, B.B.; Rodini, D.J.; van Straten, J. *J. Am. Chem. Soc.* **1980**, *102*, 5872–5880.
159. Oppolzer, W.; Robbiani, C. *Helv. Chim. Acta* **1980**, *63*, 2010–2014.
160. Oppolzer, W.; Robbiani, C.; Bättig, K. *Helv. Chim. Acta* **1980**, *63*, 2015–2018.
161. Scovell, E.G.; Sutherland, J.K. *J. Chem. Soc., Chem. Commun.* **1978**, 529–530.
162. Morgans, D.J., Jr.; Sharpless, K.B.; Traynor, S.G. *J. Am. Chem. Soc.* **1981**, *103*, 462–464.
163. Marshall, J.A.; Andersen, M.W. *J. Org. Chem.* **1993**, *58*, 3912–3918.
164. Abouabdellah, A.; Aubert, C.; Bégué, J.–P.; Bonnet–Delpon, D.; Guilhem, J. *J. Chem. Soc., Perkin Trans. 1*, **1991**, 1397–1403.
165. Nagai, T.; Kumadaki, I.; Kobayashi, Y. *Heterocycles* **1986**, *24*, 222.
166. Nagai, T.; Kumadaki, I.; Miki, T.; Kobayashi, Y.; Tomizawa, G. *Chem. Pharm. Bull.* **1986**, *34*, 1546–1552.
167. Nagai, T.; Miki, T.; Kumadaki, I. *Chem. Pharm. Bull.* **1986**, *34*, 4782–4786.

168. Abouabdellah, A.; Bégué, J.–P.; Bonnet–Delpon, D.; Lequeux, T. *J. Org. Chem.* **1991**, *56*, 5800–5808.
169. Andersen, N.H.; Hadley, S.W.; Kelly, J.D.; Bacon, E.R. *J. Org. Chem.* **1985**, *50*, 4144–4151.
170. Johnston, M.I.; Kwass, J.A.; Beal, R.B.; Snider, B.B. *J. Org. Chem.* **1987**, *52*, 5419-5424.
171. Marshall, J.A.; Andersen, M.W. *J. Org. Chem.* **1992**, *57*, 5851–5856.
172. Maruoka, K.; Ooi, T.; Yamamoto, H. *J. Am. Chem. Soc.* **1990**, *112*, 9011-9012.
173. Barrish, J.C.; Wovkulich, P.M.; Tang, P.C.; Batcho, A.D.; Uskokovic, M.R. *Tetrahedron Lett.* **1990**, *31*, 2235–2238.
174. Overmann, L.E.; Lesuisse, D. *Tetrahedron Lett.* **1985**, *26*, 4167–4170.
175. Snider, B.B.; Goldman, B.E. *Tetrahedron*, **1986**, *42*, 2951–2956.
176. Marshall, J.A.; Andersen, N.H. *Tetrahedron Lett.* **1967**, 1219–1222.
177. Marshall, J.A.; Andersen, N.H.; Johnson, P.C. *J. Org. Chem.* **1970**, *35*, 186–191.
178. Marshall, J.A.; Andersen, N.H.; Schlicher, J.W. *J. Org. Chem.* **1970**, *35*, 858–861.
179. Andersen, N.H.; Golec,.F.A., Jr. *Tetrahedron Lett.* **1977**, 3783–3786.
180. Tietze, L.F.; Bratz, M. *Synthesis* **1989**, 439–442.
181. Tietze, L.F.; Bratz, M. *Chem. Ber.* **1989**, *122*, 997–1002.
182. Tietze, L.F.; Bratz, M. *Liebigs Ann. Chem.* **1989**, 559–564.
183. Overman, L.E.; Blumenkopf, T.A.; Castaneda, A.; Thompson, A.S. *J. Am. Chem. Soc.* **1986**, *108*, 3516–3517.
184. Blumenkopf, T.A.; Bratz, M.; Castañeda, A.; Look, G.C.; Overman, L.E.; Rodriguez, D.; Thompson, A.S. *J. Am. Chem. Soc.* **1990**, *112*, 4386–4399.
185. Blumenkopf, T.A.; Look, G.C.; Overman, L.E. *J. Am. Chem. Soc.* **1990**, *112*, 4399–4403.
186. Hirst, G.C.; Johnson, T.O., Jr; Overman, L.E. *J. Am. Chem. Soc.* **1993**, *115*, 2992–2993.
187. Overman, L.E.; Thompson, A.S. *J. Am. Chem. Soc.* **1988**, *110*, 2248–2256.
188. Hug, E.; Mellor, M.; Scovell, E.G.; Sutherland, J.K. *J. Chem. Soc., Chem. Commun.* **1978**, 526–528.
189. Mellor, M.; Santos, A.; Scovell, E.G.; Sutherland, J.K. *J. Chem. Soc., Chem. Commun.* **1978**, 528–529.
190. Snider, B.B.; Philips, G.B. *J. Org. Chem.* **1984**, *49*, 183–185.
191. Mikami, K. Terada, M.; Narisawa, S.; Nakai, T. *Synlett* **1992**, 255–265.
192. Mikami, K.; Shimizu, M. *Chem. Rev.* **1992**, *92*, 1021–1050.
193. Sakane, S.; Maruoka, K.; Yamamoto, H. *Tetrahedron Lett.* **1985**, *26*, 5535–5538.
194. Sakane, S.; Maruoka, K.; Yamamoto, H. *Tetrahedron* **1986**, *42*, 2203–2209.
195. Narasaka, K.; Hayashi, Y.; Shimada, S.; Yamada, J. *Isr. J. Chem.* **1991**, *31*, 261–271.
196. Mikami, K.; Terada, M.; Sawa, E.; Nakai, T. *Tetrahedron Lett.* **1991**, *32*, 6571–6574.
197. Ziegler, F.E.; Sobolov, S.B. *J. Am. Chem. Soc.* **1990**, *112*, 2749–2758.
198. Maruoka, K.; Hoshino, Y.; Shirasaka, T.; Yamamoto, H. *Tetrahedron Lett.* **1988**, *29*, 3967–3970.
199. Mikami, K.; Terada, M.; Nakai, T. *J. Am. Chem. Soc.* **1989**, *111*, 1940–1941.

201. Mikami, K.; Terada, M.; Nakai, T. *Chemistry Express* **1989**, *4*, 589–592.
202. Terada, M.; Mikami, K.; Nakai, T. *Tetrahedron Lett.* **1991**, *32*, 935–938.
203. Terada, M.; Mikami, K.; Nakai, T. *J. Chem. Soc., Chem. Commun.* **1990**, 1623–1624.
204. Mikami, K.; Terada, M. *Tetrahedron* **1992**, *48*, 5671–5680.
205. Terada, M.; Matsukawa, S.; Mikami, K. *J. Chem. Soc., Chem. Commun.* **1993**, 327–328.
206. Mikami, K.; Matsukawa, S. *J. Am. Chem. Soc.* **1993**, *115*, 7039–7040.
207. van der Meer, F.T.; Feringa, B.L. *Tetrahedron Lett.* **1992**, *33*, 6695–6696.
208. Mikami, K.; Narisawa, S.; Shimizu, M.; Terada, M. *J. Am. Chem. Soc.* **1992**, *114*, 6566–6568.
209. Akhmedov, M.A.; Akhmedov, I.M.; Musaeva, Kh.E.; Sardarov, I.K.; Kostikov, R.R.; Menchikov, L.G. *Zh. Org. Khim.* **1991**, *27*, 2297–2300.

3. Lewis Acid-Promoted Allylsilanes and Allylstannanes Addition to Aldehydes and Ketones

I. Introduction and History

The first example of a thermal reaction between an allyltin and aldehydes was reported in 1967 by König and Neumann who reacted allyltriethyltin with aldehydes at 140°C in dibutylether solution to obtain homoallylic alcohols after protolysis (80% yield). They also showed that the reaction rate could be enhanced by adding $ZnCl_2$ (heated neat at 100°C, 80% yield after 2 h).[1]

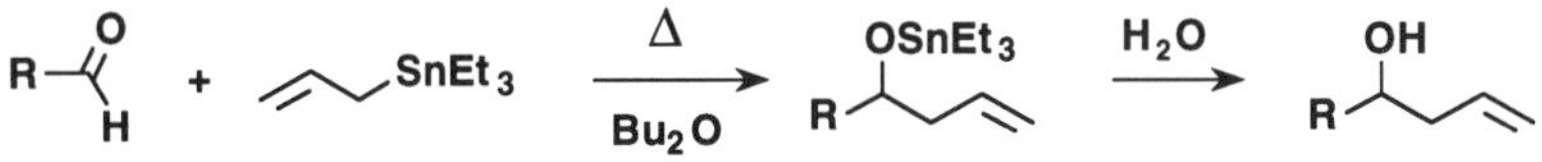

Pereyre showed in 1971 that the reaction occurs with allylic rearrangement and requires activated aldehydes when crotyltributyltin is used.[2] He was also the first to establish that uncatalyzed crotylstannation of activated aldehydes, such as chloral, occurs with remarkable stereospecificity, (*E*)-crotyltri-*n*-butyltin giving rise to *anti*-alcohols and (*Z*)-crotyltri-*n*-butyltin to *syn*-alcohols.[3]

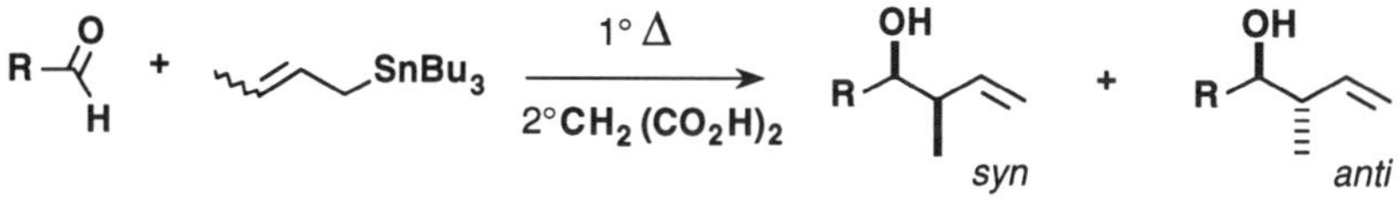

Then, Calas established in 1974 that the addition of allyltrimethylsilane or of propargyltrimethylsilane to α-chloroacetone or chloral, in the presence of Lewis acids such as $AlCl_3$, $GaCl_3$ and $InCl_3$,[4,5] leads also to homoallylic or homoallenic alcohols.

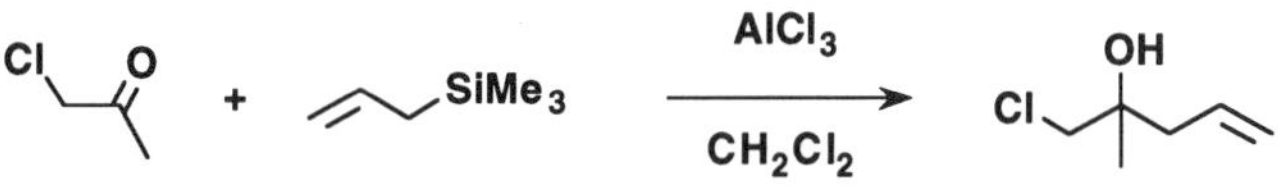

In 1975, Abel and Rowley studied the reaction between perhalogeno ketones and allylic derivatives of silicon and tin. They established that both the noncatalyzed and catalyzed (1% $AlCl_3$) addition of allyltrimethyltin to hexafluoroacetone proceeded with allylic rearrangement and led to 4,4-bis(trifluoromethyl)-4-trimethylstannoxybut-1-enes (however with a lower yield in the latter because of the formation of Me_3SnCl). The authors proposed a transition state involving an electrophilic assistance of tin on the carbonyl group.[6]

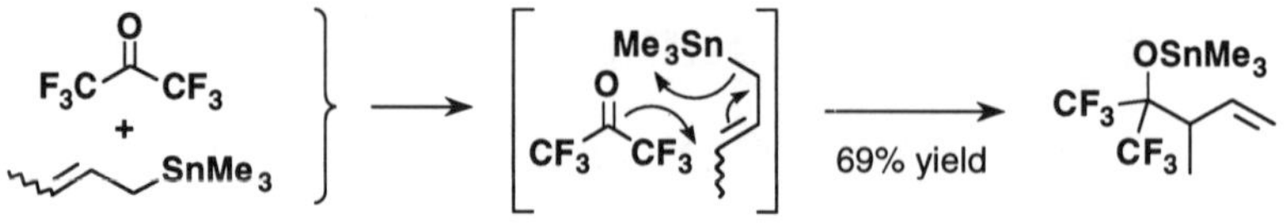

In complete contrast, the reaction between allyltrimethylsilane and hexafluoroacetone in the absence of catalyst led to a vinyl silyl alcohol. However, in the presence of 1% $AlCl_3$, significant yields of β-alkenyloxysilanes are formed.

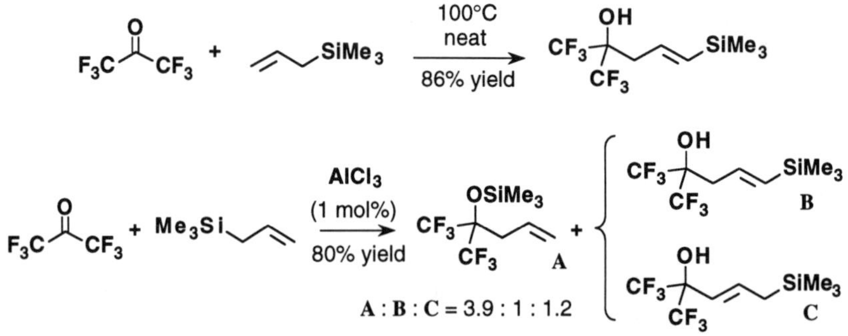

In 1976, Hosomi and Sakurai showed that a wide variety of aliphatic, alicyclic and aromatic carbonyl compounds can react with allylsilanes, provided they are activated by $TiCl_4$ The authors also confirmed that a regiospecific allylic transposition occurs.[7]

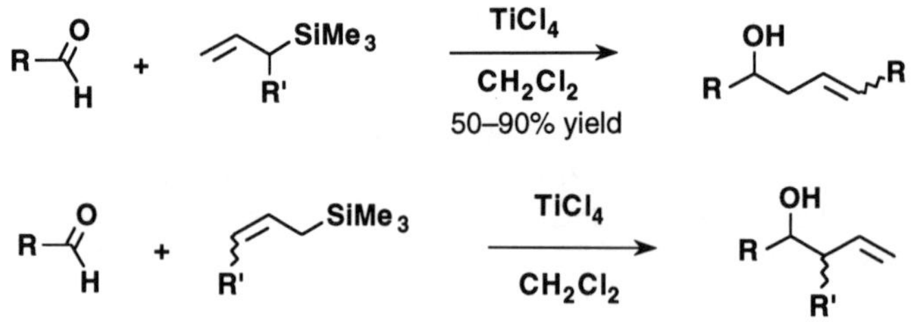

The same year, Calas and Dunoguès reported on the $AlCl_3$-mediated allylation of ordinary aldehydes.[8]

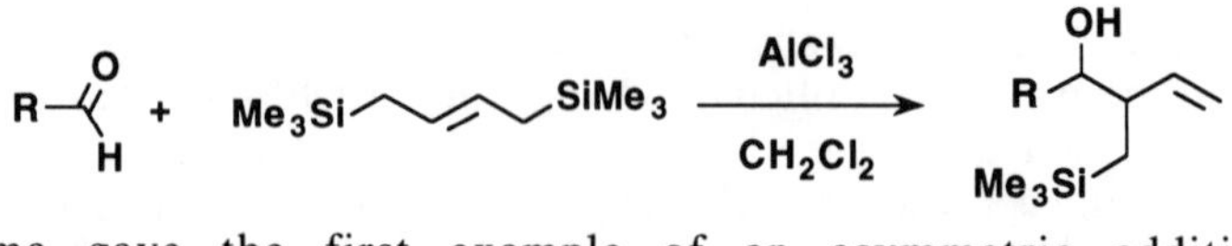

Ojima gave the first example of an asymmetric addition of allyltrimethylsilane to α-keto esters.[9]

$TiCl_4$, CH_2Cl_2, 93% yield; (*S*); 55% e.e.

This example also outlines the regioselectivity of the reaction. Other studies enable to say that allylsilanes and allyltins react preferentially with aldehydes in the presence of esters or ketones and with ketones in the presence of esters.[10–12]

Ojima then showed that 3-trimethylsilylcyclopent-1-ene can add to aldehydes, ketones and α-keto-esters in the presence of $TiCl_4$.[13]

In 1977, Tagliavini demonstrated that the noncatalyzed allylation of methyl isopropylketone could even be carried out neat by using dibutylallyl or triallyltin chlorides.[14] To account for such a reactivity of an unactivated ketone, the authors postulated its electrophilic activation by the tin atom.

$SnClBu_2$; neat, 80°C, 76% yield; H_2O

A year later, Andersen established that the Lewis acid- or fluoride ion-induced allylation of aldehydes can also occur in an intramolecular way. He even observed such a reaction while attempting to purify, by rapid chromatographic filtration through deactivated silica (moist benzene), aldehydes bearing an allylsilane moiety.[15]

silica gel, **R = $SiMe_3$** ; **A** : **B** = 1 : 2
$SnCl_4$ - CH_2Cl_2; 78% yield; **R = H**; **A** : **B** = 1 : 1
BF_3- OEt_2 ; CH_2Cl_2; R = H; **A** : **B** = 5.7 : 1

Also in 1978, Naruta and Maruyama reported on the allylstannation of quinones at low temperature in the presence of BF_3-Et_2O.[16] Similar results can also be obtained with $SnCl_4$.[17] Moreover, this reaction provides a direct route for the synthesis of isoprenyl quinones such as coenzyme Q_1 and vitamin K_2.

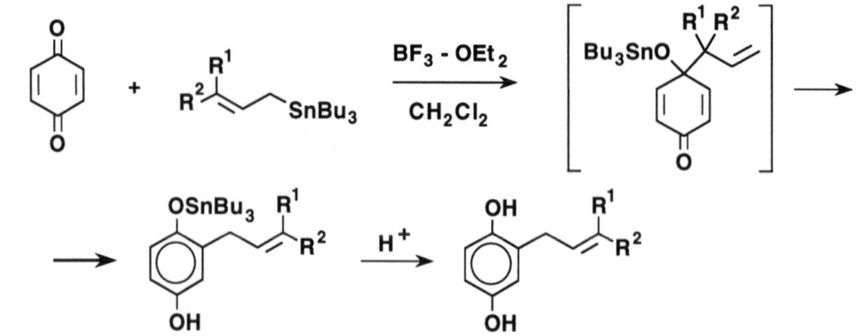

A year later, Sakurai also reported on the Lewis acid [BF_3-OEt_2 and $(Et_2Al)_2SO_4$] induced allylation of aldehydes with different allyltrimethyltin derivatives including dienic ones.[18]

Regarding the diastereoselectivity of the addition of crotylstannanes to aldehydes, Yamamoto made a milestone contribution in 1980 when showing that the BF_3-Et_2O promoted reaction leads to *syn* adducts regardless of the double-bond geometry.[19]

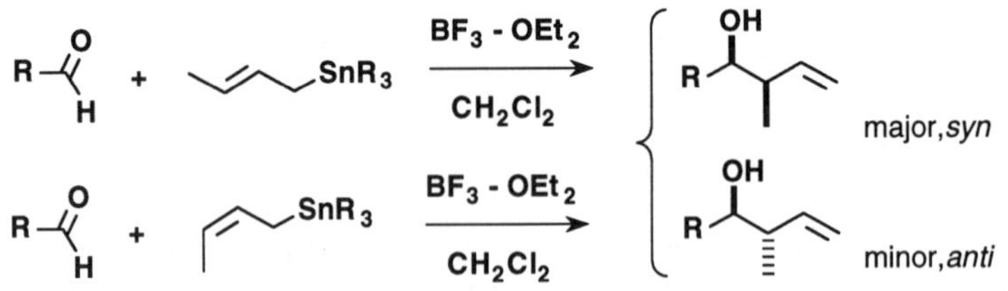

In 1983, Kumada established that the *syn* selectivity can be extended to the $TiCl_4$-catalyzed addition of γ-substituted allylsilanes to aldehydes.[20]

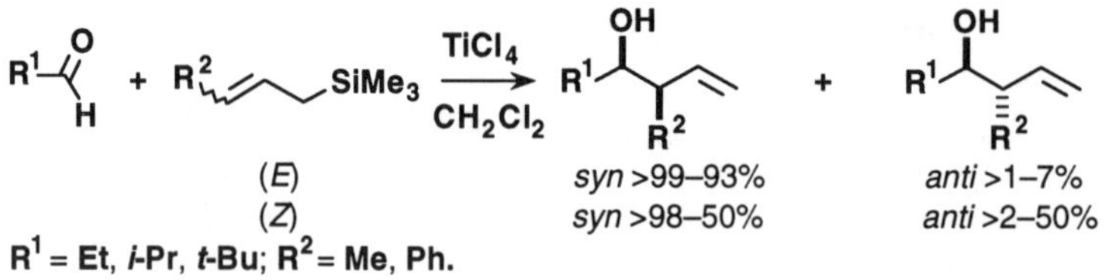

At the same time, Sakurai proposed to develop the fluoride-induced cleavage of C–Si bonds as a source of allyl anion equivalent.[21]

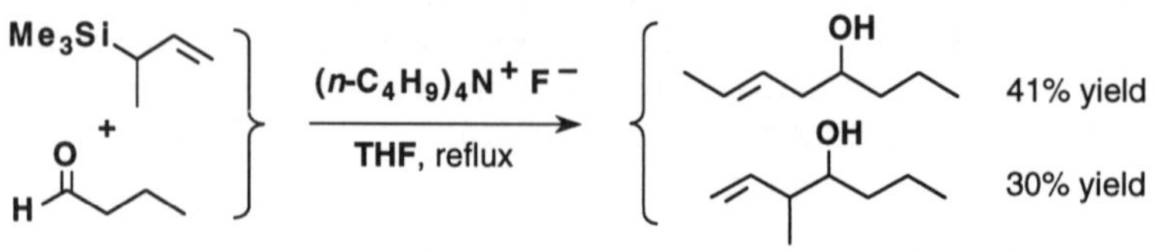

Thus, the nucleophilic attack of the fluoride anion on the silicon atom of allyltrimethylsilane results in the formation of a trimethylallyl fluorosiliconate which can then react with carbonyl compounds and lead to silyl ethers which subsequently undergo hydrolysis.

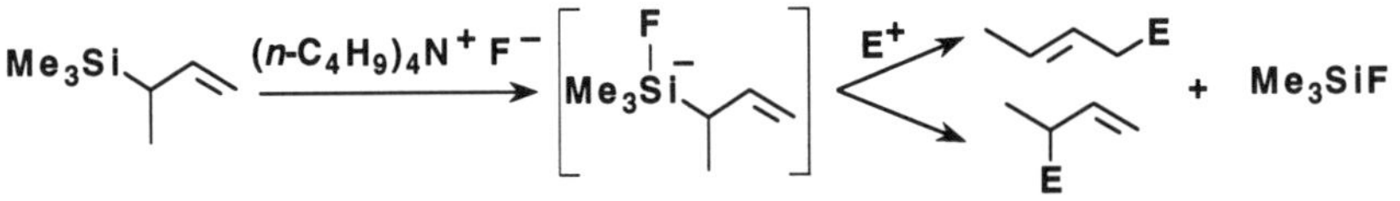

The difference between Lewis acid and n-Bu_4NF induced reactivities of propargyltrimethylsilane towards carbonyl componds was also studied: chloroprenic derivatives were obtained with $TiCl_4$[22] while α-allenic alcohols were produced with n-Bu_4NF.[23]

Over the years, the chemistry of allylstannanes[24–30] and allylsilanes[31–47] has been the subject of a number of reviews.

II. Intermolecular Addition to Achiral Aldehydes

II.1 Mechanism

The addition of allylstannanes or allylsilanes to carbonyl compounds may be considered as a nucleophilic addition. The Lewis acid is added in order to increase the reactivity of the electrophilic carbonyl compounds towards the only weakly nucleophilic stannyl or silyl reagents. However, the same reaction may also be considered as an electrophilic addition of the carbonyl-Lewis acid adduct to an allylsilane or allylstannane. It is well known, from the work of Dunitz and Bürgi, that the angle of attack of a nucleophile on a carbonyl group should be obtuse.[48–50] On the other hand, the angle of attack, of an electrophile, on a carbon-carbon double bond is predicted to be acute.[51,52]

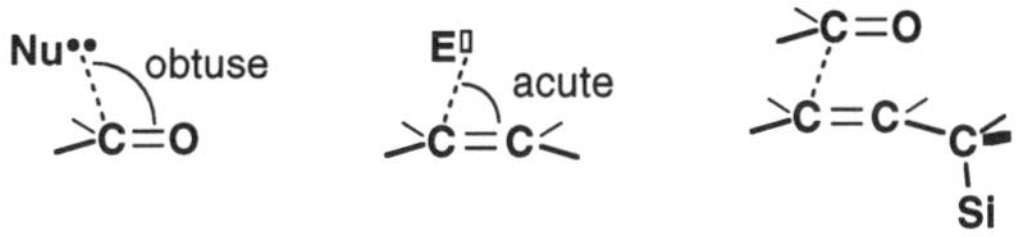

These two electronic requirements are met in the transition state proposed for the uncatalyzed (thermal or high pressure) addition of allylstannanes to aldehydes (allylstannanes can, a priori, be viewed as a more nucleophilic version of the silyl compounds[53]).

A transition state similar to the aldol reaction model of Zimmerman-Traxler[54] accounts for the observed stereochemistry (*syn*-S_E2' mechanism): (*E*)-trialkylcrotylstannanes lead to *anti* homoallylic alcohols whereas (*Z*)-trialkylcrotylstannanes lead to *syn* homoallylic alcohols.

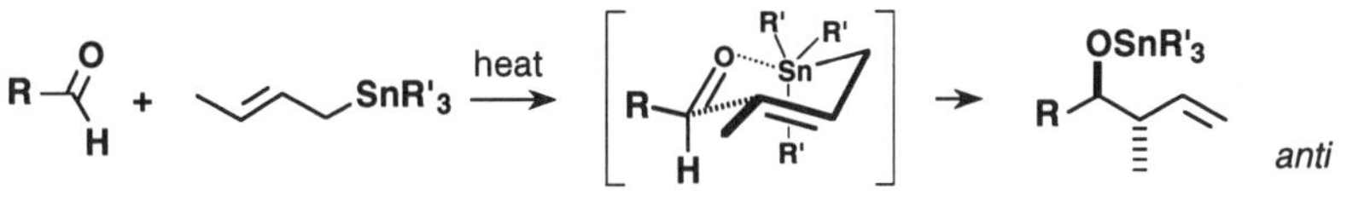

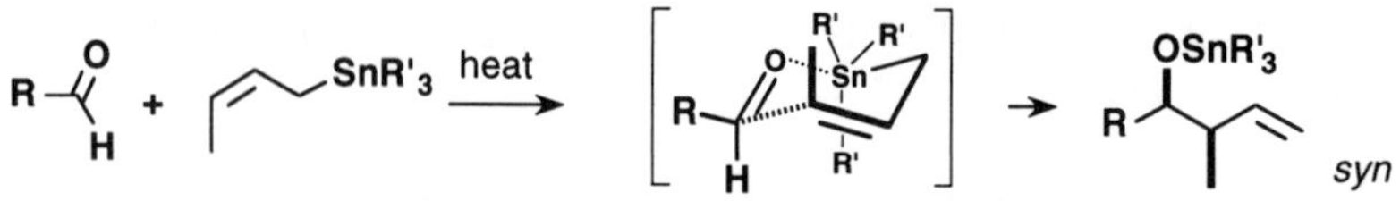

A concerted mechanism, involving a cyclic transition state, is consistent with the magnitude of the activation volume for the thermal addition of chloral to allyltri-*n*-butylstannane (-33.4 ± 0.6 $cm^3\,mol^{-1}$).[55]

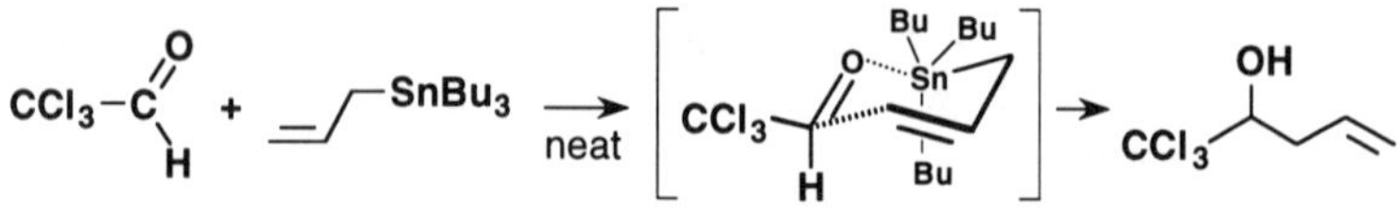

A low stereoselectivity is observed when the allylstannane is obtained *in situ*. Thus, Nokami reported that the allylation of aldehydes and ketones leading to homoallylic alcohols can be carried out successfully from allyl bromide and metallic tin in the presence of water (the reaction of allylbromide with metallic tin is known to give diallyltin bromide).[56] Oshima found that allylic phosphates, upon treatment with the reagent prepared from Bu_3SnLi and Et_2AlCl or from SnF_2 and Et_2AlCl in the presence of catalytic amounts of $Pd(PPh_3)_4$, afford allyltin compounds able to react with aldehydes to produce homoallylic alcohols in good yields.[57] Masuyama discovered that allylic acetates, allylic carbonates or allylic alcohols and $SnCl_2$ in the presence of Pd(0) lead to allyltin species which promote carbonyl allylation.[58–60] In the same way, Luche showed that aldehydes, in the presence of ketones, undergo preferential allylation by the tin mediated method (allylbromide, tin, water, THF, sonication).[61]

However, a pericyclic transition state is not involved when the reaction is catalyzed by a Lewis acid. Thus, as already mentioned, Yamamoto showed that the BF_3-promoted addition of crotyltin compound to aldehyde occurs stereoconvergently (stereoconvergence denotes the predominant formation of the same stereomer from both stereomeric precursors) to produce *syn* adducts.[19]

When the crotylstannation is performed in the presence of $TiCl_4$, Keck established that diastereoselectivity is a function of the addition order of the reactants. Thus, "normal" addition (addition of the allyltin to the Lewis acid-aldehyde complex) provokes *syn* selectivity while "inverse" addition (addition of the aldehyde to the allyltin-Lewis acid mixture) results in the predominant formation of the *anti* adduct.[62,63]

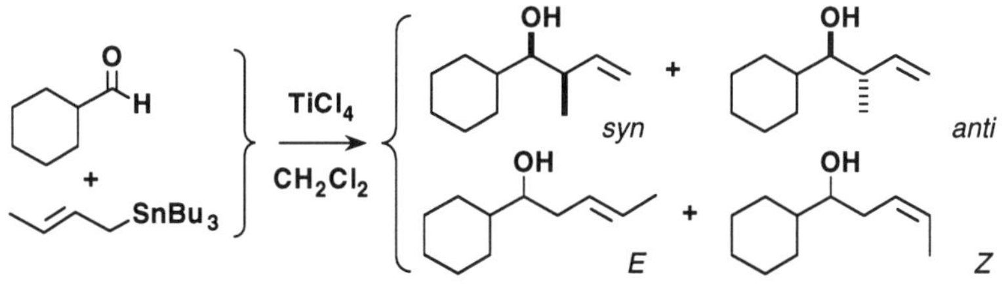

"normal" addition: *syn* : *anti* : *E* : *Z* = 181 : 14 : 1 : 4.2 (*syn* : *anti* = 13 : 1)
"inverse" addition: *syn* : *anti* : *E* : *Z* = 1 : 20.6 : 1.1 : 0 (*syn* : *anti* = 1 : 21)

This reversal of selectivity could result from a transmetallation leading to the formation of a crotyltitanium species. Indeed, these are known to yield preferentially *anti*-adducts when used alone (*vide infra*).[64]

In crotyltitanium compounds, the crotyl moiety is bonded directly to a Lewis acid center imposing therefore a chair-like geometry on the transition state. The formation of such a pericyclic transition state **A**, analogous to the Zimmerman-Traxler model[54] results in the formation of the *anti*-adducts.

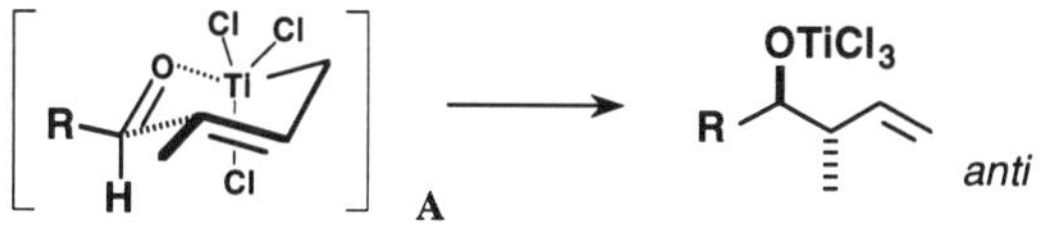

Moreover, by using an optically active allylsilane, Kumada established that its addition to formaldehyde takes place, with retention of the optical activity of the allyl moiety, following an S_E2' mechanism (studies on S_E2' reaction of optically active allylsilanes have demonstrated that the reactions proceed with *anti* stereochemistry).[65]

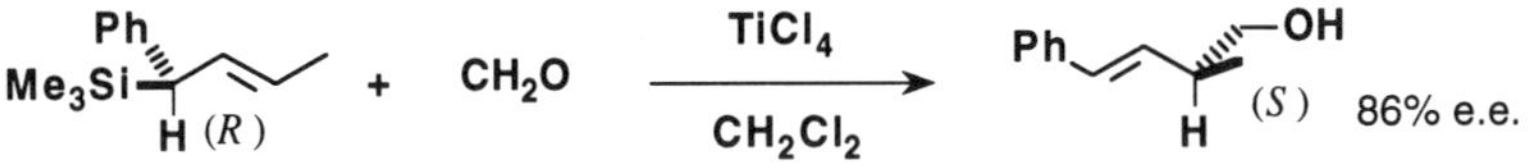

However, a transition state analogous to the Zimmerman-Traxler model[54] cannot account for these results. Thus, in order to explain the particularly high stereoselectivity of the addition of (*E*)-crotyltri-*n*-butyltin to aldehydes, Yamamoto examined eight possible transition states.[19] Out of these, antiperiplanar transition state **A1** leading to the *syn* diastereomer is favored on steric grounds. One can also consider, as Denmark does (*vide infra*), transition states of synclinal geometry. Again, transition state **S1**, leading to the *syn* diastereomer, is more favored than transition state **S2**.

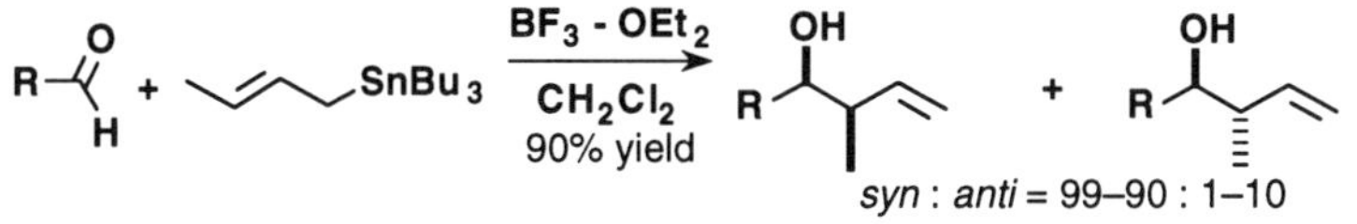

syn : *anti* = 99–90 : 1–10

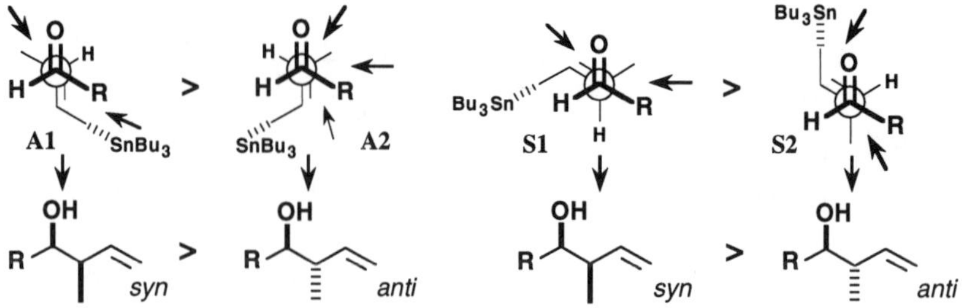

A similar result is reported with (*Z*)-crotyltributyltin (one example).[19]

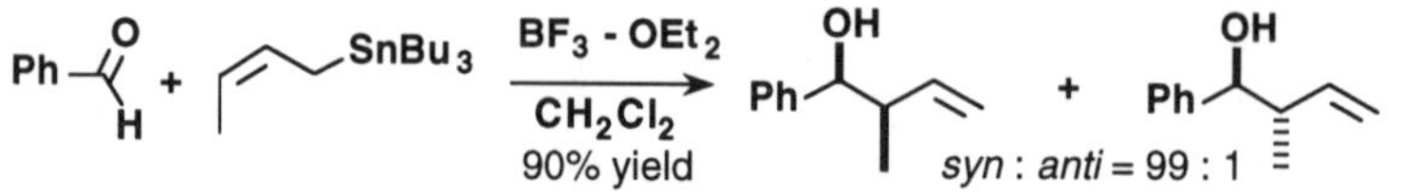

In this case, the antiperiplanar transition state **A'1** leading to the *syn* diastereomer is favored compared to transition state **A'2**. However, the most favored synclinal transition state **S'1** should lead to the *anti* diastereomer. Considering the high *syn* selectivity, reaction paths through synclinal transition states look quite improbable in this process.

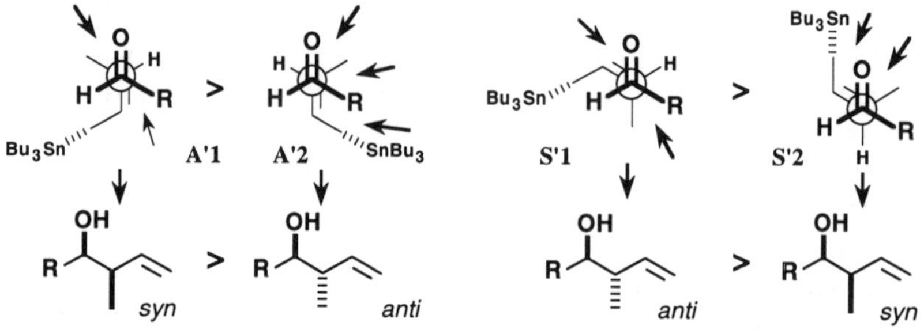

Starting from optically active allylsilanes, Kumada showed that the antiperiplanar transition state proposed by Yamamoto accounts for the stereochemistry of the alcohols resulting from the reaction.[66–68] Thus, on the one hand, the addition of the (*E*)(*R*)-isomers leads, with an excellent stereoselectivity, to the *syn* alcohols.

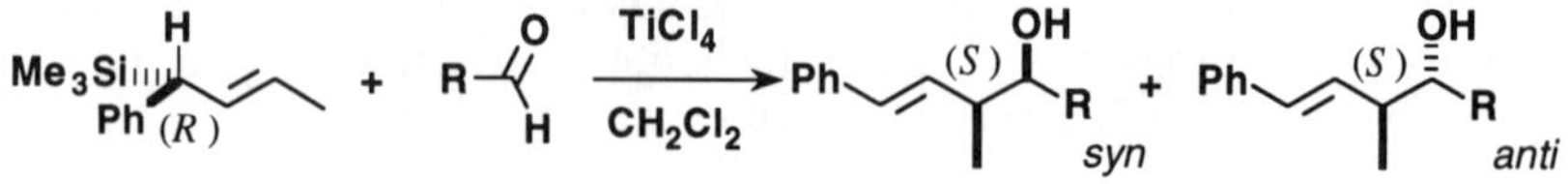

R = **Me** ; 76% yield; *syn* : *anti* = 11.5 : 1 **R** = ***t*-Bu** ; 47% yield; *syn* : *anti* = >99 : <1
R = ***i*-Pr** ; 27% yield; *syn* : *anti* = >99 : <1

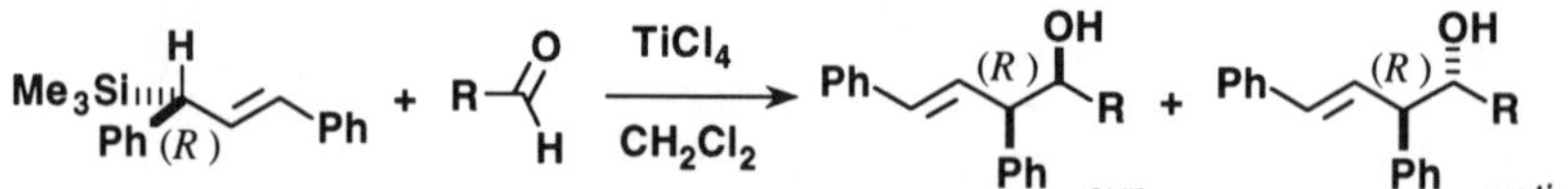

R = ***t*-Bu** ; 44% yield; *syn* : *anti* = >99 : <1

Antiperiplanar transition state **A1** accounts in a better way for the result than the synclinal transition state **S1**.

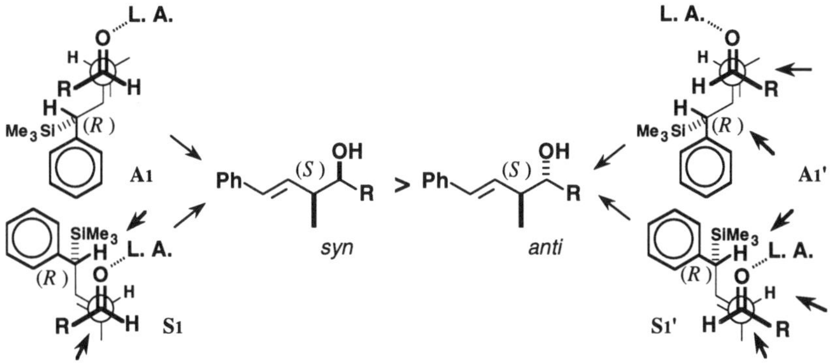

On the other hand, the addition of the (*Z*)(*R*)-isomers to various aldehydes leads also predominantly to the *syn* alcohols, enantiomers of the ones obtained from the (*E*)(*R*)-isomers, but with, in most cases, less selectivity (and not at all with ethanal).

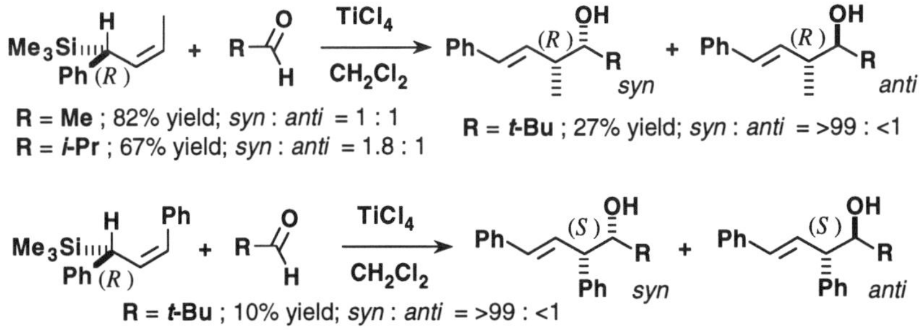

R = Me ; 82% yield; *syn* : *anti* = 1 : 1
R = *i*-Pr ; 67% yield; *syn* : *anti* = 1.8 : 1
R = *t*-Bu ; 27% yield; *syn* : *anti* = >99 : <1

R = *t*-Bu ; 10% yield; *syn* : *anti* = >99 : <1

Again, the antiperiplanar transition state **A2** seems to be the better one.

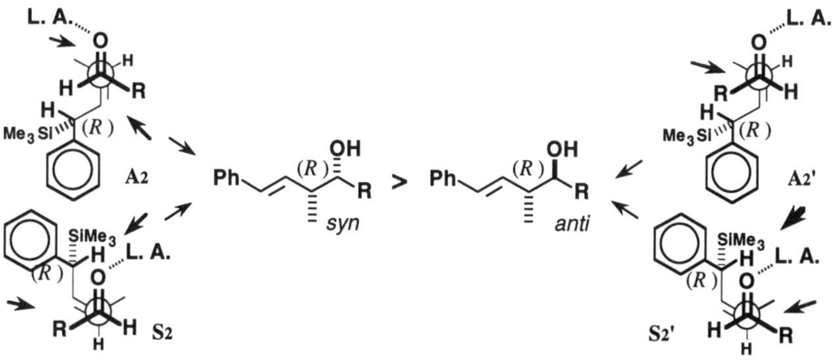

The *anti* stereochemistry of S_E2' reaction is once again confirmed by the addition of (*S*)-3-trimethylsilylcyclopentene to pivalaldehyde.[69] The major product is the *syn* isomer, of absolute configuration (*R*,*S*)(the absolute configuration of the *anti* isomer has not been established).

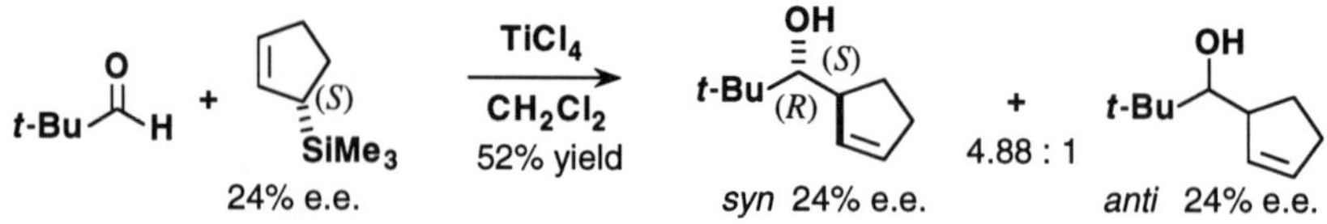

The optical activity retention and the absolute configuration of the major product are in agreement with reaction paths involving either an antiperiplanar or a synclinal transition state.

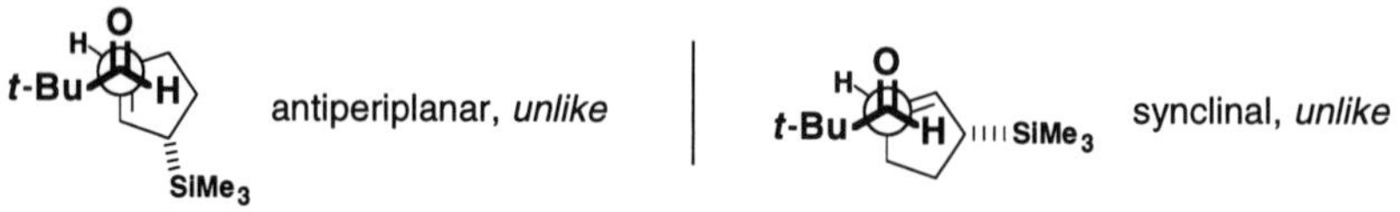

Recently, Fleming studied the addition of an allenyl silane, of high enantiomeric purity, to isobutyraldehyde and again found that the S_E2' reaction involved is stereospecifically *anti*. The major product, *syn* isomer **A** (85% yield), indicates a transfer of chirality close to 100%.[70]

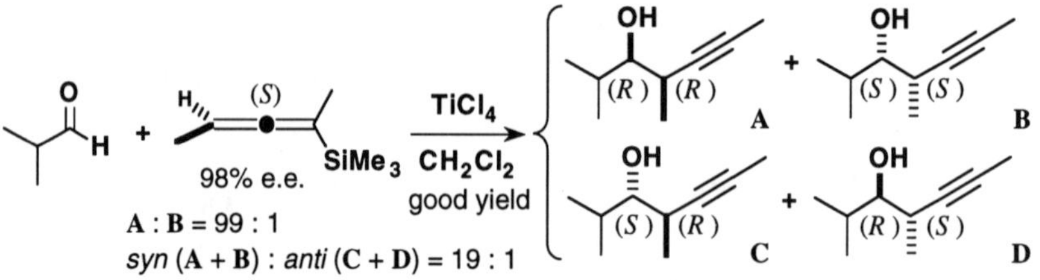

Syn (**A** + **B**) and *anti* (**C** + **D**) diastereomers are present in a 19:1 ratio. Once again, antiperiplanar transition state **A'**, leading to **A**, accounts for the observed stereoselectivity. **A'** is more probable than **C'** which would lead to the formation of alcohol **C**.

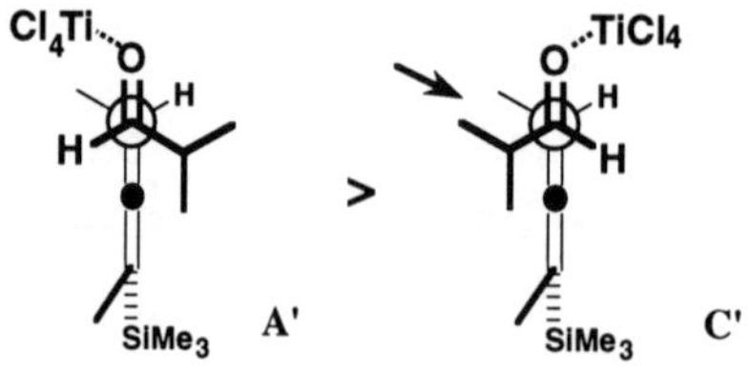

Other examples of the addition of allenylsilanes to aldehydes have also been reported by Danheiser.[71]

Fleming also described a remarkably high level of chiral information transfer in the reaction between an optically active heptadienylsilane and isobutyraldehyde. The major product arises from the attack of the diene on the face *anti* to the silyl group. The level of selectivity shows that the *anti* attack is nine times faster than the *syn* attack (*syn* to the silyl group), even though the chiral center is five atoms away from the site of the reaction.[72,73]

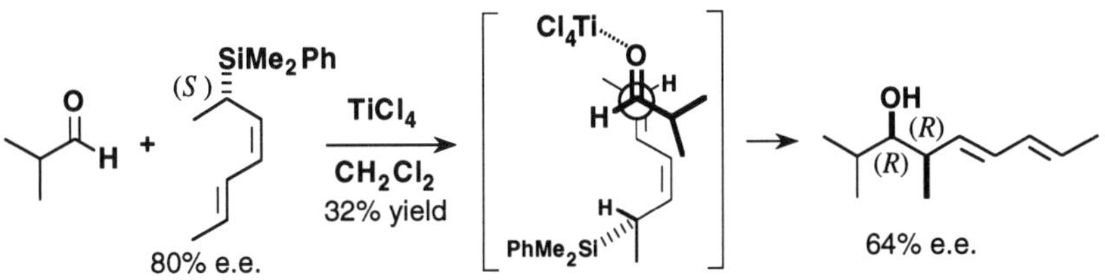

The change of mechanism resulting from the introduction of a Lewis acid is clearly evident when examining the following results. Sato showed that the (*E*)-crotyltitanium reacts with aldehydes to afford *anti* adducts preferentially,[64] but Reetz established that this diastereoselectivity can be reversed by using BF_3-OEt_2.[74]

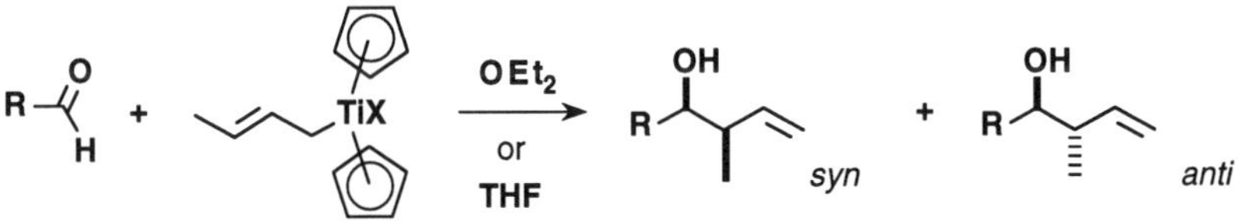

without **BF_3** (solvent **OEt_2**):

X = **Cl**; **R** = **Ph**; 96% yield; *syn*:*anti* = 1 : 1.5
X = **Br**; **R** = **Ph**; 92% yield; *syn*:*anti* = 0 : 1
X = **Br**; **R** = **Et**; 90% yield; *syn*:*anti* = 1: 24.1
X = **Cl**; **R** = **Et**; 92% yield; *syn*:*anti* = 1: 2
X = **Br**; **R** = ***i*-Pr**; 87% yield; *syn*:*anti* = 1: 99
X = **I**; **R** = **Et**; 86% yield; *syn*:*anti* = 1: 13.3
X = **I**; **R** = **Ph**; 96% yield; *syn*:*anti* = 1: 15.7

with **BF_3** (2 equiv)(solvent **THF**):

X = **Cl**; **R** = **Ph**; 98% yield; *syn*:*anti* = 6.1 : 1
X = **Br**; **R** = **Ph**; 90% yield; *syn*:*anti* = 6.1 : 1
X = **Br**; **R** = ***i*-Pr**; 75% yield; *syn*:*anti* = 10.1 : 1
X = **I**; **R** = **Ph**; 90% yield; *syn*:*anti* = 3.2 : 1

Sato's results are explained by a conventional pericyclic transition state such as **A**, whereas the stereoselectivity observed by Reetz can only be rationalized through an "open" antiperiplanar transition state similar to the one proposed by Yamamoto.

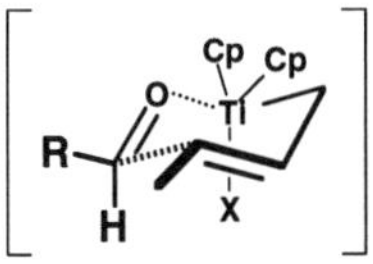

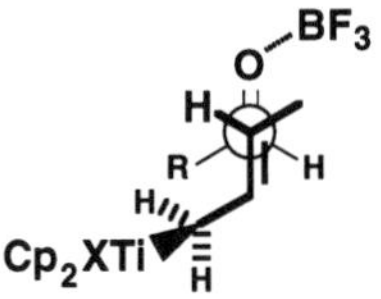

In order to make a choice between the possible transition states, Nakai cyclized a mixture of optically pure aldehydes **A** and **B** of (*E*) and (*Z*)

configuration, respectively.[75] A single optically pure, cyclopentenol of (*S*, *S*) configuration is obtained from the mixture.

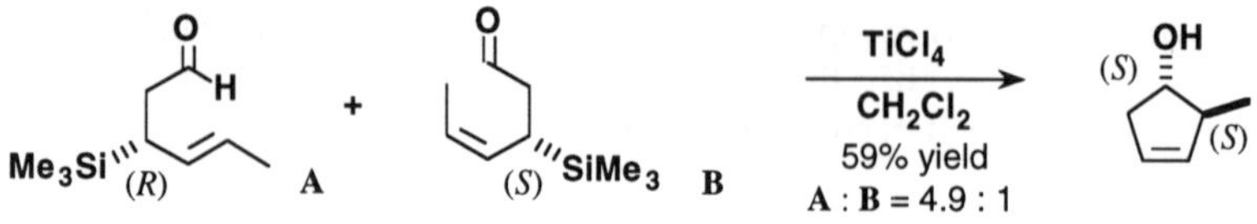

This result is compatible with the cyclization of **A** through an antiperiplanar transition state and of **B** through a synclinal one.

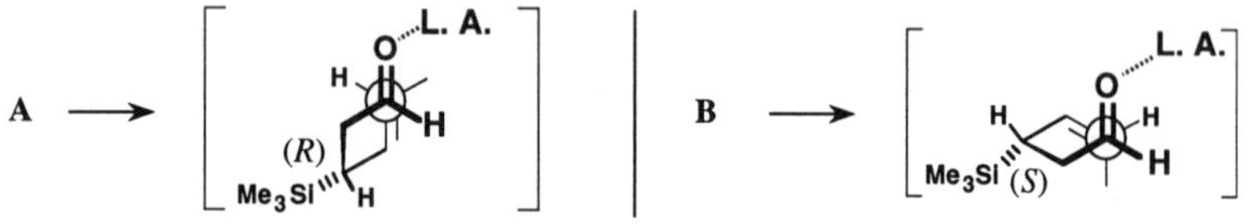

In a series of papers, Denmark studied the stereochemistry of allylmetal-aldehyde condensations. For that purpose, he studied the intramolecular condensation of allylsilane **A** and of its tin counterpart **A'**. He first showed that both *syn* and *anti* alcohols **B** and **C** are obtained from **A**, in proportions depending on the Lewis acid used.[76]

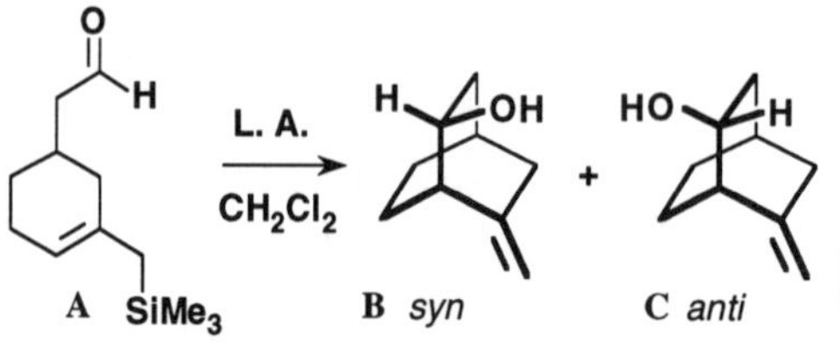

$SnCl_4$; *syn* : *anti* = 1 : 1
Et_2AlCl; *syn* : *anti* = 2 : 1
$FeCl_3$; *syn* : *anti* = 2.3 : 1
$AlCl_3$; *syn* : *anti* = 3.8 : 1
BF_3-OEt_2; *syn* : *anti* = 4 : 1
n-$Bu_4N^+ F^-$ / THF; *syn* : *anti* = 1 : 2.3

From these results, a crude correlation, between the covalent radius of the Lewis acid metal and the stereoselectivity, appears, i.e., the larger the metal, the less selective the reaction.

The *syn* alcohol **B** results from a synclinal transition state whereas the *anti* alcohol **C** results from an antiperiplanar transition state ("open-chain transition state"). Following Denmark, the association of the Lewis acid with the carbonyl group gives the *anti*-complex, the major steric contributions arising from the interaction between the Lewis acid and the (trimethylsilyl)methyl group. The very fact that *syn*-products are observed with $SnCl_4$ suggests a stereoelectronic advantage for synclinal orientation of the reactants under electrophilic conditions. The reversal of stereoselectivity, observed with the fluoride-induced cyclization, suggests a change of mechanism.

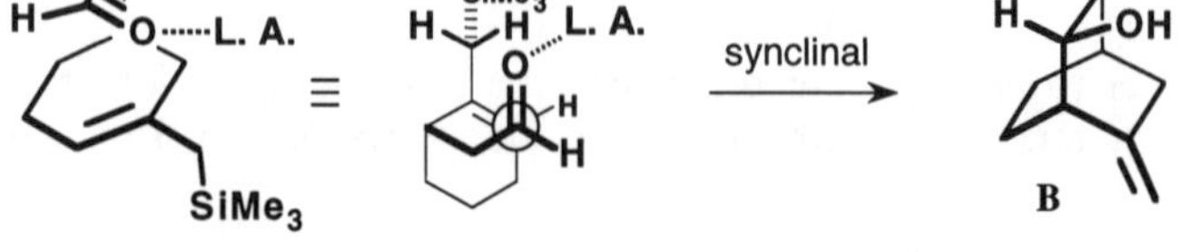

L. A. antiperiplanar C

Considering the complexation with $SnCl_4$, the well-known 2:1 (carbonyl compound:$SnCl_4$) complexation stoichiometry (see Chapter 1, Section III) is probably responsible for the greater "effective steric bulk". A high dilution should therefore increase the ratio of 1:1 complex and favor the formation of the *syn* isomer. That proved to be right.[77]

$SnCl_4$, CH_2Cl_2; A → B *syn* + C *anti*

A, 0.25 M; **L.A.** 0.275 M; **A** : **B** = 1 : 1.3
A, 0.05 M; **L.A.** 0.055 M; **A** : **B** = 1.3 : 1
A, 0.005 M; **L.A.** 0.0055 M; **A** : **B** = 1.8 : 1
A, 0.0005 M; **L.A.** 0.00055 M; **A** : **B** = 6.1 : 1

The analogous allylstannane **A'** displays an enhanced reactivity and a greater *syn* selectivity. In addition the reaction, which leads also to the *syn* alcohol when uncatalyzed, is relatively insensitive to the nature of the Lewis acid.[78,79]

A' → (L. A., CH_2Cl_2) → B (*syn*) + C (*anti*)

$TiCl_4$; 84% conversion; **A** : **B** = 4.6 : 1
BF_3-OEt_2 ; 93% conversion; **A** : **B** = 6.7 : 1
$AlCl_3$; 89% conversion; **A** : **B** = 8.1 : 1
Et_2AlCl ; 89% conversion; **A** : **B** = 9 : 1
$ZrCl_4$; 95% conversion; **A** : **B** = 9 : 1
$SnCl_4$; 85% conversion; **A** : **B** = 13.3 : 1
$FeCl_3$; 93% conversion; **A** : **B** = 49 : 1
CF_3COOH ; 95% conversion; **A** : **B** = 99 : 1
90 °C, 8h, benzene; 85% conv.; **A** : **B** = 1 : 0

Yamamoto did not obtain a good diastereoselectivity when adding crotyltri-*n*-butylstannane to various pyruvates. The increased steric bulk of the ester group brings about an inversion of the *like/unlike* ratio in favor of the former.[80]

BF_3-OEt_2, CH_2Cl_2, 84–88% yield → *like* + *unlike*

R = Me ; *like* : *unlike* = 1.5 : 1
R = Me_3CCH_2 ; *like* : *unlike* = 1.3 : 1
R = $PhCH_2$; *like* : *unlike* = 1 : 1
R = 2,6-Me_2-C_6H_3 ; *like* : *unlike* = 1 : 1.3 (**$EtAlCl_2$** was used as Lewis Acid)

These stereochemical features can be explained by the following antiperiplanar transition states.

BF_3 O H CO_2R R_3Sn → *like* | BF_3 O H CO_2R SnR_3 → *unlike*

The $SnCl_4$ catalyzed addition of allylstannanes to aldehydes displays a complex behavior due to transmetallation reactions. However, ^{119}Sn NMR proved to be a useful tool to study the dynamics of such reactions.

In early works, Tagliavini (*vide infra*), Keck[81] and Yamamoto[19] have discussed the possibility of metathetical processes due to ligand exchange between the Lewis acid and the allylic stannane. Thus, Keck reported that "normal" addition with $SnCl_4$ (i.e., addition of crotyltri-*n*-butylstannane to a solution of aldehyde and $SnCl_4$ at –78°C) gives roughly equal amounts of *syn*, *anti*, (*E*) and (*Z*) isomers. If crotyltri-*n*-butyl stannane is added to $SnCl_4$ at room temperature and the mixture then cooled to –78°C before the addition of the aldehyde ("inverse" addition), the reaction leads preferentially to the *syn* and *anti* alcohols (the *syn* isomer being the major one).

$SnCl_4$, CH_2Cl_2; products: *syn*, *anti*, *E*, *Z*

"normal" addition: *syn* : *anti* : *E* : *Z* = 1.5 : 1.8 : 2.5 : 1 (*syn* : *anti* = 1 : 1.1)
"inverse" addition: *syn* : *anti* : *E* : *Z* = 18.2 : 62.4 : 1 : 1.8 (*syn* : *anti* = 1 : 3.4)

The first result ("normal" addition) can be accounted for by a competition between the expected addition process and the addition of 3-methyl-3-trichlorostannylprop-1-ene which results from a transmetallation between $SnCl_4$ and crotyltri-*n*-butylstannane (equation 1); the so formed stannyl propene could then add to the aldehyde to afford (*E*)- and (*Z*)-homoallylic alcohols. In the case of the "reverse" addition, the formation of the stannyl propene (equation 1) is followed by an isomerization (equation 2) yielding a crotyl trichlorostannane which will react with the aldehyde to afford *syn* and *anti* alcohols.[81] Such transmetallation between allyltri-*n*-butyltin and tin tetrachloride was also shown by Naruta.[82]

$SnBu_3$ + $SnCl_4$ ⇌ $SnCl_3$ + Bu_3SnCl (1)

$SnCl_3$ → $SnCl_3$ (2)

Denmark observed, by means of a spectroscopic investigation (^{13}C and ^{119}Sn NMR), that metathesis is demonstrably faster than addition in most if not all cases.[83] However, Keck emphasized the sensitivity of such experiments to experimental details. In particular, he showed that transmetallation pathways involving the conversion of allyltri-*n*-butyl stannane to allyltrichloro stannane prior to the addition to simple aldehydes do not occur from stable 2:1 $(RCHO)_2$-$SnCl_4$ complexes at low temperature. However, in the presence of a slight excess of $SnCl_4$, ligand exchange is more rapid.[84] Such reverse addition was also reported by Yamamoto when using crotyltri-*n*-butyltin in the presence of $AlCl_3$-*i*-PrOH.[85]

II.2 Stereoselective addition

II.2.a Reaction between achiral aldehydes and unfunctionalized allylsilanes or allylstannanes

Despite the fact that the Lewis acid promoted allylation of carbonyl compounds with allylsilanes is a well-established procedure for the preparation of homoallylic alcohols, different authors evidenced the presence of by-products and yields lowering on changing the recommended experimental procedures. Taddei noticed that allyltrimethylsilane in the presence of $AlCl_3$ or $AlBr_3$ reacts with two equivalents of aldehydes to give good yields of 4-halogenotetrahydropyrans.[86] The reaction is stereospecific giving "all *cis*" substituted tetrahydropyrans (see also refs.87–90).

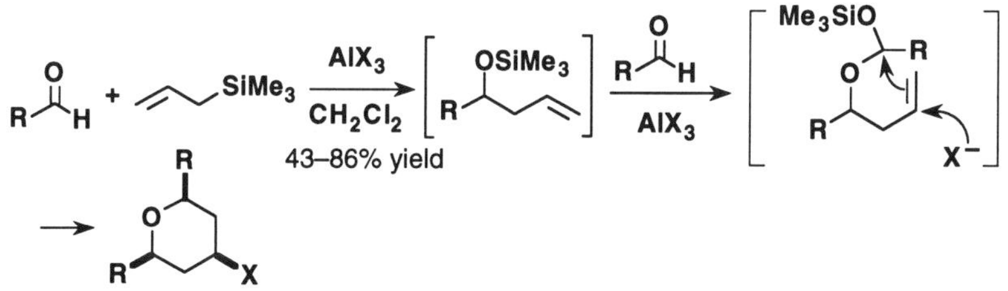

AlX_3, CH_2Cl_2, 43–86% yield

As already mentionned, the *syn* selectivity of the addition of crotyltin or silane derivatives to aldehydes is a major feature. Thus, both (*Z*) and (*E*)-crotyltri-*n*-butyltins lead predominantly to *syn* adducts upon their BF_3-OEt_2 catalyzed reaction with a wide variety of aldehydes. Particularly, Yamamoto provided a wide number of examples, thus establishing that the stereoselectivity is always greater than 90%.[91]

$BF_3 \cdot OEt_2$, CH_2Cl_2; *syn*; *anti*

90% yield; *syn* : *anti* = 49 : 1

90% yield; *syn* : *anti* = 10.1 : 1

92% yield; *syn* : *anti* = 49 : 1

90% yield; *syn* : *anti* = 9 : 1

$BF_3 \cdot OEt_2$, CH_2Cl_2

90% yield ; *syn* : *anti* = 99 : 1

Wacker-type oxidation of the obtained homoallylic alcohols provides a route to *syn*-β-ketol in high yields.[92]

$BF_3 \cdot OEt_2$, CH_2Cl_2, 92% yield; *syn*

$PdCl_2$; $CuCl_2$; O_2, DMF - H_2O, 87–92% yield

R = Me; *n*-Pr

Similar formation of predominantly *syn* homoallylic alcohols was also observed from α,β-unsaturated aldehydes.[91]

$BF_3 \cdot OEt_2$, CH_2Cl_2

88% yield ; *syn* : *anti* = 10.1 : 1

Crotyltri-*n*-butyltin also adds to aldimines in the presence of Lewis acid to give homoallylamines. In general, little or no immediate reaction is observed at –78°C. However, addition of Lewis acid to the aldimine at –78°C, brief warming to 23°C, recooling to –78°C, and addition of the allyltri-*n*-butyltin affords the desired product with, in most cases, high *syn* selectivity.[93]

$TiCl_4$, CH_2Cl_2 — *syn* + *anti*

78% yield; *syn* : *anti* = 23 : 1

60% yield; *syn* : *anti* = 20 : 1

85% yield; *syn* : *anti* = 3 : 1

84% yield; *syn* : *anti* = 30 : 1

(*E*)-Hex-2-enyltri-*n*-butyltin exhibits the same *syn* selectivity as crotyltri-*n*-butyltin.[94]

BF_3-OEt_2, CH_2Cl_2, 93% yield — *syn* 12 : 1 *anti*

However, (*E*)-cinnamyltriphenyltin only gives *anti* isomers.

BF_3-OEt_2, CH_2Cl_2 — *syn* + *anti*

R = Me; 84% yield ; *syn* : *anti* = <1 : >99
R = $PhCH_2CH_2$; 69% yield ; *syn* : *anti* = <1 : >99
R = *c*-C_6H_{11}; 65% yield ; *syn* : *anti* = <1 : >99
R = Ph; 76% yield ; *syn* : *anti* = <1 : >99

With γ,γ-dialkylallyltin systems such as (*E*)-geranyltri-*n*-butyltin, *syn* selectivity is again observed, the bulkier alkyl group being *syn* to the hydroxy group.

BF_3-OEt_2, CH_2Cl_2 — *syn* + *anti*

R = Me; 81% yield ; *syn* : *anti* = >99 : <1
R = $PhCH_2CH_2$; 60% yield ; *syn* : *anti* = >99 : <1
R = *c*-C_6H_{11}; 67% yield ; *syn* : *anti* = >99 : <1
R = Ph; 98% yield ; *syn* : *anti* = >99 : <1

Koreeda established that (*E*)-pent-3-en-2-yltri-*n*-butyltin adds, also with great stereoselectivity, to propanal leading to the *syn* alcohol which upon hydrogenation affords the *syn*-4-methylheptan-3-ol; the latter is an aggregation pheromone secreted by the European elm bark beetle, *Scolytus multistriantus* Marsham.[95]

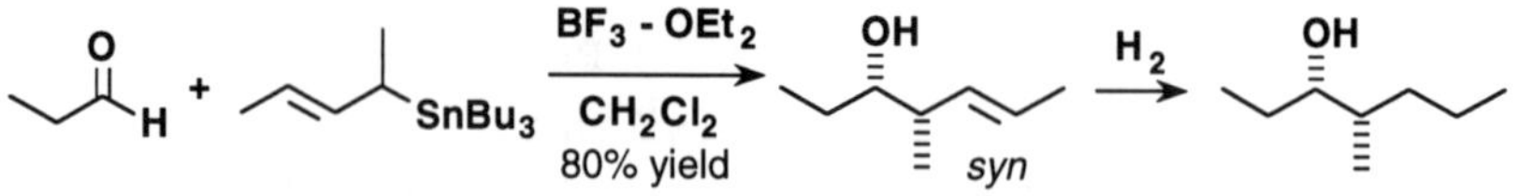

Thomas studied both the uncatalyzed and the catalyzed addition of α-methylcrotylstannanes to aldehydes. Thus, he reported that, on heating, the reaction leads stereoselectively to *anti*-(*Z*)-5-hydroxy-4-methylpent-2-enes (after flash chromatography).[96,97]

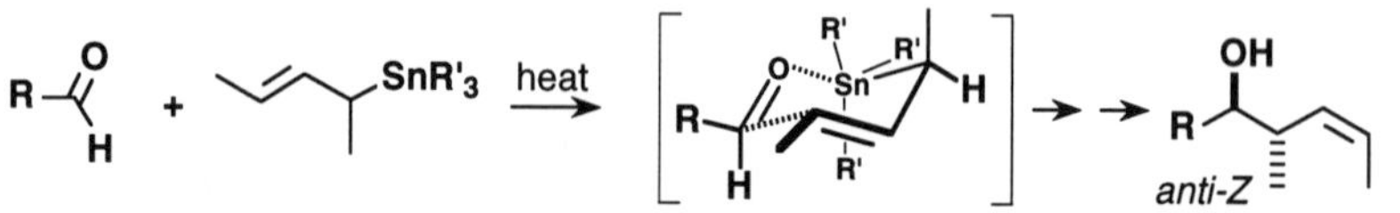

The stereoselectivity of the uncatalyzed reaction is consistent with a six-membered, chair-like, cyclic transition state in which the α-methyl substituent adopts a pseudo-axial position (to account for the (*Z*) configuration). It was tentatively suggested that tin being trigonal bipyramidal in the transition state, a steric interaction with the R' substituent (R' = *n*-Bu or Ph) would result from an equatorial α-methyl group and would therefore prevent the formation of such a transition state.

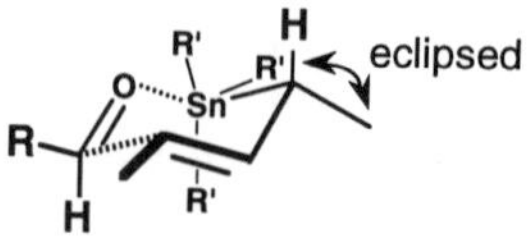

Similar results are also observed with (*E*)-3-methylpent-3-en-2-yltri-*n*-butyltin.

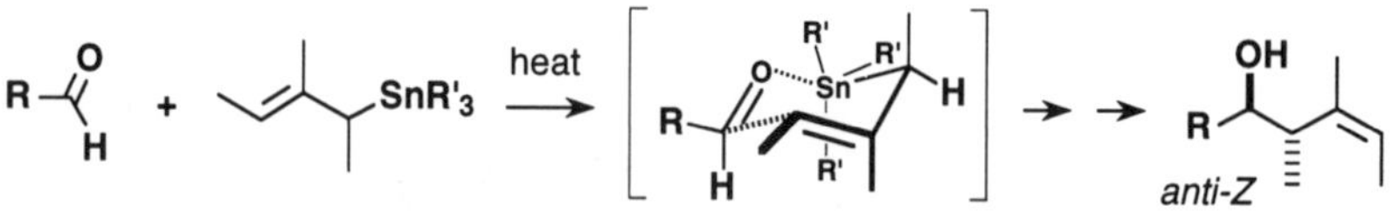

As for the catalyzed reaction, again, although less stereoselective than the uncatalyzed reaction, the Lewis acid-promoted reaction leads predominantly to the *syn*-(*E*)-isomers.

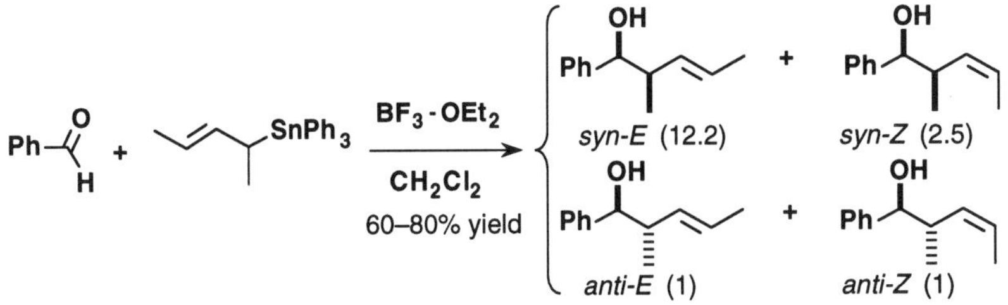

In a similar way, the BF_3-OEt_2 catalyzed reaction between (*E*)-3-methylpent-3-en-2-yltri-*n*-butyltin and benzaldehyde gives mainly the *syn*-(*E*)-adduct which accounted for 75% of the product mixture.

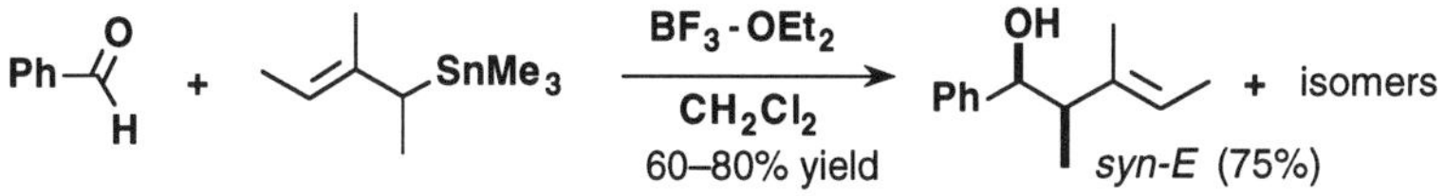

2-Cyclohexenylation of aldehydes, promoted by cyclohexenyltrimethylstannane in the presence of BF_3-OEt_2, occurred with only low to moderate diastereoselectivity as established by Kitching.[98]

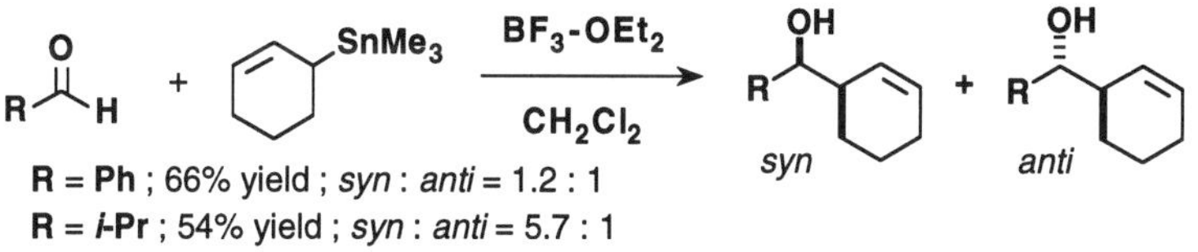

R = Ph ; 66% yield ; *syn* : *anti* = 1.2 : 1
R = *i*-Pr ; 54% yield ; *syn* : *anti* = 5.7 : 1

Conjugated dienyltin and silane derivatives have also been used to produce aldehydes. With such compounds, addition can occur in principle either at the γ- or ε-position. Thus, in the Lewis acid-promoted reaction between the (*E,E*) conjugated pentadienyltins **A** and *p*-nitrobenzaldehyde, addition at the ε-position of the dienyltins **A** preferentially occurs[99] (see also ref.100).

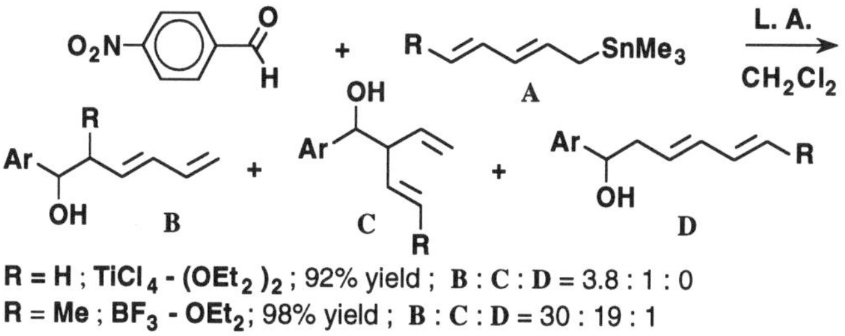

R = H ; $TiCl_4$ - $(OEt_2)_2$; 92% yield ; B : C : D = 3.8 : 1 : 0
R = Me ; BF_3 - OEt_2; 98% yield ; B : C : D = 30 : 19 : 1

In contrast, the corresponding (*Z,E*)-isomers were found to undergo, under the same conditions, preferential addition at their γ-positions.

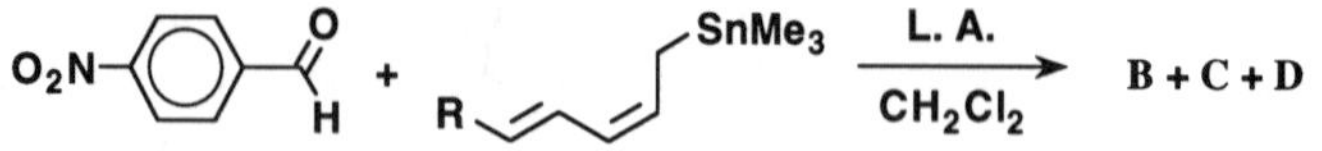

R = H ; $TiCl_4$ - $(OEt_2)_2$; 94% yield ; B : C : D = 1 : 3.2 : 0
R = Me ; BF_3 - OEt_2 ; 95% yield ; B : C : D = 25 : 74 : 1

In comparison, (*E,E*) and (*Z,E*) pentadienyltrimethylsilanes both exhibit ε-selectivity and lead to the same conjugated diene (see also ref.101).

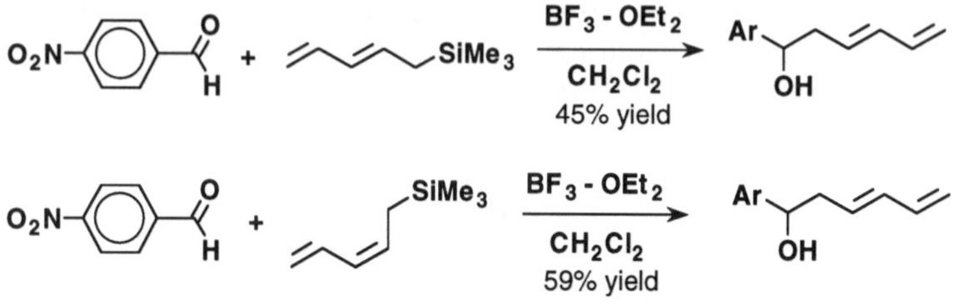

Marshall studied the addition of an allenylstannane to three representative aldehydes. It appears from that work that diastereoselectivity dramatically increases as a function of the steric bulk of the R group.[102]

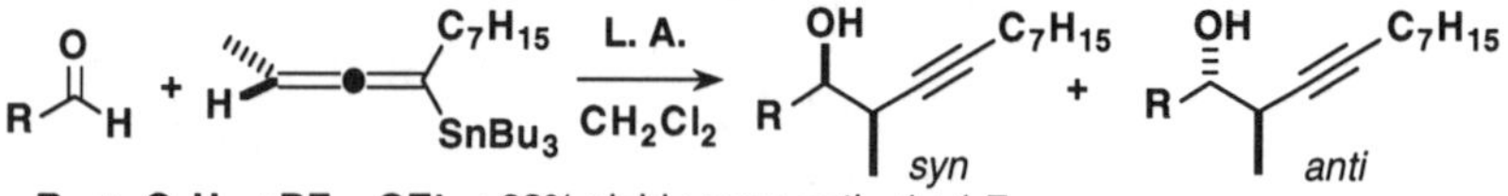

R = *n*-C_6H_{13}; BF_3- OEt_2 ; 83% yield ; *syn* : *anti* = 1 : 1.7
R = *n*-C_6H_{13}; $MgBr_2$-OEt_2 ; 56% yield ; *syn* : *anti* = 2.2 : 1
R = *i*-Pr; BF_3 - OEt_2 ; 80% yield ; *syn* : *anti* = 99 : 1
R = *i*-Pr; $MgBr_2$ - OEt_2; 48% yield ; *syn* : *anti* = 7.3 : 1
R = *t*-Bu; BF_3- OEt_2; 92% yield ; *syn* : *anti* = 99 : 1

Diallyl silanes have also been used towards aldehydes. An efficient route to (±)-muscone starting from the addition of 1,8-bis(trimethylsilyl)octa-2,6-diene to ethanal, was proposed.[103]

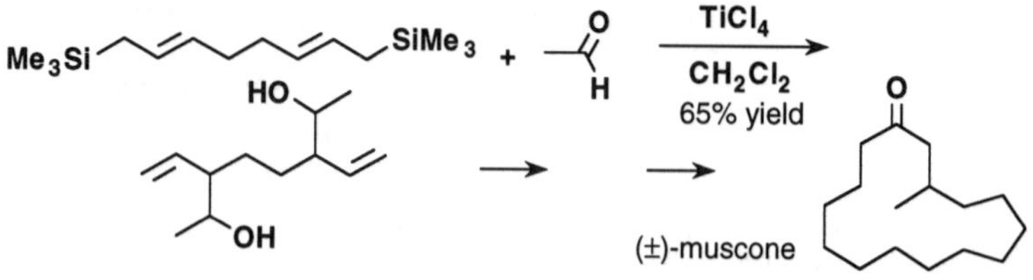

The following divinyltetrahydrofurans (stereochemistry unspecified) were also obtained from various aldehydes by Miginiac.[104]

$RCHO$ + Me_3Si–(diene)–$SiMe_3$ → (BF_3-OEt_2, CH_2Cl_2, 50–75% yield)

R = Me, Et, *n*-Pr, *n*-Hex, *i*-Bu, Benzyl

Tagliavini[105] observed an allylic rearrangement in the solvent-free redistribution reaction between (*E*)- and (*Z*)-crotyltin and dibutyltin dichloride. But-1-en-3-yldi-*n*-butylchlorotin is formed initially either from (*E*)- or (*Z*)-crotyltin (equations 1 and 2).[105]

(*E*)-crotyl-$SnBu_3$ + Bu_2SnCl_2 → (but-1-en-3-yl)$ClSnBu_2$ + Bu_3SnCl (1)

(*Z*)-crotyl-$SnBu_3$ + Bu_2SnCl_2 → (but-1-en-3-yl)$ClSnBu_2$ + Bu_3SnCl (2)

(but-1-en-3-yl)$ClSnBu_2$ + Bu_2SnCl_2 → (*E*)-crotyl-$SnClBu_2$ + Bu_2SnCl_2 (3)

(but-1-en-3-yl)$ClSnBu_2$ + Bu_2SnCl_2 → (*Z*)-crotyl-$SnClBu_2$ + Bu_2SnCl_2 (4)

(*Z*)-crotyl-$SnClBu_2$ ⇌ (*E*)-crotyl-$SnClBu_2$ (5)

Rapid addition to aldehydes gives rise to "linear" homoallylic alcohols of (*Z*)-configuration. Thus, a regioreversed addition, which is due to the formation of the allyldibutylchlorotin, occurs.[106,107]

[$RCHO$ + crotyl-$SnBu_3$] added to the neat Bu_2SnCl_2, 25 °C → [(but-1-en-3-yl)$ClSnBu_2$] → *Z* + *E* + *syn* + *anti*

Benzaldehyde: 95% yield; *Z* : *E*: *syn* : *anti* = 1 : 0 : 0 : 0

Propanal: 94% yield; *Z* : *E*: *syn* : *anti* = 17.4 : 0 : 1.2 : 1

Isobutyraldehyde: 85% yield; *Z* : *E*: *syn* : *anti* = 14.5 : 0 : 1.6 : 1

2-Methylbutanal: 82% yield; *Z* : *E*: *syn* : *anti* = 10 : 0 : 1.5 : 1

Pivalaldehyde: 82% yield; *Z* : *E*: *syn* : *anti* = 1 : 0 : 0 : 0

However, after a long period, changes occur in the distribution of isomeric organotin compounds and (*Z*)- and (*E*)-crotyldi-*n*-butylchlorotins (equations 3 and 4) are formed in a 1:1.5 ratio.[108] Again, *syn* and *anti* homoallylic alcohols are formed exclusively.

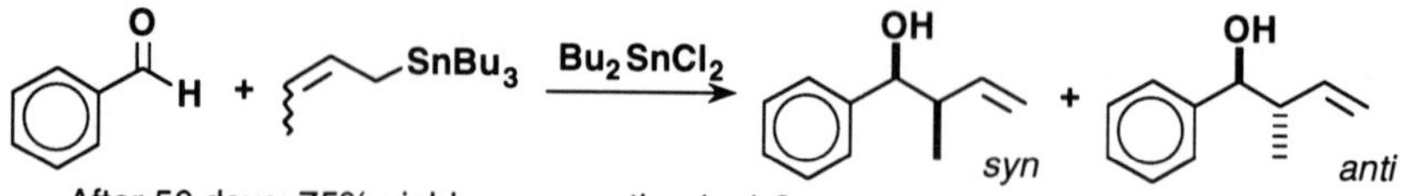

After 50 days; 75% yield ; *syn* : *anti* = 1 : 1.2

The same authors also studied the reactivity of crotyldi-*n*-butylchlorotins towards α,β-unsaturated aldehydes.[109]

Baba recently reported a reaction of allyltri-*n*-butyltins with aldehydes, catalyzed by dibutyltin dichloride together with additive (coordinative) compounds such as phosphine oxides or tetraalkylammonium halides; these compounds are known to form complexes with dialkyltin dihalides. Benzoyl chloride is added as a quencher and selective allylation of aldehydes occurs with retention of the allylic system producing linear homoallylic benzoates.[110]

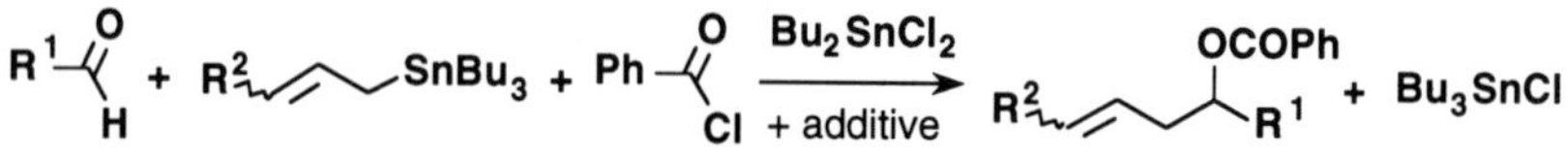

$R^1 = H$; $R^2 = Et$; additive : Et_4NCl; 84% yield
$R^1 = H$; $R^2 = Ph$; additive : Et_4NCl; 86% yield
$R^1 = Me$; $R^2 = Ph$; additive : Ph_4PI; 81% yield

The following catalytic cycle seems plausible.

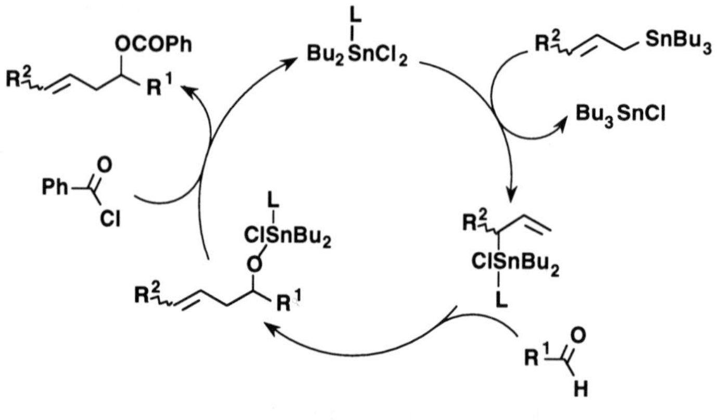

Finally, Bosnich recently reported that $Cp_2Ti(OTf)_2$ could be used in only 0.5 mol% to induce a reaction between a variety of methylallylsilanes and aldehydes.[111]

II.2.b Reaction between achiral aldehydes and functionalized allylsilanes or allylstannanes

Functionalized allylsilanes, prepared in two steps from diketene, add, in the presence of $TiCl_4$, to various aldehydes and ketones to give hydroxy acids. These undergo, upon treatment by hydrochloric acid, cyclization and double bond migration to yield the corresponding conjugated δ-lactones in fair yields.[112]

TiCl$_4$, CH$_2$Cl$_2$; HCl, 3N, 78% yield

R^1 = *n*-Pr; R^2 = H; 81% yield
R^1 = *i*-Pr; R^2 = H; 79% yield
R^1 = *i*-Bu; R^2 = H; 78% yield
R^1 = Me; R^2 = Me; 80% yield
R^1-R^2 = -(CH$_2$)$_4$; 78% yield

Evans[113] showed that the addition of methyl 2-[(dimethylphenyl)silyl]but-3-enoate to aldehydes affords the corresponding γ-functionalized (*E*)-homoallylic alcohols. Treatment of the latter with benzaldehyde, in the presence of potassium *tert*-butoxide, reveals an acetal protected *syn*-1,3-diol moiety.[113]

TiCl$_4$, CH$_2$Cl$_2$

R = Et; 66% yield
R = *i*-Pr; 70% yield
R = Ph(CH$_2$)$_2$; 65% yield

t-BuOK, THF

R = Et; 79% yield; selectiv. : >19 : 1
R = *i*-Pr; 71% yield; selectiv. : >19 : 1
R = Ph(CH$_2$)$_2$; 73% yield; selectiv. : 24 : 1

Quintard[114] established that the BF$_3$-OEt$_2$ catalyzed addition of (ethoxyalkenyl)tri-*n*-butyltins to aldehydes provides a useful way of preparing α-ethoxy homoallylic alcohols with high *syn*-selectivity.[114,115]

BF$_3$ - OEt$_2$, CH$_2$Cl$_2$, 70% yield

syn : *anti* = 13.3 : 1

The (trimethylsilyl)trifluoromethanesulfonate (TMS-OTf) promoted condensation of allylsilane **B** to aldehydes or ketones, such as **A,** smoothly gives oxonium cations **C**. Cations **C** are then intramolecularly intercepted by the allylsilane moiety to afford the dihydropyran derivatives **D**, **E** and **F**.[116,117] Spiro ethers can also be obtained by following the same sequence.

R^2 R^1 O + SiMe$_3$ Me$_3$SiO B → TMS-OTf, CH_2Cl_2 → R^1 R^2 O SiMe$_3$ TfO$^{(-)}$ (+) C → { D, E, F }

A

R^1 = *n*-pentyl, R^2 = H; 88% yield; D : E : F = 8 : 1.5 : 1

A = (cycloalkanone, ( )n) → D + E + F

n = 1 : 93% yield; D : E : F = 7.8 : 2.2 : 1; n = 2 : 97% yield; D : E : F = 11.9 : 1 : 1

In close analogy, using different ω-silyloxyallyltrimethylsilanes, Miginiac regiospecifically prepared, in the presence of BF_3-OEt_2, 2-alkyltetrahydro oxepins and 2-alkyltetrahydrooxocins.[118]

R–CHO + Me$_3$SiO( )n SiMe$_3$ → BF_3 - OEt_2, CH_2Cl_2 → R, H, O, Me$_3$Si, ( )n SiMe$_3$ → − Me$_3$SiOBF$_3^-$

R, H, O$^+$, ( )n SiMe$_3$ → − Me$_3$SiOSiMe$_3$ → R, H, O, ( )n

n = 1; 54-90% yield
n = 2; 55-77% yield

Marko[119] showed that the outcome of the addition of 2-trimethyl silyloxymethylallylsilane to aldehydes depends upon the Lewis acid used. With $TiCl_4$, the reaction leads to the normal allylation products, whereas Et_2AlCl promotes the ene reaction which affords silylenolethers.[119]

R–CHO + SiMe$_3$ OSiMe$_3$ → $TiCl_4$, CH_2Cl_2 → R OH OH

R–CHO + SiMe$_3$ OSiMe$_3$ → Et_2AlCl, CH_2Cl_2 → R OH SiMe$_3$ OSiMe$_3$

In contrast, with BF_3-OEt_2, two molecules of aldehyde are involved and the reaction yields, with moderate yield but as a single diastereoisomer, an *exo*-methylene tetrahydropyran containing three chiral centers.

The proposed mechanism for that unusual reaction starts with an ene-type reaction. Further condensation of the free hydroxy function with the unreacted aldehyde then generates an oxonium cation that undergoes intramolecular allylsilane addition. This reaction leads to an *exo*-methylene tetrahydropyran in which all the substituents occupy equatorial positions.[119]

As already mentioned, the stereo-outcome of the reaction can also depend upon the nature of the Lewis acid. Thus, Nishigaichi reported that the allylation of aldehydes by allylsilane **A**, bearing an asymmetric ethereal functionality, proceeds in divergent manners upon the Lewis acid used: $TiCl_4$ promotes the formation of *syn* isomer **B**, whereas BF_3-OEt_2 induces predominantly the formation of *anti* isomer **C**.[120]

R = *p*-$NO_2C_6H_4$; $TiCl_4$ - $(OEt_2)_2$; 98% yield ; *syn* : *anti* = 1 : 0
R = *p*-$NO_2C_6H_4$; BF_3-OEt_2 ; 67% yield ; *syn* : *anti* = 1 : 3.8
R = 2,6-$Cl_2C_6H_3$; $TiCl_4$ - $(OEt_2)_2$; 84% yield ; *syn* : *anti* = 1 : 0
R = 2,6-$Cl_2C_6H_3$; BF_3-OEt_2 ; 72% yield ; *syn* : *anti* = 1 : 8.1
R = *n*-C_7H_{15} ; $TiCl_4$ - $(OEt_2)_2$; 94% yield ; *syn* : *anti* = 1 : 0
R = *n*-C_7H_{15} ; BF_3-OEt_2 ; 28% yield ; *syn* : *anti* = 1 : 2.4

This divergent selectivity is rationalized in the way the Lewis acids coordinate differently to the reagent. $TiCl_4$, which has two acceptor sites, can coordinate to both the allylsilane **A** and the aldehyde as depicted in cyclic transition state **B1**; **B1** leading to *syn* product **B**. In contrast, BF_3-OEt_2, which bears a single acceptor site, can only coordinate to one reagent, the aldehyde, leading therefore to an acyclic transition state such as **C1** or **C2**, which both lead to *anti* product **C**.

A dramatic reversal of regioselectivity, due to the nature of the Lewis acid, was also reported by Nishigaichi. Thus, while the BF_3-OEt_2 promoted addition of the following pentadienyltin derivative to aldehydes leads to ε-adducts, the $TiCl_4$ induced reaction leads in contrast to γ-adducts with good, and in one occasion complete, selectivity.[121,122]

L. A : BF_3-OEt_2
R = C_9H_{19} ; 67% yield ; ε : γ : α = 14 : 1 : 1.7
R = c-C_6H_{11} ; 100% yield ; ε : γ : α = 9 : 1.1 : 1
R = Ph ; 74% yield ; ε : γ : α = 2.8 : 1 : 1.2

L. A : $TiCl_4$ R = C_9H_{19} ; 52% yield ; ε : γ : α = 1.8 : 22.3 : 1
R = c-C_6H_{11} ; 51% yield ; ε : γ : α = 3.7 : 9.6 : 1
R = Ph ; 61% yield ; ε : γ : α = 0 : 1 : 0
R = p-$MeO_2CC_6H_4$; 100 % yield ; ε : γ : α = 0 : 1 : 0

More interestingly, the γ-adduct obtained from the $TiCl_4$-mediated addition to *p*-nitro benzaldehyde was found to be *syn-syn* diastereomer **B**. The selectivity can be rationalized through the formation of chelated transition state **A**.

Koreeda reported that, in the presence of BF_3-OEt_2, (*E*)- and (*Z*)-γ-alkoxyallyltins, just like crotyltins, add stereoselectively to aldehydes, leading, in both cases mainly to *syn*-products.[123]

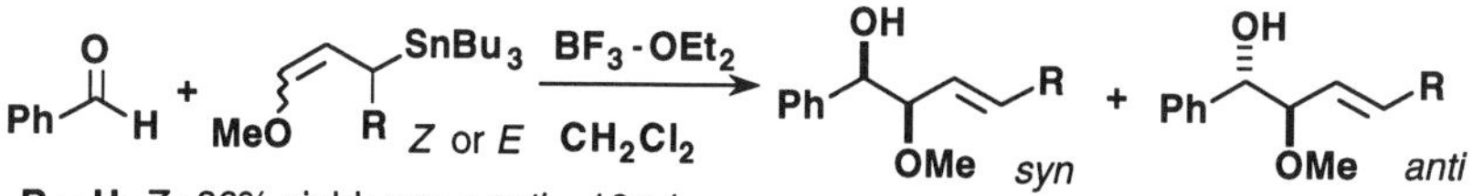

R = H; *Z*; 86% yield; *syn* : *anti* = 10 : 1
R = H; *E*; 87% yield; *syn* : *anti* = 14 : 1
R = Me; *E* : *Z* = 2 : 1; 86% yield; *syn* : *anti* = 19 : 1

This result was used to prepare a key intermediate in the synthesis of (±)-exo-brevicomin. Thus, the following diol monomethyl ether was obtained in 80% yield with a *syn* stereoselectivity greater than 20:1.

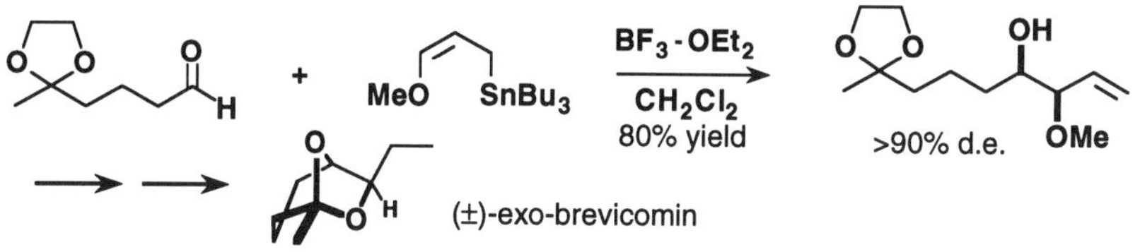

(±)-exo-brevicomin

Most studies dealing with the addition of allylsilanes or allylstannanes reported so far were performed on saturated aliphatic aldehydes or benzaldehyde derivatives. As for Marshall, he examined the effect of an α,β-ethylenic or α,β-acetylenic bond on the stereoselectivity of the reaction.[124] The BF_3-OEt_2-promoted addition of the following (*E*)- and (*Z*)-allylstannanes mixture to crotonaldehyde shows a good *syn*-selectivity. This selectivity can however be completely reversed when changing the catalyst from BF_3-OEt_2 to $TiCl_4$ and by using the "inverse" addition procedure (Keck's procedure[62]). The good *anti*-selectivity is nevertheless obtained with a lower chemical yield.

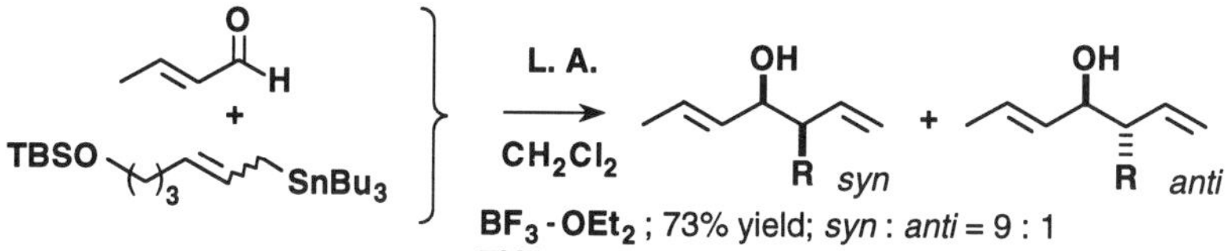

BF_3-OEt_2; 73% yield; *syn* : *anti* = 9 : 1
$TiCl_4$; 47% yield; *syn* : *anti* = 1 : 19

The addition of the same allylstannanes to the following acetylenic aldehyde gives rise to closely related results.

BF_3-OEt_2; 81% yield; *syn* : *anti* = 9 : 1
$TiCl_4$; 60% yield; *syn* : *anti* = 1 : 19

Sano and Ueno demonstrated that 1,3-bis(tri-*n*-butylstannyl)-2-methylenepropane **A** reacts with aldehydes under thermal conditions to give **B** at considerably lower temperatures than normal allylstannanes do. The addition of a second aldehyde in the presence of a Lewis acid leads to the production of bis-homoallylic alcohols **C**.[125]

R^1CHO + **A** → (110°C, toluene) → **B**

R^2CHO + **B** → (BF_3 - OEt_2, CH_2Cl_2) → **C**

The Lewis acid promoted addition of **A** to aldimine was also performed. The condensation, in the presence of a Lewis acid, of the resultant aminoallylstannane **D** to aldehydes affords aminoalcohols **E** that can subsequently undergo cyclization to yield piperidines.[126]

PhCH=N-R^1 + **A** → (BF_3 - OEt_2, CH_2Cl_2) → **D**

R^2CHO + **D** → (BF_3 - OEt_2, CH_2Cl_2, 47–74% overall yield) → **E**

The reactivity of 1-ethoxy-3-trimethylsilylpropyne **A** towards various carbonyl compounds was studied by Miginiac.[127] Thus, α-chloroketone **B** led to dienic ester **E** through intermediates **C** and **D**.

A + **B** → ($TiCl_4$, CH_2Cl_2, 71% yield) → [**C**] → [**D**] → **E**

Finally, 3-chloroallyltrimethylsilane was shown to react with aldehydes in the presence of $AlCl_3$ to give chloromethylether of homoallylic alcohols.[128]

II.2.c Reaction between achiral aldehydes and chiral and homochiral allylsilanes or allylstannanes

The presence of a stereogenic center close to the double bond of an allylsilane lowers the stereoselectivity of the addition to aliphatic aldehydes. Thus, the following reaction gave a mixture of diastereomers (76:21:2:1) in 60% chemical yield. Taddei established the relative stereochemistry of the major isomer and proposed that it was formed via antiperiplanar transition state **A**.[129]

$BF_3 \cdot OEt_2$, CH_2Cl_2, 46% yield

A

α-Methylene-γ-butyrolactones such as **C** were prepared in high yields by Tanaka. Thus, *N*-monosubstituted 2-[(tributylstannyl)methyl] propenamides such as **A** add, in the presence of a Lewis acid, to aldehydes (here isovaleraldehyde), yielding the resulting γ-hydroxy amides **B**. An acidic hydrolysis then promotes cyclization and leads to lactones **C**.[130,131]

$TiCl_4$ (4 equiv.), CH_2Cl_2, 96% yield; H^+, H_2O; 78% e.e.

A synthesis of homochiral α-(alkoxy)- and γ-(alkoxy)allyltins was developed by Marshall.[132] The addition of *n*-Bu_3SnLi to α,β-unsaturated aldehydes followed by *in situ* oxidation of the adducts affords acylstannanes. These are then reduced, with Noyori's (*R*)-BINAL-H reagent, to the optically active α-(alkoxy)allyltins (>95% e.e.) which could then undergo stereospecific BF_3-OEt_2 induced rearrangement to yield optically active (*Z*)-γ-(alkoxy)allyltins (for previous observation of that 1,3-isomerization, see refs.114,115, and 133 for a discussion of its mechanism).

Bu_3SnLi; Azodicarbonyl dipiperidide, *t*-BuOMgBr; (*R*)-BINAL-H, THF; BF_3-OEt_2, CH_2Cl_2

The following benzyloxymethyl (BOM) ether derivative adds smoothly to representative aldehydes, affording the *syn* diol derivatives with virtually complete transfer of chirality.[134]

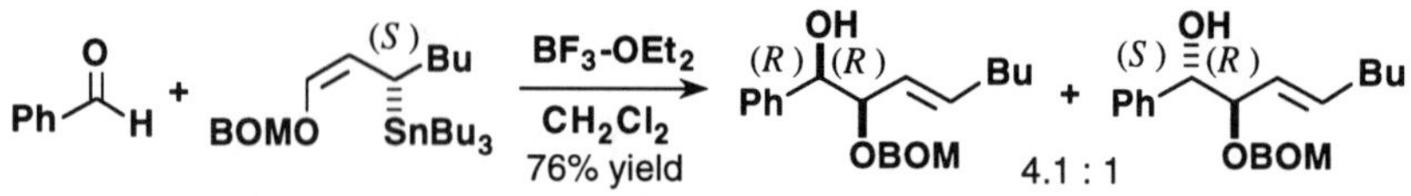

Similarly, Yamamoto proposed the asymmetric synthesis of *syn* 1,2-diols from achiral aldehydes and optically active γ-(tetrahydropyranyloxy) allylstannanes.[135]

Taddei showed that naturally occurring α-amino acids can easily be transformed into optically active allylsilanes.[136] Addition of an optically active allylsilane to a solution of 2-methylpropanal and $TiCl_4$ in CH_2Cl_2 at room temperature, followed by aqueous work-up gives directly the piperidine derivatives.[137] These could be formed via an intermediate iminium ion obtained by nucleophilic attack of the nitrogen atom on the aldehyde in the presence of Lewis acid, followed by intramolecular attack of the allylic carbon adjacent on the silicon atom.

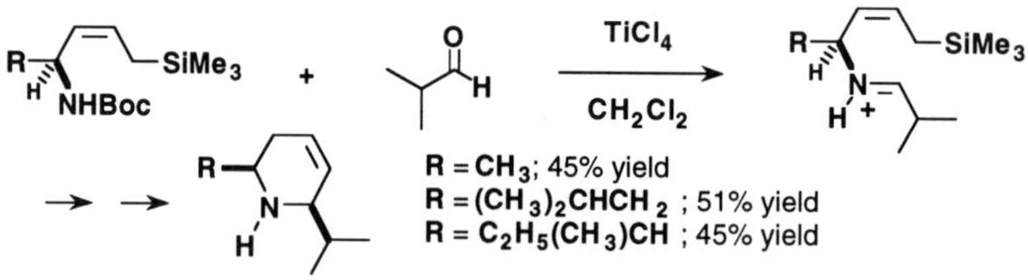

Thomas used (*E*)-1-alkoxymethoxybut-2-enyl(tri-*n*-butyl)stannanes as homo-enolate equivalents. As expected, the uncatalyzed, thermal, addition to aldehyde displays *anti*-selectivity.[138] Thus, chloromethyl (–)-menthyl ether reacts with (*E*)-1-hydroxybut-2-enyl(tri-*n*-butyl)stannane to give a separable mixture of (*E*)-1-(–)-menthoxymethoxybut-2-enyl(tri-*n*-butyl)stannanes. Heated separately with an excess of aldehyde, the (1*S*)-isomer leads to the *anti*-(*E*)-(3*R*)-enol ether. The observed stereochemistry is suggestive of a mechanism proceeding through an associated six-membered transition state.[139]

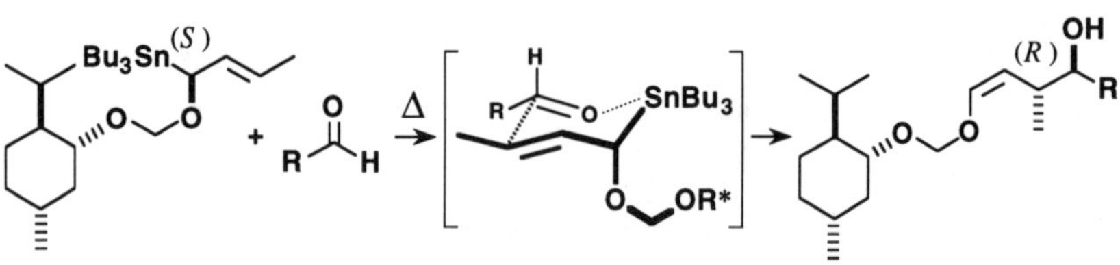

Gung prepared chiral allylstannanes from 8-phenylmenthol according to Thomas's method. After separation of the diastereomers, he established that excellent diastereofacial selectivity between (*R*)-allylstannane and benzaldehyde can be achieved under BF_3-OEt_2 catalysis; the major diastereomer being the *syn*-(*Z*)-(3*R*, 4*S*) isomer.[140]

BF_3 - OEt_2

CH_2Cl_2

75% yield

syn-Z 19 : 1 *anti-Z*

The diastereofacial selectivity is low when the (*S*)-allylstannane diastereomer is used and *syn*-(*E*)-(3*R*) isomers are predominantly obtained from most aliphatic aldehydes. However, with benzaldehyde, the facial selection is different and the *syn*-(*Z*)-(3*S*) isomer is obtained as the main isomer.

BF_3 - OEt_2

CH_2Cl_2

75 % yield

syn-Z *syn-E*

R = *n*-hexyl; *syn-Z* : *syn-E* = 1.6 : 1
R = *n*-(*E*)-2-hexenyl; *syn-Z* : *syn-E* = 1 : 1.1
R = cyclohexyl; *syn-Z* : *syn-E* = 1 : 4.6
R = phenyl; *syn-Z* : *syn-E* = 95 : <1; (*anti-Z* = 5 %)

The predominant formation of (*Z*)-enol ether, upon reaction between aromatic aldehydes and α-(alkoxy)allylstannanes, cannot be rationalized on steric grounds only. The "inside alkoxy" effect proposed by Houk[141] for electrophilic additions to chiral alkenes, may also be involved. Thus, a synclinal transition state, with an *anti* relation between the R group of the aldehyde and the allyl moiety and with an "inside alkoxy" could account for the formation of the *syn*-(*Z*)-(3*S*) isomer. With aliphatic aldehydes, the difference of stability between the *anti* and *syn* aldehyde-Lewis acid complexes is relatively small (see Chapter 1, Section III.1.b) and therefore, the antiperiplanar transition state, with BF_3 *syn* to the R group and the "outside alkoxy", becomes more favorable on steric grounds.

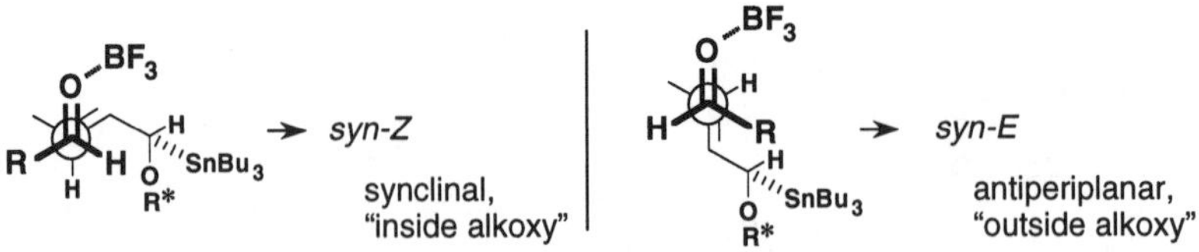

In a series of papers, Thomas investigated the transmetallation between tin tetrachloride and γ-substituted-allyltri-*n*-butylstannanes, leading to allyltrichlorostannanes (see also ref.81). These, although unstable, are very reactive towards aldehydes.[142–144] Crotylstannane was mixed with one molar equivalent of tin tetrachloride at –78°C, and after 5 minutes, a precooled solution of benzaldehyde was added. A mixture of the four possible products was obtained.

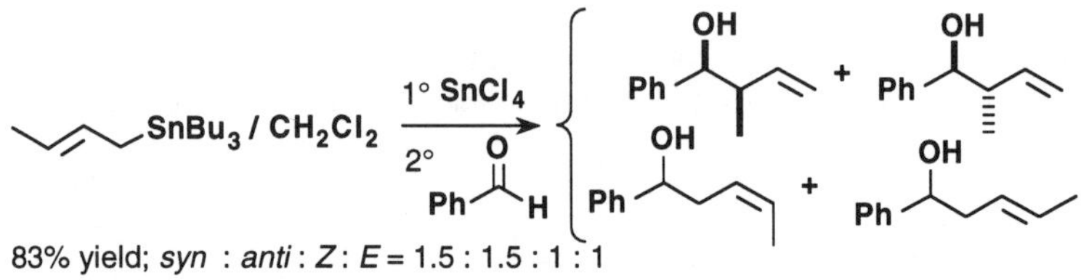

83% yield; *syn* : *anti* : *Z* : *E* = 1.5 : 1.5 : 1 : 1

However, it was found that (*S*)-4-benzyloxypentenyltri-*n*-butylstannane reacts with aldehydes, in the presence of $SnCl_4$ with excellent 1,5-asymmetric induction, to yield (*Z*)-*syn*-alcohols.

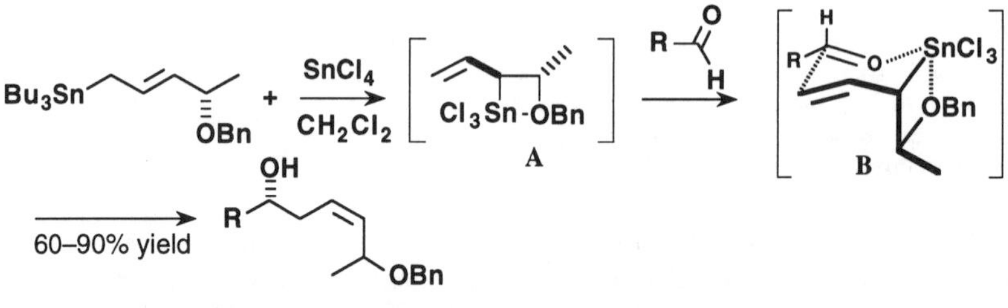

60–90% yield

It is suggested that the δ-benzyloxyallylstannane undergoes transmetallation with tin tetrachloride to generate allyltrichlorostannane **A** which may be stabilized by a hypervalent oxygen-tin interaction. The usual concerted, cyclic, transition state **B**, in which the substituent adjacent to the tin atom is in the axial position[96] would then account for the stereochemistry of the alcohols obtained. This 1,5-asymmetric induction has been used by Thomas in a remarkably stereoselective synthesis of an aliphatic 1,5,9,13-polyol.[145]

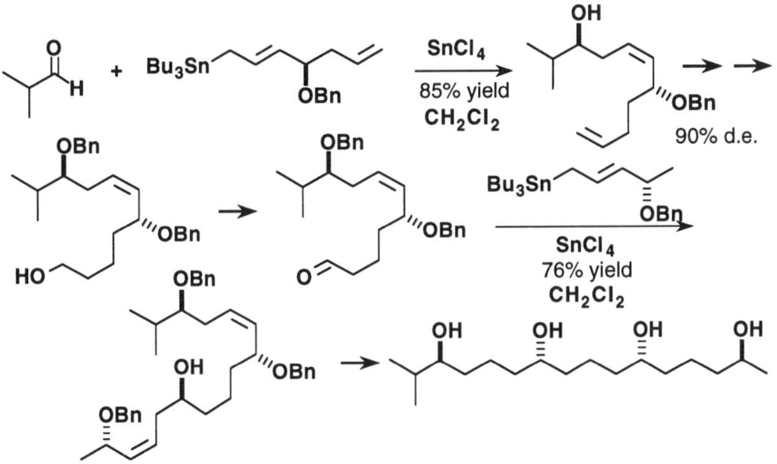

Even 1,6-asymmetric induction was observed with allylstannanes bearing an ε-hydroxy or ε-methoxy substituent.[146] The *syn*-(*Z*)-hex-3-en-1,6-diol formation is consistent with a mechanism involving transmetallation of the stannanes to give **A** which is again stabilized by oxygen-tin interaction, and which then reacts with the aldehyde via the six-membered ring, chair-like transition state **B** in which the substituent α to tin is in axial position.

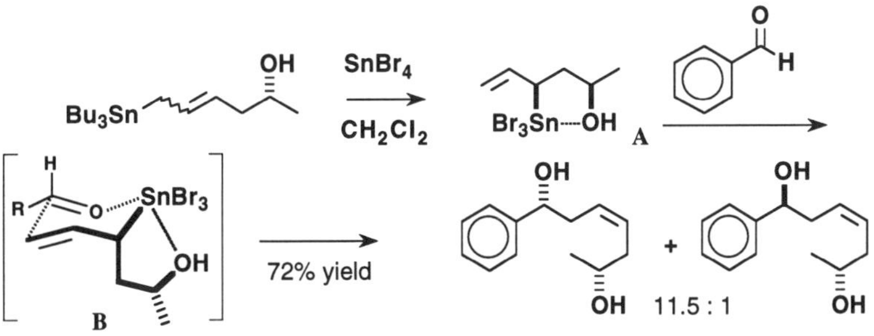

As for Panek,[147] he studied the diastereoselectivity of the addition of optically active β-methyl-substituted crotylsilane derivatives to α-benzyloxy propanal. He showed that the stereochemistry of the adduct is reversed from *syn* to *anti* by changing the Lewis acid from BF_3-OEt_2 to $MgBr_2$-OEt_2.[147]

BF_3-OEt_2 ; 68% yield ; **A** : **B** = 6.5 : 1
$ZnCl_2$; 58% yield ; **A** : **B** = 2 : 1
$AlCl_3$; 54% yield ; **A** : **B** = 1.2 : 1
$SnCl_4$; 60% yield ; **A** : **B** = 1 : 2
$TiCl_4$; 57% yield ; **A** : **B** = 1 : 4.2
$MgBr_2$ -OEt_2 ; 59% yield ; **A** : **B** = 1 : 12.2

Thus, the use of BF_3-OEt_2 affords the *syn*-homoallylic alcohol which results from a non-chelation-controlled addition. Indeed, under such conditions, the *Re* face of the aldehyde is matched with the *Si* face of the crotylsilane. The following three transition states can lead to the *syn* adduct.

A1 (*unlike*) A2 (*unlike*) A3 (*unlike*)

The minor *anti* adduct results from an attack of the crotylsilane from its *Si* face to the *Si* face of the aldehyde. All three following *like* transition states would lead to the *anti* adduct.

B1 (*like*) B2 (*like*) B3 (*like*)

Panek proposed that the *syn* adduct was formed through antiperiplanar transition state **A1.** However, the observed selectivity could also be accounted for by postulating that the reaction proceeds through a synclinal transition state; thus, when comparing *unlike* and *like* transition states, the biggest difference is clearly between the two synclinal ones, **A3** and **B3.**

The formation of the enantiomeric *syn* alcohol **C** would result from an addition of the *Re* face of the crotylsilane to the *Si* face of the aldehyde (unmatched addition).

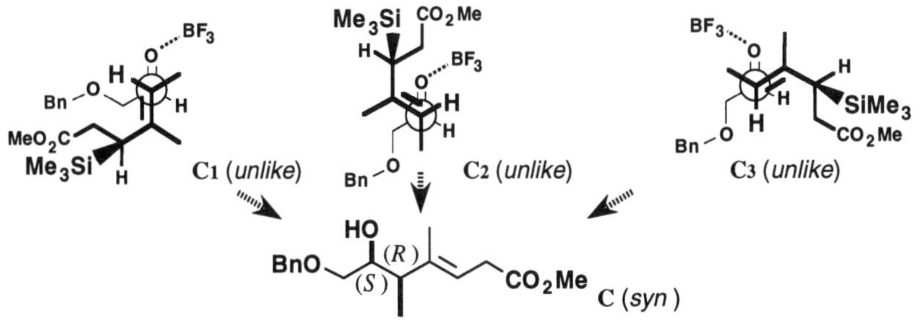

On the other hand, bidentate Lewis acids, such as $MgBr_2$, promote chelation-controlled addition. *Anti* homoallylic alcohol **B** could therefore be formed through *like* transition state **B4**, which is less hindered than **A1**.

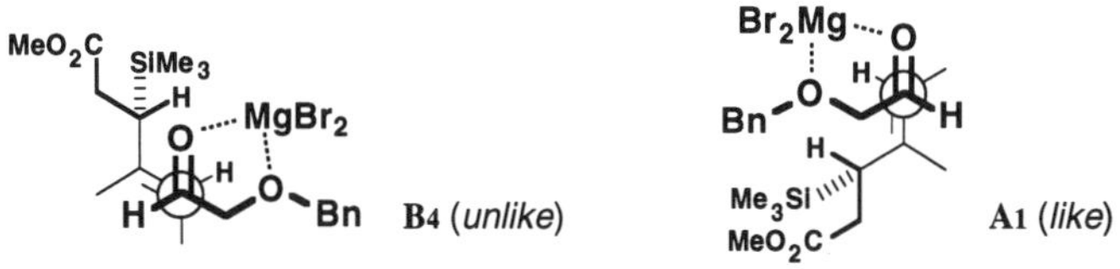

Panek also described the diastereoselective additions, mediated by catalytic amounts of TMS-OTf in the presence of trimethylsilylbenzyl ether (TMS-OBn), of homochiral (*E*)-crotylsilane derivatives to various aldehydes. The TMS-OTf-catalyzed addition of trimethylsilylbenzyl ether to aldehydes generates oxonium ions which then react with the crotylsilane derivatives. The reaction generally proceeds with a high level of diastereoselection in favor of the *syn* isomers. An antiperiplanar transition state can account for such a selectivity.[148]

The addition of the following *syn* (*R* ,*R*)-crotylsilane derivative mainly occurs on the *Re* face of the aldehydes. The diastereofacial selectivity increases as a function of the size of the alkyl R group.

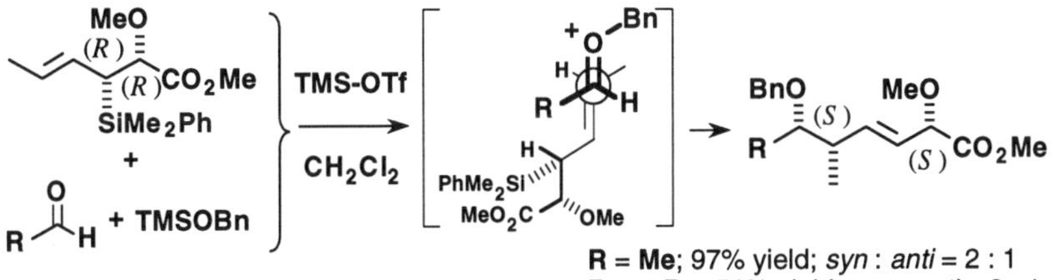

R = **Me**; 97% yield; *syn* : *anti* = 2 : 1
R = **n-Bu**; 51% yield; *syn* : *anti* = 3 : 1
R = ***i*-Pr**; 60% yield; *syn* : *anti* = 19 : 1

Addition to aromatic aldehydes is even more diastereoselective.

OAc
(*R*)
(*R*) CO_2Me
$SiMe_2Ph$
+
O
R H + TMSOBn
TMS-OTf
CH_2Cl_2
+O Bn
H
R H
H
$PhMe_2Si$
MeO_2C OAc
BnO OAc
R (*S*) (*S*) CO_2Me

O
R H ≡ MeO MeO O H
98% yield;
syn : *anti* = 30 : 1
;
MeO O
O_2N H
OMe
92% yield;
syn : *anti* = 30 : 1

Similarly, the corresponding *anti* (*R*,*S*)-crotylsilane derivative adds to the *Re* face of aldehydes, again with a higher *syn* selectivity for aromatic aldehydes.

OMe
(*R*)
(*S*) CO_2Me
$SiMe_2Ph$
+
O
R H + TMSOBn
TMS-OTf
CH_2Cl_2
+O Bn
H
R H
H
$PhMe_2Si$
MeO_2C OMe
BnO OMe
R (*S*) (*R*) CO_2Me

O
R H ≡ BnO O H
53% yield;
syn : *anti* = 20 : 1
;
MeO O
H
OMe
93% yield;
syn : *anti* = 30 : 1

By contrast, addition of the *anti* (*S*,*R*)-crotylsilane derivative occurs on the *Si* face of aldehydes.

(*S*)
(*R*) CO_2Me
$SiMe_2Ph$
+
O
R H + TMSOBn
TMS-OTf
CH_2Cl_2
+O Bn
H
R H
H
$PhMe_2Si$
MeO_2C
BnO
R (*S*) (*S*) CO_2Me

R = ***t*-Bu**; 94% yield; *syn* : *anti* = 30 : 1
R = **cyclohexyl**; 88% yield; *syn* : *anti* = 30 : 1

Taking advantage of these results, Panek proposed a highly enantioselective synthesis of tetrahydrofuran derivative **B** from homochiral (*E*)-crotylsilane **A**.[149] Oxidation of the dimethylphenylsilyl group by Hg^{2+} according to Fleming's procedure,[150] leads to the diastereomerically pure hydroxy tetrahydrofuran derivative **C**.

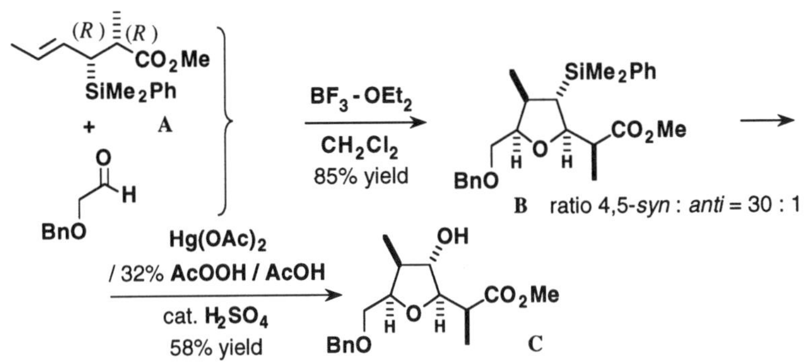

The formation of the *cis*-2,5-disubstituted tetrahydrofuran **B** involves an addition of the silane derivative to the aldehyde, followed by heterocyclization. The stereochemistry of the product requires an antiperiplanar arrangement of the participating π bonds in transition state **B1** and then a rotation about the new bond to favor the cyclization. In this case, the 1,2 migration competes favorably with the elimination of the dimethylphenylsilyl group which generally occurs after the addition to aldehydes.

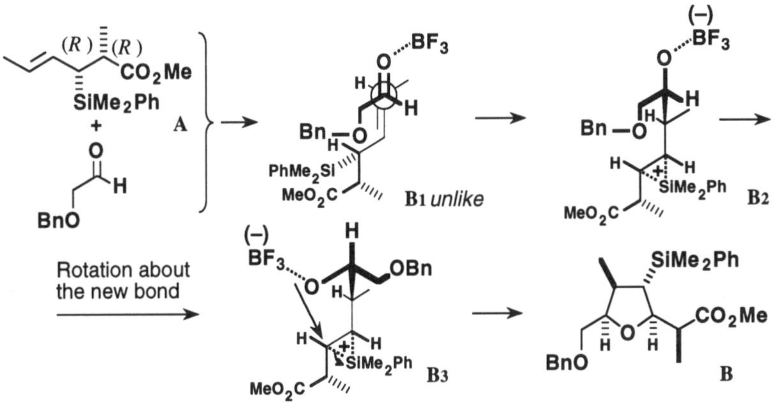

Finally, allylsilane or tin with chiral auxiliaries directly bonded to the metal atom have also been used. Optically active diallylbis(2-phenylbutyl)tin was added to various aldehydes to afford homoallylic alcohols with e.e. from 20 to 80%.[151] The (*S*)-stannane isomer transfers one of its allyl moiety to the *Re* face of the aldehyde.

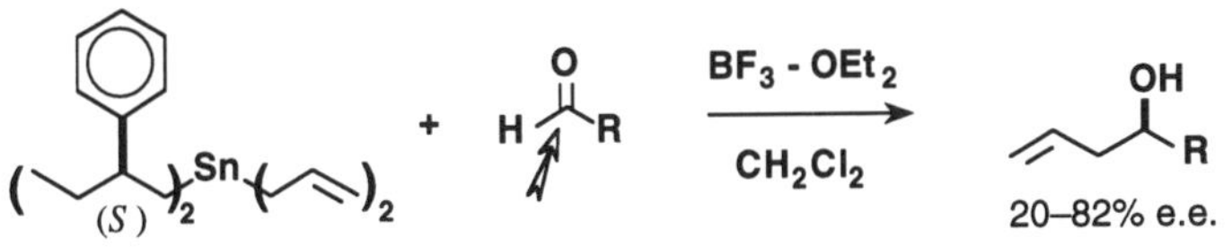

Similarly, Taddei built an allylsilane derivative, which reacts with aldehydes and ketones with low to medium enantioselectivity. The methoxy group, borne by the chiral auxiliary, presumably provides coordination with titanium.[152]

R–CHO + (MeO-pinanyl)Me$_2$Si-allyl —$TiCl_4$, CH_2Cl_2→ R–CH(OH)*–CH$_2$CH=CH$_2$ 39–61% overall yield; 21–46% e.e.

Another chiral allylsilane, bearing an alkoxy group directly attached to a bornane skeletal, was reported to react with *n*-butanal to give the corresponding homoallylic alcohol but with lower e.e.[153] Alternative approaches were provided by using alkoxyallylsilanes obtained from readily available optically active alcohols; homoallylic alcohols with e.e. in the range 18-23% were obtained.[154,155] Finally, chiral (pyrrolidinylmethyl)allylsilane, prepared from (bromomethyl)allylsilane and optically active compounds derived from (*S*)-proline, afforded homoallylic alcohols with enantiomeric excess up to 50%.[156]

II.2.d Catalytic asymmetric allylation of achiral aldehydes

Seebach prepared optically active homoallylic alcohols from achiral aldehydes and optically active dichlorodialkoxytitanium derived from (*S*)-(–)-1-phenylethanol.[157]

The enantioselective addition of allyltri-*n*-butylstannane to aldehydes, using chiral Lewis acid as catalysts, has recently received much attention. Thus, remarkably effective and simple procedures lead, with excellent yields both chemical and optical, to homoallylic alcohols, precursors of β-hydroxy aldehydes or β-hydroxy carboxylic acids.

R–CHO + allyl–MR_3 —Chiral Catalyst→ R–CH(OH)CH$_2$CH=CH$_2$ —[O]→ R–CH(OH)CH$_2$CHO ; —[O]→ R–CH(OH)CH$_2$COOH

Yamamoto used a catalyst **A**, prepared from tartaric acid, in which the Lewis acid activity is due to an (acyloxy)borane moiety.[158,159] The addition, promoted by catalyst **A** (20 mol%), of various achiral allylsilanes to benzaldehyde produced homoallylic alcohols with moderate to very good enantiomeric excess.

PhCHO + R^1CH=C(R^2)CH$_2$SiMe$_3$ —**A** (*i*-PrO, O, CO_2H, BH, O, *i*-Pr), 20 mol%, acetonitrile, 2° (*n*-Bu)$_4$NF→ PhCH(OH)CH(R^1)C(R^2)=CH$_2$

R^1 = R^2 = H ; 46% yield ; 55% e.e.
R^1 = H ; R^2 = Me ; 68% yield ; 82% e.e.
R^1 = Me ; R^2 = Me ; 63% yield ; 92% e.e.
R^1 = Me ; R^2 = Et ; 74% yield ; 96% e.e.

The boron substituent of **A** was found to have a marked influence on the chemical yield and the enantiomeric excess of the reaction. Thus, changing the hydrogen for the 3,5-bis(trifluoromethyl)phenyl group proved to be most effective as shown in the following scheme.

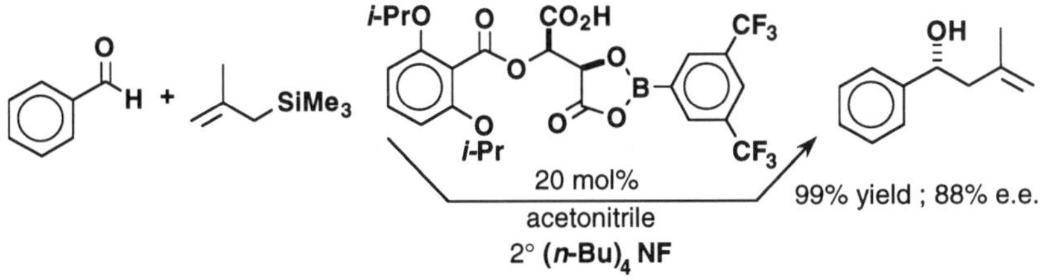

i-PrO O CO_2H CF_3 OH H + $SiMe_3$ O i-Pr B CF_3 20 mol% acetonitrile 2° $(n\text{-}Bu)_4NF$ 99% yield ; 88% e.e.

Yamamoto's chiral (acyloxy)borane catalyst was used by Marshall who replaced the allylsilane nucleophiles by their more reactive tin analogs. He found that the use of perceived promoters such as trifluoroacetic anhydride increases the efficiency of the reaction.[160]

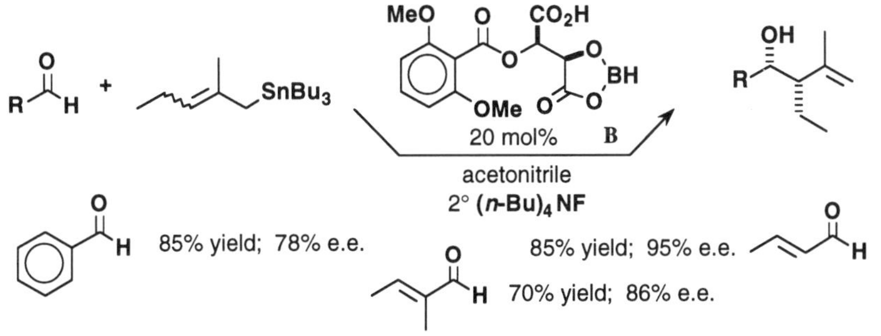

MeO O CO_2H BH OMe R H + $SnBu_3$ 20 mol% **B** acetonitrile 2° $(n\text{-}Bu)_4NF$ OH R

H 85% yield; 78% e.e.

H 70% yield; 86% e.e.

85% yield; 95% e.e. H

Nakai proposed to use the binaphthol-derived chiral titanium complex (*S*)-(BINOL-$TiCl_2$), prepared *in situ* from optically pure (*S*)-binaphthol and diisopropoxytitanium dichloride in the presence of 4 Å molecular sieves. Used in only 10 mol%, it catalyzes the addition of allylic silanes and stannanes to glyoxylates.[161] Thus, *syn*-α-hydroxy-β-methyl esters are obtained with both good diastereoselectivity and enantioselectivity, but however in moderate yield.

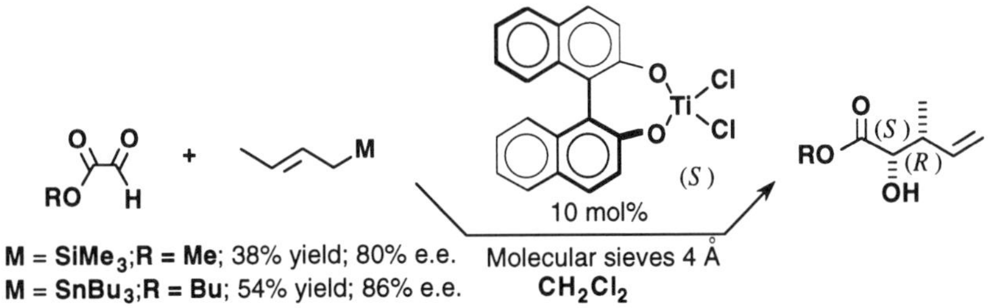

O Cl Ti Cl (*S*) RO H + M 10 mol% Molecular sieves 4 Å CH_2Cl_2 RO (*S*) (*R*) OH

M = $SiMe_3$;R = Me; 38% yield; 80% e.e.
M = $SnBu_3$;R = Bu; 54% yield; 86% e.e.

Nevertheless, the hydroxy methyl ester, obtained from the addition of crotyltrimethylsilane to methylglyoxylate, can then be transformed into verrucarinolactone, a degradation product of verrucarin A (antitumor activity).[162]

MeO ... OH → → → HO ... O (–)-Verrucarinolactone

The same catalyst, (*S*)-BINOL-$TiCl_2$, used by Tagliavini in 20 mol%, also in the presence of activated 4 Å molecular sieves, promotes the allylation of simple achiral aldehydes with good chemical yields and excellent enantiomeric excess.[163]

R–CHO + SnBu3 → (S)-BINOL-TiCl2, 10 mol%, Molecular sieves 4 Å, CH_2Cl_2 → R–CH(OH)–CH2CH=CH2

R = C_7H_{15} ; 83% yield; 97.4% e.e.
R = C_5H_{11} ; 75% yield ; 98.4% e.e.
R = *c*-C_6H_{11} ; 75% yield ; 92.6% e.e.

Keck prepared a similar chiral catalyst by two different methods: first (method A), by heating a mixture of BINOL and Ti(O*i*-Pr)$_4$ (1:1) with powdered 4 Å molecular sieves; second (method B), by heating a 2:1 mixture [BINOL:Ti(O*i*-Pr)$_4$] in the presence of catalytic amount of CF_3SO_3H or CF_3CO_2H.[164,165] Both catalysts were then employed successfully at 10 mol%.

R–CHO + SnBu3 → (R)-BINOL-Ti(O*i*-Pr)2, 10 mol%, Molecular sieves 4 Å / CH_2Cl_2 → R–CH(OH)–CH2CH=CH2

Method B :
R = Ph ; 98% yield; 92% e.e.
R = *c*-C_6H_{11}; 95% yield; 92% e.e.
R = *i*-Pr ; 97% yield; 87% e.e.
R = Ph-CH_2CH_2; 98% yield; 96% e.e.
R = (*E*)-PhCH=CH ; 78% yield; 77% e.e.

Method A :
R = Ph ; 88% yield; 95% e.e.
R = *c*-C_6H_{11}; 66% yield; 94% e.e.
R = *i*-Pr ; 89% yield; 96% e.e.
R = Ph-CH_2CH_2; 93% yield; 96% e.e.

An extension to the addition of methallyltri-*n*-butylstannane to aldehydes led to the observation of a positive nonlinear effect. In one case, the use of a (*R*)-BINOL, of only 50% e.e., gave the homoallylic alcohol with 88% e.e.[166]

R-CHO + $SnBu_3$ reagent; (R) catalyst: OH / OH + {$Ti(O\textit{i}\text{-}Pr)_4$ / CF_3COOH}

R = Ph ; 95% yield; 96% e.e.
R = c-C_6H_{11}; 65% yield; 75% e.e.
R = furyl ; 90% yield; 99% e.e.
R = Ph-CH_2-CH_2; 90% yield; 98% e.e.

10 mol%
Molecular sieves 4 Å

III. Intermolecular Addition to Chiral Aldehydes

III.1 Reaction with unfunctionalized chiral aldehydes and ketones

III.1.a Unfunctionalized α-chiral aldehydes

The two carbonyl faces of α-chiral aldehydes are diastereotopic; therefore the addition of allylsilanes or allylstannanes to such compounds can give rise to a mixture of diastereomers. When these aldehydes do not bear polar groups (Lewis bases), the qualitative diastereofacial selectivity of the addition can be predicted by Cram's rule,[167] refined by the Felkin-Anh model. The Felkin model[168] best agrees with predictions based on ab initio calculations.[169,170] This model lays particular stress on the importance of the so-called antiperiplanar effect: a transition state in which the incipient bond (between the nucleophile and the carbonyl) and the bond between the α-carbon atom and the group that provides the greater σ^*-π^* overlap with the carbonyl π^* orbital (i.e., the more electronegative group)(L = large, M = medium, S = small) are in antiperiplanar position will be favored. Cieplak proposed that transition states could be stabilized by hyperconjugative delocalization of an adjacent antiperiplanar σ-bond into the incipient σ*-bond. As a result, nucleophilic attack *anti* to the better electron donating σ-bond would be favored[171,172] (for a discussion between the Anh-Eisenstein model[169] and the Cieplak model, see the study from Paddon-Row[173]). The Felkin-Anh model can be applied to interpret the Lewis acid-induced additions of both allylsilanes and allylstannanes to α-chiral aldehydes.

L. A. + MR_3 → [Nu, L. A., M, L, S, H] → Cram addition

The non-predicted diastereomer results from what is commonly called the *anti*-Cram addition.

anti-Cram addition

The addition of allyltrimethylsilane to 2-methylbutanal is the simplest case one can think of (L = Et, M = Me, S = H). With BF_3-OEt_2, the addition takes place without selectivity, which is not really surprising because the stereo differentiation, provided by the competition between the methyl and ethyl groups, is not important. However, $TiCl_4$ induces a moderate *syn*-selectivity in accordance with the Felkin-Anh model.[174]

L. A., CH_2Cl_2

Major, *syn* + minor, *anti*

BF_3-OEt_2 ; 49% yield; *syn* : *anti* = 1.1 : 1
$TiCl_4$; 58% yield; *syn* : *anti* = 1.6 : 1

→ Major, *syn* | → minor, *anti*

Heathcock reported that the addition of allyltrimethylsilane and methallyltrimethylsilane to 2-phenylpropanal (hydratropaldehyde) — the standard substrate for testing diasteroselective reactions — affords *syn*- and *anti*-isomeric homoallylic alcohols in a 2:1 ratio which depends only slightly on the Lewis acid involved.[175] Selectivity is however better when using methallyl trimethylsilane and in this case, BF_3-OEt_2 was found to provide the best result (7:1 ratio in favor of the *syn* adduct).

L. A., CH_2Cl_2 — *syn* + *anti*

R = H, $TiCl_4$; 86% yield ; *syn* : *anti* = 1.6 : 1
R = H, BF_3-OEt_2 ; 47% yield ; *syn* : *anti* = 2 : 1
R = H, $SnCl_4$; 86% yield ; *syn* : *anti* = 2.2 : 1
R = Me, $TiCl_4$; 66% yield ; *syn* : *anti* = 2.8 : 1
R = Me, BF_3-OEt_2 ; 64% yield ; *syn* : *anti* = 7 : 1
R = Me, $SnCl_4$; 68% yield ; *syn* : *anti* = 3.2 : 1

Double diastereoselectivity occurs when adding crotyltri-*n*-butyltin to α-chiral aldehydes. Thus, Yamamoto established that, out of the four possible diastereomers, the *syn-syn* homoallylic alcohol is about all that can be obtained when crotyltri-*n*-butyltin is added to 2-phenylpropanal in the presence of BF_3-OEt_2; a remarkable selectivity compared to results obtained under high pressure conditions.[176]

syn-syn + *syn-anti* : *anti-syn* + *anti-anti* = 6.1 : 1
syn-syn + *anti-syn* : *syn-anti* + *anti-anti* = >99: 1

Hyperbaric reaction (10 Kbar) :
syn-syn + *syn-anti* : *anti-syn* + *anti-anti* = 2.3 : 1
syn-syn + *anti-syn* : *syn-anti* + *anti-anti* = 1 : 2.2

A combination of the Felkin-Anh model and of Yamamoto's antiperiplanar transition state can account for that *syn-syn* selectivity.

This *syn-syn*-selectivity was used by Koreeda to synthesize a steroid side-chain.[177]

R = H; (*Z*)-isomer
R = Me; (*E*)-isomer

R = H; 74.5% yield
R = Me; 35% yield

III.1.b Unfunctionalized chiral ketones

The addition of allyltri-*n*-butylstannane to 2-methylcyclohexanone leads to the equatorial adduct with a great selectivity.[178]

BF_3-OEt_2, CH_2Cl_2, 71% yield; 15.7 : 1

The fact that the equatorial attack is favored is illustrated by the addition of allyltri-*n*-butylstannane to 4-*tert*-butylcyclohexanone, in which the tertiobutyl substituent prevents any inversion of the cycle under the reaction conditions.

SnBu$_3$ BF$_3$-OEt$_2$ CH$_2$Cl$_2$ 93% yield OH OH 11.5 : 1

III.2 Reaction with functionalized chiral aldehydes and ketones

III.2.a Addition of allylsilanes or allylstannanes

A high degree of stereoselectivity can be realized under chelation control. Indeed, an oxygen atom of an ether function (or more generally a Lewis base), in α-, β- or possibly γ-position can serve as an anchor for the metal center of a Lewis acid. Since Cram's pioneering work on chelation-control in Grignard-type addition to chiral alkoxy compounds,[179,180] a number of reports on related matter have appeared,[181] and related transition state structures have been calculated.[182,183] Chelation control involves Cram's cyclic model and requires a Lewis acid bearing two coordination sites.

Thus, Reetz observed by ^{1}H NMR the chelation of 2-methoxy cyclohexanone with $TiCl_4$: the methoxy signal is shifted downfield by 0.65 ppm for a 1:1 mixture. However, ^{13}C NMR provides even further evidence of chelation. The carbonyl carbon atom, the α-carbon atom and the methoxy carbon atom are, all three, shifted downfield (by 13.7, 8.9 and 6.9 ppm respectively). As a result, the addition of allyltrimethylsilane to 2-methoxy cyclohexanone produces merely a single diastereomer resulting from an equatorial attack.[184]

O O Me SiMe$_3$ TiCl$_4$ CH$_2$Cl$_2$ 70% yield O TiCl$_4$ O Me

OH O Me >99 : <1 OH O Me

Comparatively, allyl Grignard reagents or allyltri(diethylamino)titanium reagents deliver only a product ratio of 3:1 (equatorial:axial).

Reetz also observed a very high 1,2 asymmetric induction when adding allyltrimethylsilane or methallyltrimethylsilane to 2-(benzyloxy)propanal. Addition takes predominantly place on the less hindered face of the chelate.[184]

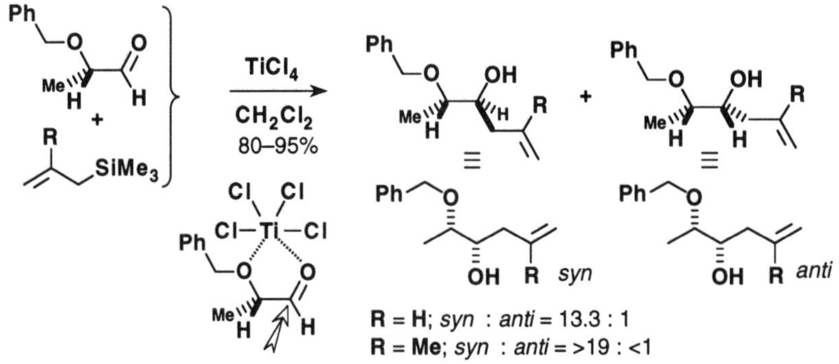

$TiCl_4$, CH_2Cl_2, 80–95%

R = H; *syn* : *anti* = 13.3 : 1
R = Me; *syn* : *anti* = >19 : <1

Heathcock reported that good to excellent *syn* selectivity can be achieved in the reactions between 2-(benzyloxy)propanal and allyltrimethylsilane or methallyltrimethylsilane, provided $SnCl_4$ is used as the Lewis acid.[185]

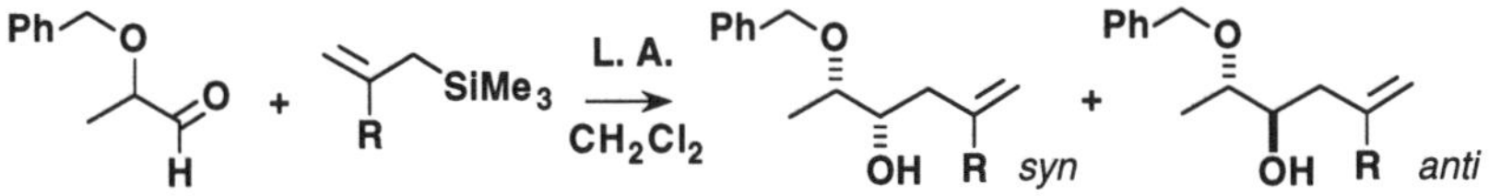

L. A., CH_2Cl_2

R = H, $SnCl_4$; 94% yield ; *syn* : *anti* = 35 : 1
R = H, BF_3-OEt_2 ; 50% yield ; *syn* : *anti* = 1 : 1.5
R = Me, $SnCl_4$; 81% yield ; *syn* : *anti* = 45 : 1
R = Me, BF_3-OEt_2 ; 40% yield ; *syn* : *anti* = 1 : 2.6

The same study was also carried out on 3-benzyloxy-2-methylpropanal and 3-(benzyloxy)butanal (see also Reetz's work[186]).

$SnCl_4$, CH_2Cl_2

R = H ; 92% yield ; *syn* : *anti* = 1 : 12
R = Me ; 83% yield ; *syn* : *anti* = 1 : 10

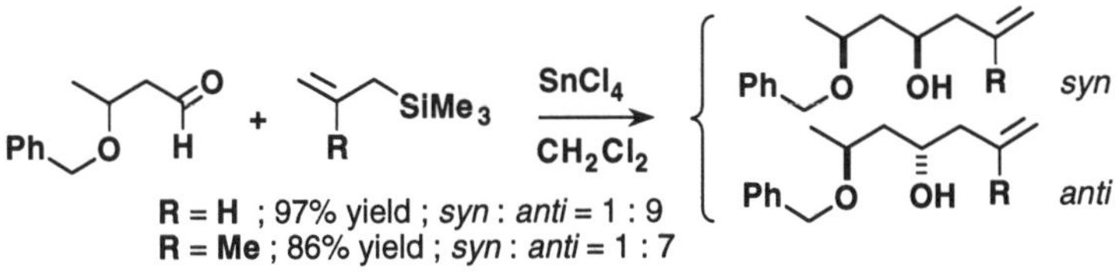

$SnCl_4$, CH_2Cl_2

R = H ; 97% yield ; *syn* : *anti* = 1 : 9
R = Me ; 86% yield ; *syn* : *anti* = 1 : 7

These results were rationalized as the consequence of a chelation of $SnCl_4$ by both oxygen atoms of the alkoxy and the carbonyl groups. The greater *syn* selectivity obtained from 2-(benzyloxy)propanal may result from a greater steric difference between the two diastereotopic faces of the carbonyl group in the five-membered chelate. Indeed, it is more rigid than the corresponding six-membered ring one.

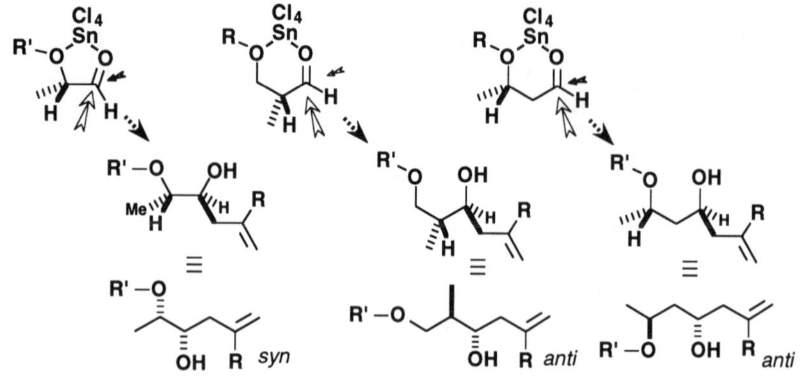

From these examples, the following stereochemical relationship can be proposed:

α-chelation: *syn* alcohol as major diastereomer
β-chelation: *anti* alcohol as major diastereomer

Reetz also studied the addition of allyltrimethylsilane to 3-(benzyloxy)butanal. If the products obtained from $TiCl_4$ or $SnCl_4$ are indeed the expected "chelation" products, those obtained from BF_3-OEt_2 or BF_3 (gas) come as a real surprise. As a matter of fact, although BF_3 has only one coordination site, and is in principle therefore incapable of chelation, the "chelation" product dominates as well.[187]

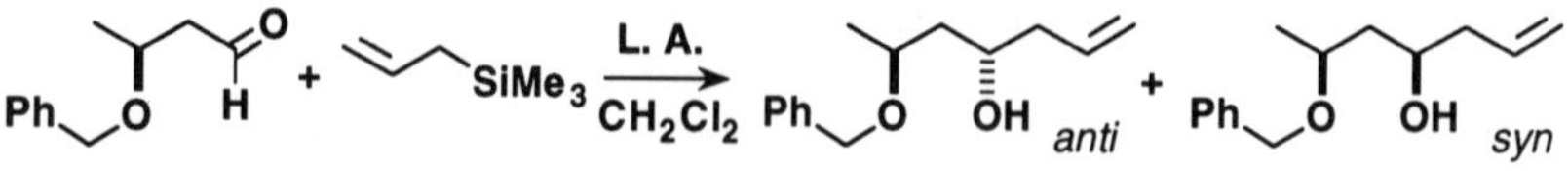

$TiCl_4$; >85% yield; *anti* : *syn* = 19 : 1
$SnCl_4$; >85% yield; *anti* : *syn* = 19 : 1
$AlCl_3$; >85% yield; *anti* : *syn* = 8.1 : 1
BF_3-OEt_2; >85% yield; *anti* : *syn* = 5.7 : 1
BF_3(gas); >85% yield; *anti* : *syn* = 10.1 : 1

It was proposed that normal complexation by BF_3 may induce a conformation in which the less hindered π-face of the carbonyl group is the same as in the chelate. Dipolar repulsion could be responsible for such a conformation.

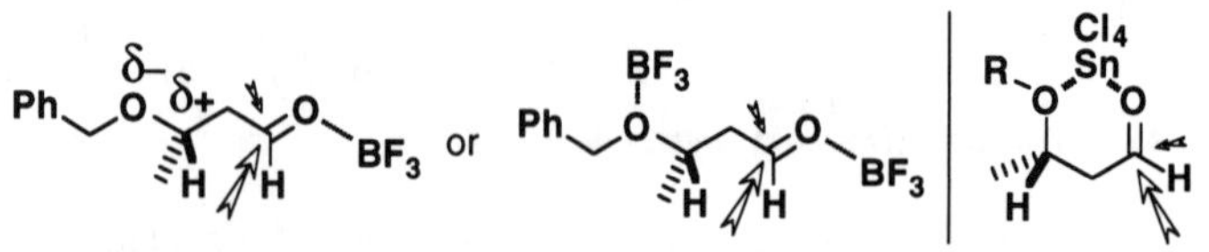

Yamamoto reported a new approach to control the acidity of strong Lewis acids. Thus, the Lewis acid-base combination $TiCl_4$-XPh_3 (X = As, Sb, Bi) generates, *in situ*, $TiCl_4$ which selectively coordinates electrophiles through its two sites just before the reaction takes place. On the other hand, the $TiCl_4$-PPh_3-mediated addition produces only very low diastereoselectivity because the phosphine complex does not liberate free $TiCl_4$ and consequently only one coordination site is available to the aldehyde.[188]

OBn, O, H + SnBu$_3$ —L. A., CH_2Cl_2→ OBn, OH *syn* + OBn, OH *anti*

$TiCl_4$; mode of add. : A; 76% yield; *syn* : *anti* = 1 : 0
$TiCl_4$-$AsPh_3$; mode of add. : A; 70% yield; *syn* : *anti* = 1 : 0
$TiCl_4$-$SbPh_3$; mode of add. : A; 76% yield; *syn* : *anti* = 19 : 1
$TiCl_4$-PPh_3; mode of add. : A; 70% yield; *syn* : *anti* = 1.1 : 1
$TiCl_4$; mode of add. : B; 76% yield; *syn* : *anti* = 1.1 : 1
$TiCl_4$-$AsPh_3$; mode of add. : B; 70% yield; *syn* : *anti* = 15.7 : 1
$TiCl_4$-$SbPh_3$; mode of add. : B; 73% yield; *syn* : *anti* = 4.6 : 1
Mode of addition A : normal addition mode
Mode of addition B : inverse addition mode

However, in many syntheses, hydroxy groups are also protected as silyl ethers. Thus, Keck evaluated the variations of diastereoselectivity in allyltri-*n*-butylstannane addition to α-(hydroxy protected) aldehydes vs. the nature of the protecting group, the Lewis acid and the solvent. Considering the nature of the Lewis acid, very high *syn*-selectivity was observed with $MgBr_2$ and $TiCl_4$ (both being capable of forming chelate structure) while a low *anti* selectivity, corresponding to the "Cram" selectivity, was observed with BF_3-OEt_2 (which only bears one coordination site).[189]

O, H, BnO + SnBu$_3$ —L. A., CH_2Cl_2→ OH, BnO *syn* + OH, BnO *anti*

BF_3-OEt_2 ; *syn* : *anti* = 1 : 1.6
$MgBr_2$; 85% yield; *syn* : *anti* = 250 : 1
$TiCl_4$; 75% yield; *syn* : *anti* = 250 : 1

The marked tendency of $MgBr_2$ and $TiCl_4$ to selectively catalyze the formation of *syn*-homoallylic alcohols from α-(benzyloxy)aldehydes is not observed from α-(*tert*-butyldimethylsiloxy)aldehydes; an excellent Cram selectivity is however observed with BF_3-OEt_2. According to Keck, this lower selectivity results from the lower basicity of silyl ethers, compared to benzyl ethers, due to oxygen-silicon interactions (see Chapter 1). Hence bidentate chelation of Lewis acids should be less effective and in addition, the electron withdrawal from the oxygen would result in a lowering of the C–O σ*. These combined phenomena should increase the stabilization of the Felkin-Anh antiperiplanar addition to the carbonyl group.

O, H, O, *t*-BuMe$_2$Si + SnBu$_3$ —BF_3-OEt_2, CH_2Cl_2, 83% yield→ OH, O, *t*-BuMe$_2$Si *syn* + OH, O, *t*-BuMe$_2$Si *anti*

1 : 10.1

Reetz observed that additions of allyltrimethylsilane or methallyl trimethylsilane to 2-*O*-benzyl-3-*O*-(*tert*-butyldimethylsilyl)-glyceraldehyde [2-(benzyloxy)-3-(*tert*-butyldimethylsilyloxy)propanal], induced by Lewis acids capable of bisligation ($TiCl_4$ or $SnCl_4$), afford almost exclusively *syn* adducts.[190]

$TiCl_4$; R = H ; >98% yield; *syn* : *anti* = >49 : <1
$SnCl_4$; R = H ; 94% yield; *syn* : *anti* = >49 : <1
BF_3-OEt_2; R = H ; 90% yield; *syn* : *anti* = 1 : 4.3
$SnCl_4$; R = Me ; >98% yield; *syn* : *anti* = >32.3 : <1
BF_3-OEt_2; R = Me ; 94% yield; *syn* : *anti* = 1 : 2.3

This result is clearly in accordance with those reported by Keck on the ability or not of benzyl and silyl ethers to chelate Lewis acids[189] (see also Chapter 1, Section III.2). With BF_3-OEt_2, the opposite diastereofacial selectivity is observed. The production of the *anti* products, a result of a non-chelation process, can be rationalized by the intervention of a Cornforth-type dipolar transition state.[191] This model which originally refers to chiral α-chloro ketones, can indeed be extended to α-alkoxy analogs.

Similarly, Guanti showed that the $MgBr_2$-catalyzed allylation of a series of diprotected asymmetrized bis(hydroxymethyl)acetaldehydes with allyltri-*n*-butylstannane proceeds with good selectivity. The stereochemistry of the adducts is in agreement with the formation of cyclic chelates in which only one ether group is involved, due to the nature of the protecting group.[192]

Reissig established that methyl β-chiral-β-formyl esters can form chelates thus making it possible to obtain "chelation" products by addition of allyltrimethylsilane. The initial products are γ-hydroxycarboxylates (*anti* and *syn* isomers) which undergo cyclization during work-up to afford 5-allyl-substituted γ-lactones (*trans*- and *cis*-isomers).[193,194]

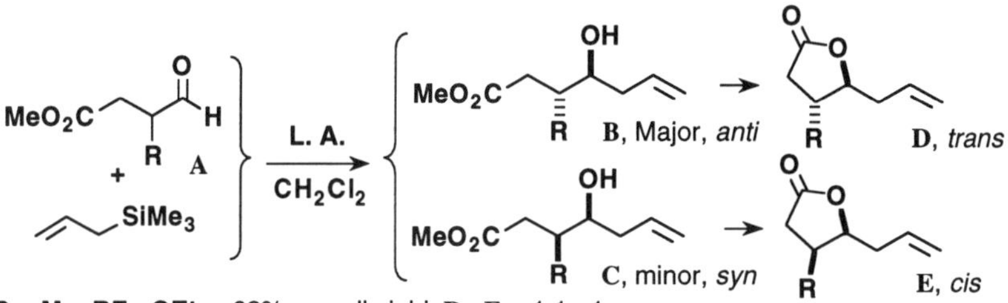

R = Me; BF_3-OEt_2 ; 62% overall yield; D : E = 1.1 : 1
R = Me; $TiCl_4$; 99% overall yield; D : E = 1.7 : 1
R = *i*-Pr; $TiCl_4$; 69% overall yield; D : E = 11.5 : 1
R = Ph; $TiCl_4$; 63% overall yield; D : E = 13.3 : 1

In this particular case, the formation of a chelate was supported by significant downfield shift of the ^{13}C NMR signals of both carbonyl carbon atoms.[194]

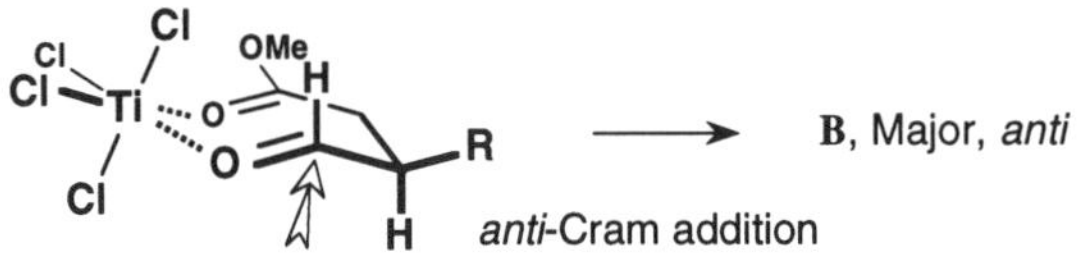

B, Major, *anti*

anti-Cram addition

1,3-Asymmetric induction was also observed when allyltrimethylsilane is added to α-chiral β-formyl esters, in the presence of $TiCl_4$.[194]

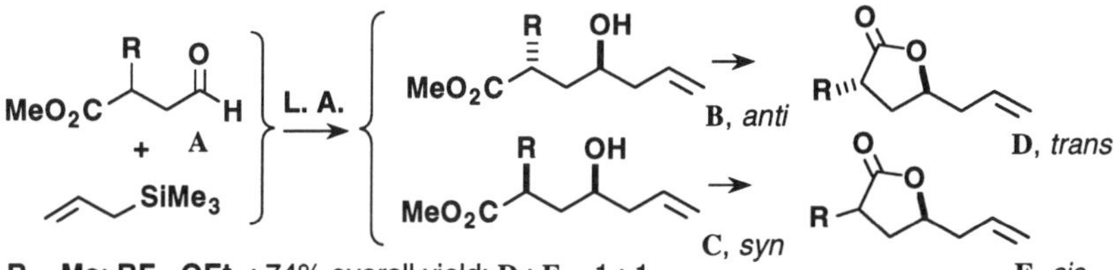

R = Me; BF_3-OEt_2 ; 74% overall yield; D : E = 1 : 1
R = Me; $TiCl_4$; 67% overall yield; D : E = 1 : 1
R = Ph; $TiCl_4$; 64% overall yield; D : E = 1 : 3.17

However, as can be seen from these results, the use of $TiCl_4$ to promote the chelate formation does not raise the stereoselectivity unless the R group is relatively large, as is the phenyl group.

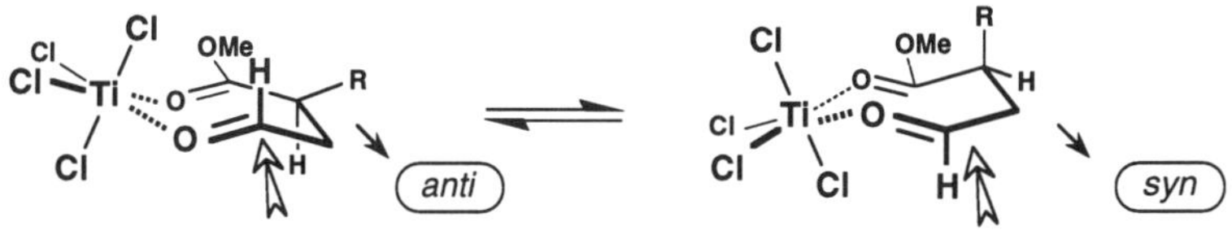

α-Chiral-β,β-dialkyl-β-formyl esters also provide, upon addition of allyltrimethylsilane, the corresponding tetrasubstituted γ-lactones with a surprisingly high excess of the *trans* isomer when $TiCl_4$ is used.

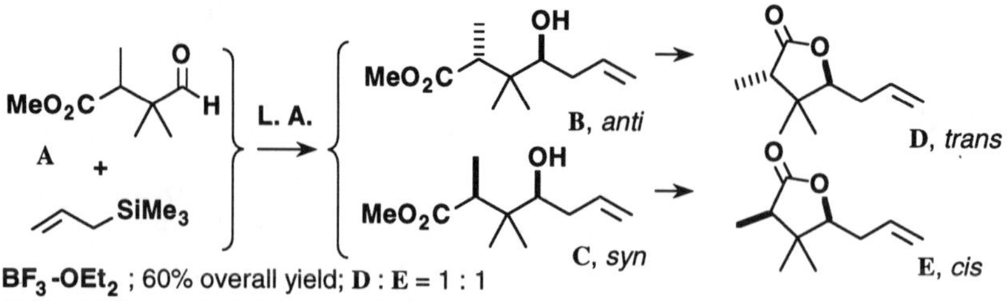

BF_3-OEt_2 ; 60% overall yield; **D** : **E** = 1 : 1
$TiCl_4$; 61% overall yield; **D** : **E** = 9 : 1

The $SnCl_4$-promoted addition of allyltrimethylsilane to α-aminoaldehydes is controlled by α-chelation and therefore predominantly leads to *syn*-alcohols.[195] These alcohols can subsequently be converted into various hydroxyethylene dipeptide isosteres.

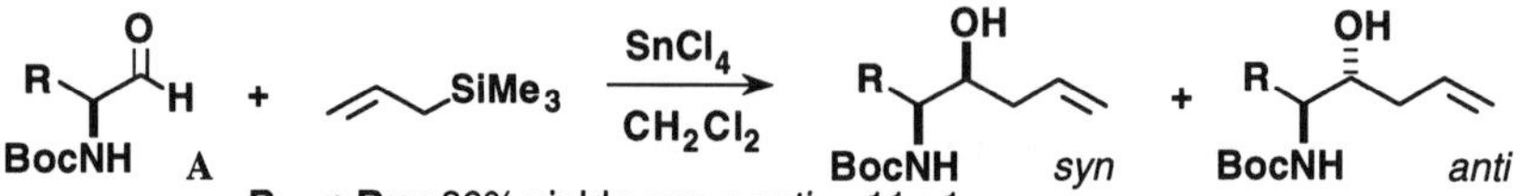

R = *n*-Bu ; 80% yield ; *syn* : *anti* = 11 : 1
R = Bn ; 65% yield ; *syn* : *anti* = 10 : 1
R = *i*-Bu ; 68% yield ; *syn* : *anti* = 20.6 : 1

In contrast, the addition of cyclopentenylmethyltrimethylsilane to Boc-leucinal occurred with only low diastereoselectivity.[196]

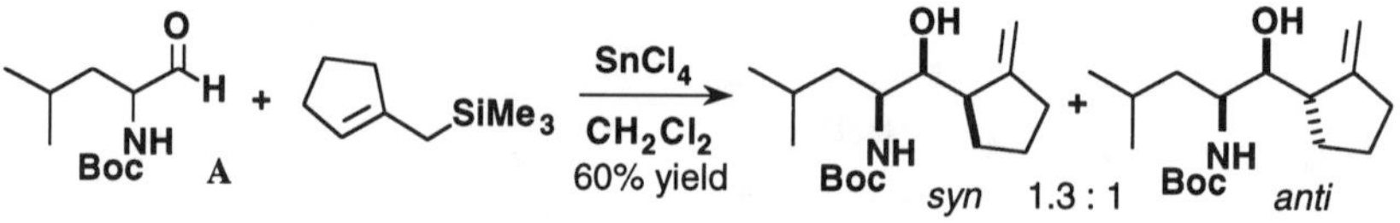

A high *syn* selectivity was again achieved by Kamimura when using $TiCl_4$ or $SnCl_4$.[197] Thus, in an important study on Lewis acid promoted stereoselective carbon-carbon bond formation of 3-formyl-Δ^2-isoxazolines, he studied the addition of allyltrimethylsilane and allyltri-*n*-butylstannane vs. the nature of the Lewis acid. $TiCl_4$ and $SnCl_4$ lead to the *syn* adduct through a postulated chelate while BF_3-OEt_2 yields the *anti* adduct through a nonchelated transition state. The reactivity of $AlCl_3$ is somehow more surprising.

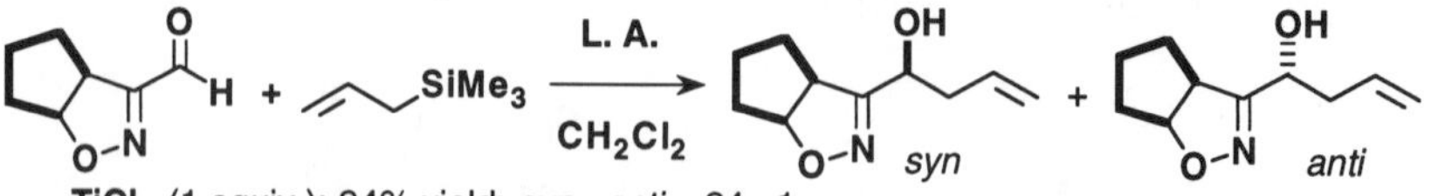

$TiCl_4$ (1 equiv.); 94% yield; *syn* : *anti* = 24 : 1
$SnCl_4$ (1 equiv.); 93% yield; *syn* : *anti* = 32.3 : 1
$AlCl_3$ (1 equiv.); 65% yield; *syn* : *anti* = 10.1 : 1
BF_3-OEt_2 (2 equiv.); 24% yield; *syn* : *anti* = 1 : 4

$TiCl_4$, CH_2Cl_2, 70% yield; *syn* 4.3 : 1 *anti*

Allylation of 3-oxo amides, in the presence of various Lewis acids, proceeds smoothly to give 3-hydroxy amides with high stereoselectivity.[198]

L. A., CH_2Cl_2

R = Ph ; $TiCl_4$; 76% yield ; **A** : **B** = >99 : <1
R = Ph ; $MeAlCl_2$; 80% yield ; **A** : **B** = 96 : 4
R = Me ; $MeAlCl_2$; 76% yield ; **A** : **B** = >99 : <1

Addition of allyltrimethylsilane to chiral α-keto-amides derived from (*S*)-proline esters gives rise to homoallylic alcohols of high diastereoisomeric excesses[199] (for studies, see refs.200, 201).

L. A., CH_2Cl_2

$TiCl_4$; R^1 = Me ; R^2 = Ph ; 65% yield; 75% d.e.
$SnCl_4$; R^1 = Bn ; R^2 = Ph ; 46% yield; 89% d.e.

Kiyooka noticed a dramatic change in diastereoselectivity depending upon the quantity of $TiCl_4$ used during the addition of allyltrimethylsilane to chiral α-*N*-(carbobenzyloxy)amino aldehydes.[202] Thus, although a high *anti* stereoselectivity was observed when adding allylsilane to *N*-Cbz-L-serinal acetonide in the presence of 1 equivalent of $TiCl_4$, the magnitude of this selectivity decreased with the concentration of Lewis acid. Even more surprising, an inversion of diastereoselectivity was observed for concentrations below 0.5 equivalent. This might be explained by assuming the possible formation of two types of Lewis acid-aldehyde complex, 1:1 and 2:1 complexes. Thus, the *syn* selectivity found with 0.5 equivalent of $TiCl_4$ was thought to arise from a reaction proceeding through a 2:1 complex, whereas the *anti* selectivity observed with 1.0 equivalent is consistent with β-chelation, which, in such a case, is more effective than α-chelation.

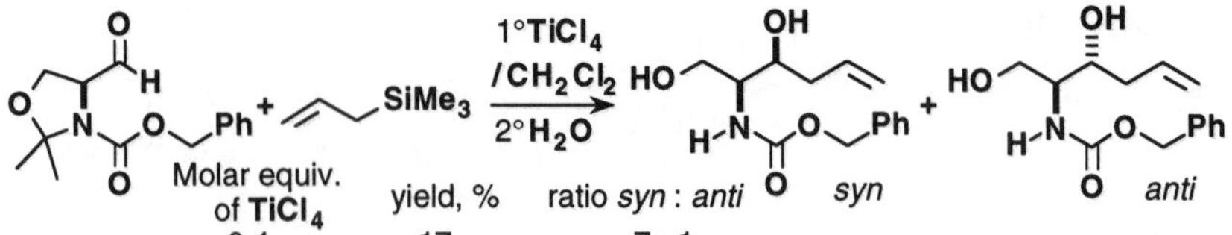

Molar equiv. of $TiCl_4$	yield, %	ratio *syn* : *anti*
0.4	17	7 : 1
0.5	47	8 : 1
0.7	56	1 : 2
0.8	72	1 : 9
1.0	84	1 : 20
1.2	80	1 : 3
1.5	78	1 : 1
2.0	69	1 : 1

The presence of a 2:1 complex seems to be confirmed by the high yield of the reaction between allylsilane and 2-phenylpropanal (80%) when only 0.5 equivalent of $TiCl_4$ is used. However, the reaction failed with 0.4 equivalent of $TiCl_4$.

In the case of *N*-Cbz-prolinal, it is worth noting that high *syn* selectivity (28:1) was obtained when the reaction was performed with 0.5 equivalent of $TiCl_4$, but that no selectivity at all (1:1) could be observed when using 1.0 equivalent. This *syn* selectivity apparently arises again from a transition state involving a 2:1 complex of the aldehyde and $TiCl_4$.

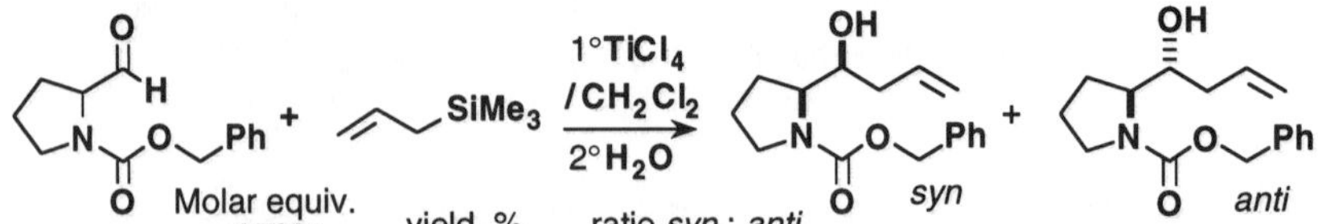

Molar equiv. of $TiCl_4$	yield, %	ratio *syn* : *anti*
0.5	63	28 : 1
0.6	70	26 : 1
0.7	73	6.5 : 1
0.8	80	1 : 1
1.0	89	1 : 1
1.5	91	1 : 1

In contrast, the reaction between acyclic *N*-Cbz-valinal and allyltrimethylsilane promoted by 1.0 equivalent of $TiCl_4$ leads to the corresponding homoallylic alcohol with high *syn* selectivity, which can result from a chelation control via a 1:1 complex.

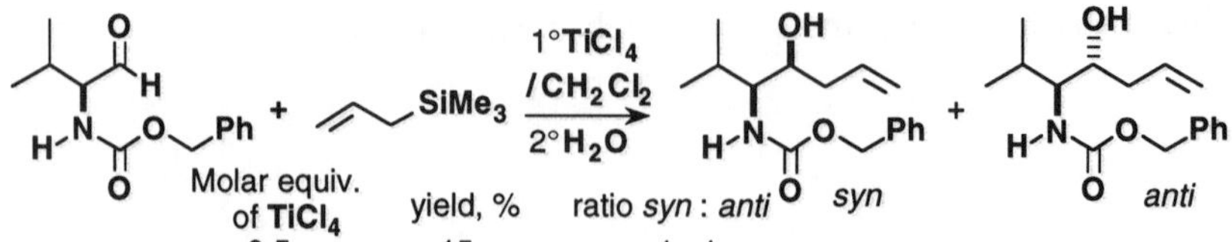

Molar equiv. of $TiCl_4$	yield, %	ratio *syn* : *anti*
0.5	15	1 : 1
0.6	46	1.5 : 1
0.7	62	3 : 1
0.8	78	12 : 1
1.0	82	14 : 1
1.2	79	15 : 1
1.5	65	13 : 1

Sugar derivatives were also used in a number of occasions as chiral aldehydes. Thus, the $TiCl_4$-mediated addition of prenyltri-*n*-butylstannane to dimethylene-L-xylose occurs with high stereoselectivity.[203]

$TiCl_4$, CH_2Cl_2, 65% yield; > 95% d.e.

The stereospecific allylation of aldose derivatives can also be performed with allyltrimethylsilane. In the following examples, due to Danishefsky, a complete reversal of selectivity occurs when changing the nature of the Lewis acid.[204]

L. A., CH_2Cl_2

BF_3 - OEt_2 ; 80% yield; (*R*) : (*S*) = 90 : 1
$TiCl_4$; 89% yield; (*R*) : (*S*) = 1 : 20

L. A., CH_2Cl_2

BF_3 - OEt_2 ; 78% yield; (*R*) : (*S*) = 20 : 1
$TiCl_4$; 64% yield; (*R*) : (*S*) = 1 : 10

In the course of the BF_3-mediated addition of (*E*)-cinnamyltrimethylsilane to arabinose derivatives, Sugimura observed the formation of a tetrahydrofuran derivative (82% yield) resulting from the cyclization of cationic intermediate **B** with migration of the oxygen atom of the acetonide.[205]

A + Ph-CH=CH-CH$_2$-$SiMe_3$ → (BF_3-OEt_2, CH_2Cl_2, 82% yield) → [**B**] →

Stereoselective allylation of the aldehyde function of L-xylose derivatives gives rise to homoallylic alcohols in variable proportion depending on the

Lewis acid involved.[206] These results brought Koomen to postulate the intervention of a chelated transition state when using $TiCl_4$.

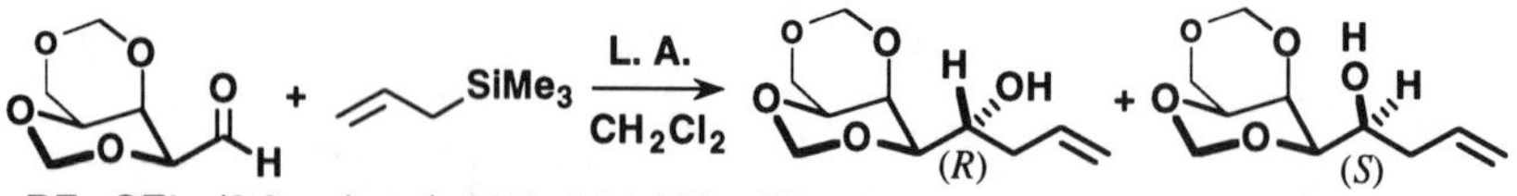

BF_3-OEt_2 (3.3 mol. eq.); 69% yield; (*R*) : (*S*) = 6.7 : 1
$TiCl_4$ (1.1 mol. eq.); 72% yield; (*R*) : (*S*) = 1 : 15.7

When using benzylidene acetal derivatives, only one diastereomer was obtained, regardless of the Lewis acid involved. At low temperature, the trimethylsilyl ether (**B**, R = $SiMe_3$) was even isolated in some cases. Moreover, tricyclic by-product **C** was isolated, in significant yield, as a single diastereomer when using BF_3-OEt_2.

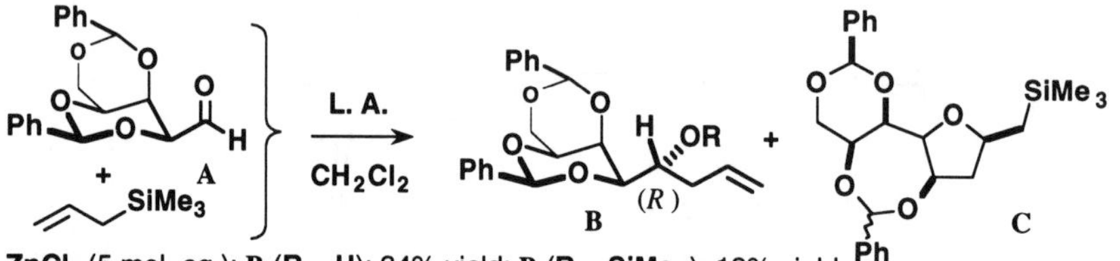

$ZnCl_2$ (5 mol. eq.); **B** (R = H): 34% yield; **B** (R = $SiMe_3$): 19% yield
BF_3-OEt_2 (3.3 mol. eq.); **B** (R = H): 40% yield; **B** (R = $SiMe_3$): 12% yield; **C**: 21% yield

The formation of the tetrahydrofuran derivative **C** can be explained by the attack of one of the acetal oxygen atoms on the carbenium ion, stabilized by the silyl group in the β position, followed by neutralization of the resulting intermediate oxonium ion by the negatively charged oxygen atom.

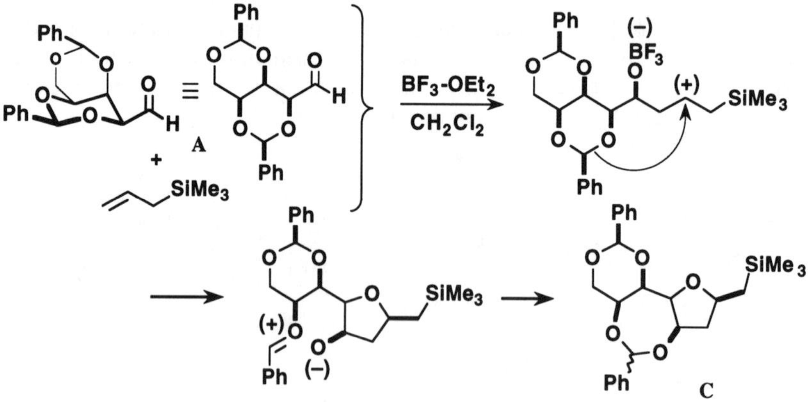

Surprisingly, tetrahydrofuran derivative **C** could be converted to homoallyl alcohol **B** upon treatment with BF_3-OEt_2 at –20°C.

Ph O O SiMe3 O O O C Ph $\xrightarrow[\text{CH}_2\text{Cl}_2,\ 34\%\ \text{yield}]{\text{BF}_3\text{-OEt}_2}$ Ph O O H OR O Ph O (*R*) B

Sugars have also been used as removable chiral auxiliaries. Thus, Kunz reported that (*S*)-configured homoallylamines, which are valuable synthons for further transformations, can be synthesized diastereoselectively by $SnCl_4$ induced addition of allylsilanes to Schiff bases of galactopyranosylamine derivatives.[207]

PivO OPiv O PivO N X OPiv H + SiMe3 $\xrightarrow[\text{CH}_2\text{Cl}_2,\ 37\text{–}82\%\ \text{yield}]{\text{SnCl}_4}$

PivO OPiv Cl Cl Cl–Sn–Cl O PivO O N X O H X → PivO OPiv O H N PivO X OPiv H

d.e. = 7 : 1 to 27 : 1

During his synthesis of (±)-kumausallene, Overman added an allylsilane derivative to a sensitive polyoxygenated bicyclic aldehyde. The $TiCl_4$ mediated addition afforded the corresponding alcohol, as a single *anti* isomer, with 91% yield. The use of BF_3-OEt_2 provided the same alcohol in lower yield but also with high stereoselectivity.[208]

O H H O O MeO2C H A + SiMe3 $\xrightarrow[\text{CH}_2\text{Cl}_2]{\text{TiCl}_4}$ OH H O O MeO2C H B 82% d.e.

As a matter of fact, crystallographic analysis indicated that simple Cram stereoselection is operative, even in the presence of bidentate Lewis acid. Therefore, it was proposed that, presumably, because of the aldehyde disposition on the concave face of the bicyclic compound, the chelation of the tetrahydrofuran oxygen atom to the titanium center of the Lewis acid is not possible on steric grounds (structure **C**, nonchelation model; structure **D**, chelation model).

TiCl4 O H H O O MeO2C H C H H O O O–TiCl4 MeO2C H D

The addition of allyltindifluoroiodide, formed *in situ* upon oxydative addition of stannous fluoride to allyliodide, to 1,2-*O*-isopropylidene-D-glyceraldehyde proceeds with *anti* selectivity (intermediates tin ethers were in fact quenched with phenoxyacetylchloride to yield the corresponding esters).[209]

SnF_2
1,2-Dimethyl-2-imidazolidinone / DMF
$SnIF_2$

74% yield
anti 4.3 : 1 syn

Anti selectivity was also observed by Mukaiyama when adding the same allyltindifluoroiodide[210] or diallyltindibromide[211] to 4-*O*-benzyl-2,3-*O*-isopropylidene-L-threose.

THF
100% yield
anti 9 : 1 syn

As part of his efforts towards the synthesis of rifamycins, Kishi studied several allylation reactions. After many unsuccessful attempts, the tin reagent, prepared according to Mukaiyama's procedure from allyliodide and $SnCl_2$, was found to react smoothly with the precursor aldehyde to yield the desired alcohol with high *syn*-stereoselectivity (20:1). Such a selectivity was rationalized by the intervention of a bicyclic-like (*trans*-decalin-like) transition state involving the complexation of the organometallic reagent with two oxygen atoms.[212]

$SnCl_2$
1,2-Dimethyl-2-imidazolidinone / benzene
$SnICl_2$

benzene
71% yield
syn : *anti* = 20 : 1

Finally, Ohno showed that, under $TiCl_4$ catalysis, allyltrimethylsilane adds to α-chiral acylsilanes to yield predominantly, after *n*-Bu_4NF induced protolysis, the corresponding *syn* homoallylic alcohols.[213]

III.2.b Addition of crotylsilanes or crotylstannanes

Asymmetric addition of crotylsilane or crotylstannane derivatives to chiral aldehydes affords diastereomeric mixtures of homoallylic alcohols resulting from a double diastereoselection.[214,215]

Keck studied the chelation controlled diastereofacial selectivity of crotyl-tri-*n*-butylstannane additions to chiral α-alkoxyaldehydes. Low diastereoselectivity was obtained with monodentate BF_3-OEt_2 but bidentate Lewis acids, such as $TiCl_4$ or $MgBr_2$, proved to be very efficient in yielding products consistent with a "chelation control" addition of crotyltri-*n*-butyl stannane.[216]

BF_3-OEt_2 ; R = cyclohexyl ;
syn-syn : *syn-anti* : *anti-syn* : *anti-anti* = 60 : 1 : 23.8 : 6; chelation : Felkin = 2 : 1
$TiCl_4$; R = cyclohexyl ;
syn-syn : *syn-anti* : *anti-syn* : *anti-anti* = 1.7 : 1 : 0 : 0; chelation : Felkin = >200 : 0
$MgBr_2$; R = cyclohexyl ;
syn-syn : *syn-anti* : *anti-syn* : *anti-anti* = 12.3 : 1 : 0 : 0; chelation : Felkin = >200 : 0

As a result of the chelation of both oxygen atoms (of the carbonyl and benzyloxy groups), the *Si* face of the carbonyl group is made clear and therefore the addition leads to the *syn-syn* diastereomer.

In the same area, Mikami and Nakai, in an important communication, studied the stereochemistry of the addition of crotylsilane and stannane derivatives to 2-(benzyloxy)propanal, as a function of the allylic moiety substitution. They first established, confirming Keck's results, that the addition of (*E*)- and (*Z*)-crotyltri-*n*-butylstannane **A** provides predominantly, under chelation conditions ($MgBr_2$), *syn*-homoallylic alcohol **B**.[217]

A, *E* : *Z* = 1.3 : 1 ; **B** : **C** = 13.3 : 1
A, *E* : *Z* = 9 : 1 ; **B** : **C** = 9 :1

The antiperiplanar transition state, proposed by Yamamoto, accounts in a satisfactory way for these results. From (*Z*)-crotylstannane, transition state **B1** leading to the *syn* adduct, is more favorable than transition state **C2** and from (*E*)-crotylstannane, transition state **B2**, leading also to the *syn* adduct, can again be regarded as more favorable than transition state **C1**.

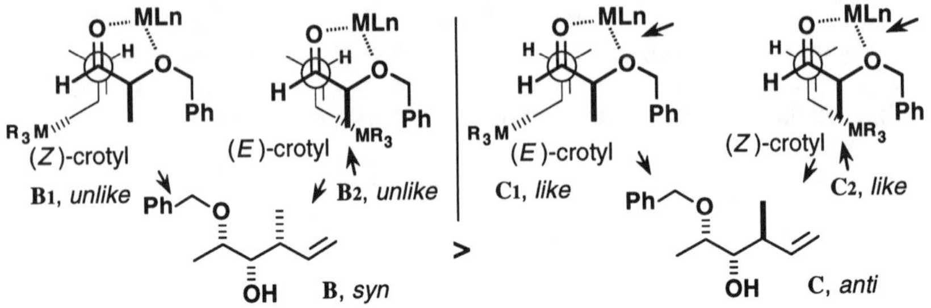

The rationalization of these results through the intervention of a synclinal transition state seems less obvious however. Thus, transition states **B3** and **B4**, that would lead to the *syn* adduct, do not look less hindered than transition states **C3** and **C4**.

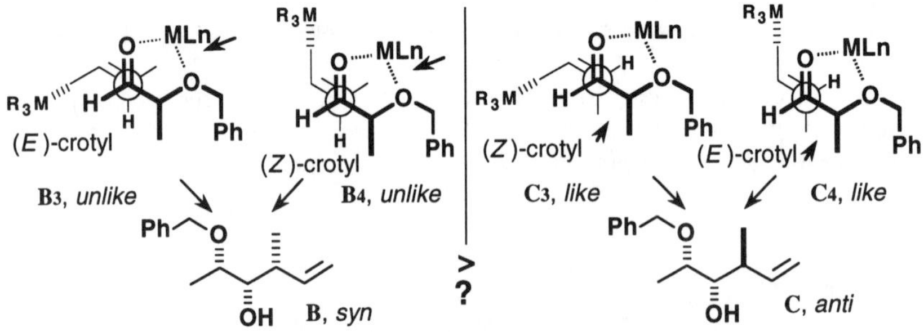

In sharp contrast with their first results, Nakai and Mikami then observed that the addition of 2-methylcrotyltri-*n*-butylstannane to the same (benzyloxy)aldehyde affords predominantly the *anti*-adduct, even under chelation conditions ($MgBr_2$).

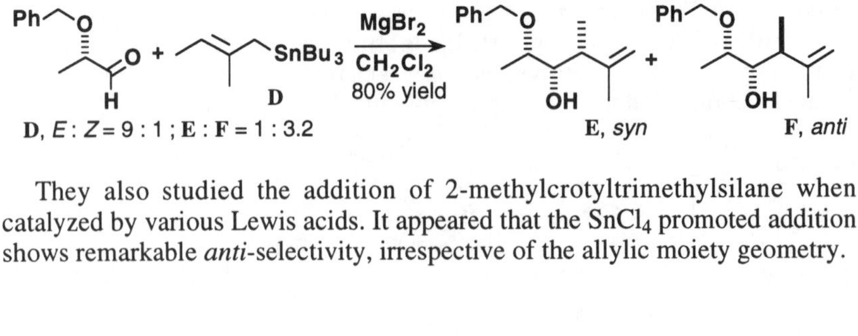

They also studied the addition of 2-methylcrotyltrimethylsilane when catalyzed by various Lewis acids. It appeared that the $SnCl_4$ promoted addition shows remarkable *anti*-selectivity, irrespective of the allylic moiety geometry.

$TiCl_4$; **G**, *E* : *Z* = 9 : 1 ; 80% yield; **E** : **F** = 1 : 1.9
$ZnBr_2$; **G**, *E* : *Z* = 9 : 1 ; 70% yield; **E** : **F** = 1 : 2
$MgBr_2$; **G**, *E* : *Z* = 9 : 1 ; 80% yield; **E** : **F** = 1 : 3.2
$SnCl_4$; **G**, *E* : *Z* = 9 : 1 ; 90% yield; **E** : **F** = 1 : 32.3

$SnCl_4$; **H**, *E* : *Z* = <1 : >19 ; 85% yield; **E** : **F** : **I** : **J** = 1 : 15.7 : 0 : 0
BF_3-OEt_2; **H**, *E* : *Z* = <1 : >19; 90% yield; **E** : **F** : **I** : **J** = 8 : 1 : 88 : 3
(2-Dimethyl-*t*-butylsiloxypropanal was used instead of 2-benzyloxypropanal)
BF_3- OEt_2 (2 equiv.) ; **H**, *E* : *Z* = <1 : >19 ; 90% yield; **E** : **F** : **I** : **J** = 3.3 : 0 : 8.3 : 1

With BF_3-OEt_2 (2 equivalents) which is a monodentate Lewis acid and cannot therefore chelate the oxygen atom of the benzyloxy group, a high yield of *syn*-selectivity is observed. This *syn*-selectivity even increases when the benzyloxy group is exchanged for a *tert*-butyldimethyl silyloxy group.

Mikami and Nakai came to the conclusion that an antiperiplanar transition state cannot account for this *anti*-selectivity (see the commentary of Fleming).[218] The presence of the β-methyl group induces new steric interactions that destabilize these transition states. Thus, regardless of the antiperiplanar transition state (**E1**, **E2**, **F1** or **F2**) there is always a steric repulsion between the chelate ring and either the β-methyl group or the $-CH_2MR_3$ group of the crotyl moiety.

The authors therefore suggested that, in that particular case, the reaction is more likely to proceed through a synclinal transition state. Hence, as can be seen from the following scheme, transition states **F3** and **F4** which both lead, from the (*Z*) and (*E*) crotyl derivative respectively, to the *syn* adduct can be regarded as more favorable than transition states **E3** and **E4.**

(*E*)-2-methylcrotyl
E3, *unlike*

(*Z*)-2-methylcrotyl
E4, *unlike*

(*Z*)-2-methylcrotyl
F3, *like*

(*E*)-2-methylcrotyl
F4, *like*

E, *syn* < F, *anti*

Antiperiplanar transition states, however, still account better for the *syn*-selectivity observed when using BF_3-OEt_2 .

E5, *unlike* → E, *syn*

E6, *like* → F, *anti*

E7, *unlike* → I, *syn*

E8, *like* → J, *anti*

The *syn*-selectivity exhibited by crotyltri-*n*-butylstannane, in the presence of BF_3-OEt_2, was applied by Yamamoto to the stereoselective synthesis of the Prelog-Djerassi lactonic acid.[219] Thus, the BF_3-OEt_2 (1 equivalent) promoted addition of crotyl tin to *meso*-4-carbomethoxy-2-methylpentanal **A** (*meso*-dimethylglutaric hemialdehyde) gives the desired *syn-anti* lactone **B** with at least 94% stereoselectivity. Greater amounts (2 or 3 equivalents) of the Lewis acid induce a loss of stereoselectivity. This result is quite surprising since it implies a chelation controlled condensation (*anti*-Cram's rule).

A + R-substituted crotyl-$SnBu_3$ → (BF_3-OEt_2, CH_2Cl_2) → **B** *syn-anti* (chelation) + **C** *anti-anti* (chelation) + **D** *anti-syn* (Felkin-Anh) + **E** *syn-syn* (Felkin-Anh)

R = Me ; 150°C, 24 h ; **B : C : D : E** = 2 : 1.1 : 1 : 0
R = H ; **BF_3 - OEt_2** ; 98% yield ; **B : C : D : E** = 94-97 : 3-4 : 1 : 1
R = H ; **BF_3 - OEt_2**; (2 equiv.); 90% yield ; **B : C : D : E** = 83-91 : 5-9 : 1-3 : 2-5
R = H ; **BF_3 - OEt_2**; (3 equiv.); 90% yield ; **B : C : D : E** = 4.1 : 1 : 1.7 : 3.2
R = Me ; **BF_3 - OEt_2**; **B : C : D : E** = 11.8 : 1 : 0 : 1.4

By contrast, (±)-4-carbomethoxy-2-methylpentanal **F**, under the same conditions, produces the *syn-syn* diastereomer **G** as the major product (Felkin-Anh). However, in this case the selectivity does not depend on the amount of BF_3-OEt_2, and is even increased when $TiCl_4$ is used. This strongly suggests a nonchelated transition state.[220]

F + crotyl-$SnBu_3$ → (L. A., CH_2Cl_2) → **G** *syn-syn* (Felkin-Anh) + **H** *syn-anti* (chelation)

BF_3 - OEt_2; 97% yield; **G : H** = 3.8 : 1
$TiCl_4$; 93% yield; **G : H** = 11.5 : 1

The selectivity observed from *meso*-diastereomer **A** would normally require the intermediacy of a chelated transition state. This is however unlikely since only one coordination site is available on BF_3. So Yamamoto undertook a ^{13}C NMR study of *meso*-diastereomer **A** in the presence of the Lewis acid in order to clarify its conformation.[221] But because BF_3-OEt_2 induces the trimerization of **A**, a phenomenon which made spectral analysis difficult, the authors had to carry on with their study using $SnCl_4$ which leads also to *syn-anti* isomer **B** with high diastereoselectivity (95:5). In order to appreciate the chemical shifts induced by the Lewis acid on **A**, isobutyraldehyde and methyl isobutyrate were used as reference compounds. It appeared that $SnCl_4$, in the presence of isobutyraldehyde (1 equivalent) and methyl isobutyrate, coordinates selectively the aldehyde. The ^{13}C NMR chemical shifts given are the values observed in the presence of $SnCl_4$; values in parentheses indicate chemical shift differences between experiments with or without $SnCl_4$.

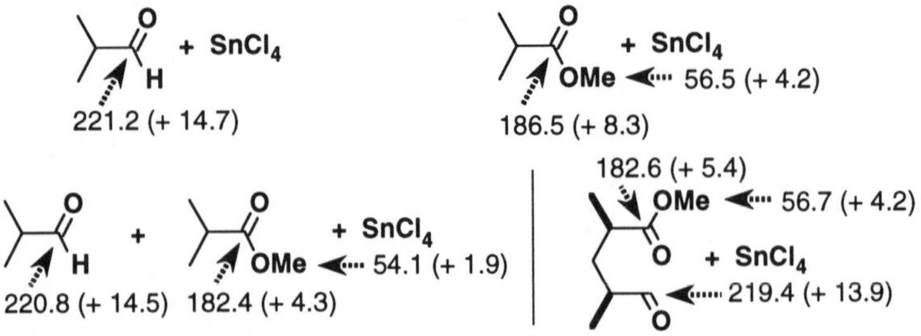

By contrast, the chemical shift differences observed on **A** with or without $SnCl_4$ clearly revealed the formation of a complex with both the ester and the aldehyde chelated to $SnCl_4$. Yamamoto concluded that the *meso*-diastereomer **A** adopts the same stable conformation regardless of the presence or absence of a Lewis acid or of its nature. **A** would then be a conformationally rigid acyclic molecule.

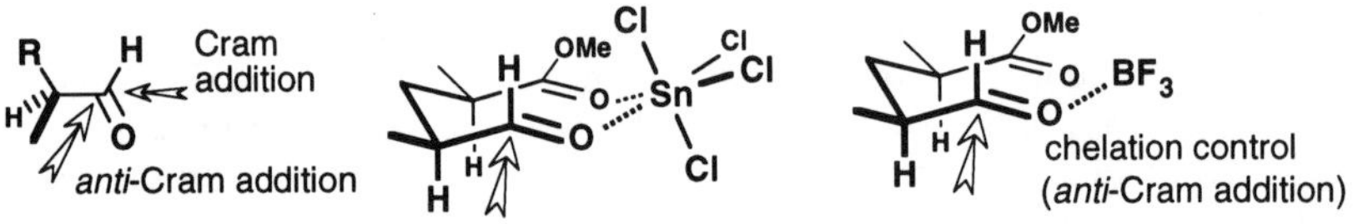

$TiCl_4$ probably also leads to a chelate. Its use by Santelli-Rouvier induced the stereoselective formation of the (±)-Prelog-Djerassi lactonic acid from *meso*-diastereomer **A** and pentenyltrimethylsilane.[222]

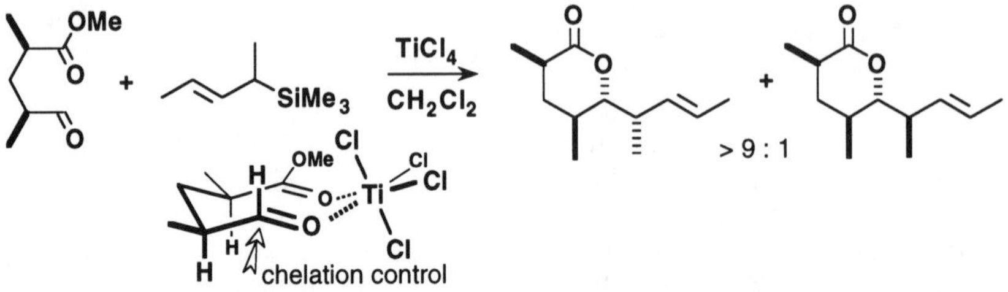

Yamamoto showed that the BF_3-OEt_2, promoted addition of (*E*)-crotyltri-*n*-butyltin to the carbonyl group of the 8-phenylmenthyl glyoxylate predominantly occurs on the *Si* face. The major product is the (*S*,*R*)-*syn*-enantiomer which is obtained with high levels of enantio- and diastereoselectivity.[223] The $TiCl_4$-mediated addition of allyltrimethylsilane to the 8-phenylmenthyl pyruvate similarly occurs on the *Si* face.[224]

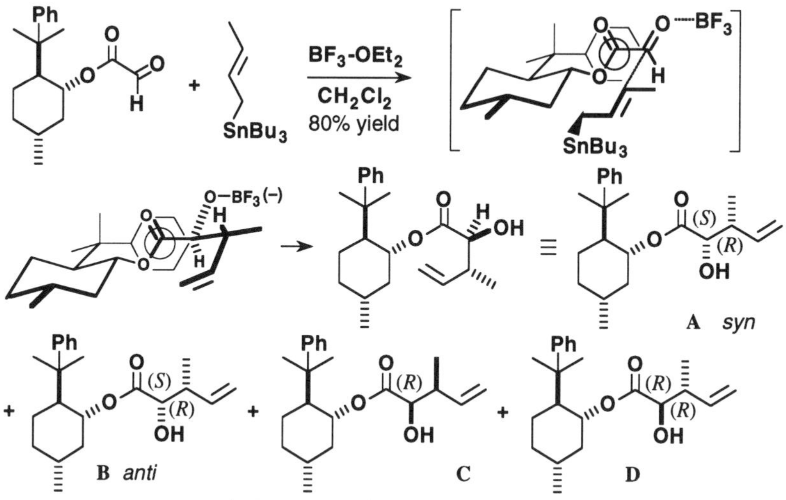

A : **B** : **C** + **D** = 12 : 1.3 : 1

The major diastereomer was then transformed, in two steps, into (–)-verrucarinolactone. A similar approach to the synthesis of (–)-verrucarinolactone was also proposed by Tamm[225] (see also ref.162).

As for Whitesell, he showed that the $SnCl_4$-catalyzed addition of (*Z*)-crotyltrimethylsilane to the same glyoxylic ester is also *syn*-stereoselective and leads to the same enantiomer however with a better selectivity (15:1) (quantitative yield). This outcome is best rationalized by assuming a stereochemical control of the addition stage through the following chelated transition state.[226]

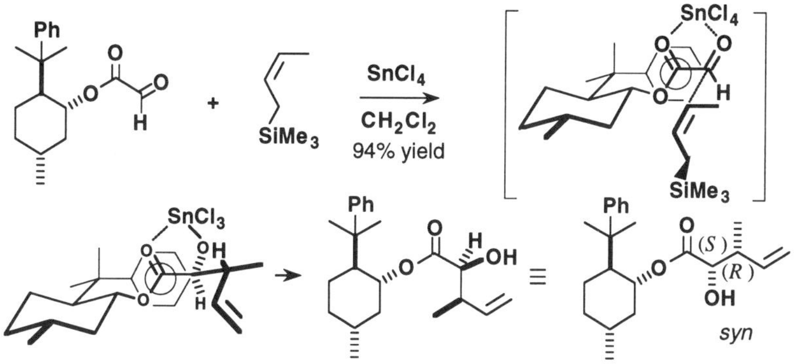

III.2.c Addition of allenylsilanes or allenylstannanes

In a study directed towards the preparation of a key intermediate in the synthesis of leucotriene B_4, Grée showed that allenylsilane **A** reacted chemo- and stereoselectively with (–)-iron tricarbonyl complex **B** in the presence of $TiCl_4$ to yield the complexed homopropargylic alcohol **C** with over 98% d.e.[227]

MeO, O, **B**, Fe $(CO)_3$, O, H + **A** C_5H_{11}, $SiMe_3$ → $TiCl_4$, CH_2Cl_2, 65% yield → MeO, O, Fe $(CO)_3$, OH, C_5H_{11} **C**

Marshall observed highly diastereoselective S_E' additions of enantio-enriched allenylstannanes to (*S*)-2-(benzyloxy)propanal. His results clearly show that diastereoselectivity depends upon chirality matching between the two reacting partners.[228]

OAc, Bu_3Sn (*S*) H + H, O (*S*) OBn → L. A., CH_2Cl_2 → (*R*) OBn OH OAc **A** *syn-syn* + (*R*) OBn OH OAc **B** *anti-anti*

BF_3-OEt_2 ; 95% yield; **A** : **B** = 2.1 : 1
$MgBr_2$; 97% yield; **A** : **B** = 100 : 1

The *syn-syn* diastereoselectivity observed when adding the (*S*)-allenyl-stannane to the (*S*)-aldehyde involves the following antiperiplanar transition states. In the BF_3-promoted addition the *S/S* pairing is mismatched and that explains the relatively low selectivity whereas, with the $MgBr_2$-chelated aldehyde, the pairing is matched since the attack can occur *anti* to the methyl substituent of the aldehyde.

[R, $SnBu_3$, (*S*), H, OBn, (*S*), H, O, BF_3] → **A** | [R, $SnBu_3$, (*S*), H, (*S*), H, OBn, O, Mg, Br, Br] → **A**

When the (*R*)-allenylstannane is added to the (*S*)-aldehyde, in the presence of BF_3-OEt_2, the pairing is matched and diastereoselectivity remarkably increased, in favor of the *anti-syn* adduct. With $MgBr_2$, diastereoselectivity stays very high but is reversed in favor of the *syn-anti* adduct.

OAc, Bu_3Sn (*R*) H + H, O (*S*) OBn → L. A., CH_2Cl_2 → (*S*) OBn OH OAc **C** *anti-syn* + (*S*) OBn OH OAc **D** *syn-anti*

BF_3- OEt_2; 97% yield; **C** : **D** = 30 : 1
$MgBr_2$; 97% yield; **C** : **D** = 1 : 92

In the BF_3-promoted reaction, the matched *S*/*R* alignment is favored by steric as well as stereoelectronic factors. However, with $MgBr_2$, a synclinal approach seems the most probable to explain the selective formation of homopropargylic alcohol **D**.

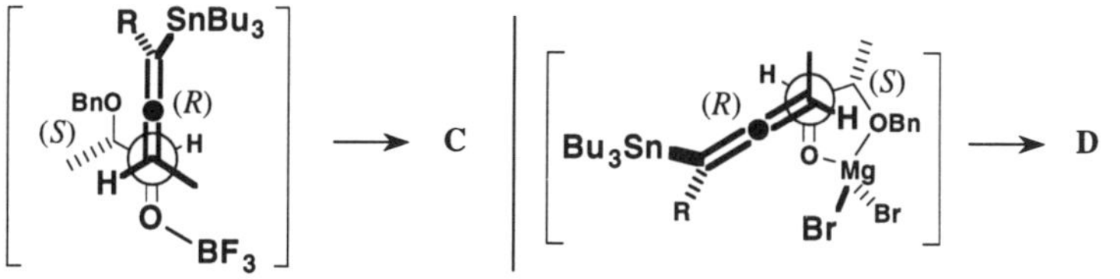

Marshall also added chiral allenylstannanes to (*R*)-3-(benzyloxy)-2-methylpropanal, in the presence of $MgBr_2$ or BF_3-OEt_2.[229] The $MgBr_2$-promoted addition of (*S*)-allenylstannanes affords the *syn-anti* adducts with a remarkably high diastereoselectivity; the following antiperiplanar chelated transition state can account for this selectivity.

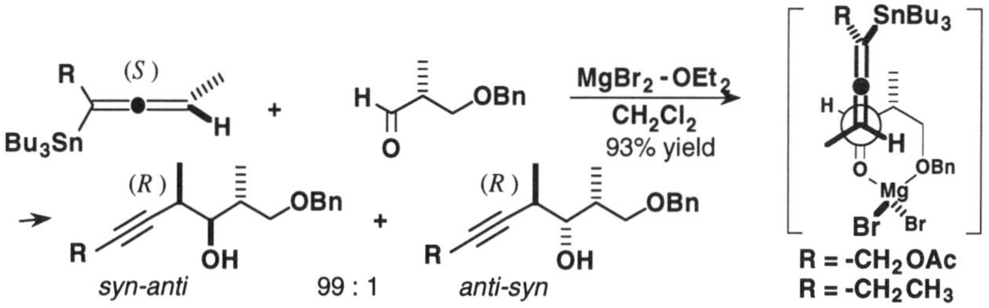

On the other hand, the $MgBr_2$-promoted addition of (*R*)-allenylstannanes affords *anti-anti* homopropargylic alcohols with even better diastereoselectivity. In this case, the stereoselectivity is again best accounted for by a synclinal chelated transition state.

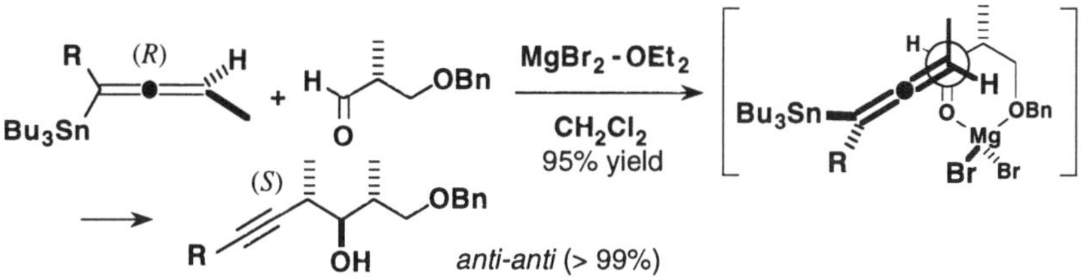

With BF_3-OEt_2, the reaction gives rise to *syn-syn* alcohols, from (*R*)-allenylstannanes, with the same very high selectivity. This is a result that can be explained by the intervention of the following antiperiplanar transition state.

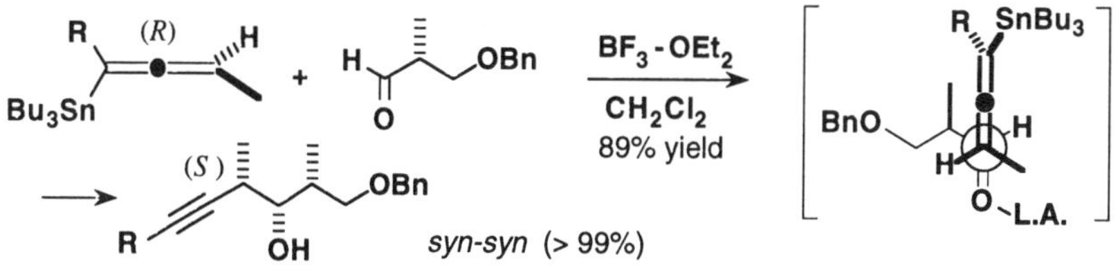

Similarly, the BF_3-OEt_2-promoted addition of an (*R*)-allenylstannane to the following α-methyl-β-alkoxy aldehyde leads to homopropargylic alcohols, bearing four stereogenic centers, in a separable 11.5:1 mixture, the Felkin-Anh product *syn-syn-anti* monoprotected diol **B**, resulting from an attack on the *Si* face of the aldehyde, being the major isomer.[230]

C_7H_{15} Bu$_3$Sn A + H O OMOM → (BF_3-OEt_2, CH_2Cl_2, 85% yield) → C_7H_{15} B OH OMOM + C_7H_{15} C OH OMOM, 11.5 : 1

[Bu$_3$Sn R, H, R, H, H, O–BF_3]

Although these additions can also be performed with $MgBr_2$-OEt_2, they were slower and less selective than when employing BF_3-OEt_2. The main isomer, homopropargylic alcohol **D**, results probably from an antiperiplanar attack of the stannane on the *Re* face of the chelated alkoxy aldehyde.

C_7H_{15} Bu$_3$Sn A + H O OMOM → ($MgBr_2$-OEt_2, CH_2Cl_2, 77% yield) → C_7H_{15} OH OMOM D

[R SnBu$_3$, H, H, MOMO, O, Mg, Br, Br]

III.2.d Addition of functionalized allylstannanes or allylsilanes

Trost prepared methylene tetrahydrofuran derivatives via a diastereoselective [3+2] type heterocyclic synthesis involving in its first step a bifunctional conjunctive stannane. Thus, 2-acetoxymethylallyltri-*n*-butylstannane adds, in the presence of BF_3-OEt_2, to a chiral aldehyde to yield the corresponding α-methylene γ-hydroxy acetate with excellent diastereoselectivity: >20:1. The so-formed acetate could then undergo transition metal-promoted cyclization to afford the desired methylene tetrahydrofuran.[231]

BnO O H + OAc SnBu$_3$ → (BF_3-OEt_2, CH_2Cl_2, 76% yield) → BnO OH OAc

→→→ BnO O or BnO O

Keck studied the addition of the following γ-(silyloxy)allyl stannane to 2- or 3-(benzyloxy)aldehydes in the presence of $MgBr_2$. He found that the stereochemistry of the adducts was consistent with chelation control, substituents about the new bond being *syn*.[232]

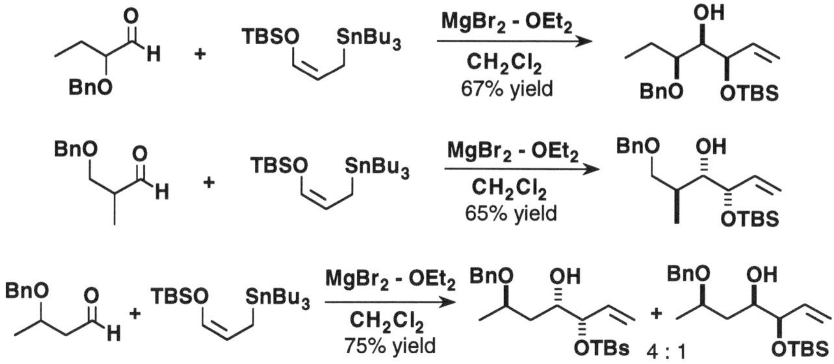

As already mentioned, Marshall proposed a synthesis of enantio-enriched γ-(alkoxy)allylstannanes by reduction of crotonoylstannanes with (*R*)-(+)- or (*S*)-(−)-BINAL-H, followed by treatment of the resulting α-(hydroxy)allylstannanes with MOMCl and 1,3-isomerization of $-SnBu_3$.[132,133] Addition of the following (*S*)-γ-(alkoxy)allylstannane to (*S*)-2-(benzyloxy)propanal, with $MgBr_2$ as catalyst, gives rise, with very high diastereoselectivity, to the *syn-syn-*(*E*) adduct.[233]

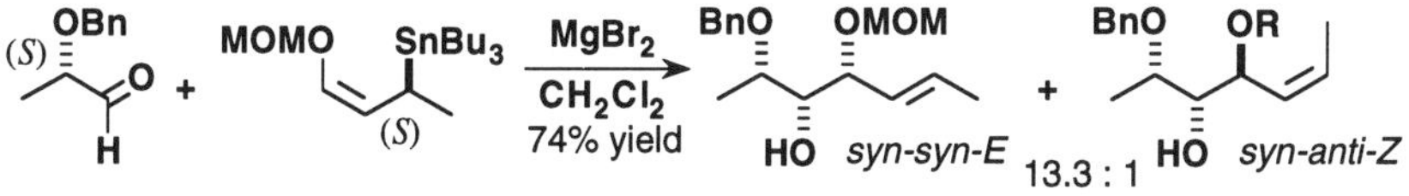

In contrast, the addition of the (*R*)-enantiomer to the same (*S*)-2-(benzyloxy)propanal was found to be mismatched; thus, the ratio of the adducts falls from 13.3:1 to 3:1.

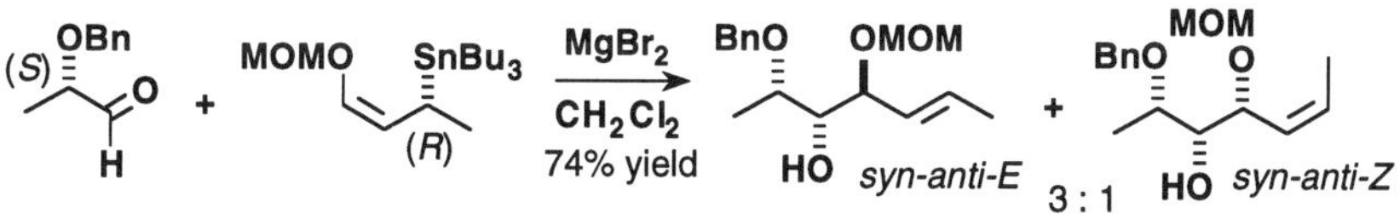

The proposed transition state for the matched addition of (*S*)-γ-(alkoxy)allylstannane to (*S*)-2-(benzyloxy)propanal is an antiperiplanar one with chelation control and pro-(*E*) carbon-carbon double bond. The corresponding transition state for the mismatched addition of the (*R*)-γ-(alkoxy)allylstannane is clearly more hindered.

syn-syn-E syn-anti-E

As expected, the matched/mismatched pairings were reversed when BF_3-OEt_2 was used as catalyst for the addition of γ-(alkoxy)allylstannanes to (*S*)-2-(benzyloxy)propanal. The (*R*)-γ-(alkoxy)allylstannane gives rise to a matched addition with high diastereoselectivity, whereas the (*S*)-γ-(alkoxy)allylstannane affords a 2:1 mixture of alcohols.

BF_3-OEt_2, CH_2Cl_2, 58% yield: anti-syn-E + syn-anti-E, 11.5 : 1

BF_3-OEt_2, CH_2Cl_2, 72% yield: syn-syn-E + anti-anti-E, 2 : 1

Several other matched pairings were examined. Thus, the BF_3-promoted addition of a (*S*)-γ-(alkoxy)allylstannane to pentabenzylglucose yielded the corresponding alcohol as a single isomer.

BF_3-OEt_2, CH_2Cl_2, 98% yield

A spectacular result was obtained by Marshall by allylation of (*R*,*R*)-tartrate derived enal **A**. The adduct **B** upon hydroxylation with catalytic OsO_4-NMO, yield a tetradecaol **C**.[234]

BF_3-OEt_2, CH_2Cl_2, 86% yield

C 40% overall yield

The addition of α-(alkoxy)allylstannanes to chiral aldehydes was also applied to the preparation of precursors of macrolide natural products such as tylosin.[235] Indeed, a remarkable stereoselectivity was observed while adding

(*R*)-allylstannane **A** to (*S*)-aldehyde **B** since the *syn-syn-*(*Z*) adduct was obtained in an 11:1 ratio.

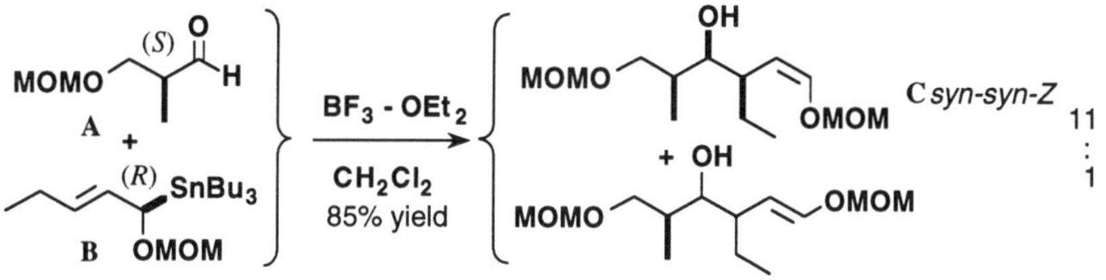

An addition according to the Felkin-Anh transition state model can explain the *syn* selectivity observed between the hydroxy and the methyl group [(*S*)-configuration for the carbinol atom].

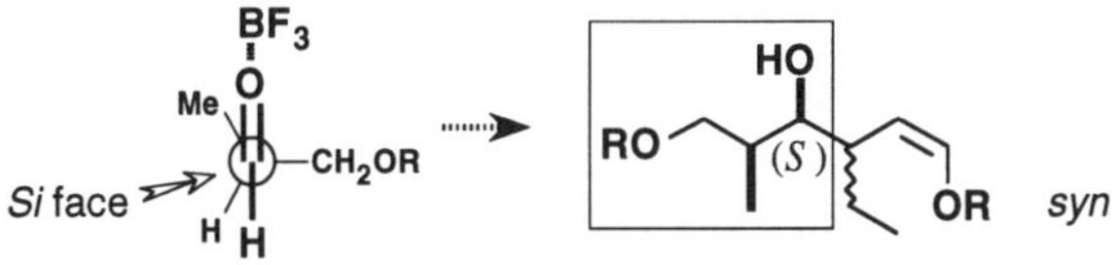

The *syn-*(*Z*) configuration of the enol ether moiety is also a consequence of the Felkin-Anh transition state (attack on the *Si* face of the carbonyl with a S_E2 mechanism). Thus, despite all of the following three transition states being hindered, the reaction probably proceeds through one of these to yield the *syn-*(*Z*) adduct.

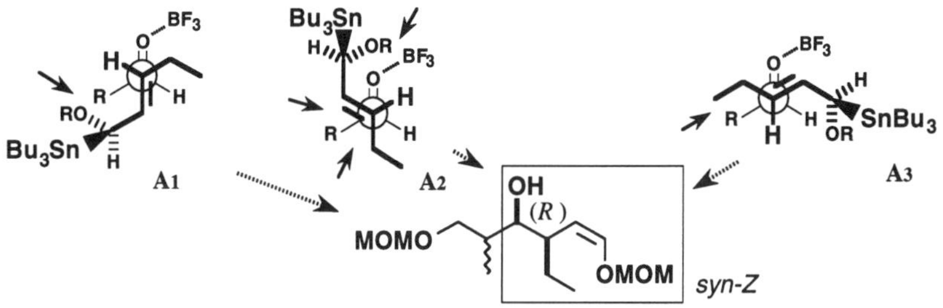

If the aldehyde involved was a simple achiral aldehyde, an addition on its *Re* face would have led, through a more favorable transition state, to the (*E*)-enol ether.[132]

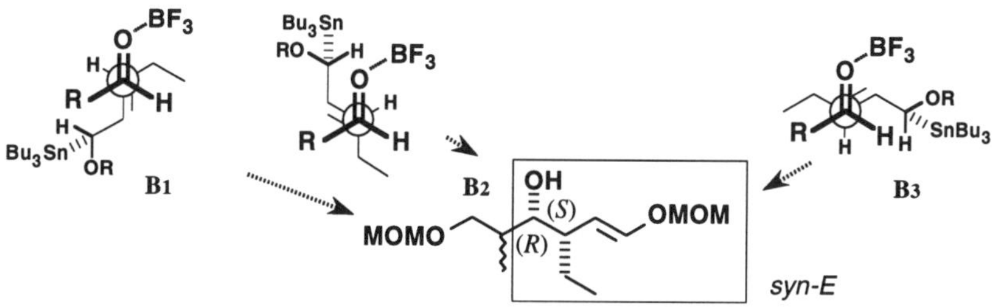

The steric requirements due to the Felkin-Anh model enable kinetic resolution. Thus, when a racemic mixture of allylstannane **B** is used towards an optically active aldehyde **A**, one notes that the (*S*)-enantiomer, which can add to the *Si* face of the aldehyde with the fewer steric interactions, is made use of first.[235]

TBSO O BnO H **A** + TBSO $SnBu_3$ OMOM **B** 1:1 mixture of diastereomers; 2.5 equiv — BF_3 - OEt_2, CH_2Cl_2, 79% yield

TBSO OH BnO OMOM OSBT **C** + TBSO (*R*) $SnBu_3$ H OMOM **B** enriched in the *R* enantiomer

Addition of one equivalent of racemic allylstannane **B** to optically active aldehyde **A** yielded, along with the expected alcohol **C** and unreacted aldehyde, optically active rearranged allylstannane **D**. Thus, the optically active aldehyde **A** shows a remarkable affinity for the (*S*)-enantiomer of allylstannane **B**.[235]

TBSO O (*S*) BnO H **A** 1 equiv + $SnBu_3$ OMOM **B** racemic; 1 equiv — BF_3 - OEt_2, CH_2Cl_2, 79% yield

TBSO OH (*R*) (*S*) BnO OMOM **C** 45% yield + (*R*) Bu_3Sn OMOM **D** 50% yield; TBSO O BnO H **A** 40% yield

The (*R*)-enantiomer of **B** undergoes a BF_3-promoted 1,3-isomerization into allylstannane **D** as proved by independent synthesis.

(*R*) $SnBu_3$ OMOM **B** — BF_3 - OEt_2, CH_2Cl_2 → (*R*) Bu_3Sn OMOM **D**

The formation of **C** results first from a Felkin-Anh model approach (addition to the *Si* face of **A**).

BF_3 O Me R CHOTBS *Si* face H H ⟶ TBSO HO BnO (*S*) OMOM *syn*

Second, as shown in the following scheme, the good matching, both in the antiperiplanar and synclinal transition states, between aldehyde **A** and the (*S*)-allystannane **B** accounts for the virtual exclusion of the (*R*)-enantiomer.

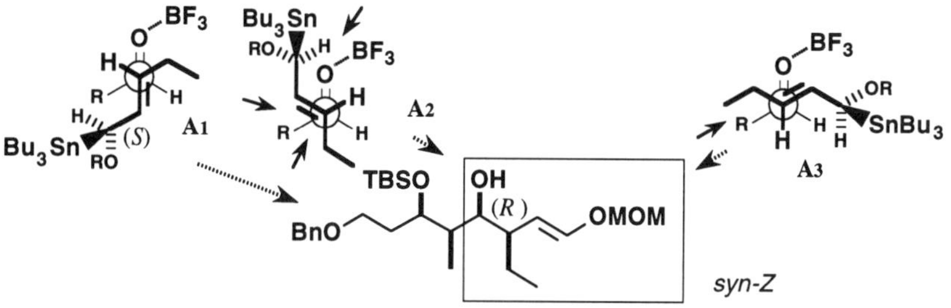

Marshall used similar strategy for the stereoselective total synthesis of bengamide E from glyceraldehyde acetonide.[236]

Panek developed an interesting strategy for the asymmetric construction of nearly diastereomerically pure 2,5-disubstituted tetrahydrofurans by adding chiral (*E*)-crotylsilanes to heterosubstituted aldehydes. The usually expected homoallylic alcohols are only produced as minor products in most cases.[237]

Under nonchelation-controlled conditions (BF_3-OEt_2 catalysis), the crotylsilane **A** was added to (*S*)-2-(benzyloxy)propanal to yield tetrahydrofuran **B**.

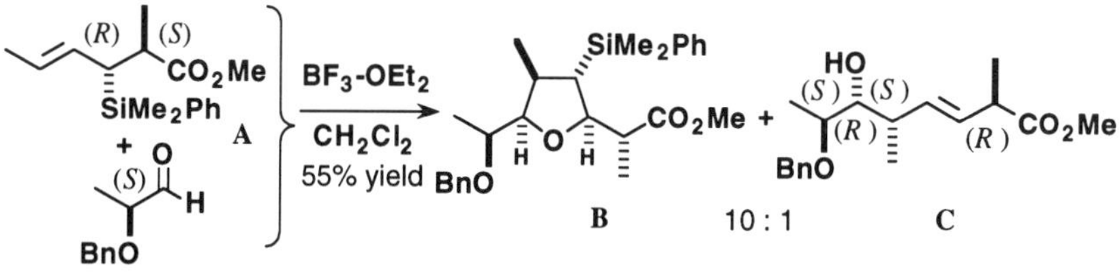

The stereochemical outcome is consistent with an anti-S_E2' pathway through an antiperiplanar arrangement **B1**, involving the creation of the incipient bond between the *Re* face of the aldehyde and the *Si* face of the allylsilane (*unlike* approach). From intermediate **B2**, a rotation about the new bond and subsequent heterocyclization, with silicon migration, gives rise to tetrahydrofuran **B**.

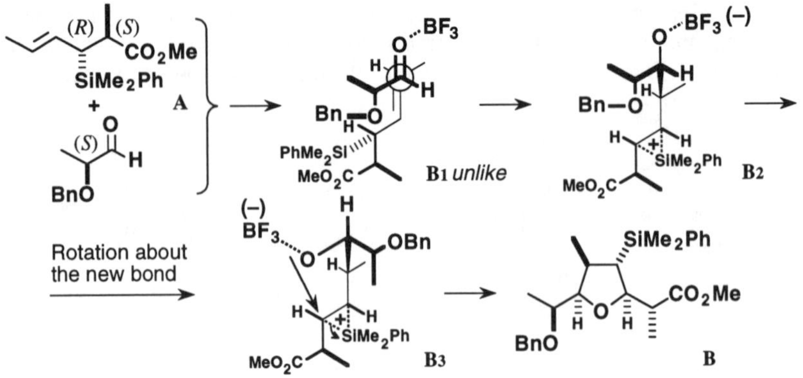

In contrast, using a bidentate Lewis acid, such as $SnCl_4$, capable of forming a chelate with the α-benzyloxy group, results in the formation of the diastereomeric tetrahydrofuran **D**.

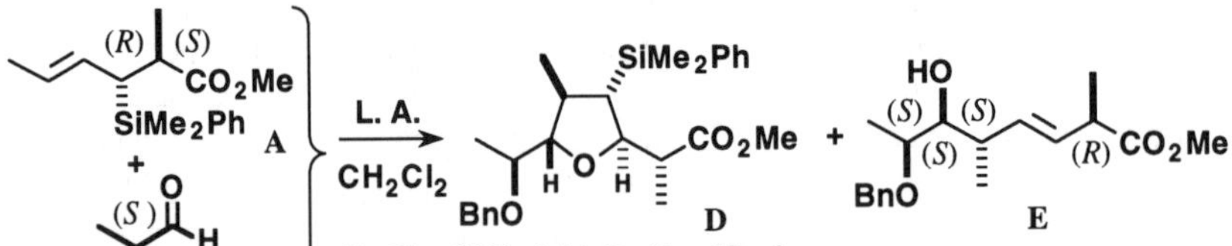

The same result is observed when using two equivalents of $AlCl_3$. This might be due to the formation of an $AlCl_2$-aldehyde complex through metathesis of two molecules of $AlCl_3$ to an $AlCl_2$-$AlCl_4$ complex prior to chelation with the aldehyde.[238] A plausible synperiplanar transition state, in which addition occurs on the *Si* face of the aldehyde, accounts for the stereochemistry of **D**.

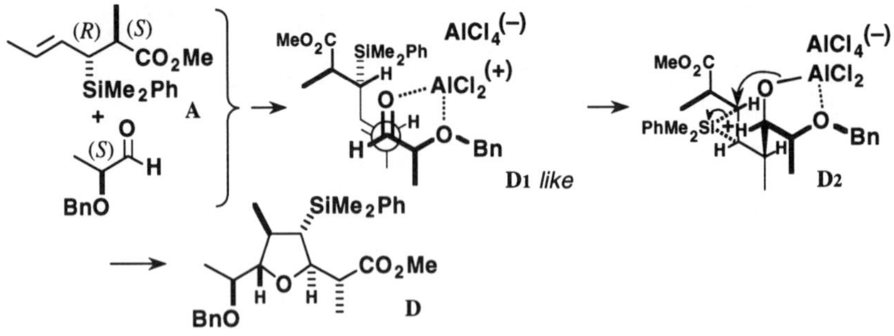

For the diastereomeric crotylsilane derivative **F**, the aldehyde π-face selection is independent of the nature of the Lewis acid employed since both BF_3-OEt_2 and $SnCl_4$ produce the *cis*-2,5-substituted tetrahydrofuran **G**.

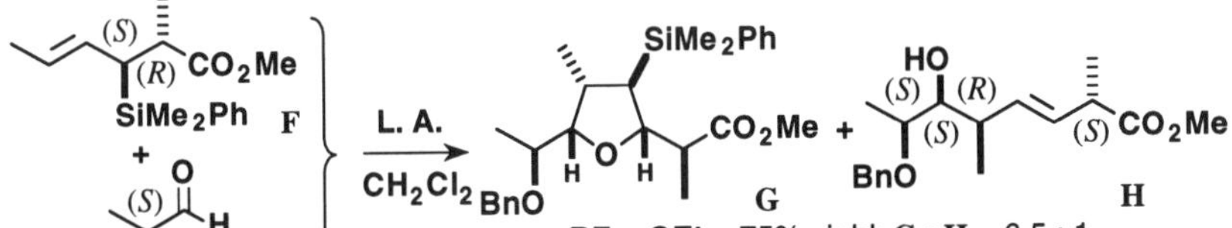

L. A., CH_2Cl_2

BF_3- OEt_2: 75% yield; **G** : **H** = 6.5 : 1
$SnCl_4$: 74% yield; **G** : **H** = 1 : 1.1
$AlCl_3$: 80% yield; **G** : **H** = 1 : 15
$TiCl_4$: 20% yield; **G** : **H** = 0 : 1

Taddei used the BF_3-promoted addition of 2-chloromethyl-3-trimethylsilylprop-1-ene **B** to *N*-Boc amino aldehydes in order to prepare hydroxyethylene dipeptide isosteres (HIV-1 protease inhibitors).[239,240]

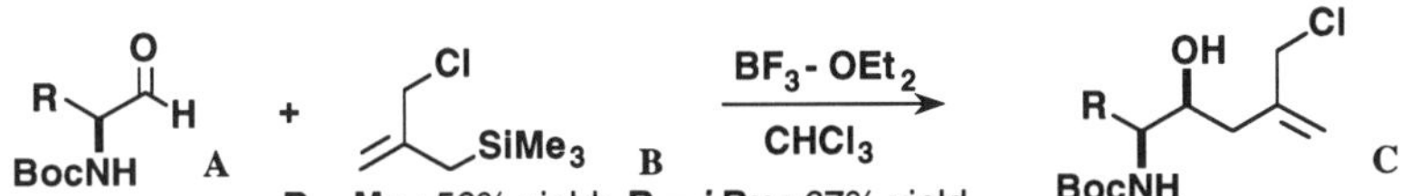

BF_3- OEt_2, $CHCl_3$

R = Me ; 56% yield; R = *i*-Bu ; 67% yield
R = Bn ; 72% yield; R = *s*-Bu ; 62% yield

IV. Intramolecular Addition of Allylstannane or Allylsilane Moieties to Aldehydes and Ketones

IV.1 Intramolecular addition without cyclization

Reetz established that the intramolecular condensation of chiral β-allylsiloxy aldehydes gives rise to homoallylic alcohols.[241]

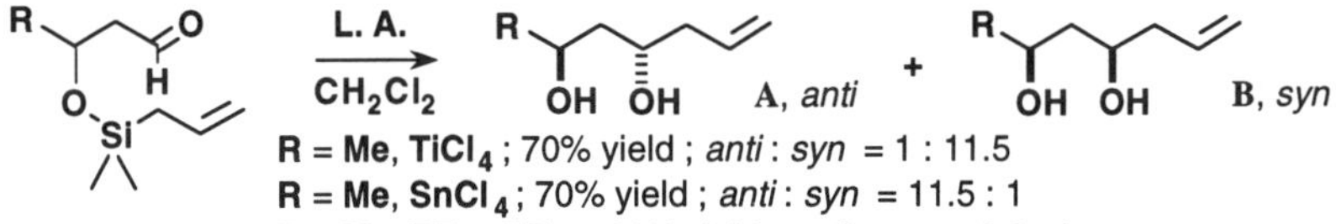

L. A., CH_2Cl_2 → **A**, *anti* + **B**, *syn*

R = Me, $TiCl_4$; 70% yield ; *anti* : *syn* = 1 : 11.5
R = Me, $SnCl_4$; 70% yield ; *anti* : *syn* = 11.5 : 1
R = Me, BF_3- OEt_2 ; 60% yield ; *anti* : *syn* = 2.3 : 1
R = *n*-Bu, $TiCl_4$; 80% yield ; *anti* : *syn* = 1 : 9

The $TiCl_4$ promoted intermolecular chelation control addition of allyltrimethylsilane to the corresponding 3-(benzyloxy)aldehydes leads to *anti* homoallylic alcohols,[186] whereas, here, *syn* alcohols are obtained as the major products of the reaction. In order to establish that an intramolecular process was responsible for such a reversal of selectivity, Reetz made cross-over experiments (D-labeled ones) which proved to be conclusive. Therefore, the *syn* selectivity can be accounted for in the following way: the chelate, formed with $TiCl_4$, promotes the intramolecular addition to the face opposite to that observed in intermolecular additions to the related chelates. That leads to the assumption that transition state **D**, which leads to *syn* diol **B**, is the reacting conformer.

B, *syn*

A, *anti*

The stereochemistry observed with BF_3-OEt_2 is in agreement with an intermolecular chelation controlled addition, similar to examples already mentioned.[187] As for $SnCl_4$, a bidentate Lewis acid, the formation of a 2:1 complex however probably induces an intermolecular process.

In contrast, when the reaction is performed on the corresponding α-chiral β-allylsilyloxy aldehyde, the stereochemistry of the adducts is the same as for the intermolecular addition of allylsilane, regardless of the Lewis acids used.

L. A., CH_2Cl_2 → A, *anti* + B, *syn*

$TiCl_4$; 95% yield ; *anti* : *syn* = 1.8 : 1
$SnCl_4$; 98% yield ; *anti* : *syn* = 3.3 : 1
$AlCl_3$; 94% yield ; *anti* : *syn* = 1.2 : 1
$EtAlCl_2$; 92% yield ; *anti* : *syn* = 5.7 : 1
(reaction in toluene)

In close analogy, Hioki showed that the intramolecular condensation of a 3-(allylsiloxy)propaniminium salt, generated *in situ*, gives rise to an allylamide in good yield and with excellent *syn* selectivity, regardless of the Lewis acid employed[242] (see also ref.137).

L. A. 2.2 equiv, CH_2Cl_2 → *anti* + *syn*

$TiCl_4$; 86% yield ; *syn* : *anti* = 48 : 1
$SnCl_4$; 93% yield ; *syn* : *anti* = 45 : 1
BF_3 - OEt_2; 86% yield ; *syn* : *anti* = 51 : 1

IV.2 Cyclization by intramolecular addition of allylsilane or allylstannane moieties

Three different modes of cyclization in which the allylmetal moiety can be linked to the carbonyl group have been studied. Type I corresponds to the formation of 3-cycloalkenols through an endo process, Type II leads to 2-vinylcycloalkanols through an exo process and Type III allows the obtention of 3-alkylidenecycloalkanols from α-alkylidene silanes or stannanes.

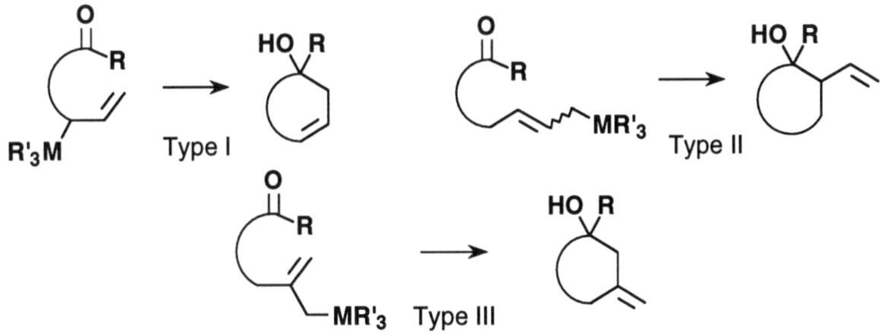

IV.2.a Preparation of 3-cycloalkenols (Type I)

As far back as 1979, Kuwajima prepared 2-substituted 3-trimethylsilyl-4-en-1-one derivatives such as **A** by a Claisen rearrangement reaction.

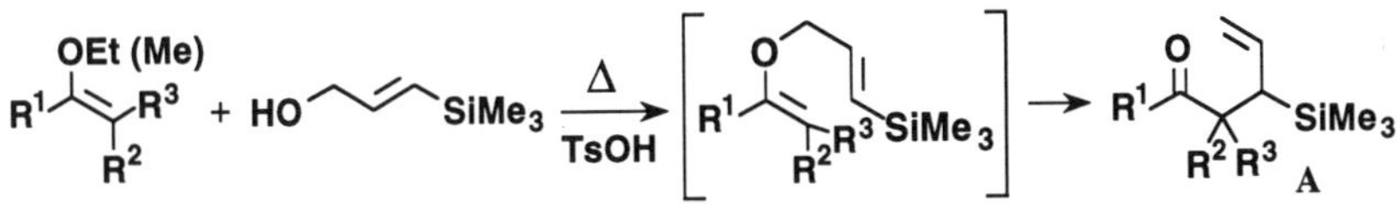

These silyl enones readily undergo Lewis acid-promoted cyclization to give 3-cyclopentenols.[243] Bicyclic alcohols can be obtained from cyclic enones.

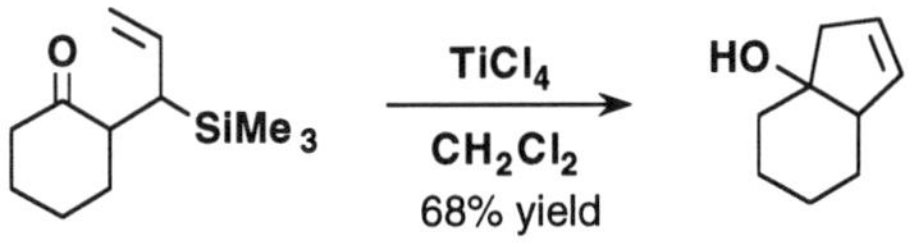

A spiro bicyclic alcohol was even prepared from the following cyclohexylallylsilyl carboxaldehyde.

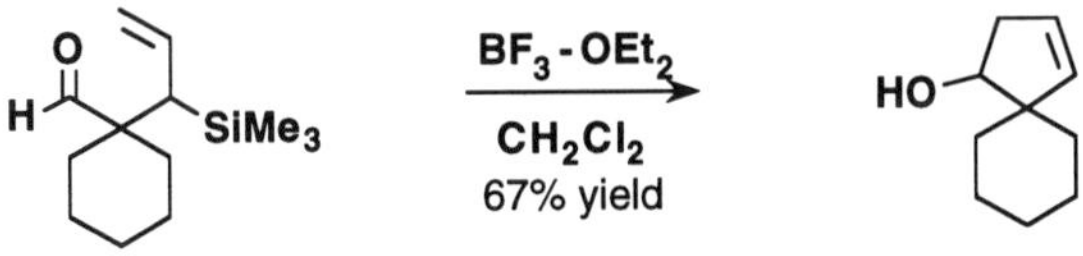

On the contrary, ketones, such as the following cyclooctanone and nonanone, do not undergo cyclization, even under more severe conditions. Thus, the cyclization reaction of 2-substituted 3-trimethylsilyl-4-en-1-ones appears to be controlled by subtle steric effects around both functionalities.

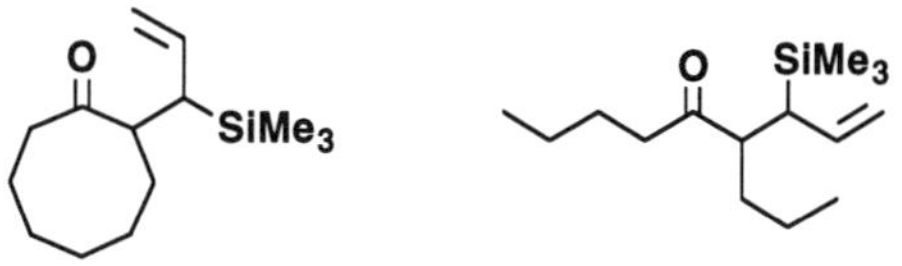

On some occasions, the suitable moiety is not isolated before cyclization. Thus, a tandem Claisen rearrangement-intramolecular allylsilane addition sequence enabled Paquette to prepare the following bromocyclopentenol.[244]

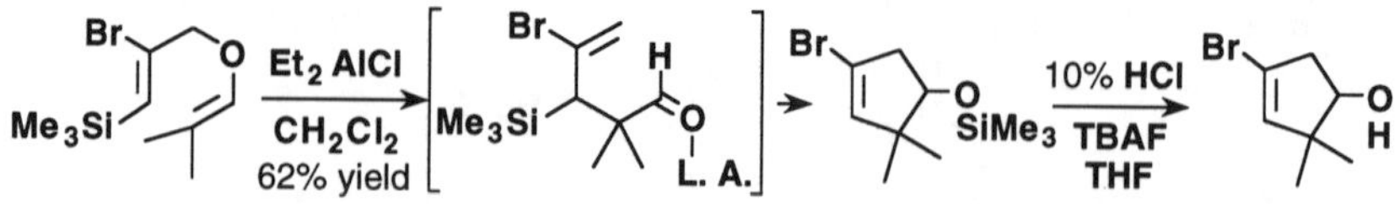

Macrocyclic homopropargyl alcohols **B**, **D** and **F** can be synthesized in high yield through the BF_3-OEt_2 promoted intramolecular condensation of allenylstannyl aldehydes **A**, **C** and **E**. Cyclization products were however obtained as mixtures of diastereomers.[245,246]

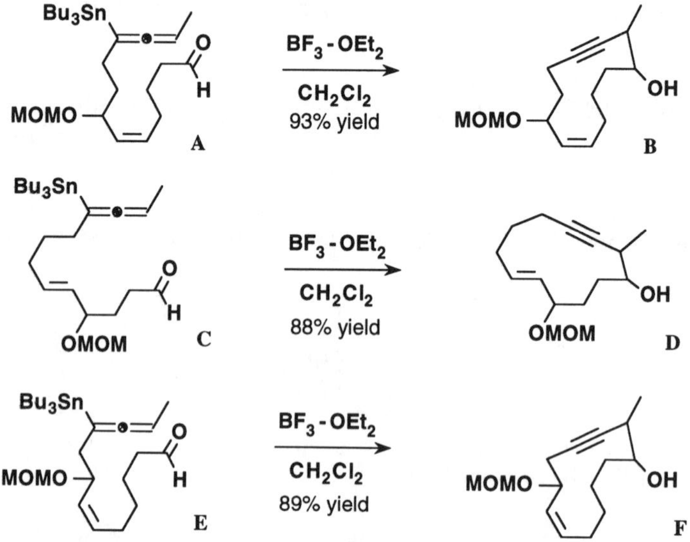

Marshall then converted these cyclic homopropargyl alcohols into 2,5-furanocyclododecenes, precursors of 2,5-furano macrocyclic diterpenes. Thus, from propargyl alcohol **F**, the bridged furan **G**, obtained upon Dess-Martin oxidation and exposure to $AgNO_3$-$CaCO_3$ in aqueous acetone, undergoes intramolecular Diels-Alder reaction at room temperature. Tetracycle **H** is thus obtained in excellent overall yield from homopropargylic alcohol **F**.

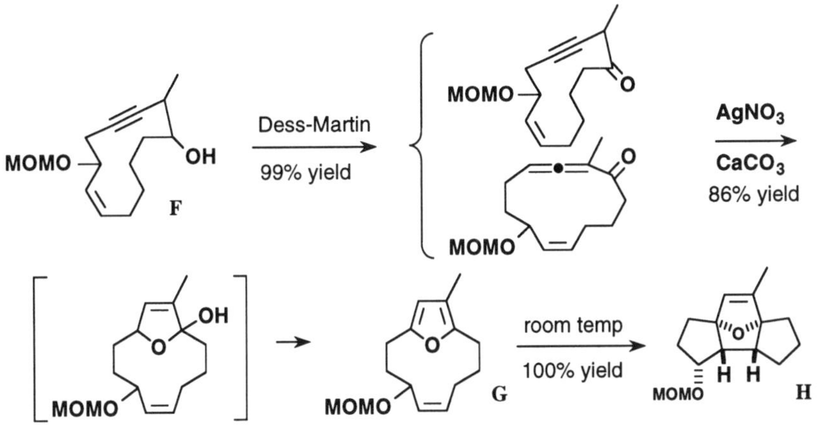

While studying macrocyclization, Marshall showed that the racemic allylstannyl aldehyde **A** exposed to BF_3-OEt_2 in CH_2Cl_2 in high dilution (0.001 M) undergoes cyclization into **C**. None of the cyclodecadienols that would have derived from an S_E' cyclization of **A** was found. These findings suggest that **A** isomerizes into **B** stereospecifically en route to **C**.[247]

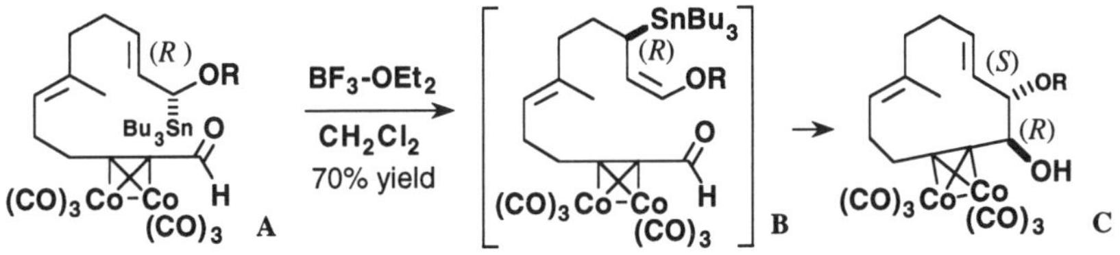

Works by Nakai on the cyclization of the optically pure (*E*)- and (*Z*)-3-trimethylsilylhex-4-enals are also relevant with this section.[75]

IV.2.b Preparation of 2-vinylcycloalkanols (Type II)

Keck observed an intramolecular condensation between allylstannane moieties and carbonyl groups separated by a single carbon atom, thus affording silyloxy vinylcyclopropanes in good yields.[248]

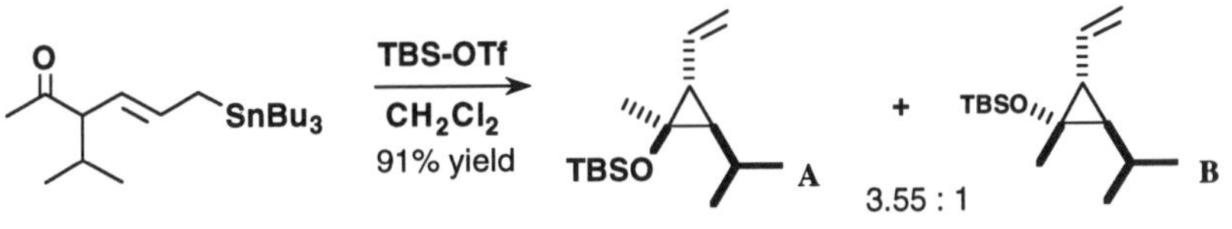

Such a cyclization must proceed through a transition state in which the carbon-carbon double bond of the allyltin moiety is eclipsed either by the methyl group, to afford isomer **A**, or by the carbon-oxygen double bond of the carbonyl group, to afford isomer **B**.

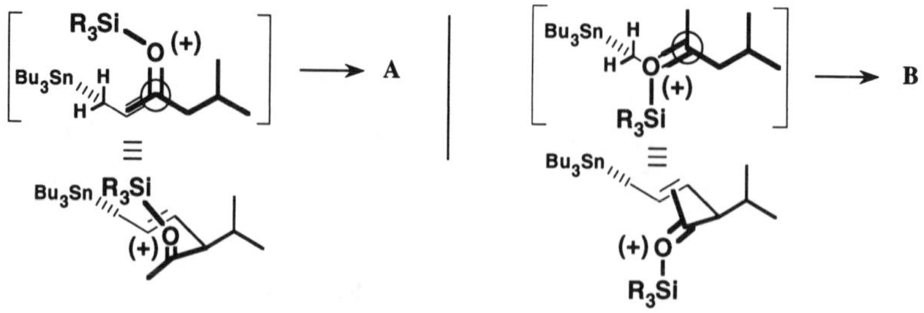

As part of a research program directed towards the synthesis of di- and triquinanes, Sarkar prepared silylated esters such as **A**. An intramolecular thermal ene-reaction gives rise to allylsilane **B** (98% yield) which was then transformed into the corresponding aldehyde **C**. **C**, upon treatment with $TiCl_4$, underwent cyclization to yield diquinane **D**, a precursor of (±)-hirsutene.[249]

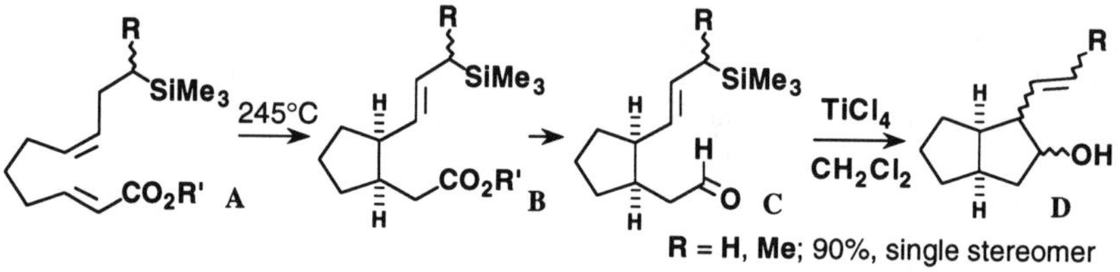

R = H, Me; 90%, single stereomer

Sakurai showed that the intramolecular condensation of 2-alkoxycarbonylallylsilane moieties to aldehydes gives α-methylene-γ-lactones, albeit with low yields.[250] Nevertheless, α-methylene-γ-lactones fused to the cyclohexyl ring should be obtainable from a suitable 8-trimethylsilyloct-6-enal derivative. Thus, Kuroda found that the following aldehyde can be cyclized, upon treatment with toluene-*p*-sulfonic acid in refluxing acetone, to yield, as the sole product of the reaction, a *cis*-fused 14,15-dinoreudesmonolide.[251,252]

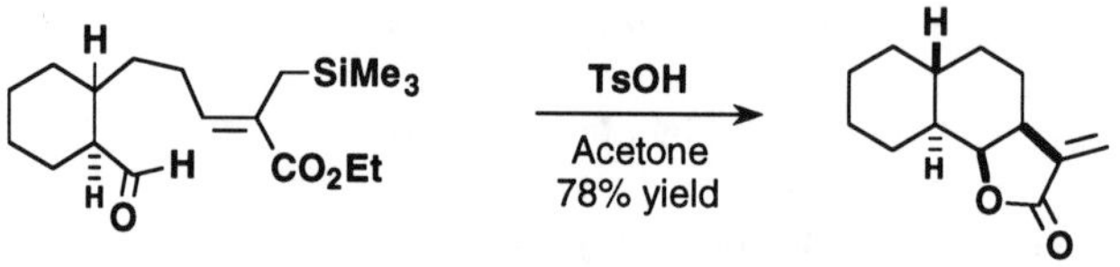

Similar results were described by Yamakawa.[253]

Analogously, cyclization of the following 9-trimethylsilylnon-7-enal derivative afforded α-methylene-γ-lactones fused to a perhydroazulene carbon framework.[254]

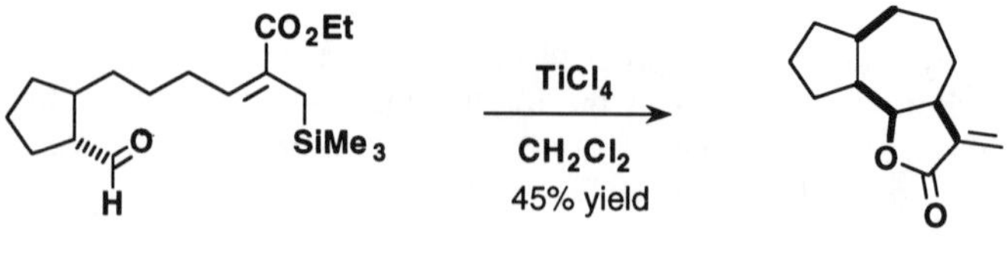

Optically active α-methylene-γ-lactones fused to cyclopentyl or cyclohexyl rings were synthesized by Yamakawa. Thus, ω-formyl-β-(chiral)alkoxycarbonylallylsilanes, such as **A**, undergo cyclization in the presence of $TiCl_4$ to yield hydroxy ester **B**. Various homochiral alkoxy groups have been tested, but best results were obtained from 8-phenylmenthyl.[255] The absolute configuration of the adducts is not specified.

A → ($TiCl_4$, CH_2Cl_2) → B → 100% yield → C

B, *cis*, 72% yield, 92% d.e.
B', *trans*, 7% yield, 53% d.e.
C, *cis*, 92% e.e.

The stereospecific formation of 2-vinylcyclohexanols was obtained upon acid catalyzed treatment of 8-trimethylsilyloct-6-enals. Best results were reached by using trifluoro acetic acid.[256]

A or B → (L. A., CH_2Cl_2) → C + D

A, $SnCl_4$; 58% yield; **C** : **D** = 1.9 : 1
A, $TiCl_4$; 60% yield; **C** : **D** = 2.3 : 1
A, BF_3 - OEt_2; 50% yield; **C** : **D** = 5.7 : 1
A, CF_3COOH ; 100% yield; **C** : **D** = 49 : 1
B, CF_3COOH ; 100% yield; **C** : **D** = 1 : 13.3

CF_3COOH, CH_2Cl_2 → Ni Raney → 50% overall yield

The cyclization of an 8-trimethylsilyloct-6-enal derivative was used by Maier in his synthesis of the cyclohexyl fragment of FK-506. However, both the BF_3-OEt_2 and $SnCl_4$ promoted reactions failed to produce valuable diastereoselectivity. The similar ene-reaction (H instead of trimethylsilyl group) was unsuccessful.[257]

BF_3-OEt_2, CH_2Cl_2, 82% yield; 1.13 : 1

The diastereoselectivities observed in the Lewis acid-mediated cyclization of (*E*)- or (*Z*)-octenals **A**, bearing a stereogenic center α to the carbon-carbon double bond, were better. As expected, the diastereofacial selection in the allylsilane moiety is *trans* with reference to the α-methyl group, regardless of the double bond geometry of **A**.[258]

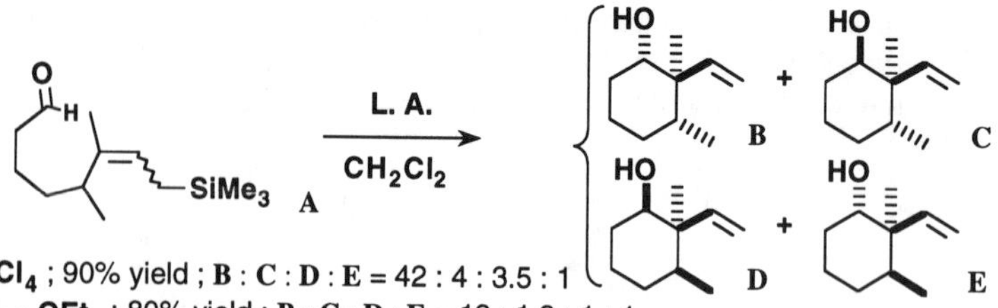

L. A.
CH_2Cl_2

Δ = *E*; $SnCl_4$; 90% yield; **B** : **C** : **D** : **E** = 42 : 4 : 3.5 : 1
Δ = *E*; BF_3 - OEt_2; 80% yield; **B** : **C** : **D** : **E** = 13 : 1.3 : 1 : 1
Δ = *Z*; $SnCl_4$; 90% yield; **B** : **C** : **D** : **E** = 1.6 : 1 : 0 : 0
Δ = *Z*; BF_3 - OEt_2; 80% yield; **B** : **C** : **D** : **E** = 2.3 : 1 : 0 : 0

In an important paper, Molander used chelation control to direct the stereochemical outcome of intramolecular additions of allylsilane moieties to β-dicarbonyl substrates.[259] Treatment of either β-ketoester **A** or **B** with four equivalents of $TiCl_4$ provided **C**, in nearly quantitative yield, with no detectable amounts of the other three possible diastereomers.

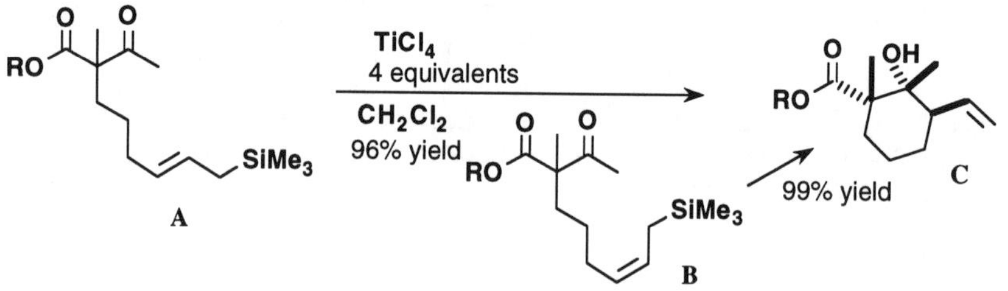

$TiCl_4$
4 equivalents
CH_2Cl_2
96% yield

99% yield

A reaction path through synclinal transition states **A'** or **B'** can account for the exclusive formation of vinylcyclohexanol **C**. These transition states correspond to a *like* approach of the unsaturated moieties.

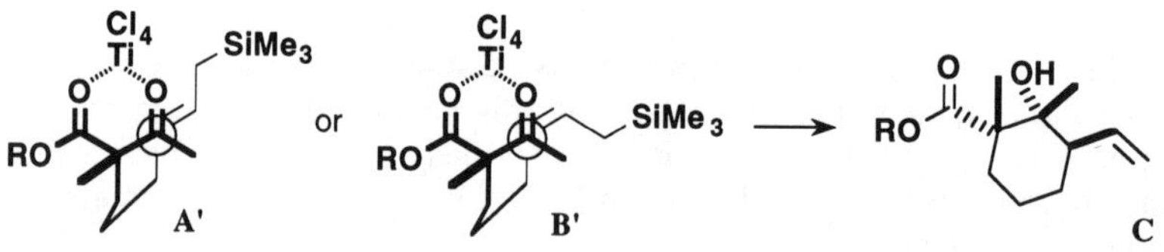

The *unlike* approach leading to alcohol **D** is clearly unfavored due to steric interactions between the approaching allylsilane and the chelated ring.

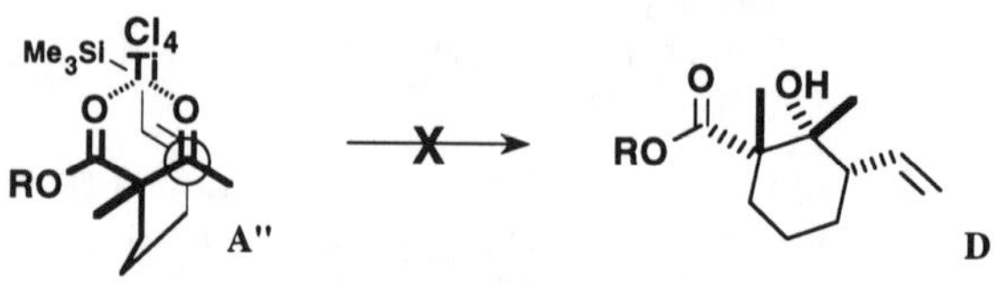

In a series of papers, Marshall developed a methodology towards the synthesis of cembranolides, based on the intramolecular addition of (α-alkoxyallyl)stannane moieties to carbonyl groups. Treatment of the racemic precursor, with BF_3-OEt_2 in CH_2Cl_2 at high dilution (0.009 M), afforded, in 88% yield, a mixture of four isomeric hydroxy enol ethers in which the major one bears the correct 1,2-*syn* relationship between the hydroxy group and the enol ether moiety.[260,261] Moreover, it turned out that the major isomer **B** is obtained, with high enantiomeric integrity, when the same reaction is performed on homochiral aldehyde **A**. It is worth noting that the thermal cyclization attempted on the same compound has so far failed.[262]

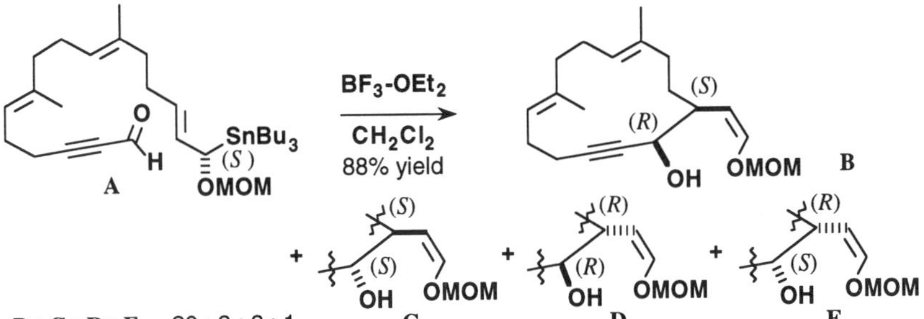

B : C : D : E = 20 : 2 : 2 : 1

The high level of chirality transfer, observed during the reaction, can be accounted for when assuming that the reaction proceeds through a *synclinal* transition state which is favored by "inside alkoxy effect".[141]

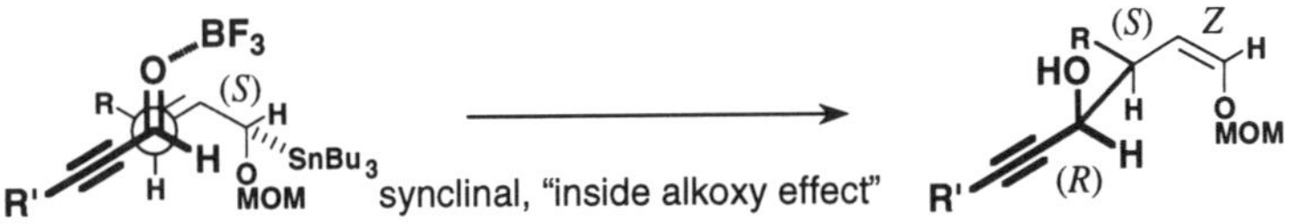

Yamamoto studied the acid catalyzed cyclization of (*E*)- and (*Z*)-5-oxa-8-(tri-*n*-butylstannyl)oct-6-enals. He observed that the stereochemical outcome of the protic acid induced cyclization depends strongly upon the double bond geometry whereas the Lewis induced cyclization affords *trans* tetrahydropyranol regardless of the double bond geometry.[263]

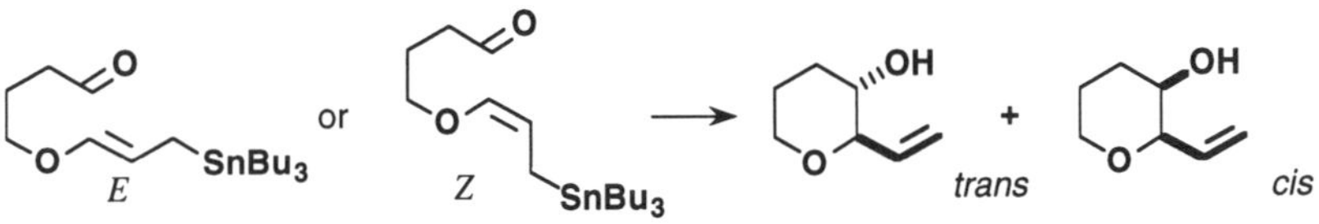

E-isomer; $\mathbf{TiCl_4}$; >95% yield; *trans* : *cis* = 5.3 : 1
E-isomer; $\mathbf{Bu_4\ NF - TiCl_4}$; 87% yield; *trans* : *cis* = 13.3 : 1
E-isomer; **TfOH** ; >95% yield; *trans* : *cis* = 1 : 0
E-isomer; $\mathbf{BF_3 - OEt_2}$; >95% yield; *trans* : *cis* = 6.7 : 1
Z-isomer; **TfOH** ; 94% yield; *trans* : *cis* = 1 : 13.3
Z-isomer; $\mathbf{TiCl_4}$; 86% yield; *trans* : *cis* = 6.7 : 1
Z-isomer; $\mathbf{BF_3 - OEt_2}$; 80% yield; *trans* : *cis* = 2.1 : 1
Z-isomer; $\mathbf{Bu_4\ NF - TiCl_4}$; 90% yield; *trans* : *cis* = 1 : 2.4

Yamamoto proposed a "push-pull" mechanism via a cyclic transition state to account for the stereochemistry of the products obtained under Bronsted acid catalysis.

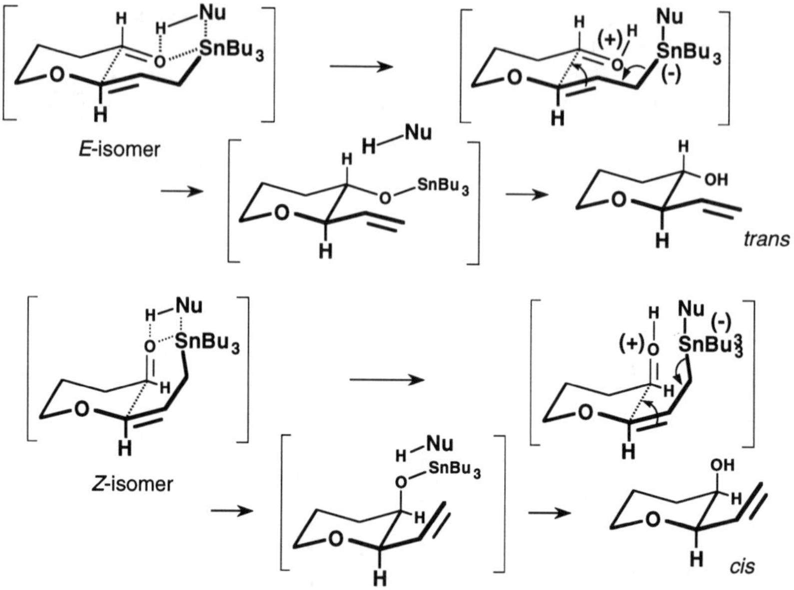

Yamamoto also established that the intramolecular cyclization of ω-tri-*n*-butylstannyl ether aldehydes can proceed with high stereocontrol. Thus, upon treatment with BF_3-OEt_2, the following tin derivative leads to the corresponding 3-hydroxyoxepanes **A** and **B** in a 13.3:1 ratio.[264]

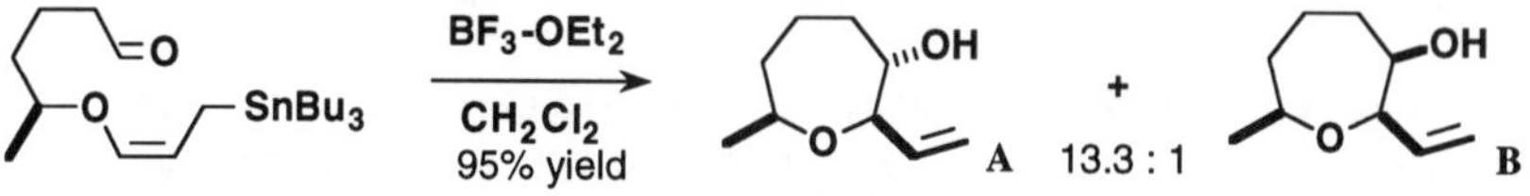

This selectivity may result from the preference, on steric grounds, for a synclinal transition state (leading to **A**) over a synperiplanar one (leading to **B**).

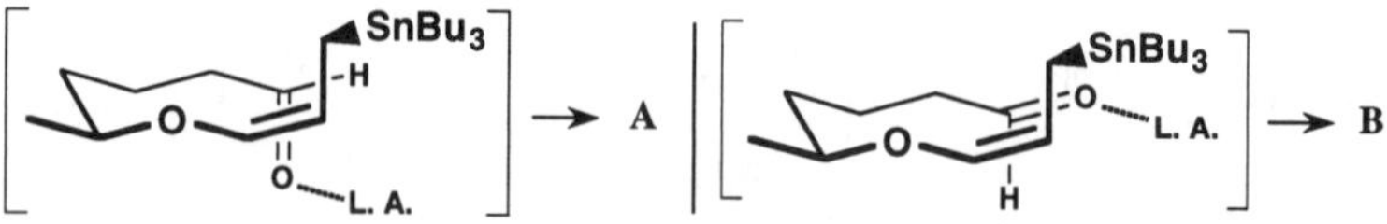

Yamamoto then first applied this method to the synthesis of a 6.7.7.6 ring system, as a model of the polycyclic framework of brevetoxin B.

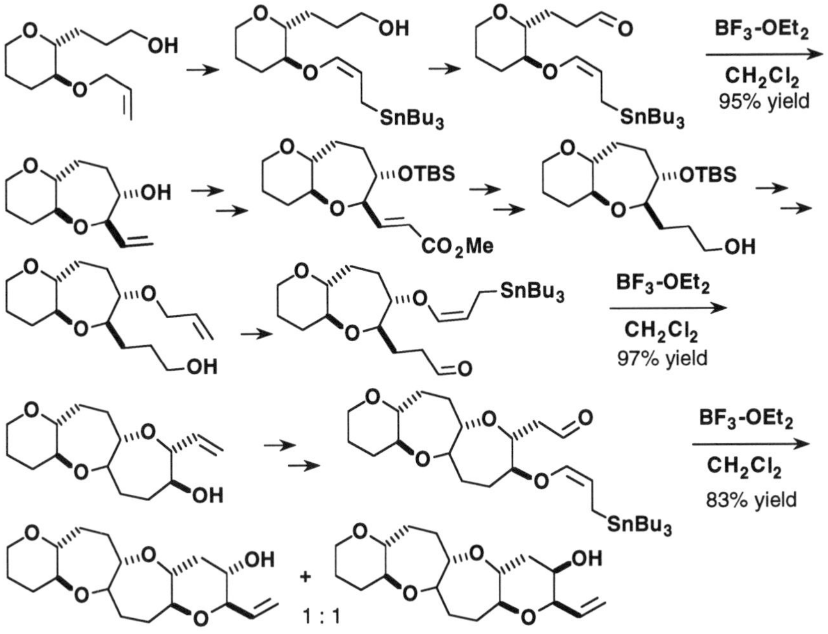

More recently, the same methodology was used to synthesize the hemibrevetoxin ring system (7,7,6,6-tetracyclic ether skeleton).[265]

Transannular cyclization with high control of the stereochemistry has been used by White in the course of the total synthesis of (±)-africanol and (±)-isoafricanol. From trimethylcyclohexenone, a three step sequence led to alcohol **B**. An anionic oxy-Cope rearrangement gave rise to cyclodec-5-enone **C** which, upon treatment with $SnCl_4$, underwent cyclization to afford hydroazulenols **D** and **E**.[266] The related n-Bu_4NF mediated cyclization of 7-trimethylsilylcyclodec-5-enone led also to the corresponding hydroazulenol with a *cis* ring junction.[267]

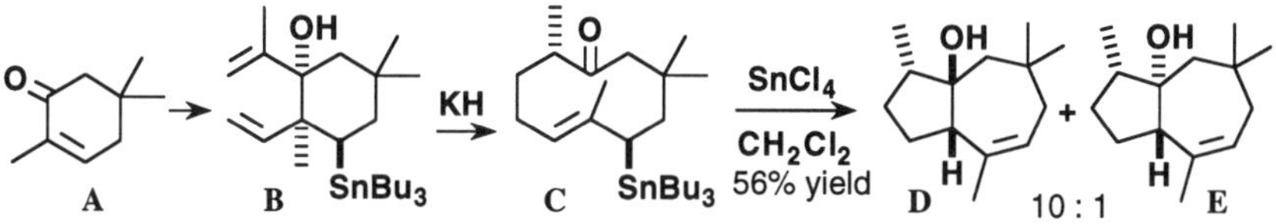

IV.2.c Preparation of 3-methylenecycloalkanols (Type III)

In the course of his use of bifunctional conjunctive reagents, Trost prepared 2-bromo-3-(trimethylsilyl)prop-1-ene, a potentially binucleophilic reagent. Copper-catalyzed Grignard addition to conjugated enones generates 4-trimethylsilylmethylpent-4-enone derivatives. Fluoride ion and $TiCl_4$ induced cyclizations of these adducts suffered from protodesilylation. However, the use of $EtAlCl_2$ avoided this complication, which presumably arises from the adventitious presence of HX. Thus, subjecting the three following adducts to two equivalents of $EtAlCl_2$ in toluene at 0°C resulted in the formation of bicyclo [n.2.1.] systems.[268]

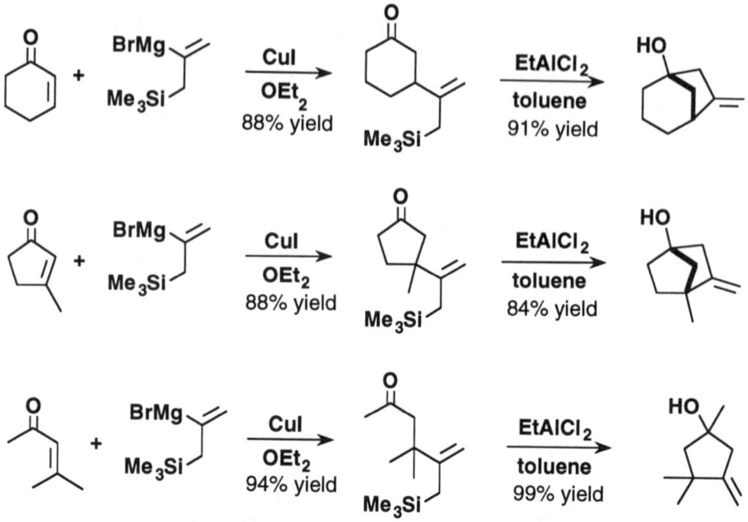

CuI, OEt_2, 88% yield; $EtAlCl_2$, toluene, 91% yield

CuI, OEt_2, 88% yield; $EtAlCl_2$, toluene, 84% yield

CuI, OEt_2, 94% yield; $EtAlCl_2$, toluene, 99% yield

The following bifunctional conjunctive reagent **B**, that contains both a nucleophilic and an electrophilic center, was obtained by Trost from 2,3-bis(trimethylsilyl)buta-1,3-diene **A**. Alkylation of β-ketoesters or ketosulfones gives rise to 4-methylene-5-trimethylsilylmethylhex-5-en-2-one derivatives **C**. Cyclization into 2,3-dimethylenebicyclic compounds occurs smoothly in the presence of $EtAlCl_2$.[269,270] The formation of the 2,3-dimethylenecyclohexane unit sets the stage for a second annulation, the Diels-Alder reaction.

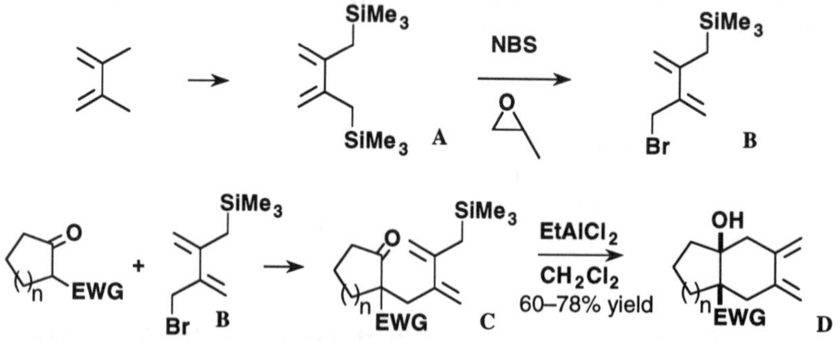

n = 1 or 2 **EWG** : $-CO_2Me$; $-SO_2Ph$

In 1978, Andersen published the first intramolecular allylsilane addition to aldehyde.[15] The cyclization reaction occurred with a better yield than the corresponding ene reaction (hydrogen atom instead of the trimethylsilyl group) which afforded significant quantities of endocyclic olefin products.[271]

SnCl$_4$; R = SiMe$_3$; 78% yield; A : B = 1.4 : 1
BF$_3$-OEt$_2$; R = H ; quant. yield; A : B = 5.7 : 1
(n-Bu)$_4$NF / THF ; R = H ; quant. yield; A : B = 1 : 4.6

The selectivity in favor of axial alcohol **A** increases when changing $SnCl_4$ for BF_3-Et_2O. Stereoelectronic factors appear to favor the synclinal transition state (precursor of **A**), but steric bulk, due to $SnCl_4$, seems to induce the preferential formation of the antiperiplanar transition state.

As a part of his work on intramolecular allylsilane condensation,[259] Molander observed a high diastereoselectivity for the cyclization of the following ethyl acetylacetate derivatives.

TiCl$_4$ 4 equiv. CH$_2$Cl$_2$ 97% yield; >200 : 1

Extension of the chelation-controlled cyclization reaction to unsymmetrical β-diketones in some instances provided excellent regioselectivity between the two distinct ketone groups.

TiCl$_4$ 4 equiv. CH$_2$Cl$_2$
R = Me; 81% yield
R = Et; 71% yield; A : B = 1 : 1.5
R = i-Pr; 66% yield; A : B = 5.3 : 1
R = t-Bu; 71% yield; A : B = >20 : 1
R = Ph; 87% yield; A : B = >20 : 1

$TiCl_4$-promoted cyclization was also effective in generating a spirocyclic ring system with interesting diastereoselectivity.

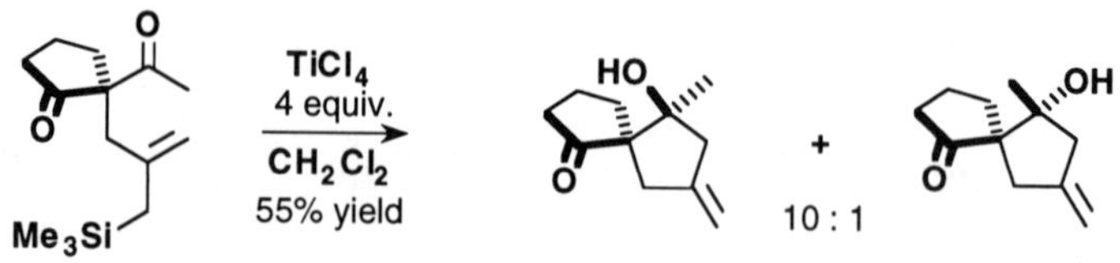

Lee used a tandem Mukaiyama addition - allylsilane cyclization to afford 3-methylenebicyclo[5.4.0]undeca-1,6-diol **D** through intermediate **C**. Unfortunately, the overall yield was low.[272]

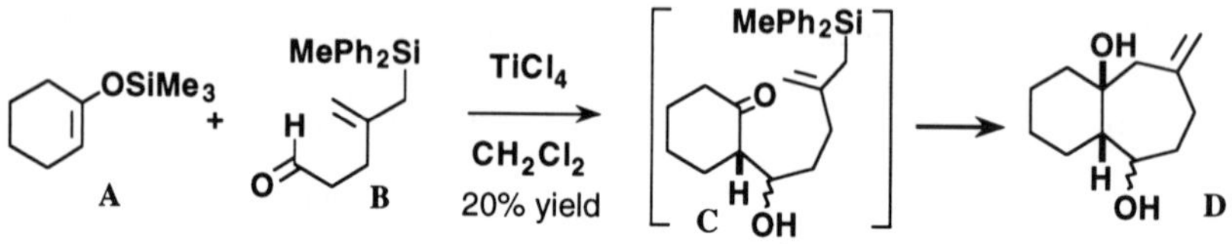

Finally, the milestone contribution from Denmark on the stereochemistry of transition states and particularly, his proposition of a synclinal one, was based on results observed during the cyclization of a 5-trimethylsilylmethyl hex-5-enal unit.[76–79]

V. Conclusion

The stereoselective allylmetallation of aldehydes is a synthetically very useful reaction. Indeed, it provides a powerful means for preparing acyclic structures with one or two new stereogenic centers, and can, in some cases, be substituted with advantages to the aldol reaction.

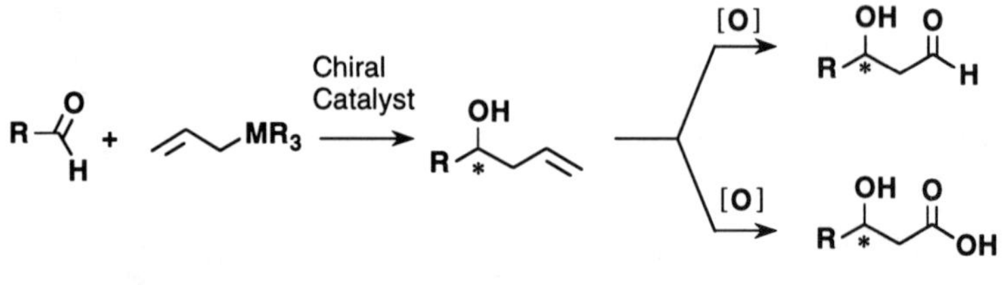

Among the numerous allyl metallic reagents (M = B, Cr, Mg, Ti, Sn, Si) which have been developed over the last ten to fifteen years, allylsilanes and allylstannanes used in conjunction with Lewis acids are of particular interest. Indeed, these reagents, which are easy to prepare and to handle, lead, in the presence of aldehydes, to the formation of homoallylic alcohols with both high chemical yields and high stereoselectivities (or even enantioselectivities).

Finally, recent works have shown that the use of homochiral Lewis acids, used in only 10–20% mol, can lead to the formation of alcohols of high optical purity. Great attention will certainly focus on this area in the future.

References

1. König, K.; Neumann, W.P. *Tetrahedron Lett.* **1967**, 495–498.
2. Servens, C.; Pereyre, M. *J. Organomet. Chem.* **1971**, *26*, C4–C6.
3. Servens, C.; Pereyre, M. *J. Organomet. Chem.* **1972**, *35*, C20–C22.
4. Calas, R.; Dunoguès, J.; Deleris, G.; Pisciotti, F. *J. Organomet. Chem.* **1974**, *69*, C15–C17.
5. Deleris, G.; Dunoguès, J.; Calas, R. *J. Organomet. Chem.* **1975**, *93*, 43–50.
6. Abel, E.W.; Rowley, R.J. *J. Organomet. Chem.* **1975**, *84*, 199–229.
7. Hosomi, A.; Sakurai, H. *Tetrahedron Lett.* **1976**, 1295–1298.
8. Deleris, G.; Dunoguès, J.; Calas, R. *Tetrahedron Lett.* **1976**, 2449–2450.
9. Ojima, I.; Miyazawa, Y.; Kumagai, M. *J. Chem. Soc., Chem. Commun.* **1976**, 927–928.
10. Takuwa, A.; Nishigaichi, Y.; Yamashita, K.; Iwamoto, H. *Chem. Lett.* **1990**, 1761–1764.
11. Craven, A.; Tapolczay, D.J.; Thomas, E.J.; Whitehead, J.W.F. *J. Chem. Soc., Chem. Commun.* **1985**, 145–147.
12. Aono, T.; Hesse, M. *Helv. Chim. Acta* **1984**, *67*, 1448–1452.
13. Ojima, I.; Kumagai, M.; Miyazawa, Y. *Tetrahedron Lett.* **1977**, 1385–1388.
14. Tagliavini, G.; Peruzzo, V.; Plazzogna, G.; Marton, D. *Inorg. Chim. Acta* **1977**, *24*, L47–L48.
15. Sarkar, T.K.; Andersen, N.H. *Tetrahedron Lett.* **1978**, 3513–3516.
16. Maruyama, K.; Naruta, Y. *J. Org. Chem.* **1978**, *43*, 3796–3798.
17. Naruta, Y. *J. Am. Chem. Soc.* **1980**, *102*, 3774–3783.
18. Hosomi, A.; Iguchi, H.; Endo, M.; Sakurai, H. *Chem. Lett.* **1979**, 977–980.
19. Yamamoto, Y.; Yatagai, H.; Naruta, Y. ; Maruyama, K. *J. Am. Chem. Soc.* **1980**, *102*, 7107–7109.
20. Hayashi, T.; Kabeta, K.; Hamachi, I.; Kumada, M. *Tetrahedron Lett.* **1983**, *24*, 2865–2868.
21. Hosomi, A.; Shirahata, A.; Sakurai, H. *Tetrahedron Lett.* **1978**, 3043–3046.
22. Pornet, J. *Tetrahedron Lett.* **1981**, *22*, 453–454.
23. Pornet, J. *Tetrahedron Lett.* **1981**, *22*, 455–456.
24. Pereyre, M.; Quintard, J.P.; Rahm, A. *Tin in Organic Synthesis*, Butterworths, London, **1987**.
25. Yamamoto, Y.; Maruyama, K. *Heterocycles* **1982**, *18*, 357–386.
26. Hoffmann, R.W. *Angev. Chem., Int. Ed. Engl.* **1982**, *21*, 555–642.
27. Yamamoto, Y. *Aldrichimica Acta* **1987**, *20*, 45–49.
28. Yamamoto, Y. *Acc. Chem. Res.* **1987**, *20*, 243–249.
29. Roush, W.R. in *Comprehensive Organic Synthesis*, Trost, B.M.; Fleming, I., Eds., Pergamon, Oxford, **1991**, Vol. 2, Chap. 1.1, pp. 1–53.
30. Marshall, J.A. *Chemtracts–Org. Chem.* **1992**, 75–98.
31. Colvin, E.W. *Chem. Soc. Rev.* **1978**, *7*, 15–64.
32. Chan, T.H.; Fleming, I. *Synthesis* **1979**, 761–786.
33. Fleming, I. in *Comprehensive Organic Synthesis*, Barton, D.H.R., Ollis, W.D., Eds, Pergamon, Oxford, **1979**, Vol. 3, p 541.
34. Colvin, E.W. *Silicon in Organic Chemistry*, Butterworths, London, **1981**.
35. Calas, R. *J. Organomet. Chem.* **1980**, *200*, 11–36.
36. Sakurai, H. *Pure Appl. Chem.* **1982**, *54*, 1–22.
37. Weber, W.P. *Silicon Reagents for Organic Synthesis*, Springer Verlag, Berlin, **1983**.

38. Fleming, I.; Terrett, N.K. *Pure Appl. Chem.* **1983**, *55*, 1707–1713.
39. Schinzer, D. *Janssen Chim. Acta* **1988**, *6*, 11–14.
40. Fleming, I.; Dunoguès, J.; Smithers, R. *Org. React.* **1989**, *37*, 57–575.
41. Birkofer, L.; Stuhl, O. in *The Chemistry of Organic Silicon Compounds*, Patai, S.; Rappoport, Z., Eds.; Wiley and Sons, New York, **1989**; Chap.10.
42. Larson, G.L. in *The Chemistry of Organic Silicon Compounds*, Patai. S.; Rappoport, Z., Eds.; Wiley and Sons, New York, **1989**; Chap. 11, pp. 763–808.
43. Majetich, G. *Organic Synthesis: Theory and Applications*, Hudlicky, T., Ed.; JAI Press, Greenwich, CT, **1989**; Vol. 1, pp. 173–240.
44. Yamamoto, T.; Sasaki, N. *Stereochem. Organomet. Inorg. Compounds*, **1989**, *3*, 363–441. *Chem. Abstr.* **1990**, *113*, 151509.
45. Nishigaichi, Y.; Takuwa, A.; Naruta, Y.; Maruyama, K. *Tetrahedron* **1993**, *49*, 7395–7426.
46. Hosomi, A. *Acc. Chem. Res.* **1988**, *21*, 200–206.
47. Nativi, C.; Ricci, A.; Taddei, M. *Front. Organosilicon Chem. (Proc. Int. Symp. Organosilicon Chem.)* 9th, **1990** (Publ. 1991), 332–343.
48. Bürgi, H.B.; Dunitz, J.D.; Shefter, E. *J. Am. Chem. Soc.* **1973**, *95*, 5065–5067.
49. Bürgi, H.B.; Dunitz, J.D.; Lehn, J.M.; Wipff, G. *Tetrahedron* **1974**, *30*, 1563–1572.
50. Bürgi, H.B. *Angew. Chem. Int. Ed. Engl.* **1975**, *14*, 460–473.
51. Anh, N.T.; Thanh, B.T. *Nouv. J. Chim.* **1986**, *10*, 681–683.
52. Anh, N.T.; Elkaïm, L.; Thanh, B.T.; Maurel, F.; Flament, J.P. *Bull. Soc. Chim. Fr.* **1992**, *129*, 468–477.
53. Andersen, N.H.; McCrae, D.A.; Grotjahn, D.B.; Gabhe, S.Y.; Theodore, L.J.; Ippolito, R.M.; Sarkar, T.K. *Tetrahedron*, **1981**, *37*, 4069–4079.
54. Zimmerman, H.E.; Traxler, M.D.; *J. Am. Chem. Soc.* **1957**, *79*, 1920–1923.
55. Isaacs, N.S.; Maksimovic, L.; Rintoul, G.B.; Young, D.J. *J. Chem. Soc., Chem. Commun.* **1992**, 1749–1750.
56. Nokami, J.; Otera, J.; Sudo, T.; Okawara, R. *Organometallics* **1983**, *2*, 191–193.
57. Matsubara, S.; Wakamatsu, K.; Morizawa, Y.; Tsuboniwa, N.; Oshima, K.; Nozaki, H. *Bull. Chem. Soc. Jpn* **1985**, *58*, 1196–1199.
58. Masuyama, Y.; Hayashi, R.; Otake, K.; Kurusu, Y. *J. Chem. Soc., Chem. Commun.* **1988**, 44–45.
59. Masuyama, Y.; Otake, K.; Kurusu, Y. *Tetrahedron Lett.* **1988**, *29*, 3563–3566.
60. Masuyama, Y.; Takahara, J.P.; Kurusu, Y. *J. Am. Chem. Soc.* **1988**, *110*, 4473–4474.
61. Petrier, C.; Einhorn, J.; Luche, J.L. *Tetrahedron Lett.* **1985**, *26*, 1449–1452.
62. Keck, G.E.; Abbott, D.E.; Boden, E.P.; Enholm, E.J. *Tetrahedron Lett.* **1984**, *25*, 3927–3930.
63. Keck, G.E.; Castellino, S.; Andrus, M.B. in *Selectivities in Lewis Acid Promoted Reactions*, Schinzer, D., Ed.; Kluwer Academic Publishers, Dordrecht, Holland, **1989**, pp. 73–105.
64. Sato, F.; Iida, K.; Iijima, S.; Moriya, H.; Sato, M. *J. Chem. Soc., Chem. Commun.* **1981**, 1140–1141.
65. Hayashi, T.; Konishi, M.; Ito, H.; Kumada, M. *J. Am. Chem. Soc.* **1982**, *104*, 4962–4963.

66. Hayashi, T.; Konishi, M.; Kumada, M. *J. Am. Chem. Soc.* **1982**, *104*, 4963–4965.
67. Hayashi, T.; Kabeta, K.; Hamachi, I.; Kumada, M. *Tetrahedron Lett.* **1983**, *24*, 2865–2868.
68. Hayashi, T.; Konishi, M.; Kumada, M. *J. Org. Chem.* **1983**, *48*, 281–282.
69. Hayashi, T.; Kabeta, K.; Yamamoto, T.; Tamao, K.; Kumada, M. *Tetrahedron Lett.* **1983**, *24*, 5661–5664.
70. Buckle, M.J.C.; Fleming I. *Tetrahedron Lett.* **1993**, *34*, 2383–2386.
71. Danheiser, R.L.; Carini, D.J.; Kwasigroch, C.A. *J. Org. Chem.* **1986**, *51*, 3870–3878.
72. Fleming, I.; Kindon, N.D.; Sarkar, A.K. *Tetrahedron Lett.* **1987**, *28*, 5921–5924.
73. Fleming, I.; Higgins, D.; Sarkar, A.K. in *Selectivities in Lewis Acid Promoted Reactions*, Schinzer, D., Ed.; Kluwer Academic Publishers, Dordrecht, Holland, **1989**, pp. 265–280.
74. Reetz, M.T.; Sauerwald, M. *J. Org. Chem.* **1984**, *49*, 2293–2295.
75. Mikami, K.; Maeda, T.; Kishi, N.; Nakai, T. *Tetrahedron Lett.* **1984**, *25*, 5151–5154.
76. Denmark, S.E.; Weber, E.J. *Helv. Chim. Acta* **1983**, *66*, 1655–1660.
77. Denmark, S.E.; Henke, B.R.; Weber, E. *J. Am. Chem. Soc.* **1987**, *109*, 2512–2514.
78. Denmark, S.E.; Weber, E.J. *J. Am. Chem. Soc.* **1984**, *106*, 7970–7971.
79. Denmark, S.E.; Weber, E.J.; Wilson, T.M.; Willson, T.M. *Tetrahedron* **1989**, *45*, 1053–1065.
80. Yamamoto, Y.; Komatsu, T.; Maruyama, K. *J. Chem. Soc., Chem. Commun.* **1983**, 191–192.
81. Keck, G.E.; Abbott, D.E. *Tetrahedron Lett.* **1984**, *25*, 1883–1886.
82. Naruta, Y.; Nishigaichi, Y.; Maruyama, Y. *Tetrahedron* **1989**, *45*, 1067–1078.
83. Denmark, S.E.; Wilson, T.; Willson, T.M. *J. Am. Chem. Soc.* **1988**, *110*, 984–986.
84. Keck, G.E.; Andrus, M.B.; Castellino, S. *J. Am. Chem. Soc.* **1989**, *111*, 8136–8141.
85. Yamamoto, Y.; Maeda, N.; Maruyama, K. *J. Chem. Soc., Chem. Commun.* **1983**, 742–743.
86. Coppi, L.; Ricci, A.; Taddei, M. *Tetrahedron Lett.* **1987**, *28*, 973–976.
87. Gambaro, A.; Boaretto, A.; Marton, D.; Tagliavini, G. *J. Organomet. Chem.* **1983**, *254*, 293–304.
88. Boaretto, A.; Furlani, D.; Marton, D.; Tagliavini, G. *J. Organomet. Chem.* **1986**, *299*, 157–167.
89. Wei, Z.Y.; Wang, D.; Li, J.S.; Chan, T.H. *J. Org. Chem.* **1989**, *54*, 5768–5774.
90. Boaretto, A.; Marton, D.; Tagliavini, G.; Gambaro, A. *Inorg. Chim. Acta* **1983**, *77*, L153–L154.
91. Yamamoto, Y.; Yatagai, H.; Ishihara, Y.; Maeda, N.; Maruyama, K. *Tetrahedron* **1984**, *40*, 2239–2246.
92. Yatagai, H.; Yamamoto, Y.; Maruyama, K. *J. Am. Chem. Soc.* **1980**, *102*, 4548–4550.
93. Keck, G.E.; Enholm, E.J. *J. Org. Chem.* **1985**, *50*, 146–147.
94. Koreeda, M.; Tanaka, Y. *Chem. Lett.* **1982**, 1299–1302.
95. Koreeda, M.; Tanaka, Y. *Chem. Lett.* **1982**, 1297–1298.
96. Hull, C.; Mortlock, S.V.; Thomas, E.J. *Tetrahedron Lett.* **1987**, *28*, 5343–5346.

97. Hull, C.; Mortlock, S.V.; Thomas, E.J. *Tetrahedron* **1989**, *45*, 1007–1015.
98. Young, D.; Kitching, W. *Aust. J. Chem.* **1985**, *38*, 1767–1777.
99. Nishigaichi, Y.; Fujimoto, M.; Takuwa, A. *J. Chem. Soc., Perkin Trans 1* **1992**, 2581–2582.
100. Naruta, Y.; Nagai, N.; Arita, Y.; Maruyama, K. *Chem. Lett.* **1983**, 1683–1686.
101. Bailey, T.R.; Garipati, R.S.; Morton J.A.; Weinreb, S.M. *J. Am. Chem. Soc.* **1984**, *106*, 3240–3245.
102. Marshall, J.A.; Wang, X.-j. *J. Org. Chem.* **1990**, *55*, 6246–6248.
103. Sakurai, H.; Eriyama, Y.; Kamiyama, Y.; Nakadaira, Y. *J. Organomet. Chem.* **1984**, *264*, 229–237.
104. Brouard, C.; Pornet, J.; Miginiac, L. *Tetrahedron* **1992**, *48*, 2385–2400.
105. Gambaro, A.; Marton, D.; Tagliavini, G. *J. Organomet. Chem.* **1981**, *210*, 57–62.
106. Gambaro, A.; Ganis, P.; Marton, D.; Peruzzo, V.; Tagliavini, G. *J. Organomet. Chem.* **1982**, *231*, 307–314.
107. Gambaro, A.; Boaretto, A.; Marton, D.; Tagliavini, G. *J. Organomet. Chem.* **1984**, *260*, 255–262.
108. Boaretto, A.; Marton, D.; Tagliavini, G.; Ganis, P. *J. Organomet. Chem.* **1987**, *321*, 199–207.
109. Boaretto, A.; Marton, D.; Tagliavini, G.; Gambaro, A. *Inorg. Chim. Acta* **1983**, *77*, L196–L197.
110. Yano, K.; Baba, A.; Matsuda, H. *Bull. Chem. Soc. Jpn* **1992**, *65*, 66–70.
111. Hollis, T.K.; Robinson, N.P.; Whelan, J.; Bosnich, B. *Tetrahedron Lett.* **1993**, *34*, 4309–4312.
112. Itoh, K.; Fukui, M.; Kurachi, Y. *J. Chem. Soc., Chem. Commun.* **1977**, 500–501.
113. Evans, D.A.; Gauchet–Prunet, J.A. *J. Org. Chem.* **1993**, *58*, 2446–2453.
114. Quintard, J.–P.; Elissondo, B.; Pereyre, M. *J. Org. Chem.* **1983**, *48*, 1559–1560.
115. Quintard, J.–P.; Dumartin, G.; Elissondo, B.; Rahm, A.; Pereyre, M. *Tetrahedron* **1989**, *45*, 1017–1028.
116. Mekhalfia, A.; Marko, I.E.; Adams, H. *Tetrahedron Lett.* **1991**, *32*, 4783–4786.
117. Marko, I.E.; Mekhalfia, A.; Bayston, D.J.; Adams, H. *J. Org. Chem.* **1992**, *57*, 2211–2213.
118. Guyot, B.; Pornet, J.; Miginiac, L. *J. Organomet. Chem.* **1989**, *373*, 279–288.
119. Marko, I.E.; Bayston, D.J. *Tetrahedron Lett.* **1993**, *34*, 6595–6598.
120. Nishigaichi, Y.; Takuwa, A.; Jodai, A. *Tetrahedron Lett.* **1991**, *32*, 2383–2386.
121. Naruta, Y.; Nishigaichi, Y.; Maruyama, K. *Chem. Lett.* **1988**, 225–228.
122. Nishigaichi, Y.; Fujimoto, M.; Nakayama, K.; Takuwa, A.; Hamada, K.; Fujiwara, T. *Chem. Lett.* **1992**, 2339–2342.
123. Koreeda, M.; Tanaka, Y. *Tetrahedron Lett.* **1987**, *28*, 143–146.
124. Marshall, J.A.; DeHoff, B.S. *J. Org. Chem.* **1986**, *51*, 863–872.
125. Sano, H.; Okawara, M.; Ueno, Y. *Synthesis* **1984**, 933–935.
126. Keck, G.E.; Palani, A. *Tetrahedron Lett.* **1993**, *34*, 3223–3224.
127. Pornet, J.; Rayadh, A.; Miginiac, L. *Tetrahedron Lett.* **1988**, *29*, 3065–3068.
128. Ochiai, M.; Fujita, E. *J. Chem. Soc., Chem. Commun.* **1980**, 1118.

129. Nativi, C.; Palio, G.; Taddei, M. *Tetrahedron Lett.* **1991**, *32*, 1583–1586.
130. Tanaka, K.; Yoda, H.; Isobe, Y.; Kaji, A. *Tetrahedron Lett.* **1985**, *26*, 1337-1340.
131. Tanaka, K.; Yoda, H.; Isobe, Y.; Kaji, A. *J. Org. Chem.* **1986**, *51*, 1856–1866.
132. Marshall, J.A.; Gung, W.Y. *Tetrahedron* **1989**, *45*, 1043–1052.
133. Marshall, J.A.; Welmaker, G.S.; Gung, W.Y. *J. Am. Chem. Soc.* **1991**, *113*, 647-656.
134. Marshall, J.A.; Gung, W.Y. *Tetrahedron Lett.* **1989**, *30*, 2183–2186.
135. Yamamoto, Y.; Kobayashi, K.; Okano, H.; Kadota, I. *J. Org. Chem.* **1992**, *57*, 7003-7005.
136. Franciotti, M.; Mordini, A.; Taddei, M. *Synlett* **1992**, 137–138.
137. Franciotti, M.; Mann, A.; Mordini, A.; Taddei, M. *Tetrahedron Lett.* **1993**, *34*, 1355-1358.
138. Pratt, A.J.; Thomas, E.J. *J. Chem. Soc., Chem. Commun.* **1982**, 1115–1117.
139. Jephcote, V.J.; Pratt, A.J.; Thomas, E.J. *J. Chem. Soc., Chem. Commun.* **1984**, 800–802.
140. Gung, B.W.; Peat, A.J.; Snook, B.M.; Smith, D.T. *Tetrahedron Lett.* **1991**, *32*, 453–456.
141. Houk, K.N.; Duh, H.–Y.; Wu, Y.–D.; Moses, S.R. *J. Am. Chem. Soc.* **1986**, *108*, 2754–2755.
142. McNeill, A.H.; Thomas, E.J. *Tetrahedron Lett.* **1990**, *31*, 6239–6242.
143. McNeill, A.H.; Thomas, E.J. *Tetrahedron Lett.* **1992**, *33*, 1369–1372.
144. Teerawutgulrag, A.; Thomas, E.J. *J. Chem. Soc., Perkin Trans. 1* **1993**, 2863–2864.
145. McNeill, A.H.; Thomas, E.J. *Tetrahedron Lett.* **1993**, *34*, 1669–1672.
146. Carey, J.S.; Thomas, E.J. *Tetrahedron Lett.* **1993**, *34*, 3935–3938.
147. Panek, J.S.; Cirillo, P.F. *J. Org. Chem.* **1993**, *58*, 999–1002.
148. Panek, J.S.; Yang, M.; Xu, F. *J. Org. Chem.* **1992**, *57*, 5790–5792.
149. Panek, J.S.; Yang, M. *J. Am. Chem. Soc.* **1991**, *113*, 9868–9870.
150. Fleming, I.; Sanderson, P.E.J. *Tetrahedron Lett.* **1987**, *28*, 4229–4232.
151. Otera, J.; Kawasaki, Y.; Mizuno, H.; Shimizu, Y. *Chem. Lett.* **1983**, 1529–1532.
152. Coppi, L.; Mordini, A.; Taddei, M. *Tetrahedron Lett.* **1987**, *28*, 969–972.
153. Nativi, C.; Ravida, N.; Ricci, A.; Seconi, G.; Taddei, M. *J. Org. Chem.* **1991**, *56*, 1951–1955.
154. Wei, Z.Y.; Li, J.S.; Wang, D.; Chan, T.H. *Tetrahedron Lett.* **1987**, *28*, 3441–3444.
155. Wei, Z.Y.; Wang, D.; Li, J.S.; Chan, T.H. *J. Org. Chem.* **1989**, *54*, 5768–5774.
156. Chan, T.H.; Wang, D. *Tetrahedron Lett.* **1989**, *30*, 3041–3044.
157. Imwinkelried, R.; Seebach, D. *Angew. Chem., Int. Ed. Engl.* **1985**, *24*, 765–766.
158. Furuta, K.; Mouri, M.; Yamamoto, H. *Synlett* **1991**, 561–562.
159. Ishihara, K.; Mouri, M.; Gao, Q.; Maruyama, T.; Furuta, K.; Yamamoto, H. *J. Am. Chem. Soc.* **1993**, *115*, 11490–11495.
160. Marshall, J.A.; Tang, Y. *Synlett* **1992**, 653–654.
161. Aoki, S.; Mikami, K.; Terada, M.; Nakai, T. *Tetrahedron* **1993**, *49*, 1783–1792.
162. Roush, W.R.; Blizzard, T.A.; Basha, F.Z. *Tetrahedron Lett.* **1982**, *23*, 2331–2334.

163. Costa, A.L.; Piazza, M.G.; Tagliavini, E.; Trombini, C.; Umani–Ronchi, A. *J. Am. Chem. Soc.* **1993**, *115*, 7001–7002.
164. Keck, G.E.; Geraci, L.S. *Tetrahedron Lett.* **1993**, *34*, 7827–7828.
165. Keck, G.E.; Tarbet, K.H.; Geraci, L.S. *J. Am. Chem. Soc.* **1993**, *115*, 8467–8468.
166. Keck, G.E.; Krishnamurthy, D.; Grier, M.C. *J. Org. Chem.* **1993**, *58*, 6543–6544.
167. Cram, D.J.; Abd Elhafez, F.A. *J. Am. Chem. Soc.* **1952**, *74*, 5828–5835.
168. Cherest, M.; Felkin, H.; Prudent, N. *Tetrahedron Lett.* **1968**, 2199–2204.
169. Anh, N.T.; Eisenstein, O. *Nouv. J. Chim.* **1977**, *1*, 61–70.
170. Anh, N.T. *Top. Curr. Chem.* **1980**, *88*, 145–162.
171. Cieplak, A.S. *J. Am. Chem. Soc.* **1981**, *103*, 4540–4552.
172. Cieplak, A.S.; Tait, B.D.; Johnson, C.R. *J. Am. Chem. Soc.* **1989**, *111*, 8447–8462.
173. Wong, S.S.; Paddon–Row, M.N. *J. Chem. Soc., Chem. Commun.* **1990**, 456–458.
174. Kunz, T.; Janowitz, A.; Reissig, H.–U. *Chem. Ber.* **1989**, *122*, 2165–2175.
175. Heathcock, C.H.; Kiyooka, S.–i.; Blumenkopf, T.A. *J. Org. Chem.* **1984**, *49*, 4214–4223.
176. Yamamoto, Y.; Yatagai, H.; Ishihara, Y.; Maeda, N.; Maruyama, K. *Tetrahedron* **1984**, *40*, 2239–2246.
177. Koreeda, M.; Tanaka, Y. *Tetrahedron Lett.* **1987**, *28*, 143–146.
178. Naruta, Y.; Ushida, S.; Maruyama, K. *Chem. Lett.* **1979**, 919–922.
179. Cram, D.J.; Kopecky, K.R. *J. Am. Chem. Soc.* **1959**, *81*, 2748–2755.
180. Leitereg, T.J.; Cram, D.J. *J. Am. Chem. Soc.* **1968**, *90*, 4019–4026.
181. Nogradi, M. *Stereoselective Synthesis*, VCH: Weinheim, **1987**, pp 160–220.
182. Frenking, G.; Köhler, K.F.; Reetz, M.T. *Tetrahedron* **1993**, *49*, 3971–3982.
183. Frenking, G.; Köhler, K.F.; Reetz, M.T. *Tetrahedron* **1993**, *49*, 3983–3994.
184. Reetz, M.T.; Kesseler, K.; Schmidtberger, S.; Wenderoth, B.; Steinbach, R. *Angew. Chem.* **1983**, *95*, 1007–1008; *Angew. Chem. Suppl.* **1983**, 1511–1526.
185. Kiyooka, S.–i; Heathcock, C.H. *Tetrahedron Lett.* **1983**, *24*, 4765–4768.
186. Reetz, M.T.; Jung, A. *J. Am. Chem. Soc.* **1983**, *105*, 4833–4835.
187. Reetz, M.T.; Kesseler, K.; Jung, A. *Tetrahedron Lett.* **1984**, *25*, 729–732.
188. Suzuki, I.; Yamamoto, Y. *J. Org. Chem.* **1993**, *58*, 4783–4784.
189. Keck, G.E.; Boden, E.P. *Tetrahedron Lett.* **1984**, *25*, 265–268.
190. Reetz, M.T.; Kesseler, K. *J. Org. Chem.* **1985**, *50*, 5434–5436.
191. Cornforth, J.W.; Cornforth, R.H.; Mathew, K.K. *J. Chem. Soc.* **1959**, 112–127.
192. Guanti, G.; Banfi, L.; Narisano, E. *Tetrahedron Lett.* **1991**, *32*, 6939–6942.
193. Kunz, T.; Reissig, H.–U. *Angew. Chem., Int. Ed. Engl.* **1988**, *27*, 268–270.
194. Kunz, T.; Janowitz, A.; Reissig, H.–U. *Chem. Ber.* **1989**, *122*, 2165–2175.
195. Vara Prasad, J.V.N.; Rich, D.H. *Tetrahedron Lett.* **1990**, *31*, 1803–1806.
196. Vara Prasad, J.V.N.; Rich, D.H. *Tetrahedron Lett.* **1991**, *32*, 5857–5860.
197. Kamimura, A.; Yoshihara, K.; Marumo, S.; Yamamoto, A.; Nishiguchi, T.; Kakehi, A.; Hori, K. *J. Org. Chem.* **1992**, *57*, 5403–5413.
198. Taniguchi, M.; Oshima, K.; Utimoto K. *Chem. Lett.* **1992**, 2135–2138.
199. Soai, K.; Ishizaki, M. *J. Chem. Soc., Chem. Commun.* **1984**, 1016–1017.
200. Kawanami, Y.; Katayama, K. *Chem. Lett.* **1990**, 1749–1752.

201. Soai, K.; Ishizaki, M. *J. Org. Chem.* **1986**, *51*, 3290–3295.
202. Kiyooka, S.–i.; Nakano, M.; Shiota, F.; Fujiyama, R. *J. Org. Chem.* **1989**, *54*, 5409–5411.
203. Hoffmann, R.W.; Schlapbach, A. *Tetrahedron Lett.* **1993**, *34*, 7903–7906.
204. Danishefsky, S.; DeNinno, M. *Tetrahedron Lett.* **1985**, *26*, 823–824.
205. Sugimura, H. *Tetrahedron Lett.* **1990**, *31*, 5909–5912.
206. Veloo, R.A.; Wanner, M.J.; Koomen, G.–J. *Tetrahedron* **1992**, *48*, 5301–5316.
207. Lashat, S.; Kunz, H. *J. Org. Chem.* **1991**, *56*, 5883–5889.
208. Grese, T.A.; Hutchinson, K.D.; Overman, L.E. *J. Org. Chem.* **1993**, *58*, 2468–2477.
209. Harada, T.; Mukaiyama, T. *Chem. Lett.* **1981**, 1109–1110.
210. Mukaiyama, T.; Suzuki, K.; Yamada, T. *Chem. Lett.* **1982**, 929.
211. Mukaiyama, T.; Yamada, T.; Suzuki, K. *Chem. Lett.* **1983**, 5–8.
212. Nagoaka, H.; Kishi, Y. *Tetrahedron* **1981**, *37*, 3873–3888.
213. Nakada, M.; Urano, Y.; Kobayashi, S.; Ohno, M. *J. Am. Chem. Soc.* **1988**, *110*, 4826–4827.
214. Nakajima, N.; Hamada, T.; Tanaka, T.; Oikawa, Y.; Yonemitsu, O. *J. Am. Chem. Soc.* **1986**, *108*, 4645–4647.
215. Boeckman, R.K.; Charette, A.B.; Asberom, T.; Johnson, B.H. *J. Am. Chem. Soc.* **1987**, *109*, 7553–7555.
216. Keck, G.E.; Boden, E.P. *Tetrahedron Lett.* **1984**, *25*, 1879–1882.
217. Mikami, K.; Kawamoto, K.; Loh, T.–P.; Nakai, T. *J. Chem. Soc., Chem. Commun.* **1990**, 1161–1163.
218. Fleming, I. *Chem. Tracts–Org. Chem.* **1991**, 21–25.
219. Maruyama, K.; Ishihara, Y.; Yamamoto, Y. *Tetrahedron Lett.* **1981**, *22*, 4235–4238.
220. Yamamoto, Y.; Taniguchi, K.; Maruyama, K. *J. Chem. Soc., Chem. Commun.* **1985**, 1429–1431.
221. Yamamoto, Y.; Nemoto, H.; Kikuchi, R.; Komatsu, H.; Suzuki, I. *J. Am. Chem. Soc.* **1990**, *112*, 8598–8599.
222. Santelli–Rouvier, C. *Tetrahedron Lett.* **1984**, *25*, 4371–4374.
223. Yamamoto, Y.; Maeda, N.; Maruyama, K. *J. Chem. Soc., Chem. Commun.* **1983**, 774–775.
224. Chen, M.–Y.; Fang, J.–M. *J. Chem. Soc., Perkin Trans. 1* **1993**, 1737–1741.
225. Grossen, P.; Herold, P.; Mohr, P.; Tamm, C. *Helv. Chim. Acta* **1984**, *67*, 1625–1629.
226. Whitesell, J.K.; Bhattacharya, A.; Buchanan, C.M.; Chen, H.H.; Deyo, D.; James, D.; Liu, C.–L.; Minton, M.A. *Tetrahedron* **1986**, *42*, 2993–3001.
227. Nunn, K.; Mosset, P.; Grée, R.; Saalfrank, R.W. *Angew. Chem., Int. Ed. Engl.* **1988**, *27*, 1188–1189.
228. Marshall, J.A.; Wang, X.–j. *J. Org. Chem.* **1991**, *56*, 3211–3213.
229. Marshall, J.A.; Wang, X.–j. *J. Org. Chem.* **1992**, *57*, 1242–1252.
230. Marshall, J.A.; Wang, X.–j. *J. Org. Chem.* **1990**, *55*, 6246–6248.
231. Trost, B.M.; Bonk, P.J. *J. Am. Chem. Soc.* **1985**, *107*, 1778–1781.
232. Keck, G.E.; Abbott, D.E.; Wiley, M.R. *Tetrahedron Lett.* **1987**, *28*, 139–142.
233. Marshall, J.A.; Luke, G.P. *J. Org. Chem.* **1991**, *56*, 483–485.
234. Marshall, J.A.; Beaudoin, S.; Lewinski, K. *J. Org. Chem.* **1993**, *58*, 5876–5877.
235. Marshall, J.A.; Yashunsky, D.V. *J. Org. Chem.* **1991**, *56*, 5493–5495.

236. Marshall, J.A.; Luke, G.P. *J. Org. Chem.* **1993**, *58*, 6229–6234.
237. Panek, J.S.; Beresis, R. *J. Org. Chem.* **1993**, *58*, 809–811.
238. Lehmkuhl, H.; Kobs, H.–D. *Liebigs Ann. Chem.* **1968**, *719*, 11–20.
239. D'Aniello, F.; Taddei, M. *J. Org. Chem.* **1992**, *57*, 5247–5250.
240. D'Aniello, F.; Géhanne, S.; Taddei, M. *Tetrahedron Lett.* **1992**, *33*, 5621–5624.
241. Reetz, M.T.; Jung, A.; Bolm, C. *Tetrahedron*, **1988**, *44*, 3889–3898.
242. Hioki, H.; Okuda, M.; Miyagi, W.; Itô, S. *Tetrahedron Lett.*, **1993**, *34*, 6131–6134.
243. Kuwajima, I.; Tanaka, T.; Atsumi, K. *Chem. Lett.* **1979**, 779–782.
244. Borelly, S.; Paquette, L.A. *J. Org. Chem.* **1993**, *58*, 2714–2717.
245. Marshall, J.A.; Wang, X.–j. *J. Org. Chem.* **1991**, *56*, 6264–6266.
246. Marshall, J.A.; Wang, X.–j. *J. Org. Chem.* **1992**, *57*, 3387–3396.
247. Marshall, J.A.; Gung, W.Y. *Tetrahedron Lett.* **1989**, *30*, 2183–2186.
248. Keck, G.E.; Tonnies, S.D. *Tetrahedron Lett.* **1993**, *34*, 4607–4610.
249. Sarkar, T.K.; Ghosh, S.K.; Subba Rao, P.S.V.; Satapathi, T.K.; Mamdapur, V.R. *Tetrahedron* **1992**, *48*, 6897–6908.
250. Hosomi, A.; Hashimoto, H.; Sakurai, H. *Tetrahedron Lett.* **1980**, *21*, 951–954.
251. Kuroda, C.; Shimizu, S.; Satoh, J.Y. *J. Chem. Soc., Chem. Commun.* **1987**, 286–288.
252. Kuroda, C.; Shimizu, S.; Satoh, J.Y. *J. Chem. Soc., Perkin Trans. 1* **1990**, 519–524.
253. Nishitani, K.; Yamakawa, K. *Tetrahedron Lett.* **1987**, *28*, 655–658.
254. Kuroda, C.; Inoue, S.; Kato, S.; Satoh, J.Y. *J. Chem. Res. (S)*, **1993**, 62–63; *J. Chem. Res. (M)*, **1993**, 458–471.
255. Nishitani, K.; Yamakawa, K. *Tetrahedron Lett.* **1991**, *32*, 387–390.
256. Itoh, A.; Oshima, K.; Nozaki, H. *Tetrahedron Lett.* **1979**, 1783–1786.
257. Maier, M.E.; Schöffling, B. *Tetrahedron Lett.* **1990**, *31*, 3007–3010.
258. Asao, K.; Iio, H.; Tokoroyama, T. *Tetrahedron Lett.* **1989**, *30*, 6397–6400.
259. Molander, G.A.; Andrews, S.W. *Tetrahedron* **1988**, *44*, 3869–3888.
260. Marshall, J.A.; DeHoff, B.S.; Crooks, S.L. *Tetrahedron Lett.* **1987**, *28*, 527–530.
261. Marshall, J.A.; Crooks, S.L.; DeHoff, B.S. *J. Org. Chem.* **1988**, *53*, 1616–1623.
262. Marshall, J.A.; Gung, W.Y. *Tetrahedron Lett.* **1988**, *29*, 1657–1660.
263. Gevorgyan, V.; Kadota, I.; Yamamoto, Y. *Tetrahedron Lett.* **1993**, *34*, 1313–1316.
264. Yamamoto, Y.; Yamada, J.–i.; Kadota, I. *Tetrahedron Lett.* **1991**, *32*, 7069–7072.
265. Kadota, I.; Matsukawa, Y.; Yamamoto, Y. *J. Chem. Soc., Chem. Commun.* **1993**, 1638–1641.
266. Fan, W.; White, J.B. *J. Org. Chem.* **1993**, *58*, 3557–3562.
267. Li Jisheng; Gallardo, T.; White, J.B. *J. Org. Chem.* **1990**, *55*, 5426–5428.
268. Trost, B.M.; Coppola, B.P. *J. Am. Chem. Soc.* **1982**, *104*, 6879–6881.
269. Trost, B.M.; Remuson, R. *Tetrahedron Lett.* **1983**, *24*, 1129–1132.
270. Trost, B.M.; Adams, B.R. *J. Am. Chem. Soc.* **1983**, *105*, 4849–4850.
271. Andersen, N.H.; McCrae, D.A.; Grotjahn, D.B.; Gabhe, S.Y.; Theodore, L.J.; Ippolito, R.M.; Sarkar, T.K. *Tetrahedron* **1981**, *37*, 4069–4079.
272. Lee, T.V.; Roden, F.S. *Tetrahedron Lett.* **1990**, *31*, 2067–2068.

4. Lewis Acid-Promoted Acetal Substitution Reaction

I. Introduction and History

Due to their stability under basic conditions, acetals are generally used as protecting groups for ketone and aldehyde functions against nucleophiles. Nevertheless, it has been known for some time that acetals can also undergo carbon-carbon bond-forming reactions through nucleophilic addition at the functional carbon atom.[1] Moreover, Müller-Cunradi and Pieroh discovered in 1939 that Lewis acids promote such a reaction between enol ethers and acetals[2] (other nucleophiles have also been used[3,4]).

However, enol ethers are only weak nucleophiles and, therefore, the substitution of acetal alkoxy group was explored mainly in two directions:

- In participation reactions analogous to the cyclization of epoxy-squalene into lanosterol[5] (see Section VI.2).
- By using activated ethylenic derivatives such as allylsilanes, allyltins, silyl enol ethers, etc.

In 1966, Johnson discovered that acetals with appropriately juxtaposed olefinic double bonds undergo facile Lewis acid-catalyzed ring closure to form mono- and bicyclic products in high yields.[6] Moreover, tricyclic products were formed in high yields and stereospecifically with respect to the ring fusions (*trans, anti, trans* configuration).

$SnCl_4$
benzene
89% yield

The stereo-outcome of these biomimetic polyolefin cyclizations is a demonstration of the Stork-Eschenmoser postulate: the cyclization occurs via chairlike conformations of the nascent rings and the addition of each double bond takes place in an antiparallel fashion.[7]

In 1974, Mukaiyama reported that $TiCl_4$ promotes the reaction of silyl enol ethers with acetals to afford the corresponding aldol products.[8] Two years later, he also showed that the same reaction could also lead to β-alkoxy esters providing ketene silyl acetals were used as nucleophiles.[9]

Also in 1976, Hosomi and Sakurai reported that allylsilanes react with various acetals including aliphatic, alicyclic and aromatic acetals in the presence of $TiCl_4$ to afford the corresponding homoallylic ethers.[10] The reaction takes place regiospecifically with allylic transposition.

In 1978, Fleming reported the cyclization of allylsilane acetal **A** into methylene methoxy cyclohexane **B**.[11]

The chemoselectivity of the reaction of allylsilanes with α- and β-keto acetals was studied by Ojima.[12] Thus, the reaction between β-keto acetal **A** and allylsilane **B** proceeded selectively on the acetal moiety to yield β-alkoxy ketone **C**.

However, it appeared that the chemoselectivity of the reaction with α-keto acetal **D** depended dramatically upon the nature of the Lewis acid employed; thus, either α-alkoxy ketone **E** or α-alkoxy alcohol **F** were exclusively obtained depending on whether the reaction was carried out with $AlCl_3$ or $TiCl_4$ respectively.

This difference of reactivity between $AlCl_3$ and $TiCl_4$ was also observed by Hosomi and Sakurai with α,β-unsaturated acetals. Thus, monoallylated compounds were obtained with $AlCl_3$ while the reaction led to diallylated compounds with $TiCl_4$.[13]

OEt OEt + SiMe3 — $AlCl_3$, CH_2Cl_2, 70% yield → OEt

OMe OMe + SiMe3 — $TiCl_4$, CH_2Cl_2, 100% yield → + (1.1 : 1)

The same authors also established that pentadienyltrimethylsilanes react smoothly with acetals with regiospecific transposition of the pentadienyl group[14] and that allyltrimethyltin reacted also with acetals in the presence of Lewis acids to yield the corresponding homoallylic ethers.[15]

Ph OMe OMe + SiMe3 — BF_3 - OEt_2, CH_2Cl_2, 75% yield → Ph OMe

Finally, Itoh studied the reaction of unsymmetrical acetals and showed that during the $TiCl_4$ promoted addition of allyltrimethylsilane to methoxy alkoxy acetals, the methoxy group was selectively substituted.[16]

The substitution of nucleophiles to alkoxy groups of acetals has therefore become over the years an important method for carbon-carbon bond formation in organic synthesis and has been covered in a number of reviews.[17–25]

II. Mechanism

The mechanism of the nucleophilic substitution of acetals has been the object of a number of studies. A priori, two extreme mechanisms can be proposed: (a) direct nucleophilic displacement of Lewis acid-acetal complex **B**, a S_N2 mechanism, (b) prior formation of oxocarbenium ion **C** which then undergoes nucleophilic attack, an S_N1 mechanism.

A (R, O-R', O-R') — L. A. → S_N2: B (L. A.–O-R', O-R') / S_N1: C (R–O(+)-R') — Nu → D (R, O-R', Nu)

Paquette, who studied chirality transfer from silicon to carbon in the course of the allylation of benzaldehyde dimethyl acetal with (–)-α-naphthylphenylmethylallylsilane, established that at least some direct displacement occurs; however, the chemical yield was low and enantiomeric excess only around 5%.[26] Again in favor of a S_N2 mechanism, Yamamoto reported in 1986 evidence that bond making and bond breaking were concerted in a reaction involving an acetal template and directed towards the synthesis of a steroidal side chain (see Section IV.1.a).[27] Finally, Johnson, who developed a procedure for the coupling of acetals with α-silyl ketones or silyl enol ethers which enables the formation of aldol ethers with high d.e., also proposed a S_N2 transition state to account for the results.[28]

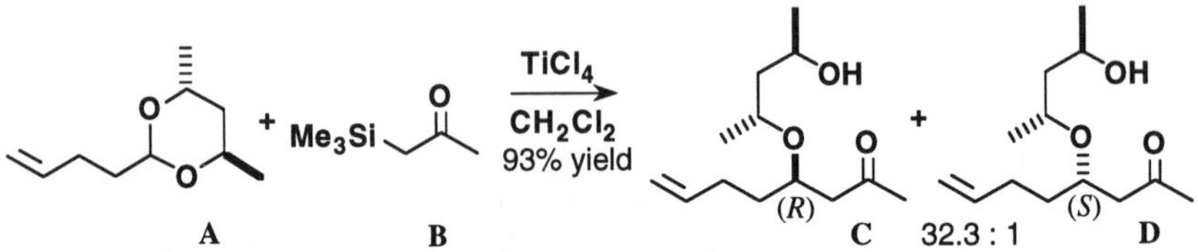

A + Me_3Si B —($TiCl_4$, CH_2Cl_2, 93% yield)→ C (*R*) + D (*S*) 32.3 : 1

S_N2 transition state **B** is stabilized by a lengthening of the 2,3-bond of the ground state **A** with consequently the release of the relatively large 2,4-diaxial H/Me interaction.[29] In the transition state, one of the electron pairs on oxygen atom 1 can participate in the stabilization of the developing positive charge (anomeric effect).[30]

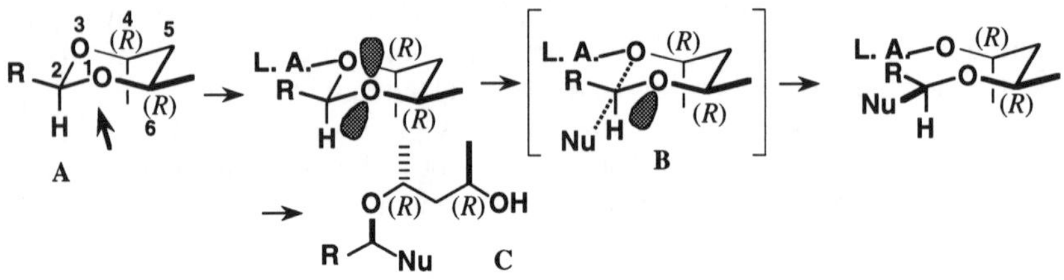

No such interaction is released in the alternative process involving the lengthening of bond 1,2 in transition state **D**, which is therefore less favored.

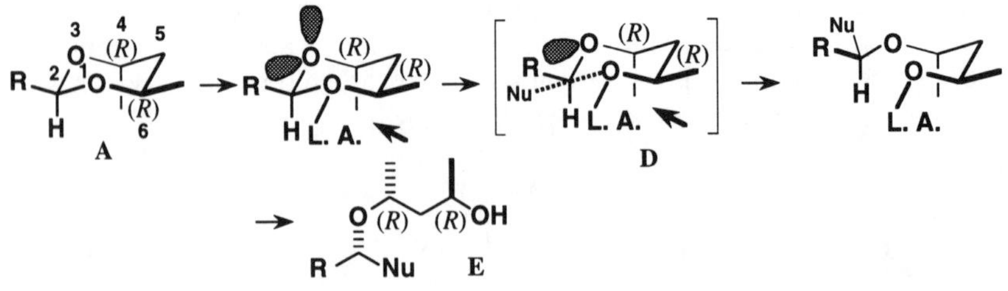

In contrast, a study carried by Mayr proved to be more in favor of an S_N1 type mechanism. Indeed, he determined the relative reactivities of various acetals in BF_3-OEt_2-catalyzed reactions with methylvinyl ether (a weak nucleophile) by competition experiments.[31] The k_{rel} values of the para-substituted benzaldehyde acetals follow a Hammett σ correlation (ρ = –4.6).

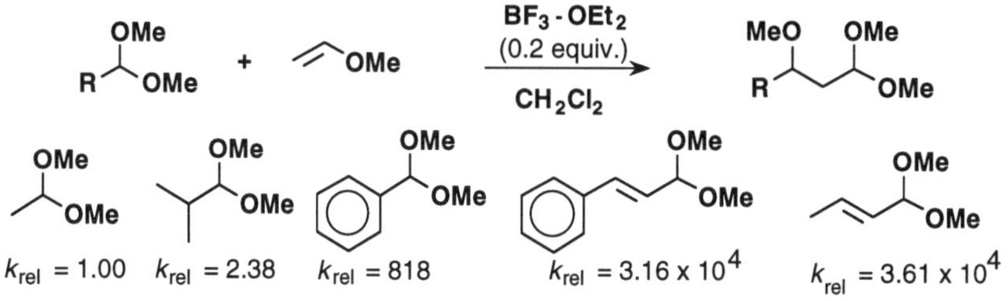

$BF_3 \cdot OEt_2$ (0.2 equiv.), CH_2Cl_2

k_{rel} = 1.00 k_{rel} = 2.38 k_{rel} = 818 k_{rel} = 3.16 x 10^4 k_{rel} = 3.61 x 10^4

Mayr also studied the kinetics of the reactions of allylsilanes, allylgermanes and allylstannanes with carbenium ions.[32]

However, the most significant contributions to a better understanding of the mechanism were made by Denmark, Heathcock and Sammakia. Thus, Denmark used acetal **A**, as an intramolecular model, assuming that it should be possible to deduce the geometry of the reaction predominant transition state from the stereochemistry of the major bicyclic ether **B** or **C**. Moreover, he compared its reactivity with the reactivity of a similar enol ether (*vide infra*).[33]

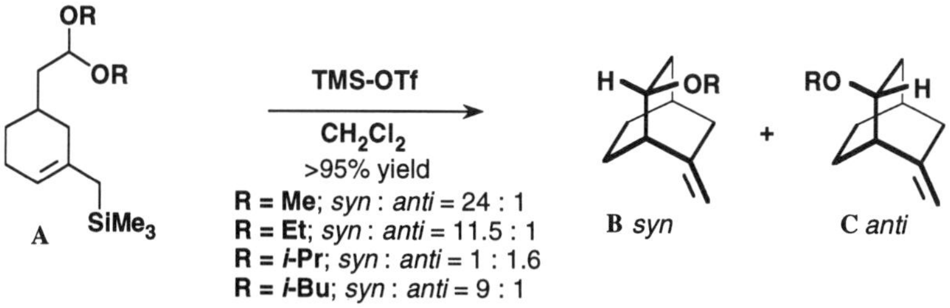

TMS-OTf, CH_2Cl_2, >95% yield

R = Me; *syn* : *anti* = 24 : 1
R = Et; *syn* : *anti* = 11.5 : 1
R = *i*-Pr; *syn* : *anti* = 1 : 1.6
R = *i*-Bu; *syn* : *anti* = 9 : 1

B *syn* **C** *anti*

As already mentioned, acetal **A** may cyclize either via nucleophilic attack on oxocarbenium ion **A'**, obtained by Lewis acid-induced ionization or via direct displacement of Lewis acid complex **A''**. In each case a synclinal or antiperiplanar transition state may occur.

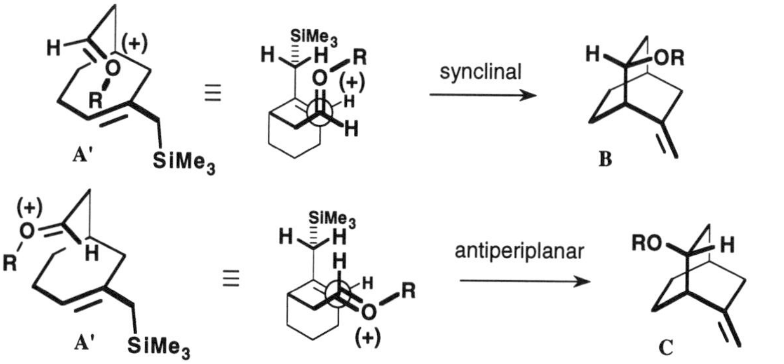

A' ≡ synclinal → B

A' ≡ antiperiplanar → C

A'' synclinal → B

A'' antiperiplanar → C

With trimethylsilyl trifluoromethanesulfonate (trimethylsilyl triflate: TMS-OTf), the *syn*:*anti* ratio (**B**:**C**) did change with acetal structure but not in a regular way. The reactions were very *syn*-selective except for R = *i*-Pr, in which case, it was weakly *anti*-selective.

On the other hand, the course of the reaction was found to be highly dependent on the Lewis acid employed. Monocoordinate Lewis acids gave predominantly *syn* ether **E** while bicoordinate Lewis acids, such as $SnCl_4$, gave nearly equal amounts of *syn* and *anti* products. However, the use of only 0.5 equivalent of $SnCl_4$ restored the *syn*-selectivity but at a lower level.

D → (L. A., CH_2Cl_2) → E *syn* + F *anti*

$SnCl_4$(1 equiv.); 35% yield; *syn* : *anti* = 1 : 1.2
$SnCl_4$(0.5 equiv.); 81% yield; *syn* : *anti* = 2.4 : 1
$AlCl_3$; 33% yield; *syn* : *anti* = 6.1 : 1
BCl_3; 57% yield; *syn* : *anti* = 4.6 : 1
BF_3-OEt_2; 95% yield; *syn* : *anti* = 3.3 : 1
TMS-OTf; 100% yield; *syn* : *anti* = 24 : 1

A ^{13}C NMR study of the Lewis acid complexes of heptanal dimethyl acetal established that addition of BF_3-OEt_2 to the acetal solution induced only partial displacement of Et_2O by the acetal; the new signals being attributable to a 1:1 complex **G**. Addition of 1.0 equivalent of $SnCl_4$ to the acetal solution gave rise to a 1:1 complex **H**. In contrast, addition of 0.5 equivalent of $SnCl_4$ to the acetal solution resulted in the quantitative formation of a single 1:2 complex **I**. The authors concluded that the range of cyclization selectivities is a consequence of the various modes of acetal complexation observed.

+ BF_3-OEt_2 ⇌ (CD_2Cl_2 - $CDCl_3$) **G**

+ $SnCl_4$ (1 equiv.) ⇌ (CD_2Cl_2 - $CDCl_3$) **H**

+ $SnCl_4$ (0.5 equiv.) ⇌ (CD_2Cl_2 - $CDCl_3$) **I**

R = *n*-C_6H_{13}

The triflic acid induced cyclization of (*Z*) and (*E*) enol ethers **J**, which is assumed to modelize an S_N1 process, was then studied; it appeared that (a) the stereoselectivity of the reaction (**B**:**C** ratio) is independent of the configuration of the enol ether double bond, (b) when R = Me or Et, the stereochemical course of the cyclization is radically different from that of the corresponding acetal **A**, being either unselective or weakly *anti*-selective, (c) when R = *i*-Pr, it is closely related to the cyclization of the corresponding acetal **A**. The authors concluded that the cyclization of acetal **A** does not occur via an S_N1 mechanism when R = Me or Et but this might be the case to a large extent when R = *i*-Pr.

J → CF_3SO_3H, CH_2Cl_2 → **B** *syn* + **C** *anti*

R = Me; J(*E*) : J(*Z*) = 24 : 1; 25% yield; *syn* : *anti* = 1.5 : 1
R = Me; J(*E*) : J(*Z*) = 0 : 1; 79% yield; *syn* : *anti* = 1.5 : 1
R = Et; J(*E*) : J(*Z*) = 1 : 0; 76% yield; *syn* : *anti* = 1 : 2.2
R = Et; J(*E*) : J(*Z*) = 1 : 3.3; 74% yield; *syn* : *anti* = 1 : 1.6
R = *i*-Pr; J(*E*) : J(*Z*) = 1 : 0; 84% yield; *syn* : *anti* = 1 : 3
R = *i*-Pr; J(*E*) : J(*Z*) = 0 : 1; 76% yield; *syn* : *anti* = 1 : 2.7

With allylstannane derivative **K**, the stereoselectivity of the reaction also depended on the nature of the Lewis acid. The same *syn*-selectivity, than when using allylsilane **D**, was observed with BF_3-OEt_2 and TMS-OTf.

K → L. A., CH_2Cl_2 → **E** *syn* + **F** *anti*

$SnCl_4$; 25% yield; *syn* : *anti* = 1 : 1.6
$TiCl_4$; 44% yield; *syn* : *anti* = 1 : 1.4
BF_3-OEt_2; 63% yield; *syn* : *anti* = 15.7 : 1
TMS-OTf; 100% yield; *syn* : *anti* = 13.3 : 1

The ^{13}C NMR analysis of the reaction between allylstannane **K** and $SnCl_4$ showed the metathesis of the allylstannane unit into allyltrichloro stannane **L** in which the acetal moiety is intramolecularly complexed.

OMe OMe + $SnCl_4$ (1 equiv.) $\xrightarrow[CD_2Cl_2 - CDCl_3]{}$ OMe OMe $SnCl_3$ + n-Bu_3SnCl

K $SnBu_3$ L

From this study, it appeared that (a) methyl, ethyl and isobutyl acetals **A** react via an S_N2 mechanism with monocoordinate Lewis acids, (b) isopropyl acetals **A** react via an S_N1 mechanism with mono- and bicoordinate Lewis acids, (c) methyl acetals **A** react slowly via an S_N1 mechanism with bicoordinate Lewis acids.

Denmark also worked on the intermolecular reaction between acetal **A** and allyltrimethylsilane. Indeed, the opening of *meso* acetal **A**, in which all three substituents adopt an equatorial position, offers an interesting opportunity to test the timing of bond breaking and making in the substitution reaction. Thus, a single racemic diastereomer **B/B'** should be formed if the reaction occurs through a concerted pathway.[34,35]

A + $SiMe_3$ — S_N2-like inversion → B ≡ (R)(S) OH ; B' ≡ (S)(R) OH

It appeared that the reaction is not stereospecific and exhibits a strong dependence on the nature of R. With the "Ti-blend" catalyst, the reaction was *syn*-selective and *n*-hexyl gave the best result. With $TiCl_4$, selectivity was similar but the reaction was insensitive to the nature of the R group. In contrast, a poor selectivity in favor of diastereomer **C** was observed with BF_3 regardless of the nature of the R group.

R–acetal **A** (S), (R) + allyl-$SiMe_3$ → (L. A., CH_2Cl_2) → **B** [O (R), (S) OH; R (R)] + **C** [O (R), (S) OH; R (R)]

L. A. : $TiCl_4$: $Ti(O\text{-}i\text{-}Pr)_4$ = 6 : 5
R = n-C_6H_{13} ; 100% yield; **B** : **C** = 11.1 : 1
R = *cyclo*-C_6H_{11} ; 98% yield; **B** : **C** = 6.2 : 1
R = n-$C_4H_9C{\equiv}C$; 100% yield; **B** : **C** = 2.5 : 1
R = C_6H_5 ; 30% yield; **B** : **C** = 2.8 : 1
R = p-$CF_3C_6H_4$; 45% yield; **B** : **C** = 3.8 : 1
R = p-$NO_2C_6H_4$; 70% yield; **B** : **C** = 3.4 : 1
L. A. : $TiCl_4$
R = n-C_6H_{13} ; 92% yield; **B** : **C** = 5.1 : 1
R = *cyclo*-C_6H_{11} ; 96% yield; **B** : **C** = 4.8 : 1
R = t-C_4H_9 ; 65% yield; **B** : **C** = 4.2 : 1
R = p-$CF_3C_6H_4$; 96% yield; **B** : **C** = 4.6 : 1
L. A. : BF_3 (gas)
R = n-C_6H_{13} ; 84% yield; **B** : **C** = 1 : 1.9
R = *cyclo*-C_6H_{11} ; 67% yield; **B** : **C** =1 : 2.8
R = p-$CF_3C_6H_4$; 72% yield; **B** : **C** = 1 : 1.7

With the related chiral acetal **D**, in which a methyl group is in axial position and the other in equatorial position (see Section IV), the stereoselectivity remained the same regardless of the Lewis acid. However, from excellent with the "Ti-blend", it dropped to low with BF_3.

C_6H_{13}–acetal **D** (R), (R) + allyl-$SiMe_3$ → (L. A., CH_2Cl_2) → **E** [C_6H_{13}, HO (R), (R), (R); ≡ O (R), (R) OH; C_6H_{13} (R)] + **F** [C_6H_{13} (S), HO, (R), (R); ≡ O (R), (R) OH; C_6H_{13} (S)]

L. A. : $TiCl_4$: $Ti(O\text{-}i\text{-}Pr)_4$ = 6 : 5 ; 100% yield; **E** : **F** = 57.7 : 1
L. A. : $TiCl_4$; 100% yield; **E** : **F** = 6.7 : 1
L. A. : BF_3 (gas) ; 78% yield; **E** : **F** = 2.5 : 1

The same authors also examined, under the "Ti-blend" catalysis, the reaction of different allyl metal (Si, Ge, Sn) reagents with acetal **D**. They found that although adduct **E** was consistently predominant, the **E**:**F** ratio was strongly dependent upon the reagent and varied from 10:1 with allyl triphenylsilane, to 100:1 with allyltrimethylgermane and even over 300:1 with allyltri-*n*-butyltin.[36]

A unified mechanistic scheme involving three distinct ion species (intimate ion pair, external ion pair and separated ions) was proposed to explain the dependence of allylation selectivity on structural factors and on the nature of the Lewis acid and the nucleophile.[35]

L. A. complex intimate ion pair inversion external ion pair oxocarbenium ion

Heathcock showed that the reaction of atropaldehyde acetals **A** with enol ether **B**, derived from pinacolone, gave rise to *syn* ketones **C** as major isomers.[37] He proposed that a steric effect, due to the alkoxy group, was responsible for the stereo-outcome of the reaction. Moreover, since a solvent effect also exists, the *syn*:*anti* (**C**:**D**) ratio increased with solvent polarity, the authors concluded in favor of an S_N1 mechanism.

A + **B** ($TiCl_4$, solvent) → **C** + **D**

R = Me
solv.: CH_2Cl_2; 84% yield; **C** : **D** = 2.5 : 1
solv.: acetonitrile; 92% yield; **C** : **D** = 4.4 : 1
solv.: toluene; 83% yield; **C** : **D** = 1.3 : 1

R = *i*-Pr
solv.: CH_2Cl_2; 82% yield; **C** : **D** = 7.3 : 1
solv.: toluene; 82% yield; **C** : **D** = 2.2 : 1

Acetal **E** reacts with enol ether **B** or ketene silylthioacetal **G** to give essentially Cram products **F** and **H** resulting from an S_N1 mechanism (the dissociation into an oxocarbenium ion being favored by a steric repulsion of the alkoxy groups).

E + **B** ($TiCl_4$, CH_2Cl_2, 94% yield) → **F**

E + **G** ($TiCl_4$, CH_2Cl_2, 94% yield) → **H**

Acetals **I** also undergo clean addition from silyl enol **B** with high stereoselectivity.

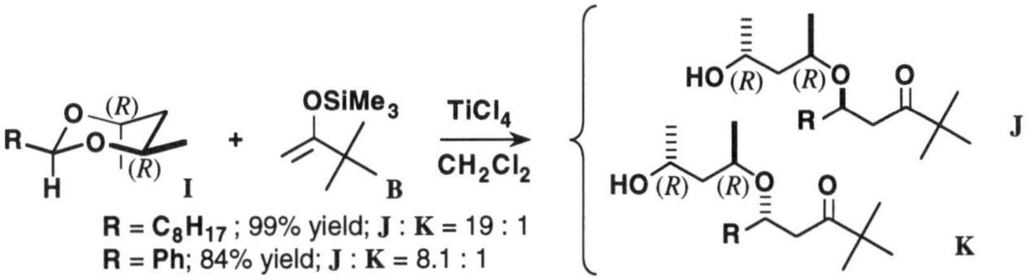

$R = C_8H_{17}$; 99% yield; **J** : **K** = 19 : 1
R = Ph; 84% yield; **J** : **K** = 8.1 : 1

Finally, Sammakia also studied in detail the mechanism of the reaction.[38] Particularly, he provided direct evidence of the existence of an oxocarbenium ion intermediate through experiments involving deuterium labeled acetal **A**.[39] Thus, he showed that, regardless of the experimental conditions, the ratios **B**:**E** and **C**:**D**, corresponding to the proportions of deuterium position in diastereomers resulting from an attack on the *Si* face or the *Re* face of the oxonium ion, were roughly identical. Such a cross-over is clearly incompatible with a pathway involving exclusively a direct displacement of the alkoxy group.

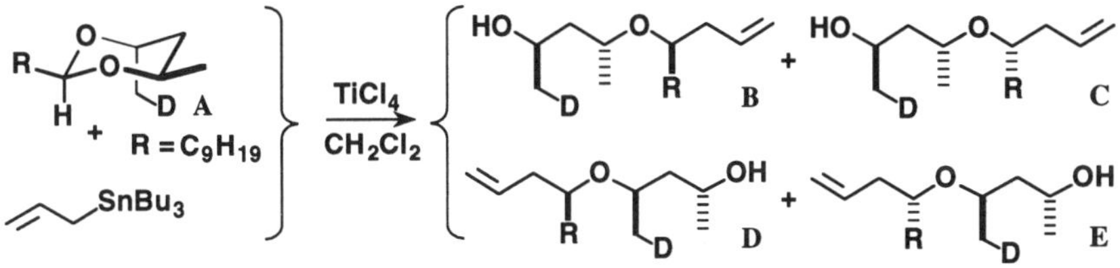

$TiCl_4$: 10 equiv.; rapid add.; **B** + **E** : **C** + **D** = 5 : 1; **B** : **E** = 1.5 : 1; **C** : **D** = 1.2 : 1
$TiCl_4$: 10 equiv.; 2h add.; **B** + **E** : **C** + **D** = 16 : 1; **B** : **E** = 1.7 : 1; **C** : **D** = 1.5 : 1
$TiCl_4$: 0.3 equiv.; 3h add.; **B** + **E** : **C** + **D** = 44 : 1; **B** : **E** = 1.4 : 1; **C** : **D** = 1.2 : 1

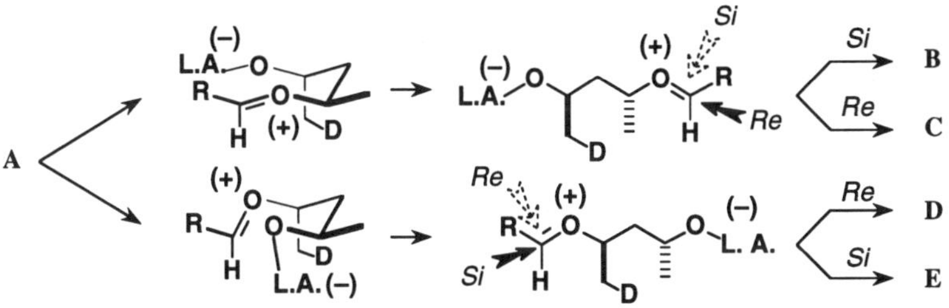

In conclusion, it appears that both mechanisms, the S_N2 and the S_N1, are likely to occur, the predominance of one or the other being highly dependent upon various parameters: the nature (inter- or intramolecular) of the reaction, and the strength of the Lewis acid and of the nucleophile being of particular importance. Indeed, a very acidic Lewis acid would favor the formation of an oxocarbenium ion prior to the attack of the nucleophile while a strong nucleophile would favor a direct displacement of an alkoxy group complexed to a Lewis acid.

III. Intermolecular Reaction

III.1 Reaction of allylsilanes or allylstannanes with achiral acetals

In addition to examples reported in Section I, the following are worth noting. Kuwajima[40] reported that 3-(trimethylsilyl)methylcyclohex-2-enone **A** reacts with various acetals and that, interestingly, the reaction takes place exclusively on the carbon attached to silicon (α-alkylation). Thus, allylsilane **A** reacted with acetal **B** to yield methoxydiketone **C**.[40]

A + B $\xrightarrow[\text{CH}_2\text{Cl}_2,\ 71\%\ \text{yield}]{\text{SnCl}_4}$ C

^{1}H NMR studies suggested the formation of stannyl derivative **D** which can be seen as equivalent to a tin enolate and would in fact be the reactive species.

A $\xrightarrow[\text{CH}_2\text{Cl}_2]{\text{SnCl}_4}$ D + Me_3SiCl

The diastereoselectivity of the reaction of crotylsilanes with acetals was studied by Hosomi and Sakurai.[41] Reactions with aliphatic acetals proceed in a regiospecific and highly *syn*-selective mode, irrespectively of the geometry of crotylsilanes.

$\xrightarrow[\text{CH}_2\text{Cl}_2]{\text{TiCl}_4}$

Z : *E* = 8.1 : 1; 99% yield; *syn* : *anti* = 10.1 : 1
Z : *E* = 1 : 15.7; 90% yield; *syn* : *anti* = 10.1 : 1

With aromatic acetals, stereoselectivity was not very good and depended heavily upon both the nature of the para-substituent and the stereochemistry of the crotylsilane. Thus, with (*Z*)-crotylsilane (*Z*:*E* = 8.1:1) — respectively (*E*)-crotylsilane, *Z*:*E* = 1:15.7 — the *anti*-selectivity (respectively *syn*-selectivity) was enhanced when increasing the electron-withdrawal character of the substituent.

E-crotylsilane :
X = *p*-MeO, 56% yield, *syn* : *anti* = 1 : 1.22
X = *p*-Me, 77% yield, *syn* : *anti* = 1.86 : 1
X = H, 94% yield, *syn* : *anti* = 3 : 1
X = *p*-NC, 84% yield, *syn* : *anti* = 4 : 1

Z-crotylsilane :
X = *p*-MeO, 66% yield, *syn* : *anti* = 1 : 1.17
X = *p*-Me, 87% yield, *syn* : *anti* = 1 : 2.12
X = H, 78% yield, *syn* : *anti* = 1 : 2.57
X = *p*-NC, 92% yield, *syn* : *anti* = 1 : 4

Regioselective allylation of trimethylsilyl substituted mixed acyclic acetal **A** provided selectively homoallylic ether **C**.[42] The formation of oxocarbenium ion **B** stabilized by the presence of a trimethylsilyl group was invoked to rationalize the observed diastereoselectivity.

$MR_3 = SiMe_3$; 65% yield; 5.2 : 1 diastereoselectivity
$MR_3 = SnBu_3$; 66% yield; 2.2 : 1 diastereoselectivity

Phenylthioacetals react, in the presence of $AlCl_3$, with several allylsilanes to give phenylthioethers. In the following example, the reaction is even *syn*-selective.[43]

2-Substituted tetrahydrofurans and tetrahydropyrans were shown to undergo regioselective ring cleavage with simultaneous carbon-carbon bond formation at the highly substituted α-carbon atom upon treatment with allyl or alkenyl silanes and $TiCl_4$.[44]

Disilanes can also be used. Thus, Santelli reported the synthesis of vinyl estratrienones based on the reactivity of 1,8-bis(trimethylsilyl)octa-2,6-diene **A** towards nitroketone ethylene ketals **B**. Indeed, the $TiCl_4$-promoted reaction between **A** and **B** led to a mixture of adducts **C**, **D** and **E** which underwent further transformation towards vinyl estratrienones (with n = 1).[45]

n = 1; 35% yield; **C** : **D** : **E** = 1.7 : 5.7 : 1
n = 2; 45% yield; **C** : **D** : **E** = 2.8 : 1.8 : 1

The same disilane **A** was also used for the preparation of bicyclic alcohol **C**, obtained as a single stereomer, from pentan-2,4-dione mono ketal **B**. A deuterium labeling established the mechanism of the reaction and, particularly, its regioselectivity[46] (see also ref.12).

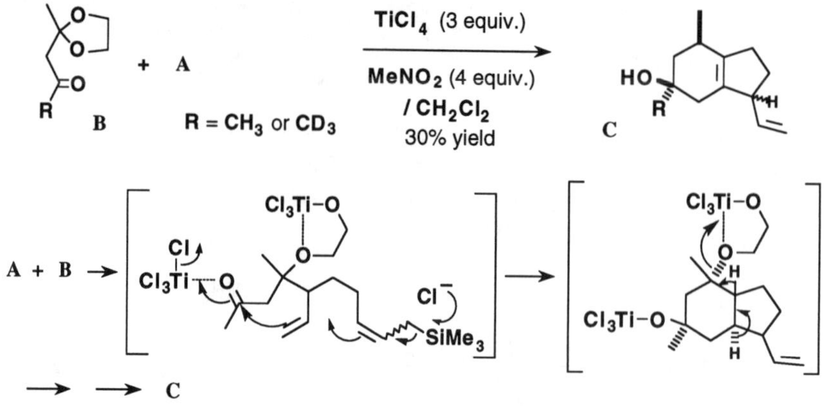

Another disilane, 2-trimethylsilylmethyl allyltrimethylsilane **A**, was used by Miginiac with various acetals or bis-acetals. Thus, diethoxy methylene cycloheptane **C** was prepared from bis-acetal **B**.[47]

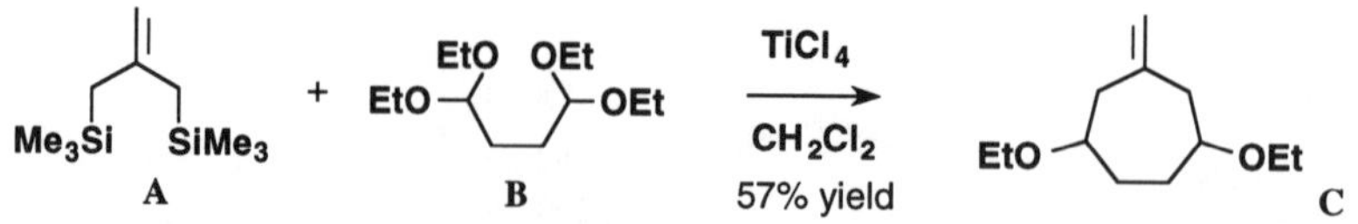

Even more interesting is the possibility of promoting the reaction under catalytic conditions. Thus, significant advances were due to Noyori who found that use of TMS-OTf as promoter makes the allylation process catalytic.[48,49]

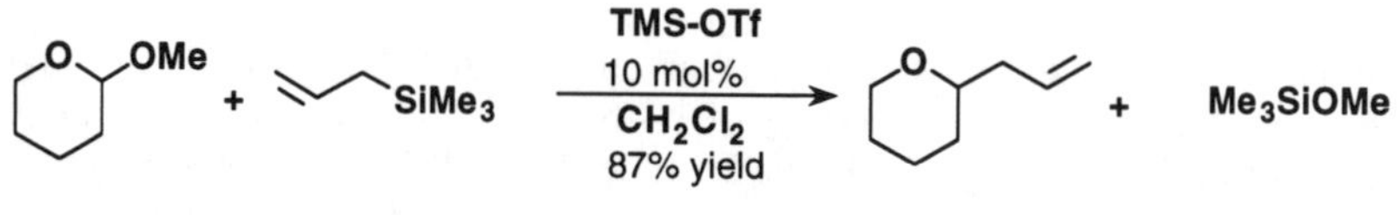

The catalyzed reaction between 4-*tert*-butylcyclohexanone dimethyl acetal **A** and allyltrimethylsilane is highly stereoselective as it produces a 13:1 mixture of ethers **B** and **C**.

Apparently the trimethylsilyl moiety, which has a strong oxophile character, serves as the chain carrier and the following mechanism was proposed.

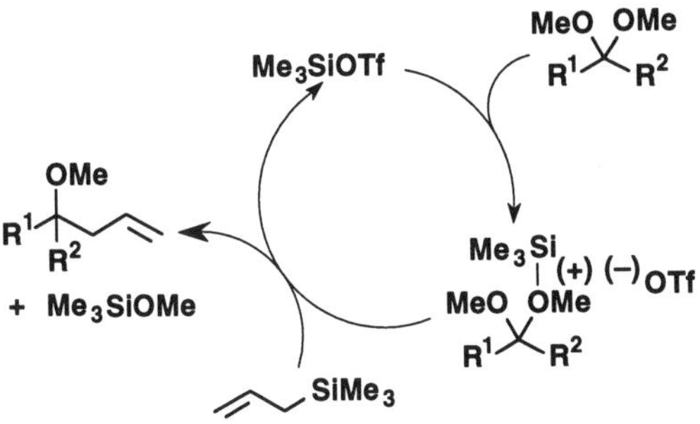

In a study directed towards the understanding of the competition of the reaction of allylsilane with protodesilylation, Fredj also reported the reaction of cyclic allylsilane with acetaldehyde diethyl acetal under TMS-OTf catalysis.[50]

Hosomi and Sakurai obtained homoallylic ethers, with regiospecific transposition of the allyl group when using catalytic amounts of iodotrimethylsilane (10 mol%).[51] Mukaiyama established that catalytic amounts (5 mol%) of trityl perchlorate ($Ph_3C^+ClO_4^-$) or diphenylboryl triflate ($Ph_2B^+OTf^-$) were enough to induce the substitution reaction. With diphenylboryl triflate, it was presumed that the boron cation stabilized by two phenyl groups acts as an activator.[52]

More recently, Kobayashi showed that scandium trifluoromethanesulfonate ($Sc(OTf)_3$) used in 5 mol% could catalyze the condensation of silyl enol ether on acetals[53] and Bosnich established that $Cp_2Ti(OTf)_2$, used in only 0.5 mol%, could catalyze the addition of allylsilanes to acetals.[54]

Finally, Westerlund demonstrated that 1,3-dithienium tetrafluoroborate **A**, a formyl cation equivalent, prepared by hydride abstraction from 1,3-dithiane, reacted with allylsilanes to give β,γ-unsaturated dithianes such as **B**.[55]

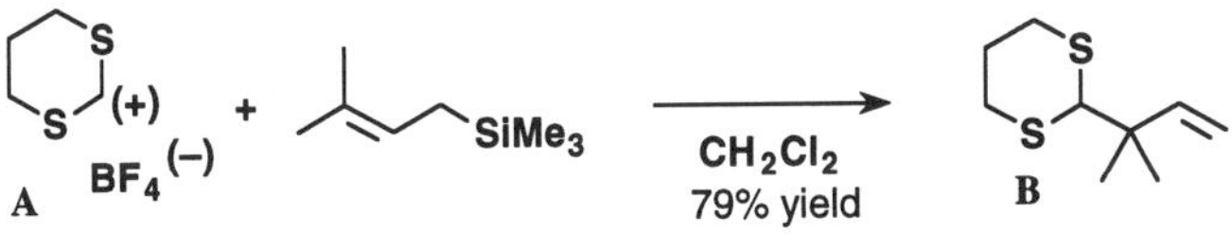

III.2 Reaction of allylsilanes or allylstannanes with chiral acetals

The reaction of allyltrimethylsilane with 2-phenylpropanal dimethylacetal occurred with a stereochemistry predictable by Cram's rule for asymmetric induction (see Chapter 3, Section III).

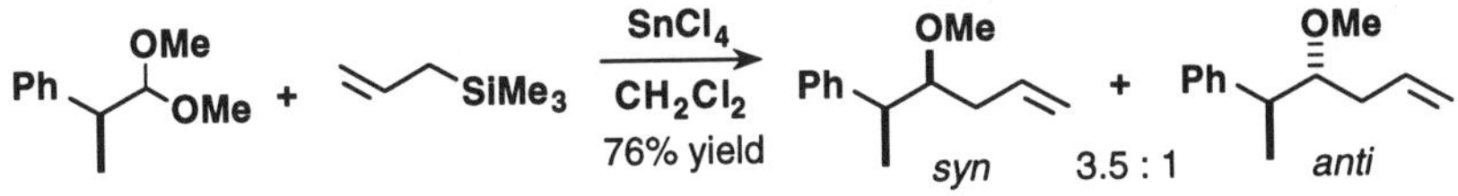

Only modest 1,3-*syn*-selectivity was obtained in reaction of allyltrimethylsilane with β-alkoxy acetals.[56]

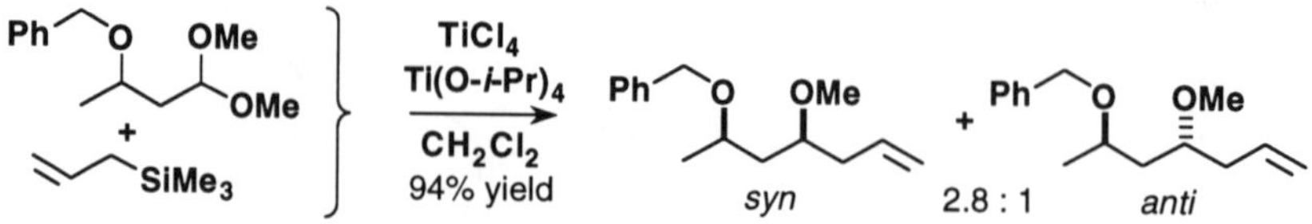

Panek showed that high diastereo- and enantioselectivity could be reached for the preparation of homoallylic ethers resulting from the TMS-OTf catalyzed addition of optically active functionalized allylsilane **B** and α- or β-alkoxy acetals such as **A**[57] or aryl acetal.[58]

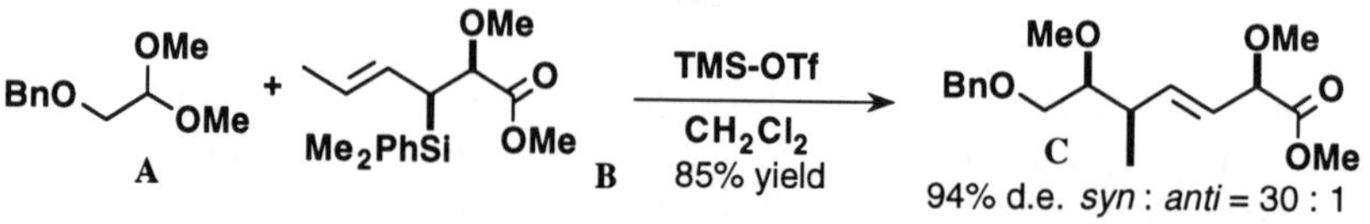

The Lewis acid promoted addition of crotyltri-*n*-butylstannane to the nor-ephedrine-derived 2-methoxy oxazolidine **A** was studied by Scolastico.[59,60] He found that the stereochemistry of the substitution depends on the nature of the Lewis acid and that either oxazolidines **B** or **C** could be selectively prepared.

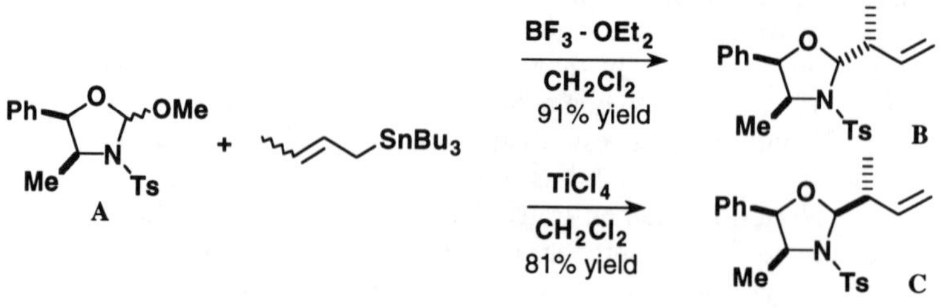

The stereochemical diversity between the BF_3-OEt_2 and the $TiCl_4$ promoted condensations may be due to an equilibration step following the C–C bond formation when $TiCl_4$ is used. Thus, in both cases, the addition of the nucleophile would occur on the *Re* face of the oxocarbenium ion.

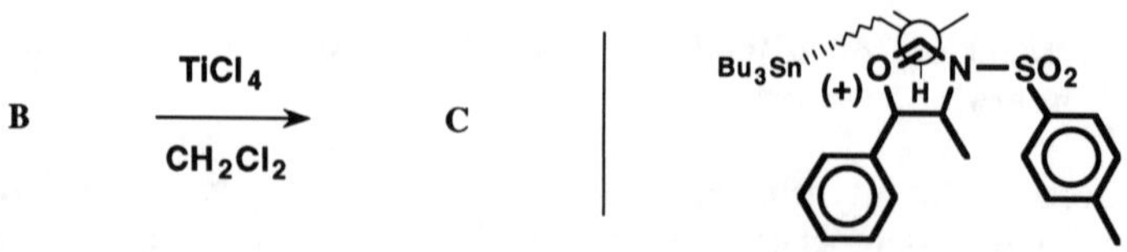

More recently, Zhou reported that the allylation of chiral 6-ethoxy-2-methyl-*N*-tosyl-piperidin-3-ol **A** with allyltrimethylsilane in the presence of $TiCl_4$ led to the predominant formation of adduct **B**.[61] The reaction was then applied to a total synthesis of (+)-azimic acid.[62]

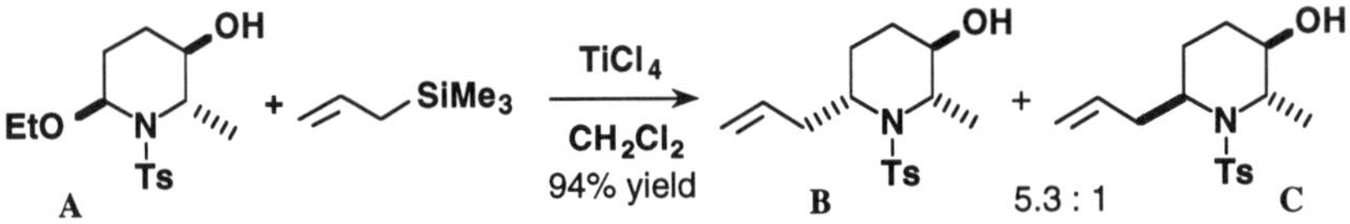

The allylation of cyclic α-hydroxy or α-methoxy amide was also studied.[63–65] With chiral cyclic α-acyloxy amides such as **A**, it was shown that the diastereofacial selectivity of the allylation of the intermediate *N*-acyliminium ions, such as **B**, was strongly dependent upon the Lewis acid involved.[66]

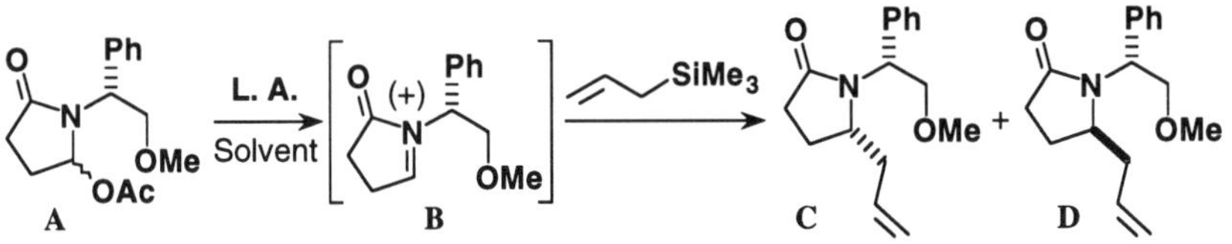

L. A. : BF_3 - OEt_2; solv. CH_2Cl_2; 66% yield; **C** : **D** = 2.8 : 1
L. A. : $TiCl_4$; solv. CH_2Cl_2; 64% yield; **C** : **D** = 4.6 : 1
L. A. : $SnCl_4$; solv. CH_2Cl_2; 91% yield; **C** : **D** = 1 : 5.3
L. A. : $SnCl_4$; solv. $CHCl_3$; 71% yield; **C** : **D** = 1 : 11.5

The stereochemistry of the TMS-OTf catalyzed allylation of 2-acetoxy-2-phenylacetaldehyde monothioacetals **A** or **B** is also dependent on the nature of the metal.[67] With allyltrimethylsilane, the *syn*:*anti* (**C**:**D** = 4:1) ratio remains constant irrespectively of the stereochemistry of the reactant.

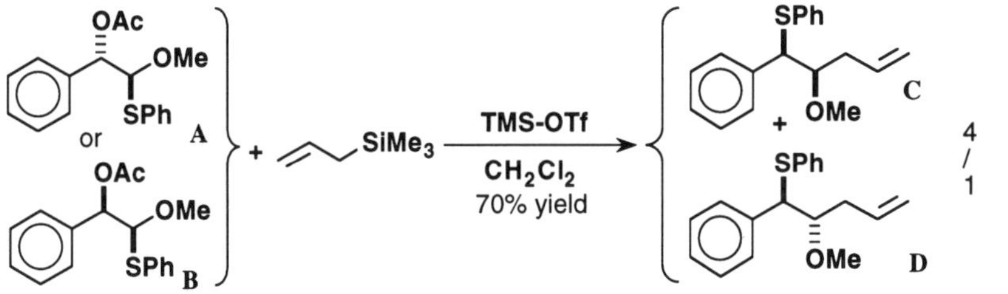

Apparently, in both cases, the reaction proceeds via a common oxonium intermediate **G** which results from the migration of the thiophenyl group of sulfonium ions **E** and **F**.

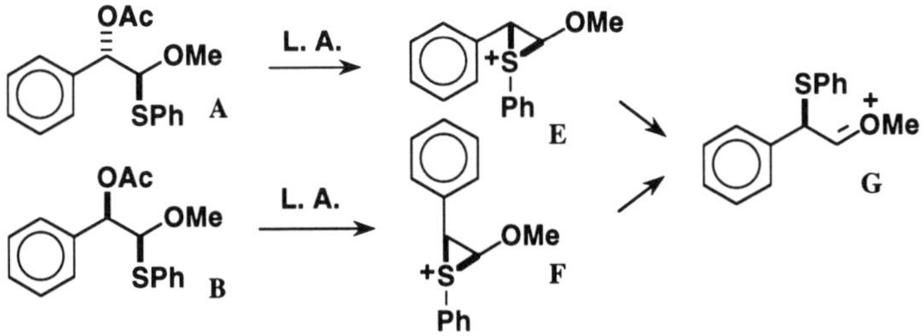

In contrast, with allyltri-*n*-butyltin, the greater nucleophilic character of the reagent enables an attack prior to the formation of the oxonium intermediate **G** (see Section II). An S_N2 mechanism therefore occurs and the **C**:**D** ratio depends upon the stereochemistry of the starting thioacetal.

A + $\diagup\!\!\diagdown$SnBu$_3$ —(TMS-OTf, CH_2Cl_2, 89% yield)→ C + D (8.1 : 1)

B + $\diagup\!\!\diagdown$SnBu$_3$ —(TMS-OTf, CH_2Cl_2, 72% yield)→ C + D (1 : 7.3)

Finally, Mukaiyama showed that optically active homoallylic ethers could be prepared, via the *in situ* formation of an optically active acetal, from achiral aldehydes, allyltrimethylsilane and (*S*)-1-phenyl-1-trimethylsilyloxyethane in the presence of catalytic amount of Ph_2BOTf.[68]

III.3 Reaction of enoxysilanes with acetals

Lee described a new strategy for the synthesis of five-, six-, and seven-membered rings, based on the addition of enoxysilanes such as **A** to functionalized acetals such as **B**.[69,70] **A** and **B** give rise to electrophilic intermediate **C** which undergoes intramolecular allylsilane addition to give bicyclic compound **D**, a precursor of hydrindenone **E**.

A (OSiMe$_3$) + B (MeO, OMe, SiMe$_3$) —(TiCl$_4$, TMS-OTf, CH_2Cl_2, 65% yield)→ [C: (+)O, SiMe$_3$, OMe] → D (OH, H, OMe) → E (O, H, OMe)

Similarly bicyclo[5.3.0]decane derivative **G** was prepared in fair yield from enoxysilane **A** and acetal **F**.

A (OSiMe$_3$) + F (MeO, MeO, SiMe$_3$) —(TiCl$_4$, TMS-OTf, CH_2Cl_2, 65% yield)→ G (OH, H, OMe)

Ishida showed that dienoxysilanes such as **A** react with acetals in the presence of $TiCl_4$ to give δ-alkoxy-α,β-unsaturated aldehydes such as **B** in good yields.[71] Further treatment of **B** with 1,8-diazabicyclo[5.4.0]undec-7-ene (DBU) afforded the corresponding conjugated polyenal **C**.

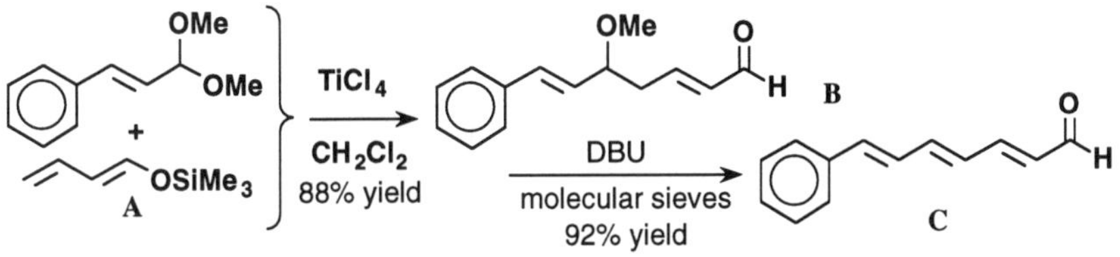

Similarly, from β-cyclocitral dimethyl acetal **D** and dienoxysilane **E**, trienal **G**, a useful intermediate in the syntheses of vitamin A and β-carotene, was obtained through δ-methoxy aldehyde **F**.

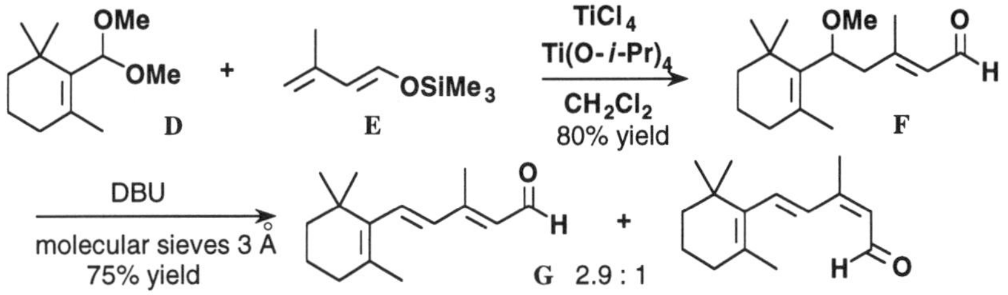

Saigo established that ortho substituted benzaldehyde dimethyl acetal **A** reacts with ketene silyl acetal **B** in the presence of $AlCl_3$ to give the corresponding aldol adduct **C** with high diastereoselectivity.[72]

The authors proposed that this remarkable 1,4-remote asymmetric induction could result from the intramolecular formation of cyclic sulfonium ion **D**, which then underwent an S_N2 inversion to selectively yield β-methoxy ester **C**.

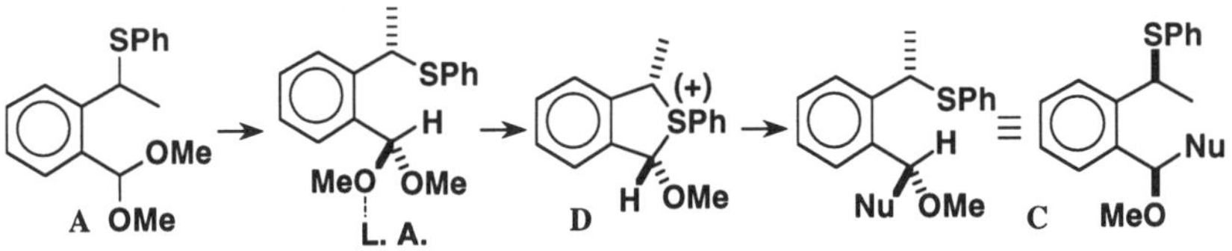

Trost reported that vinylcyclopropyloxysilanes such as **A** react with acetals, in the presence of catalytic amounts of TMS-OTf, to give cyclobutanones such as **B** in very high yield.[73]

Finally, *N*-acyl iminium ions,[74] derived from α-methoxy glycine esters, and norephedrine derived 2-methoxy oxazolidines[60] undergo nucleophilic attack from various silyl nucleophiles including silyl enol ethers and ketene silyl acetals.

III.4 Reaction of other silyl nucleophiles with acetals

α-Allenyl ethers such as **C** or **E** can be obtained upon reaction of acetals such as **A** with propargyltrimethylsilane **B**[75] or enyne silane **D**[76] in the presence of $TiCl_4$.

N-Alkoxycarbonyliminium ions derived from carbamates **A** reacted also with propargyltrimethylsilane to yield oxazinones **B** and allenes **C**.[77] The formation of oxazinones **B** results from the cyclization of vinylic carbocations **D**.

X = H; L. A. : $SnCl_4$; CH_2Cl_2; 48% yield; **B** : **C** = 1 : 0
X = H; L. A. : $EtAlCl_2$; benzene; 45% yield; **B** : **C** = 1 : 2
X = OMe; L. A. : $SnCl_4$; CH_2Cl_2; 56% yield; **B** : **C** = 1 : 0

Finally the reaction of propargyltrimethylsilane with ω-ethoxy lactams in the presence of BF_3-Et_2O afforded ω-allenyl lactams.[78]

Various other silyl nucleophiles have been used. Thus, β-alkoxy carboxylic acids, precursors to optically active alcohols, were obtained from chiral 1,3-dioxan-4-ones under Lewis catalysis in both excellent yield and diastereomeric excess.[79] In connection with results reported in Sections III.2 and III.3, Hoppe also showed that 2-methoxy-1,3-oxazolidines undergo TMS-OTf induced substitution from Me_3SiCN to afford the corresponding 2-cyano-1,3-oxazolidines.[80] Mukaiyama reported that the combined use of catalytic amounts (10 mol%) of $SnCl_4$ and $ZnCl_2$ induces the reaction of acetals such as **A** with 1-trimethylsilyl alkynes and leads to the formation of secondary propargylic ethers such as **B**.[81]

OMe
n-C8H17 OMe
A
+ Me3Si-C≡C-n-C5H11
SnCl4/ZnCl2
CH2Cl2
80% yield
OMe
n-C8H17
n-C5H11
B

Finally, [3+2] annulation reactions were performed by Danheiser who prepared, for example, pyrrolizinone **C** from allenylsilane **B** and pyrrolidinone **A**.[82]

O
NH
OEt
A
+
SiMe2t-Bu
B
TiCl4
CH2Cl2
63% yield
O
N
SiMe2t-Bu
C

IV. Intermolecular Reaction With Acetal Templates

Optically active homoallylic ethers or alcohols can also be prepared from acetal templates derived from achiral aldehydes and homochiral diols. Over the last fifteen years such acetals have proved all their usefulness in asymmetric synthesis.[17] Of particular interest are acetals prepared from diols having a C2 symmetry axis, and among these, acetals prepared from pentan-2,4-diols and butan-2,3-diols. Acetals derived from pentan-2,4-diols are particularly useful since, not only do they induce high stereoselectivity but the chiral auxiliary is easy to remove (vide infra). For a detailed discussion on acetal templates and their cleavage by organometallic reagents, such as organoaluminum compounds,[83] which are not covered here, see the review from Alexakis and Mangeney.[17]

IV.1 Acetals derived from homochiral pentan-2,4-diols

In the course of his study towards the biomimetic cyclization of polyenals (see Section VI.2), Johnson introduced the use of chiral acetals derived from (*R*,*R*)-pentan-2,4-diol and established their usefulness.[84] Johnson also used chiral acetals, derived from (*R*,*R*)-pentan-2,4-diol in intramolecular reactions with various nucleophiles. The reaction leads to hydroxy ethers in which the new chiral center is formed with high enantioselectivity. Both enantiomeric forms of pentan-2,4-diol are readily available through asymmetric hydrogenation of acetylacetone.[85]

IV.1.a Reaction with allylsilanes and allylstannanes

Homoallylic alcohol **B** was produced with 96% d.e. from allyltrimethylsilane and acetal **A**.[86]

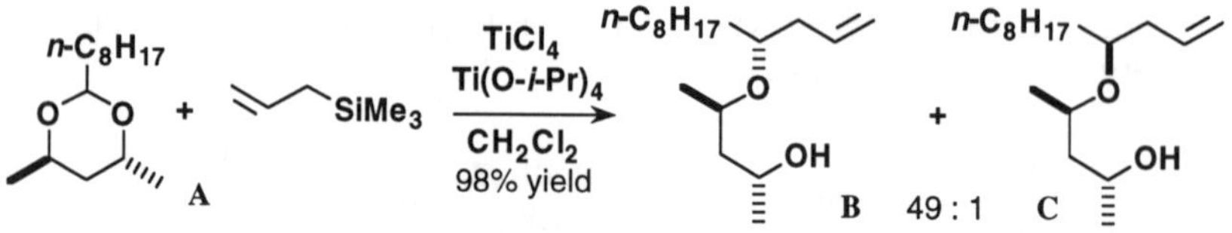

Removal, via ketone **D**, of the chiral auxiliary by an easy two step sequence (oxidation – β-elimination) affords secondary homoallylic alcohol **E** in high yield.

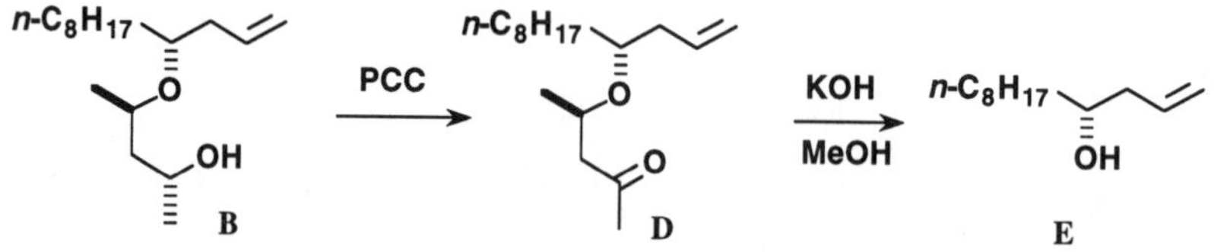

Similarly the coupling of acetal **A** with methallyltrimethylsilane, in the presence of a "Ti-blend", led to alcohol **B** with 97% d.e.[87] Keto lactone **C**, a key intermediate in the synthesis of calcitriol lactone, a major metabolite of vitamin D_3, was synthesized from alcohol **B**.

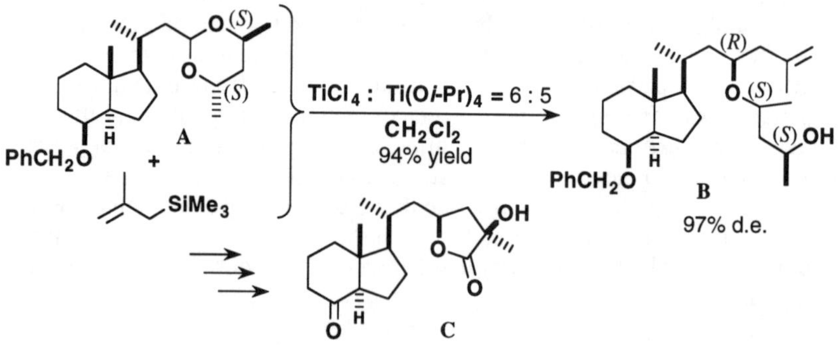

An enantiodivergent synthesis of steroidal side chains was reported by Yamamoto. Hence, on one hand, treatment of acetal **A**, prepared from (*R*,*R*)-pentan-2,4-diol, with allyltrimethylsilane or allyltri-*n*-butyltin followed by removal of the chiral auxiliary (oxidation – β-elimination) gave exclusively or almost exclusively homoallylic alcohol **B**.[27]

R_3M : Me_3Si ; 93% yield ; **B** : **C** = >99 : <1
R_3M : n-Bu_3Sn ; 85% yield ; **B** : **C** = 24 : 1

On the other hand, the reaction of acetal **D**, prepared from (*S*,*S*)-pentan-2,4-diol, produced, after the same work-up, a mixture of alcohols **B** and **C**, with in one case **C** as the major isomer.

R_3M : Me_3Si ; 93% yield ; **B** : **C** = 9 : 1
R_3M : n-Bu_3Sn ; 84% yield ; **B** : **C** = 1 : 2.33
R_3M : Ph_3Sn ; 80% yield ; **B** : **C** = 3.17 : 1

With **A**, the chirality dictated by the acetal template is in the same direction as the chirality predicted by the Felkin-Anh model (in the Felkin-Anh model, the bond with the largest group should provide the greater overlap with the bond breaking, see Chapter 3, Section III). **B** is therefore overwhelmingly produced.

With **D**, the chiralities dictated by the Cram rule and the template are opposite. The fact that **B** is still the major product in two cases indicates that the stereo-outcome of the reaction is governed by the Cram rule and that the Lewis acid was coordinated to the more hindered oxygen atom 3.

These results indicate that the direction of asymmetric induction was dictated primarily by Cram's rule and that the influence of the template was negligible except with allyltri-*n*-butylstannane and **D**. In that particular case,

C was produced predominantly, pointing out that violation of Cram's rule had occurred and the chiral induction was dictated essentially by the template.

Yamamoto also studied this competition, chelation control effect vs. acetal template effect, on more simple molecules.[88] Thus, from (*S*-*R*, *R*) acetal **A**, he observed, regardless of the nucleophilicity of the reagent, a high 1,3-asymmetric induction leading to the formation of chelation adduct **B** with 88% d.e.; indeed a synergy between both effects which operate in the same direction can account for such a result. However, from (*R*-*R*, *R*) isomeric acetal **D**, although the chelation adduct **E** is still produced predominantly, the d.e. fell to 40%. Indeed, in this case, the acetal template should have induced the formation of **F**. The formation of **E** as the major isomer establishes that, here again, the chelation control is more important than the template effect.

(*S*) (*R*) *t*-BuMe$_2$Si (*R*) **A** + MR_3 → $TiCl_4$, CH_2Cl_2 → (*S*) (*S*) OH *t*-BuMe$_2$Si *anti* **B** + (*S*) (*R*) OH *t*-BuMe$_2$Si *syn* **C**

MR_3 = $SiMe_3$; 82% yield; *anti* : *syn* = 15.6 : 1
MR_3 = $SnBu_3$; 88% yield; *anti* : *syn* = 15.6 : 1

(*R*) (*R*) *t*-BuMe$_2$Si (*R*) **D** + $SnBu_3$ → $TiCl_4$, CH_2Cl_2, 88% yield → (*R*) (*R*) OH *t*-BuMe$_2$Si *anti* **E** + (*R*) (*S*) OH *t*-BuMe$_2$Si *syn* **F**

anti : *syn* = 2.3 : 1

Another example of excellent diastereoselectivity is provided with spiroketal **A**, derived from menthone, which underwent $TiCl_4$ promoted selective ring cleavage to yield **B**.[89]

C_6H_{13} C_6H_{13} ≡ C_6H_{13} C_6H_{13} **A** + $SiMe_3$ → $TiCl_4$, CH_2Cl_2, 100% yield → $H_{13}C_6$ C_6H_{13} OH

B >95% d.e.

Finally, Quintard reported the stereoselective cleavage of α-tri-*n*-butylstannylacetal **A**, with allyltri-*n*-butylstannane in the presence of $TiCl_2(Oi\text{-}Pr)_2$, leading to homoallylether **B**.[90]

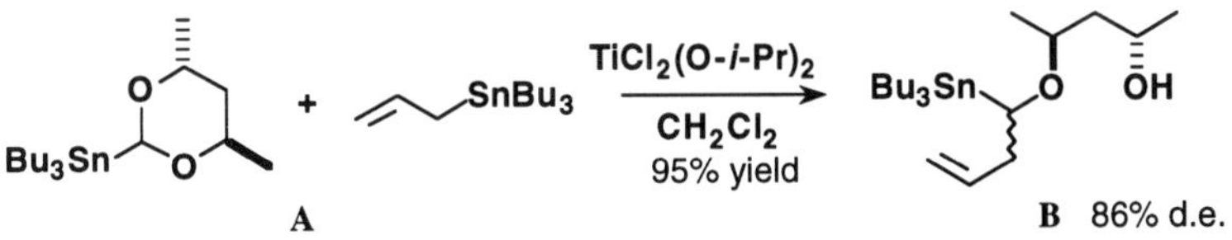

IV.1.b Reaction with other silyl nucleophiles

The coupling of acetal **A** with ditrimethylsilylacetylene led to chiral propargyl alcohol **D** through hydroxy ether **B** and β-alkoxyketone **C**[91] (see also reference 92 for a related example).

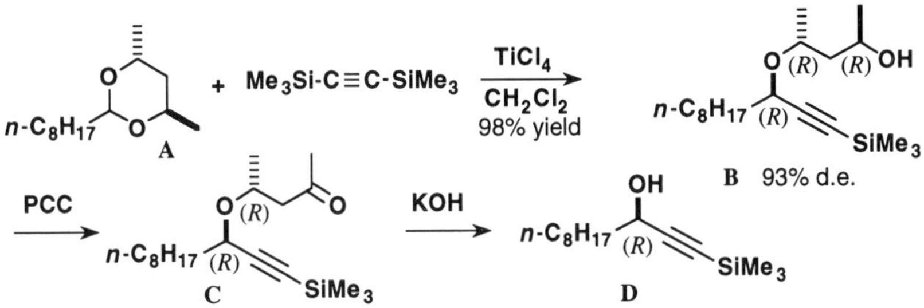

In a similar way, cyanohydrin ethers like **B** were prepared from acetal such as **A** and cyanotrimethylsilane. **B** was then converted without racemization into the corresponding cyanohydrin by oxidation and acid catalyzed elimination.[93] This reaction was used as a route to β-adrenergic blocking agents.[94]

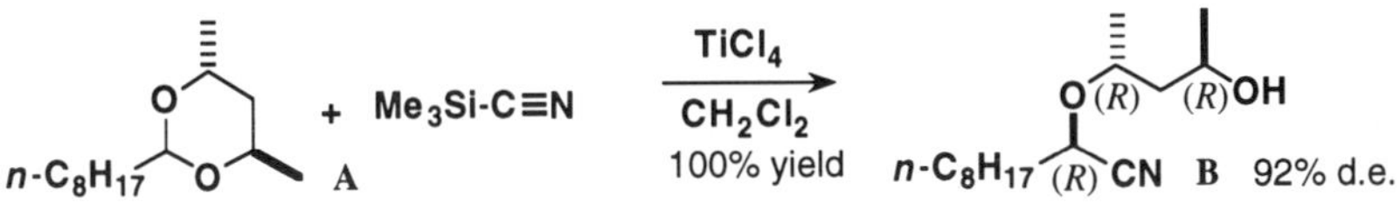

Silyl ketene acetals can also be used as nucleophiles. Thus, β-alkoxy esters, such as **B**, can be prepared in high d.e. providing therefore an access, through hydrolysis and removal of the chiral auxiliary, to β-hydroxy carboxylic acids with high e.e.[95] The procedure was applied to the synthesis of (*R*)-α-lipoic acid.

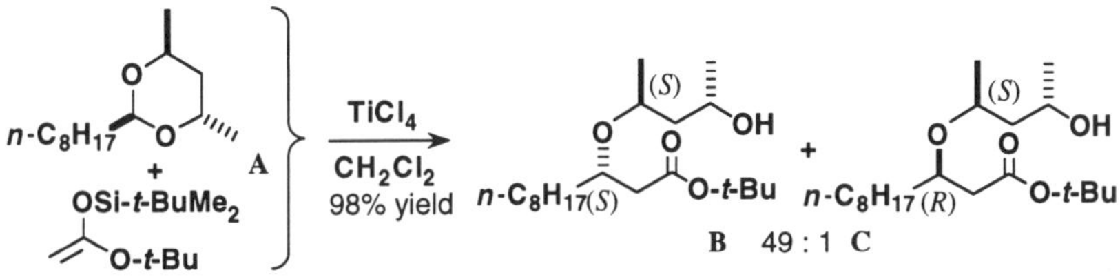

Finally, as already mentioned in Section II, Johnson studied the coupling of acetal templates derived from pentan-2,4-diols with α-silyl ketones and silyl enol ethers.[28]

IV.2 Acetals derived from homochiral butan-1,3-diols

(*R*)-Butan-1,3-diol is available, inexpensively, by $LiAlH_4$ reduction of commercial (*R*)-polyhydroxybutanoate.[96,97]

Acetals derived from butan-1,3-diol present an especially interesting situation, because the 2-position (derived from the aldehyde carbon) becomes chiral in the acetalization, and the product can exist in two diastereomeric forms if no other chiral center exists in the molecule. Moreover, a given acetal can undergo ring opening either to give a primary alcohol or a secondary alcohol.

The addition of cyanotrimethylsilane to acetal **A** led to cyano ethers **B** and **C**; the **B**:**C** ratio can vary from 1:1 to 99:1 depending on the rate of addition of the reagents and the temperature of the reaction.[29]

A + $Me_3Si\text{-}C{\equiv}N$ → ($TiCl_4$, CH_2Cl_2, 92% yield) **B** + **C**; 88% d.e. **B** 32.3 : 1 **C**

Free aldol adducts can be obtained by coupling of butane-1,3-diol acetals and silyl enol ethers. Hence, acetone trimethylsilyl enol ether, in the presence of acetal **A** and $TiCl_4$, led to cetol **B** with over 98% e.e.[98]

A + $OSiMe_3$ enol ether → ($TiCl_4$, CH_2Cl_2, 84% yield) **B** + **C**; **B** >99 : <1 **C**

Removal of the chiral auxiliary, leading to β-hydroxy ketone **E**, was achieved by oxidation, leading to ketoaldehyde **D**, followed by selective β-elimination.

B → (PCC, CH_2Cl_2) **D** → (dibenzylammonium trifluoroacetate, benzene, 81% yield from **B**) **E**

The selective opening of acetal A, leading to the formation of primary alcohol **B** results from the complexation of the Lewis acid to oxygen atom 1 which is the less hindered. It appears therefore that selectivity mainly depends upon the coordination site of the Lewis acid.

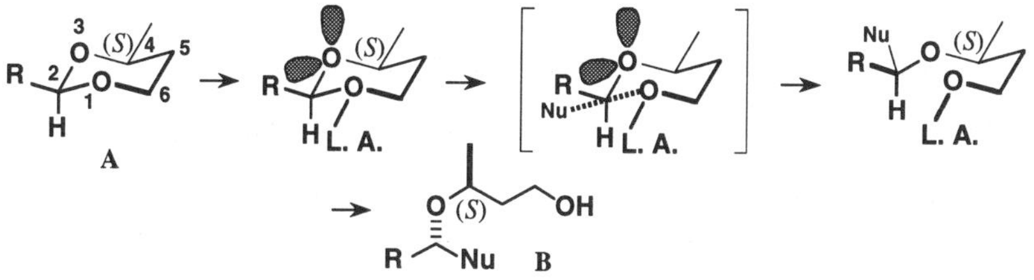

Johnson even extended the scope of the reaction to the use of ethylenic ketene acetal **B**.[99] Thus, he obtained δ-alkoxy-β-keto ester **C**, as a single diastereomer, from acetal **A**. **C** is a key intermediate in the synthesis of a mevinolin analog.

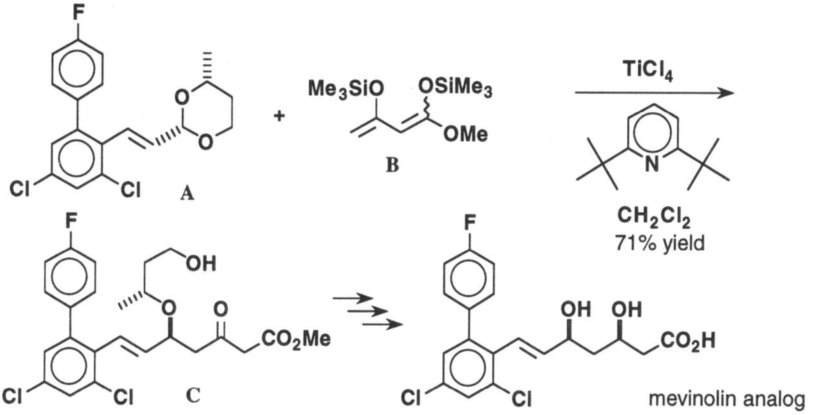

Finally, Johnson described, from leucinal derivative **A**, the preparation of ether **B**, a key intermediate in the synthesis of enantiomerically pure statine derivatives.[100]

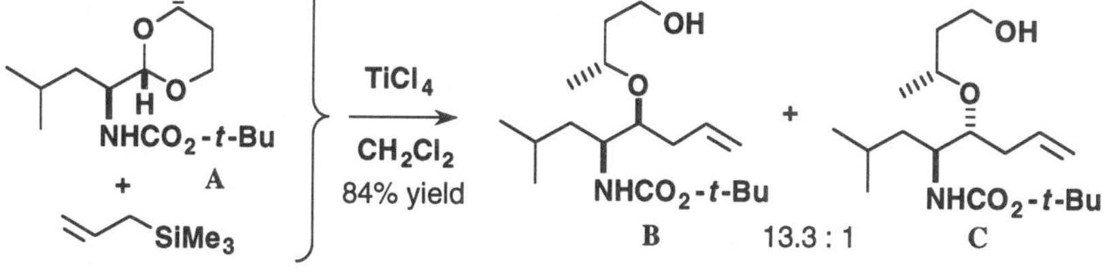

IV.3 Acetals derived from homochiral butan-2,3-diols

Johnson first used an optically active acetal derived from (*R*,*R*)-(–)-butan-2,3-diol in asymmetric polyene cyclizations (see Section VI.2).

Kishi reported an asymmetric synthesis of aklavinone using an efficient asymmetric crossed aldol reaction.[101] Acetal **A** reacted smoothly with silyl ketone **B** in the presence of $SnCl_4$ in acetonitrile (no reaction in CH_2Cl_2)

yielding predominantly crossed aldol product **C**. **C** was then transformed into optically active aklavinone.

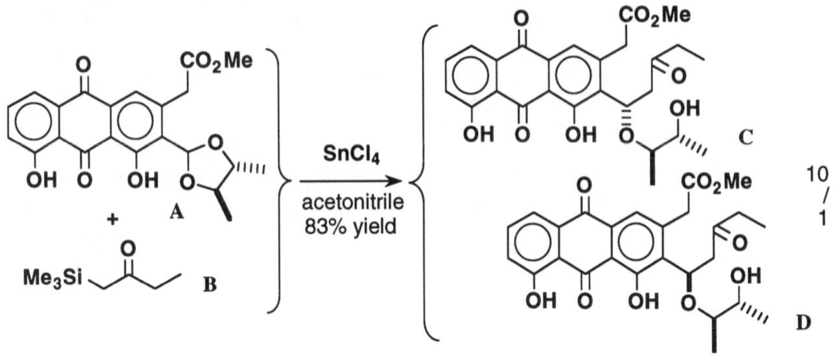

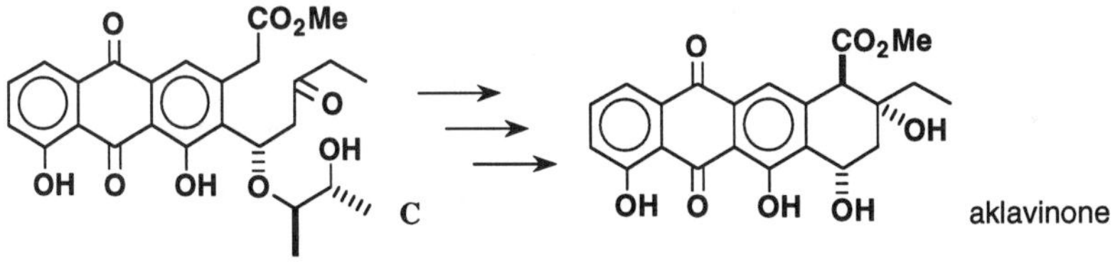

SnCl4
acetonitrile
83% yield

A, B, C, D — C : D = 10 / 1

aklavinone

Kishi also subjected other acetals including **E** to the crossed aldol reaction in the presence of silyl ketone **B** or to allylation in the presence of allyltrimethylsilane.

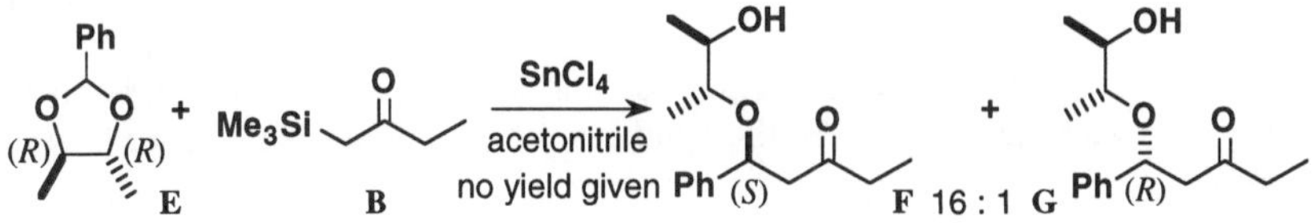

SnCl4
acetonitrile
no yield given

F 16 : 1 G

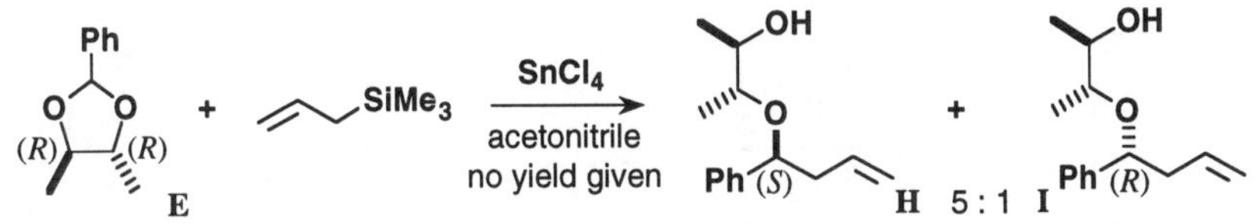

SnCl4
acetonitrile
no yield given

H 5 : 1 I

He eventually extended this approach to an asymmetric synthesis of 11-deoxydaunomycinone.[102] This approach involves, as the key step of the synthesis, the crossed aldol reaction of unsaturated silyl ketone **B** with acetal **A** to give mainly **C**.

BF_3 (gas), toluene, 80–85% yield

C : D = 13 : 1

Although the use of butan-2,3-diols induces good selectivity, it suffers from the difficulty of removing the chiral auxiliary. Indeed the α-elimination involved proceeds through a moderately convenient procedure (Na/Et_2O).[17]

IV.4 Acetals derived from other homochiral alcohols

Corcoran described a regioselective cleavage of chiral acetals derived from butan-2,4-diols.[103] However, the diastereoselectivity of the cleavage reaction is uniformly poor.

A + Me_3SiCN → (L. A., CH_2Cl_2, 90% yield) B + C

L. A. : $TiCl_4$; B : C = 250 : 1
L. A. : $SnCl_4$; B : C = 1 : 2.2
L. A. : BF_3 - OEt_2 ; B : C = 1 : 3
L. A. : $ZnBr_2$; B : C = 1 : 250

Chelation control cleavage occurred with $TiCl_4$; however, a complete reversal of the regioselectivity was observed in the use of $ZnBr_2$.

1,4-Asymmetric induction was reported by Molander in the course of the addition of various nucleophiles to 4-methoxy aldehyde acetals.[104] He showed that the stereochemistry of the substitution reaction is irrespective of the chirality of the *sec*-phenylethyl alcohol derived acetal.

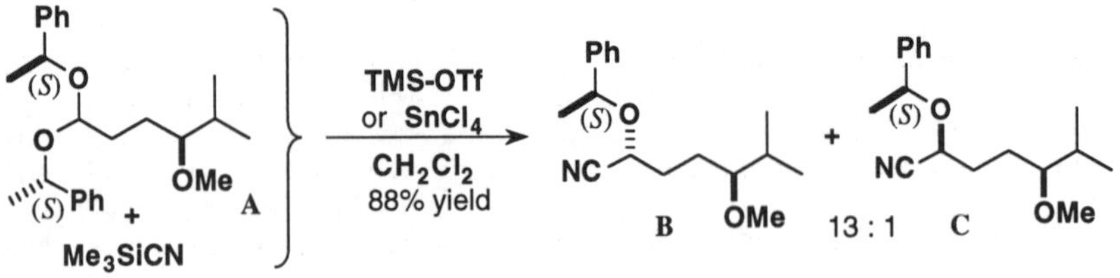

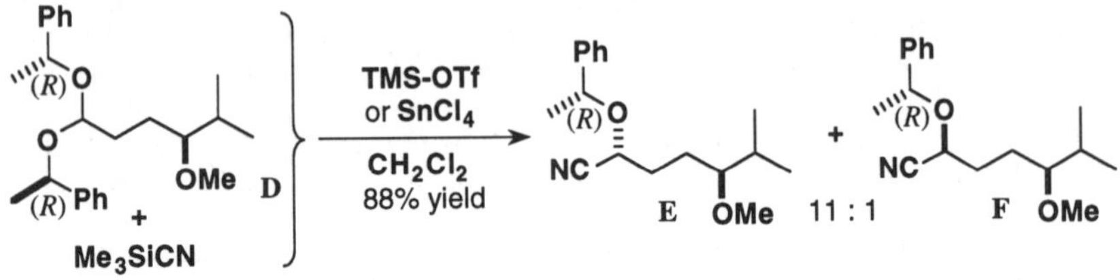

All the asymmetric induction is derived from the remote stereogenic center. The cyclic oxocarbenium ion **H** generated from **G** by participation reaction and subsequent S_N2-type displacement of the oxonium ion oxygen atom leads to the observed diastereomer **I**.

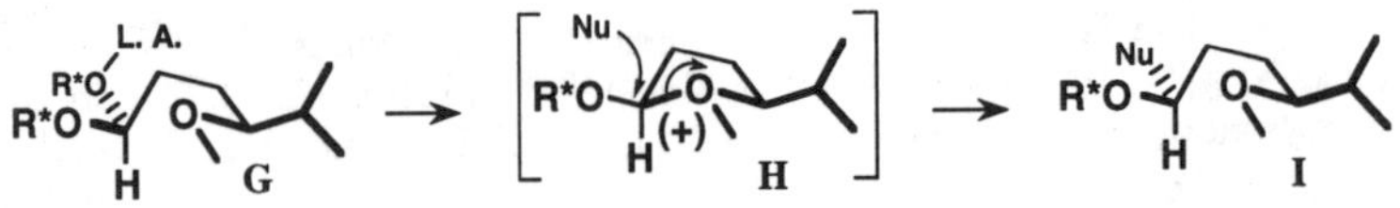

Finally, chiral orthoesters derived from diethyl or di-isopropyl (*R*,*R*)-tartrates undergo nucleophilic substitution from silyl enol ethers leading therefore to chiral 2-substituted-1,3-dicarbonyl compounds.[105]

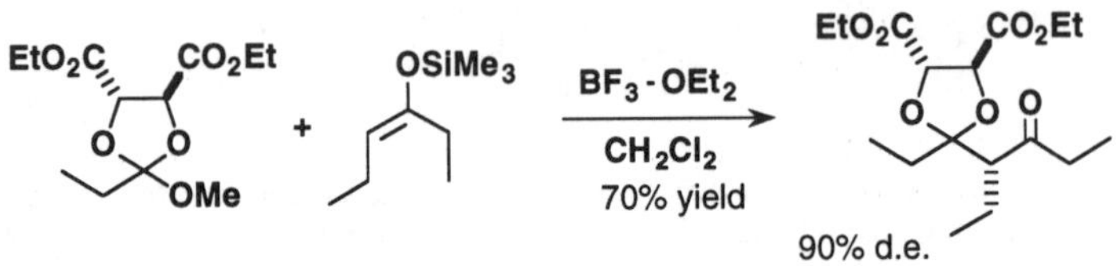

V. Anomeric Allylation of Carbohydrate Derivatives

Allylsilanes can be used under Lewis acid catalysis, for the stereocontrolled synthesis of anomerically allylated *C*-glycopyranosides, *C*-glycofuranosides and *C*-glycals.

V.1 Allylation of glycopyranosides and glycofuranosides

Kozikowski studied the allylation of ribofuranose derivative **A** and showed that the reaction appeared to have taken on more of the character of an S_N2 process as **B** was obtained as the predominant isomer.[106,107]

L. A. : $ZnBr_2$; neat; 85% yield; **B** : **C** = 4 : 1
L. A. : BF_3- OEt_2; acetonitrile; 93% yield; **B** : **C** = 7 : 1

With lyxose derivative **D**, the major product **E**, resulting from an axial addition mode, was even more predominant (formation of α-product).

1° L. A.
2° $Bu_4N^+ F^-$

L. A. : $ZnBr_2$; neat; 84% yield; **E** : **F** = 4 : 1
L. A. : BF_3- OEt_2; acetonitrile; 88% yield; **E** : **F** = 19 : 1

Highly stereoselective *C*-allylation of glycopyranosides was also reported by Hosomi and Sakurai when using TMS-OTf or iodotrimethylsilane as catalysts.[108] Thus, α-anomers **B** and **D** are obtained overwhelmingly or exclusively from α-D-glucopyranoside **A**

TMS-OTf (0.5 equiv.), MeCN, 86% yield; **B** : **C** = 10 : 1

TMS-OTf (0.5 equiv.), MeCN, 71% yield

The choice of the solvent considerably affects the stereoselectivity of the reaction of allylation of acetylated glycopyranoses. Thus, running the reaction between **A** and allyltrimethylsilane in acetonitrile produced a selective abundance of the α-anomer (**B**:**C** = 19:1) while a 1:1 mixture was obtained in 1,2-dichloroethane.[109]

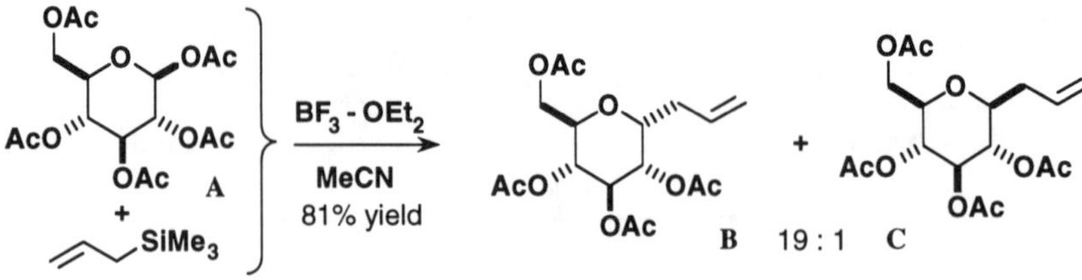

However, although the addition of penta-2,4-dienyltrimethylsilane to **A** also afforded the expected pentedienyl substituted compounds **D** and **E**, diastereomeric excess and yield were low.[110] Acton also used the same dienylsilane in synthetic studies directed towards anthracyclinone *C*-glycosides.[111]

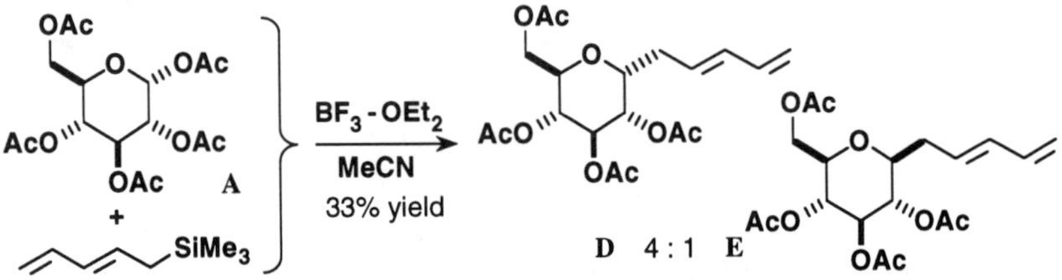

Panek described the use of C1-oxygenated allylic silane, such as **B**, as homoenolate equivalents, and used them in alkylation reactions of glucopyranose derivatives such as **A**.[112] In that case, allylated product **C** was formed predominantly providing BF_3-OEt_2 and 1,2-dichloroethane were used.

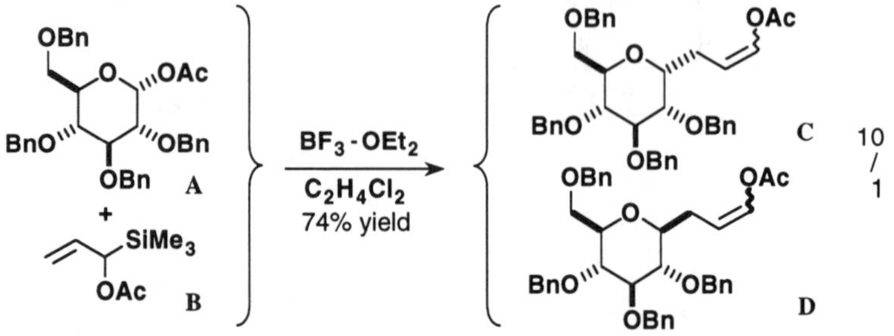

Interestingly, the addition of allylsilanes to acetals can also be promoted successfully by Bronsted acids such as perchloric acid. Thus, in the course of the synthesis of halichondrin B, from common sugars, Salomon found that the best results for allylation of **A** were obtained with $HClO_4$ in acetonitrile.[113]

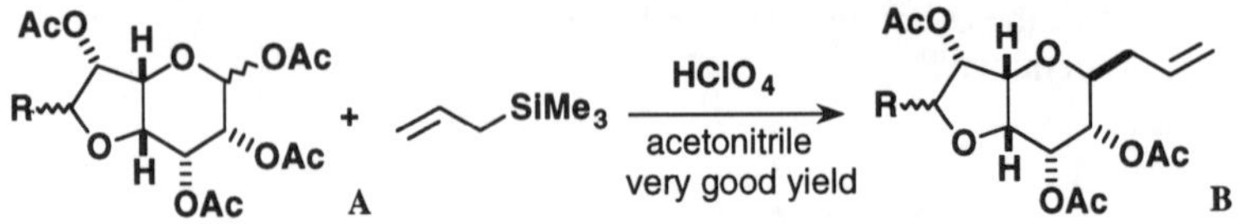

The formation of α-anomer **C** is favored by the anomeric effect which stabilizes carbocation **B** and therefore induces axial attack.

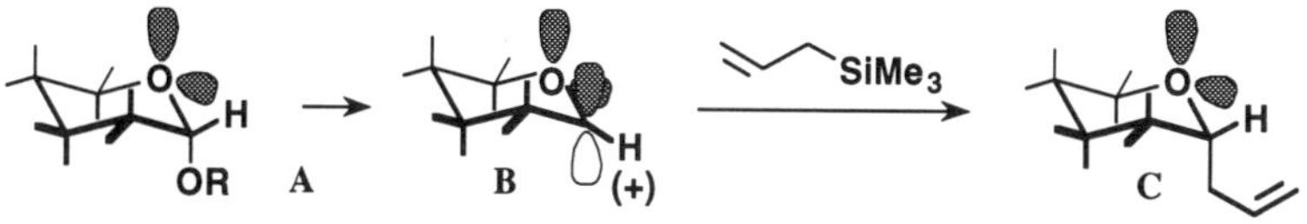

Martin showed that various xylofuranosides, glucopyranosides or hemiacetals such as **A** can lead to acyclic compounds such as **B** with a high degree of stereoselectivity upon $TiCl_4$ mediated allylation.[114]

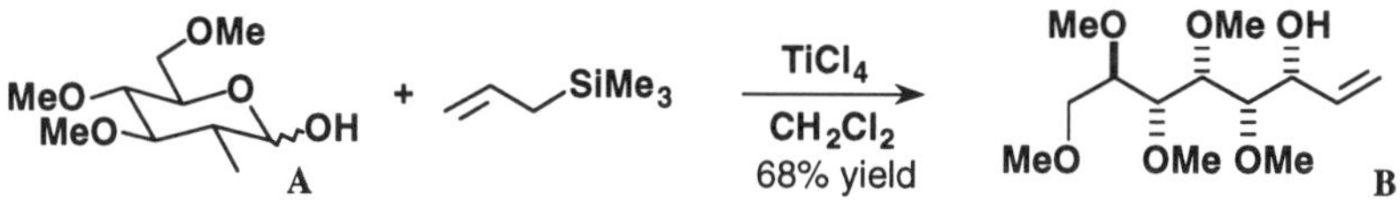

Bartlett studied the addition of allyl- and enolsilane reagents to bicyclic acetals as an approach to the synthesis of polyether nigericin.[115]

Finally, an oxetanosyl-*N*-glycoside was converted into a furanosyl-*C*-allylglycoside with allyltrimethylsilane in the presence of BF_3-OEt_2.[116]

V.2 Allylation of glycals

In 1982 Danishefsky demonstrated that activated glycals, such as glycal acetates, undergo regio- and stereoselective S_N2' nucleophilic attack at C1 by allyltrimethylsilane in the presence of a Lewis acid.[117] This "carbon-Ferrier" displacement[118] which provides an easy access to C1 allylated glycosides bearing C2–C3 unsaturation can also be obtained with (*E*)- and (*Z*)-crotylsilanes. However, the reaction exhibited modest stereoselectivity which is responsive to crotylsilane geometry.[119]

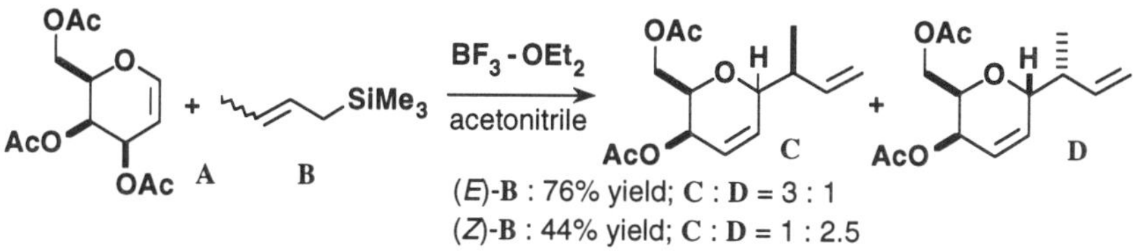

(*E*)-**B** : 76% yield; **C** : **D** = 3 : 1
(*Z*)-**B** : 44% yield; **C** : **D** = 1 : 2.5

Nevertheless, the authors proposed a synthesis of the optically active dihydropyranoid segment of indanomycin[119] and a synthesis of zincophorin, a zinc-binding antibiotic.[120,121]

More recently, Sabol showed that the Lewis acid catalyzed allylation of a diacetyl-D-xylal provided a useful entry to the synthesis of leukotriene D_4 analogs.[122]

Finally, Herscovici showed that simple olefins react with glycal in the presence of Lewis acids[123] and reported a route to 1-aldo and 2-keto *C*-glycosides, such as **C**, by condensation of glycals, such as **A**, with allylic ethers, such as **B**, which are used as homoenolates.[124,125] A hydride migration, occurring after the first step of the alkylation process, is probably responsible for this result.

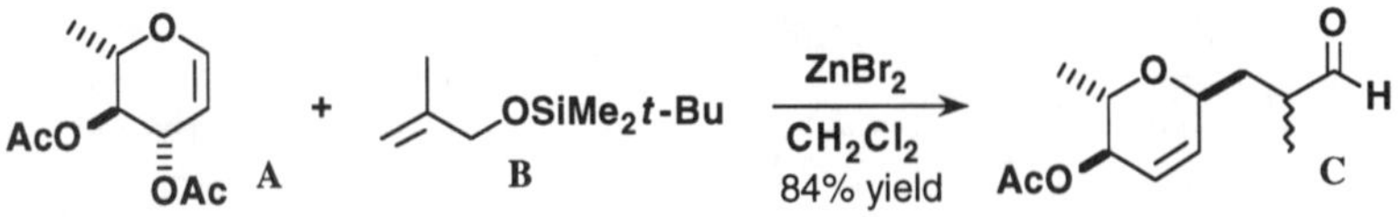

VI. Intramolecular Reaction

The nucleophilic substitution of acetal can also take place intramolecularly providing therefore a convenient access to cyclic or polycyclic derivatives. A particularly spectacular example is the biomimetic cyclization of acyclic polyenals into tetracyclic compounds (see Section VI.2).

VI.1 Reaction of silyl nucleophiles with acetals

Itoh reported that allylsilanes react with monothioacetals on treatment with $SnCl_4$ to give predominantly the corresponding homoallylethers with selective cleavage of the C–S bond. In contrast, $TiCl_4$ induced mainly the formation of homoallylic sulfides.[126] As a demonstration of possible applications, vinyltetrahydrofuran **B** and methylenetetrahydropyran **D** were synthesized from allylsilane monothioacetals **A** and **C**.

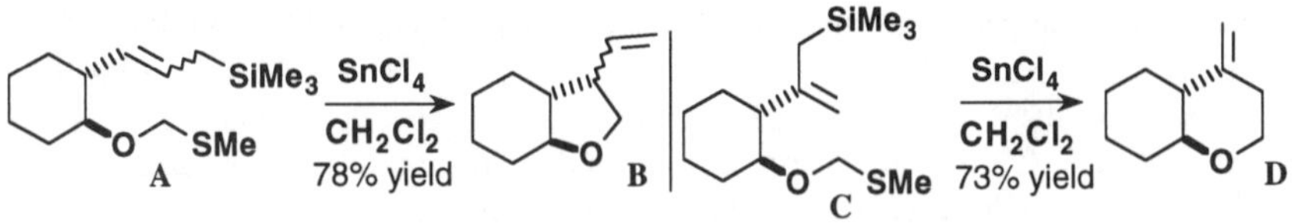

While only modest diastereoselectivity was observed during the cyclization of α-amino aldehyde dimethyl acetal **A** in the presence of $TiCl_4$,[127] the cyclization of ethylene cetal derivative **D**, in the presence of $TiCl_3(Oi\text{-}Pr)$, afforded **E** as a single 2,3-*cis*-diastereomer. This approach provides an efficient route to functionalized piperidines.

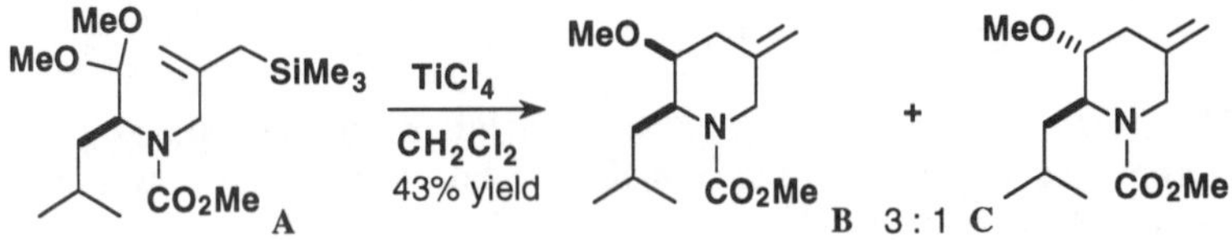

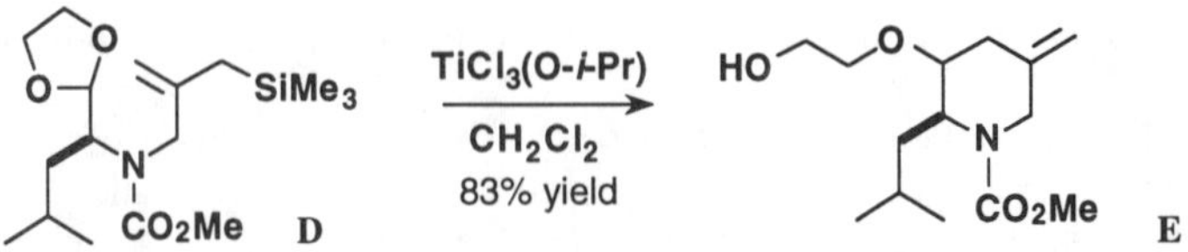

Similar results were also reported by Mann on the use of TMS-OTf as catalyst.[128]

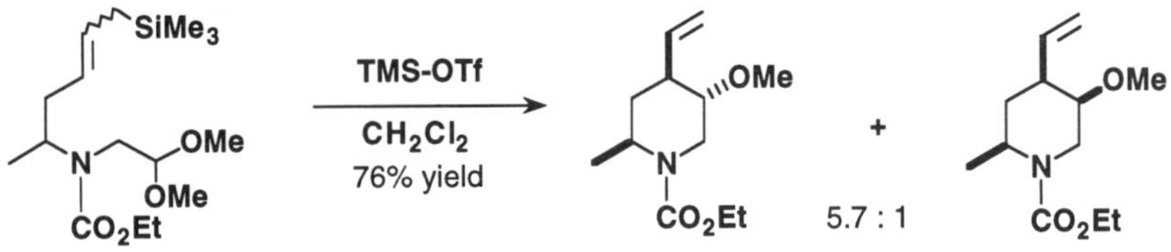

β-Alkoxycyclic ethers were obtained by Yamamoto via the cyclization of ω-organometallic ether acetals such as trialkyllead, trialkyltin,[129] trialkylsilyl or trialkylgermyl derivatives.[130] The stereoselectivity was found to depend strongly on the Lewis acid used. Hence, with acetal **A**, $TiCl_4$-PPh_3 gave both the best yield and diastereoselectivity.

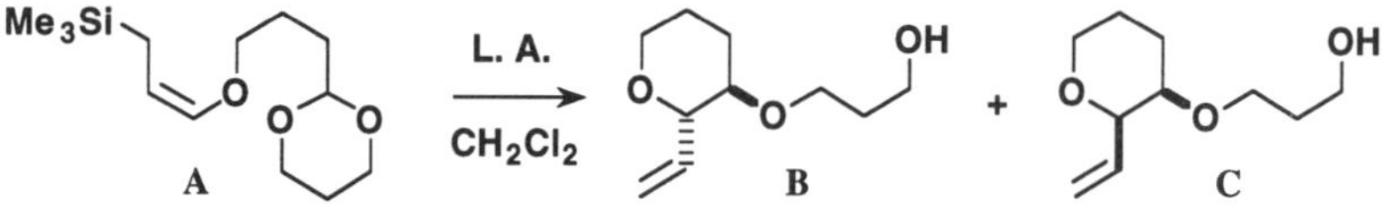

L. A. : **$TiCl_4$**; 16% yield; **B** : **C** = 3.17 : 1 **L. A.** : **$TiCl_4$ - PPh_3**; 95% yield; **B** : **C** = 49 : 1
L. A. : **$TiCl_3$(O-*i*-Pr)**; 53% yield; **B** : **C** = 1 : 3

As already mentioned,[63–66] α-acyl iminium ions are versatile electrophiles towards silicon nucleophiles (for an overview on the reactivity of such compounds, see the review from Speckamp[131]). Thus, Gramain proposed the synthesis of a key intermediate in the synthesis of (±)-mesembrine based on the intramolecular cyclization of α-acylium ion **B** generated from hydroxy lactam **A**.[132]

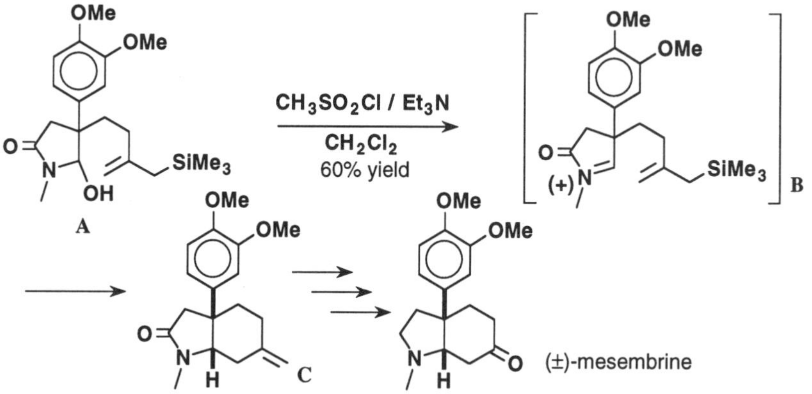

N-Acyliminium ions can also be obtained by treatment of lactams such as **A** with formic acid. Thus, on treatment with formic acid, an intramolecular alkylation with the propargylsilane moiety afforded bicyclic compound **B**, a key intermediate in the synthesis of alkaloid peduncularine.[133]

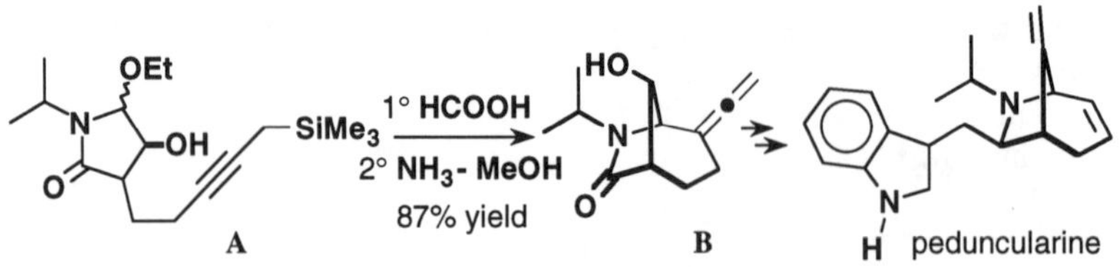

Vinylcyclopropyl ether silyl acetals, such as **A**, can undergo TMS-OTf induced cyclization with good diastereoselectivity.[73]

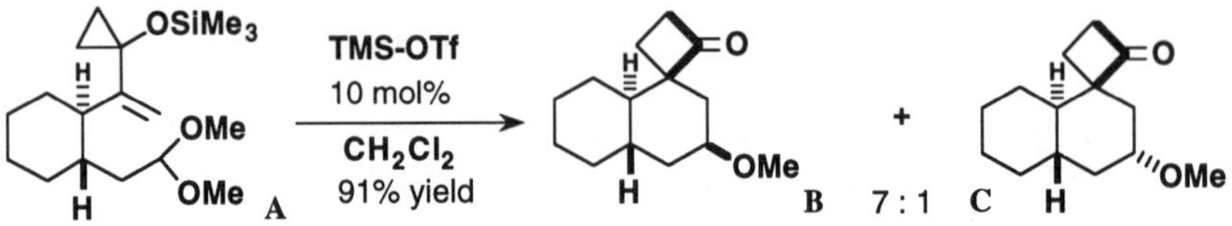

Transition state **D** can account for the formation of major diastereomer **B**.

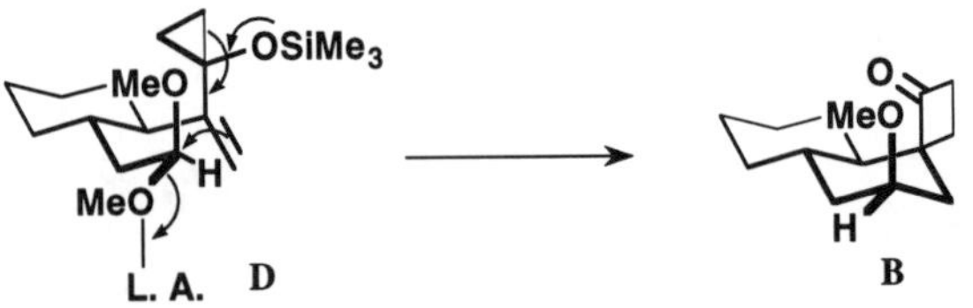

Kuo and Money reported a diastereoselective synthesis of longiborneol and longifolene from methoxy ketones **B** and **C** resulting from an intramolecular substitution reaction.[134]

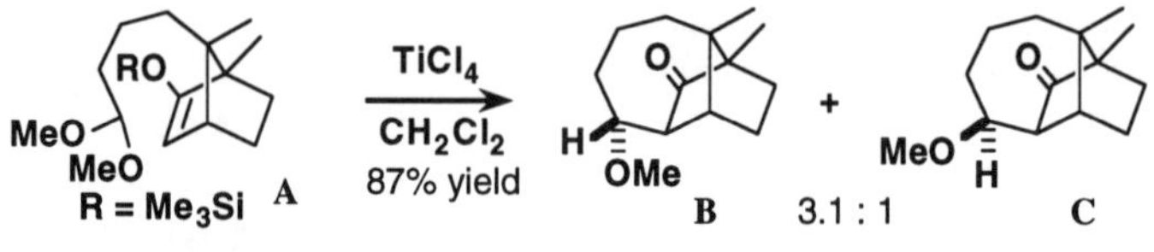

An invertive substitution process can account for the formation of major diastereomer **B**. Interestingly, the diastereoselectivity of the reaction decreased with ethylene ketal **D**.

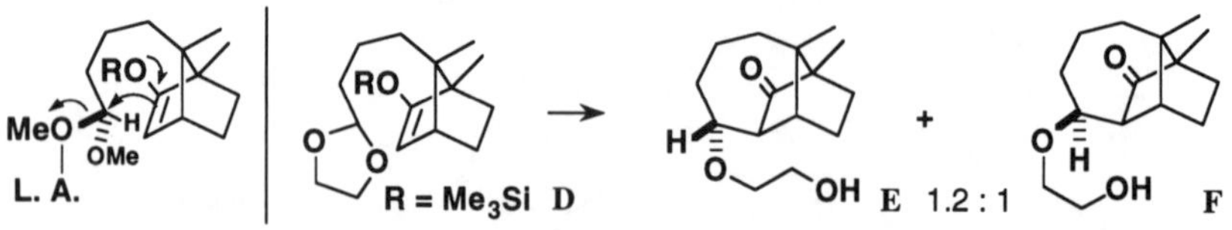

Schinzer showed that Lewis acids can induce the cyclization of allylsilane ene-ketals. Thus, he prepared 3a-hydroxy-6-hydrindanones **B** and **C** (**B**:**C** = 10:1) and **E** from ene-ketals **A** and **D** respectively.[135]

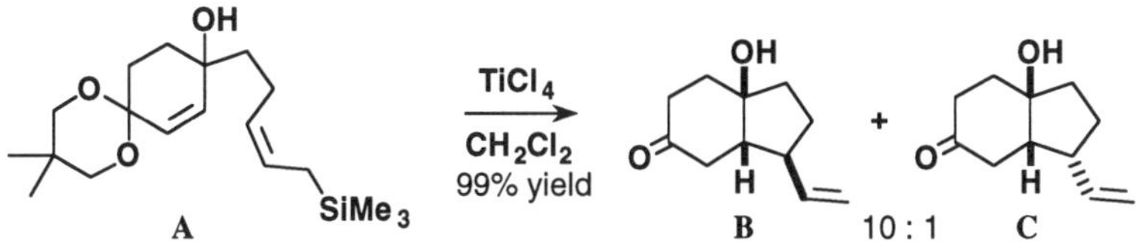

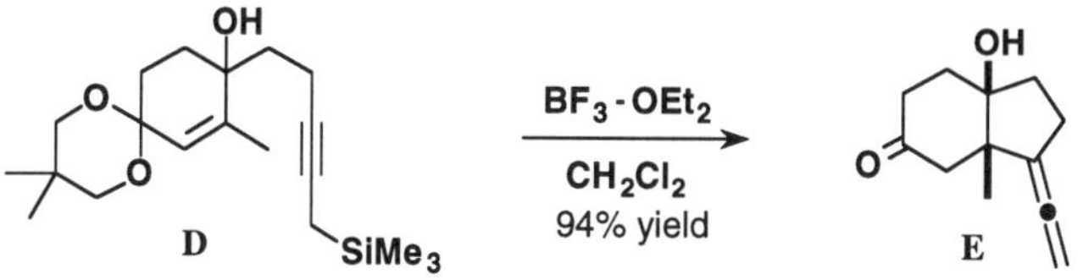

Finally, Mohr reported the acid catalyzed preparation of vinyl tetrahydrofurans resulting from the ring closure of an allylsilane acetal formed *in situ* from a hydroxy allylsilane and an acetal.[136]

VI.2 Biomimetic cationic polyolefin cyclization

Having established that acetals with appropriately juxtaposed olefinic double bonds undergo facile Lewis acid-catalyzed ring closure,[6] Johnson demonstrated that the stereochemical outcome of the reaction was dictated by the configuration of the olefinic bonds in the substrate.[137] Thus, *trans* and *cis* dienic acetals **A** and **C** undergo rapid stereospecific cyclization to give the corresponding *trans* and *cis* angularly methylated octalin derivatives **B** and **D**.

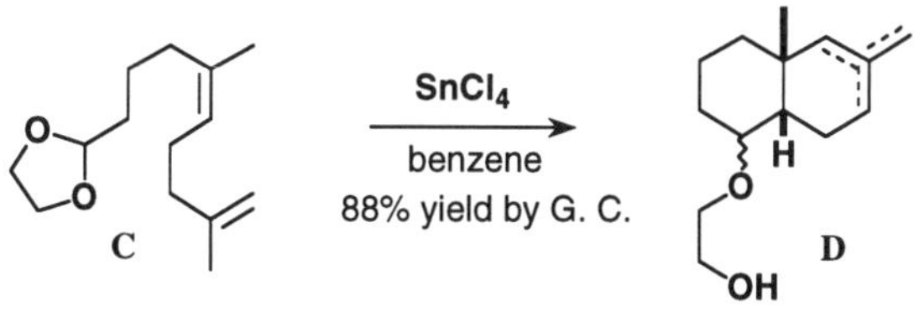

Since the process is stereospecific, it must either be synchronous, or at least concerted, or involve cationic intermediates that maintain the stereochemical integrity of the starting acetals **A** and **C**.

The introduction of acetal templates enabled the formation of optically active compounds[138,139] (see also the preparation of novel chloroanthracyclines by Rutledge[140]). Thus, the cyclization of chiral dienic acetal **A**, derived from (−)-butan-2,3-diol, led to octalin alcohols **B** and **D** in high yield. By removal of the side chain, axial octalol **C** and equatorial octalol **E** were easily obtained.[84]

SnCl$_4$ benzene 73% yield

B, 96% e.e. **B** : **D** = 2.48 : 1

D, 96% e.e.

The asymmetric induction observed affords a clue to the geometry of the transition states involved in the two cyclization reactions leading to **B** and **D**: "As the acetals open, the configuration of the oxocarbonium ion must approach a planar, sp2-hybridized geometry, with the departing oxygen atom being in a roughly perpendicular direction. On the assumption that attack of the participating double bond takes place at some point along this sequence, the four diastereomeric structures **F–I** can be considered. The least sterically congested would appear to be **F**, in which the *pro-R* oxygen atom of the acetal is the leaving group".[84] Attack of the double bond then occurs as shown by the arrow and leads to the formation of **B**.

Cyclization terminated by a propargylsilane function occurs with both better yields and greater selectivity. Thus, acetal **A** led to steroidal substructures **B** and **C**, both with *trans*-fused rings (**B**:**C** = 9.1:1), with 91% yield.[141]

TiCl$_4$ CH$_2$Cl$_2$ 91% yield

B 9.1 : 1 **C**

An S_N2-like transition state can be proposed, and a fully concerted process is needed to account for the high asymmetric induction observed. Thus, the following transition state rationalizes the production of **B** according to a concerted process.

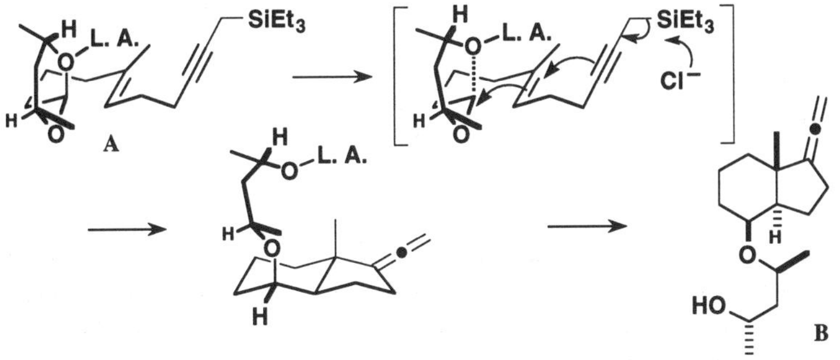

The Inhoffen-Lythgoe diol **H**, a key intermediate in the preparation of a vitamin D metabolite can be obtained from **B**. In particular, a Lewis acid-catalyzed ene reaction between **F** and paraformaldehyde gave rise to alcohol **G**.[142]

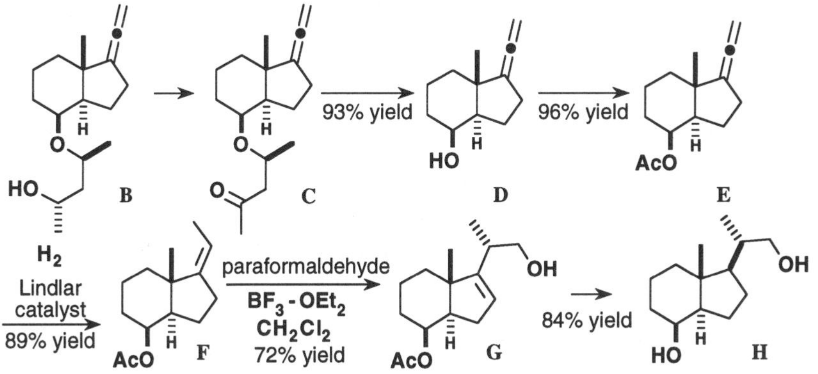

Johnson also studied cyclizations bearing a formal resemblance to the enzymic production of tetracyclic triterpenoids from squalene. Thus, he prepared, in the presence of $SnCl_4$, tetracyclic derivative **B** from tetraenic acetal **A**.[143,144]

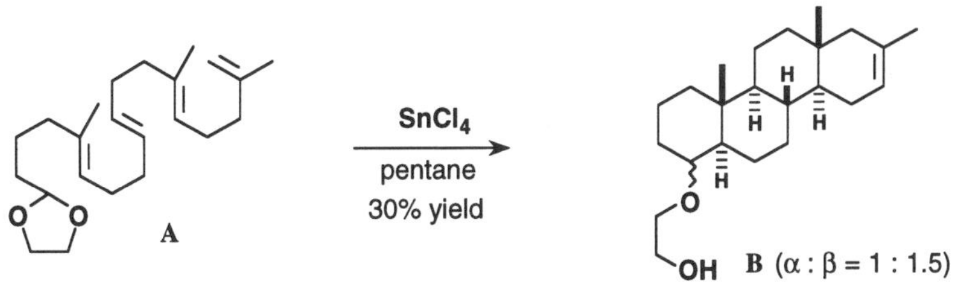

Although the reaction was highly stereoselective, yields were modest. After unsuccessful attempts to improve them, including the use of solid supports or ultrahigh pressures, Johnson proposed a new approach to this problem: a modification of one of the internal double bonds so as to enhance the propagation of the cyclization process. Indeed, the formation of **B** from **A** could arise from two bicyclization reactions occurring in tandem and involving the formation of secondary cation **C**. He therefore postulated that the introduction of a cation-stabilizing auxiliary, such as an isobutenyl group, at *pro*-C8 in polyene **D**, should enhance the propagation of the cyclization process.[145]

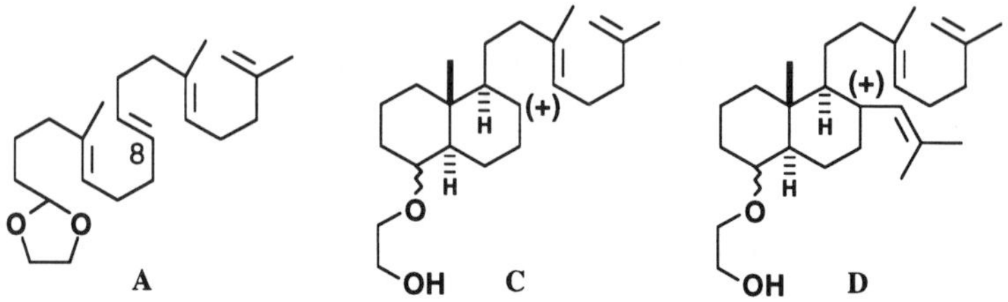

A **C** **D**

Indeed, a dramatic increase of the combined yield, from 30 to 77%, of tetracyclic products was observed .

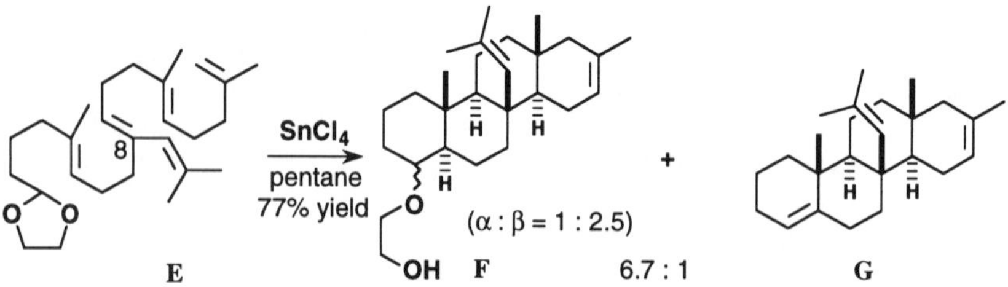

E → $SnCl_4$, pentane, 77% yield → **F** (α : β = 1 : 2.5) + **G** (6.7 : 1)

Similarly, while the use of an allylsilane group as termination of cyclization allowed the preparation of 17-vinyl steroid in only moderate yields (34%),[146] beneficial effects on polyene cyclizations were obtained by introducing a cation-stabilizing auxiliary and a propargyl function as nucleophilic group; moreover, the preparation of chiral acetal **A** allowed the formation of tetracyclic adduct **B** as a single diastereomer.[147,148]

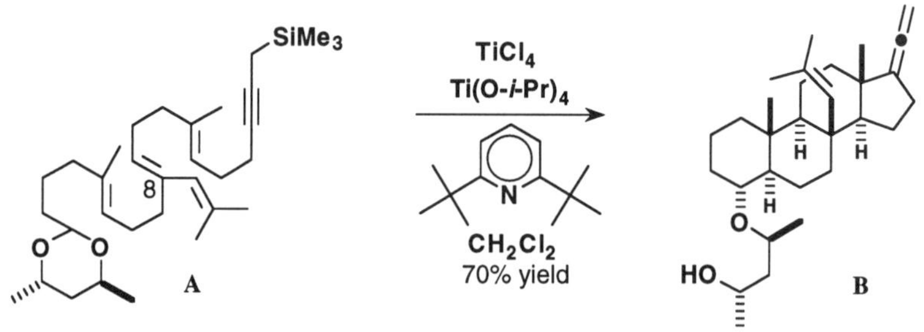

More recently, Johnson developed the use of the fluorine atom, instead of the isobutenyl group, as a cation stabilizing auxiliary and in that way prepared fluorosteroids in good to high yields.[149,150] Using that approach, he proposed a total synthesis of *dl*-β-amyrin.[151]

VII. Conclusion

The reactivity of acetals towards nucleophiles under Lewis acid catalysis coupled with the use of acetal templates provides one of the finest methods to introduce chirality on achiral carbonyl compounds.

Moreover, it appears that acetals, which in the past were only considered as protective groups for carbonyl compounds, can in fact activate such compounds under Lewis acid catalysis; a dramatic demonstration is the biomimetic cyclization of polyene compounds.

References

1. Schmitz, E.; Eichorn, I. in *The Chemistry of the Ether Linkage*, Patai, S., Ed.; Wiley, New York, **1967**, Chap. 7, p 309.
2. Müller–Cunradi, M.; Pieroh, K. U.S. Patent 2 165 962; *Chem. Abstr.* **1939**, *33*, 8210.
3. Lindell, S.D.; Elliott, J.D.; Johnson, W.S. *Tetrahedron Lett.* **1984**, *25*, 3947–3950.
4. Normant, J.F.; Alexakis, A.; Ghribi, A.; Mangeney, P. *Tetrahedron* **1989**, *45*, 507–516.
5. Abe, I.; Rohmer, M.; Prestwich, G.D. *Chem. Rev.* **1993**, *93*, 2189–2206.
6. Johnson, W.S.; Kinnel, R.B. *J. Am. Chem. Soc.* **1966**, *88*, 3861–3862.
7. Bartlett, P.A. *Olefin Cyclization Processes that Form Carbon-Carbon Bonds* in *Asymmetric Synthesis*, Vol. 3, Chap. 5, p. 341.
8. Mukaiyama, T.; Hayashi, M. *Chem. Lett.* **1974**, 15–16.
9. Saigo, K.; Osaki, M.; Mukaiyama, T. *Chem. Lett.* **1976**, 769–772.
10. Hosomi, A.; Endo, M.; Sakurai, H. *Chem. Lett.* **1976**, 941–942.
11. Fleming, I. *Chimia* **1978**, *32*, 219–220.
12. Ojima, I.; Kumagai, M. *Chem. Lett.* **1978**, 575–578.
13. Hosomi, A.; Endo, M.; Sakurai, H. *Chem. Lett.* **1978**, 499–502.
14. Hosomi, A.; Saito, M.; Sakurai, H. *Tetrahedron Lett.* **1980**, *21*, 3783–3786.
15. Hosomi, A.; Iguchi, H.; Endo, M.; Sakurai, H. *Chem. Lett.* **1979**, 977–980.

16. Nishiyama, H.; Itoh, K. *J. Org. Chem.* **1982**, *47*, 2496–2498.
17. Alexakis, A.; Mangeney, P. *Tetrahedron: Asymmetry*, **1990**, *1*, 477–511.
18. Meerwein, H. in *Houben-Weyl, Methoden der Organischen Chemie*; Thieme, Stuttgart, **1965,** Vol. VI–3, p. 199–294.
19. Effenberger, F. *Angew. Chem., Int. Ed. Engl.* **1969**, *8*, 295–312.
20. Povarov, L.S. *Russ. Chem. Rev. (Engl. Transl.)* **1965**, *34*, 639.
21. Mathieu, J.; Weill–Raynal, J. *Formation of C–C Bonds*, Thieme, Stuttgart, **1979**, Vol. III, p. 196.
22. Makin, S.M. *Russ. Chem. Rev. (Engl. Transl.)* **1965**, *38*, 237.
23. Makin, S.M. *Pure Appl. Chem.* **1976**, *47*, 173–181.
24. Mukaiyama, T.; Murakami, M. *Synthesis* **1987**, 1043–1054.
25. Johnson, W.S. *Angew. Chem., Int. Ed. Engl.* **1976**, *15*, 9–17.
26. Hathaway, S.J.; Paquette, L.A. *J. Org. Chem.* **1983**, *48*, 3351–3353.
27. Yamamoto, Y.; Nishii, S.; Yamada, J.–i. *J. Am. Chem. Soc.* **1986**, *108*, 7116–7117.
28. Johnson, W.S.; Edington, C.; Elliott, J.D.; Silverman, I.R. *J. Am. Chem. Soc.* **1984**, *106*, 7588–7591.
29. Choi, V.M.F.; Elliott, J.D.; Johnson, W.S. *Tetrahedron Lett.* **1984**, *25*, 591–594.
30. Deslongchamps, P. *Stereoelectronic Effects in Organic Chemistry*, Pergamon, Oxford, **1983**, p. 162.
31. von der Brüggen, U.; Lammers, R.; Mayr, H. *J. Org. Chem.* **1988**, *53*, 2920–2925.
32. Hagen, G.; Mayr, H. *J. Am. Chem. Soc.* **1991**, *113*, 4954–4961.
33. Denmark, S.E.; Willson, T.M. in *Selectivities in Lewis Acid Promoted Reactions*, Schinzer, D., Ed.; Kluwer Academic Publishers, **1989**, pp. 247–263.
34. Denmark, S.E.; Almstead, N.G. *J. Am. Chem. Soc.* **1991**, *113*, 8089–8110.
35. Denmark, S.E.; Almstead, N.G. *J. Org. Chem.* **1991**, *56*, 6458–6467.
36. Denmark, S.E.; Almstead, N.G. *J. Org. Chem.* **1991**, *56*, 6485–6487.
37. Mori, I.; Ishihara, K.; Flippin, L.A.; Nozaki, K.; Yamamoto, H.; Bartlett, P.A.; Heathcock, C.H. *J. Org. Chem.* **1990**, *55*, 6107–6115.
38. Sammakia, T.; Smith, R.S. *J. Org. Chem.* **1992**, *57*, 2997–3000.
39. Sammakia, T.; Smith, R.S. *J. Am. Chem. Soc.* **1992**, *114*, 10998–10999.
40. Hatanaka, Y.; Kuwajima, I. *J. Org. Chem.* **1986**, *51*, 1932–1934.
41. Hosomi, A.; Ando, M.; Sakurai, H. *Chem. Lett.* **1986**, 365–368.
42. Linderman, R.J.; Anklekar, T.V. *J. Org. Chem.* **1992**, *57*, 5078–5080.
43. Mann, A.; Ricci, A.; Taddei, M. *Tetrahedron Lett.* **1988**, *29*, 6175–6176.
44. Oku, A.; Homoto, Y.; Harada, T. *Chem. Lett.* **1986**, 1495–1498.
45. Ouvrard, P.; Tubul, A.; Santelli, M. *Bull. Soc. Chim. Fr.* **1993**, *130*, 772–778.
46. Pellissier, H.; Tubul, A.; Santelli, M. *Tetrahedron Lett.* **1993**, *34*, 827–830.
47. Guyot, B.; Pornet, J.; Miginiac, L. *Tetrahedron* **1991**, *47*, 3981–3988.
48. Tsunoda, T.; Suzuki, M.; Noyori, R. *Tetrahedron Lett.* **1980**, *21*, 71–74.
49. Murata, S.; Suzuki, M.; Noyori, R. *Tetrahedron* **1988**, *44*, 4259–4275.
50. Polla, M.; Fredj, T. *Acta Chem. Scand.* **1993**, *47*, 716–720.
51. Sakurai, H.; Sasaki, K.; Hosomi, A. *Tetrahedron Lett.* **1981**, *22*, 745–748.

52. Mukaiyama, T.; Nagaoka, H.; Murakami, M.; Ohshima, M. *Chem. Lett.* **1985**, 977–980.
53. Kobayashi, S.; Hachiya, I.; Ishitani, H.; Araki, M. *Synlett* **1993**, 472–474.
54. Hollis, T.K.; Robinson, N.P.; Whelan, J.; Bosnich, B. *Tetrahedron Lett.* **1993**, *34*, 4309–4312.
55. Westerlund, C. *Tetrahedron Lett.* **1982**, *23*, 4835–4838.
56. Kiyooka, S.–i.; Sasaoka, H.; Fujiyama, R.; Heathcock, C.H. *Tetrahedron Lett.* **1984**, *25*, 5331–5334.
57. Panek, J.S.; Yang, M. *J. Org. Chem.* **1991**, *56*, 5755–5758.
58. Panek, J.S.; Yang, M. *J. Am. Chem. Soc.* **1991**, *113*, 6594–6600.
59. Pasquarello, A.; Poli, G.; Potenza, D.; Scolastico, C. *Tetrahedron: Asymmetry* **1990**, *1*, 429–432.
60. Bernardi, A.; Piarulli, U.; Poli, G.; Scolastico, C.; Villa, R. *Bull. Soc. Chim. Fr.* **1990**, *127*, 751–757.
61. Lu, Z.–H.; Zhou, W.–S. *J. Chem. Soc., Perkin Trans. 1* **1993**, 593–596.
62. Lu, Z.–H.; Zhou, W.–S. *Tetrahedron* **1993**, *49*, 4659–4664.
63. Takacs, J.M.; Weidner, J.J.; Takacs, B.E. *Tetrahedron Lett.* **1993**, *34*, 6219–6222.
64. Meyers, A.I.; Burgess, L.E. *J. Org. Chem.* **1991**, *56,* 2294–2296.
65. Polniaszek, R.P.; Belmont, S.E.; Alvarez, R. *J. Org. Chem.* **1990**, *55,* 215–223.
66. Ukaji, Y.; Tsukamoto, K.; Nasada, Y.; Shimizu, M.; Fujisawa, T. *Chem. Lett.* **1993**, 221–224.
67. Sato, T.; Otera, J.; Nozaki, H. *J. Org. Chem.* **1990**, *55*, 6116–6121.
68. Mukaiyama, T.; Ohshima, M.; Miyoshi, N. *Chem. Lett.* **1987**, 1121–1124.
69. Lee, T.V.; Boucher, R.J.; Rockell, C.J.M. *Tetrahedron Lett.* **1988**, *29*, 689–692.
70. Lee, T.V.; Richardson, K.A.; Taylor, D.A. *Tetrahedron Lett.* **1986**, *27*, 5021–5024
71. Ishida, A.; Mukaiyama, T. *Bull. Chem. Soc. Jpn* **1977**, *50*, 1161–1168.
72. Hashimoto, Y.; Sato, Y.; Kudo, K.; Saigo, K. *Tetrahedron Lett.* **1993**, *34*, 7623–7626.
73. Trost, B.M.; Lee, D.C. *J. Am. Chem. Soc.* **1988**, *110*, 6556–6558.
74. Roos, E.C.; Hiemstra, H.; Speckamp, W.N.; Kaptein, B.; Kamphuis, J.; Schoemaker, H.E. *Recl. Trav. Chim. Pays-Bas* **1992**, *111*, 360–364.
75. Pornet, J.; Miginiac, L.; Jaworski, K.; Randrianoelina, B. *Organometallics* **1985**, *4*, 333–338.
76. Pornet, J. *Tetrahedron Lett.* **1980**, *21*, 2049–2050.
77. Esch, P.M.; Hiemstra, H.; Speckamp, W.N. *Tetrahedron Lett.* **1988**, *29*, 367–370.
78. Hiemstra, H.; Fortgens, H.P.; Speckamp, W.N. *Tetrahedron Lett.* **1984**, *25*, 3115–3118.
79. Seebach, D.; Imwinkelried, R.; Stucky, G. *Angew. Chem., Int. Ed. Engl.* **1986**, *25*, 178–180.
80. Friboes, K.C.; Haeder, T.; Aubert, D.; Strahringer, C.; Bolte, M.; Hoppe, D. *Synlett*, **1993**, 921–923.
81. Hayashi, M.; Inubushi, A.; Mukaiyama, T. *Chem. Lett.* **1987**, 1975–1978.
82. Danheiser, R.L.; Kwasigroch, C.A.; Tsai, Y.–M. *J. Am. Chem. Soc.* **1985**, *107*, 7233–7235.

83. Ishiara, K.; Hanaki, N.; Yamamoto, H. *J. Am. Chem. Soc.* **1993**, *115*, 10695–10704.
84. Bartlett, P.A.; Johnson, W.S.; Elliott, J.D. *J. Am. Chem. Soc.* **1983**, *105*, 2088–2089.
85. Ito, K.; Harada, T.; Tai, A. *Bull. Chem. Soc. Jpn.* **1980**, *53*, 3367–3368.
86. Johnson, W.S.; Crackett, P.H.; Elliott, J.D.; Jagodzinski, J.J.; Lindell, S.D.; Natarajan, S. *Tetrahedron Lett.* **1984**, *25*, 3951–3954.
87. Johnson, W.S.; Chan, M.F. *J. Org. Chem.* **1985**, *50*, 2598–2600.
88. Yamamoto, Y.; Yamada, J.–i. *J. Chem. Soc., Chem. Commun.* **1987**, 1218–1219.
89. Harada, T.; Ikemura, Y.; Nakajima, H.; Ohnishi, T.; Oku, A. *Chem. Lett.* **1990**, 1441–1444.
90. Parrain, J.–L.; Cintrat, J.–C.; Quintard, J.–P. *J. Organomet. Chem.* **1992**, *437*, C19–C22.
91. Johnson, W.S.; Elliott, R.; Elliott, J.D. *J. Am. Chem. Soc.* **1983**, *105*, 2904–2905.
92. Fujioka, H.; Kitagawa, H.; Yamanaka, T.; Kita, Y. *Chem. Pharm. Bull.* **1992**, *40*, 3118–3120.
93. Elliott, J.D.; Choi, V.M.F.; Johnson, W.S. *J. Org. Chem.* **1983**, *48*, 2294–2295.
94. Solladie–Cavallo, A.; Suffert, J.; Gordon, M. *Tetrahedron Lett.* **1988**, *29*, 2955–2958.
95. Elliott, J.D.; Steele, J.; Johnson, W.S. *Tetrahedron Lett.* **1985**, *26*, 2535–2538.
96. Seebach, D.; Züger, M. *Helv. Chim. Acta* **1982**, *65*, 495–503.
97. Seebach, D.; Züger, M. *Tetrahedron Lett.* **1985**, *26*, 2747–2750.
98. Silverman, I.R.; Edington, C.; Elliott, J.D.; Johnson, W.S. *J. Org. Chem.* **1987**, *52*, 180–183.
99. Johnson, W.S.; Kelson, A.B.; Elliott, J.D. *Tetrahedron Lett.* **1988**, *29*, 3757–3760.
100. Andrew, R.G.; Conrow, R.E.; Elliott, J.D.; Johnson, W.S.; Ramezani, S. *Tetrahedron Lett.* **1987**, *28*, 6535–6538.
101. McNamara, J.M.; Kishi, Y. *J. Am. Chem. Soc.* **1982**, *104*, 7371–7372.
102. Sekizaki, H.; Jung, M.; McNamara, J.M.; Kishi, Y. *J. Am. Chem. Soc.* **1982**, *104*, 7372–7374.
103. Corcoran, R.C. *Tetrahedron Lett.* **1990**, *31*, 2101–2104.
104. Molander, G.A.; Haar, Jr., J.P. *J. Am. Chem. Soc.* **1993**, *115*, 40–49.
105. Basile, T.; Longobardo, L.; Tagliavini, E.; Trombini, C.; Umani–Ronchi, A. *J. Chem. Soc., Chem. Commun.* **1990**, 759–760.
106. Kozikowski, A.P.; Sorgi, K.L. *Tetrahedron Lett.* **1982**, *23*, 2281–2284.
107. Kozikowski, A.P.; Sorgi, K.L.; Wang, B.C.; Xu, Z.–b. *Tetrahedron Lett.* **1983**, *24*, 1563–1566.
108. Hosomi, A.; Sakata, Y.; Sakurai, H. *Tetrahedron Lett.* **1984**, *25*, 2383–2386.
109. Giannis, A.; Sandhoff, K. *Tetrahedron Lett.* **1985**, *26*, 1479–1482.
110. Horton, D.; Miyake, T. *Carbohydrate Research* **1988**, *184*, 221–229.
111. Acton, E.M.; Ryan, K.J.; Tracy, M. *Tetrahedron Lett.* **1984**, *25*, 5743–5746.
112. Panek, J.S.; Sparks, M.A. *J. Org. Chem.* **1989**, *54*, 2034–2038.
113. Cooper, A.J.; Pan, W.; Salomon, R.G. *Tetrahedron Lett.* **1993**, *34*, 8193–8196.
114. Martin O.R.; Rao, S.P.; Yang, T.–F.; Fotia, F. *Synlett* **1991**, 702–704.
115. Holmes, C.P.; Bartlett, P.A. *J. Org. Chem.* **1989**, *54*, 98–108.

116. Kato, K.; Minami, T.; Takita, T.; Nishiyama, S.; Ohgiya, T.; Yamamura, S. *J. Chem. Soc., Chem. Commun.* **1989**, 1037–1038.
117. Danishefsky, S.; Kerwin, Jr., J.F. *J. Org. Chem.* **1982**, *47*, 3803–3805.
118. Ferrier, R.J. *J. Chem. Soc.* **1964**, 5443–5449.
119. Danishefsky, S.J.; DeNinno, S.; Lartey, P. *J. Am. Chem. Soc.* **1987**, *109*, 2082–2089.
120. Danishefsky, S.J.; Selnick, H.G.; DeNinno, M.P.; Zelle, R.E. *J. Am. Chem. Soc.* **1987**, *109*, 1572–1574.
121. Danishefsky, S.J.; Selnick, H.G.; Zelle, R.E.; DeNinno, M.P. *J. Am. Chem. Soc.* **1988**, *110*, 4368–4378.
122. Sabol, J.S.; Cregge, R.J. *Tetrahedron Lett.* **1989**, *30*, 6271–6274.
123. Herscovici, J.; Muleka, K.; Antonakis, K. *Tetrahedron Lett.* **1984**, *25*, 5653–5656.
124. Herscovici, J.; Delatre, S.; Antonakis, K. *J. Org. Chem.* **1987**, *52*, 5691–5695.
125. Herscovici, J.; Muleka, K.; Boumaiza, L.; Antonakis, K. *J. Chem. Soc., Perkin Trans. 1* **1990**, 1995–2009.
126. Nishiyama, H.; Narimatsu, S.; Sakuta, K.; Itoh, K. *J. Chem. Soc., Chem. Commun.* **1982**, 459–460.
127. Kano, S.; Yokomatsu, T.; Iwasawa, H.; Shibuya, S. *Heterocycles* **1987**, *26*, 2805–2809.
128. Mann, A.; Nativi, C.; Taddei, M. *Tetrahedron Lett.* **1988**, *29*, 3247–3250.
129. Yamada, J.–i.; Asano, T.; Kadota, I.; Yamamoto, Y. *J. Org. Chem.* **1990**, *55*, 6066–6068.
130. Kadota, I.; Gevorgyan, V.; Yamada, J.–i.; Yamamoto, Y. *Synlett* **1991**, 823–824.
131. Speckamp, W.N.; Hiemstra, H. *Tetrahedron* **1985**, *41*, 4367–4416.
132. Gramain, J.–C.; Remuson, R. *Tetrahedron Lett.* **1985**, *26*, 4083–4086.
133. Klaver, W.J.; Hiemstra, H.; Speckamp, W.N. *J. Am. Chem. Soc.* **1989**, *111*, 2588–2595.
134. Kuo, D.L.; Money, T. *Can. J. Chem.* **1988**, *66*, 1794–1804.
135. Schinzer, D.; Ruppelt, M. *Chem. Ber.* **1991**, *124*, 247–248.
136. Mohr, P. *Tetrahedron Lett.* **1993**, *34*, 6251–6254.
137. Johnson, W.S.; van der Gen, A.; Swoboda, J.J. *J. Am. Chem. Soc.* **1967**, *89*, 170–172.
138. Johnson, W.S.; Harbert, C.A.; Stipanovic, R.D. *J. Am. Chem. Soc.* **1968**, *90*, 5279–5280.
139. Johnson, W.S.; Harbert, C.A.; Ratcliffe, B.E.; Stipanovic, R.D. *J. Am. Chem. Soc.* **1976**, *98*, 6188–6193.
140. Brown, E.G.; Cambie, R.C.; Holroyd, S.E.; Johnson, M.; Rutledge, P.S.; Woodgate, P.D. *Tetrahedron Lett.* **1989**, *30*, 4735–4736.
141. Johnson, W.S.; Elliott, J.D.; Hanson, G.J. *J. Am. Chem. Soc.* **1984**, *106*, 1138–1139.
142. Batcho, A.D.; Berger, D.E.; Davoust, S.G.; Wovkulich, P.M.; Uskokovic, M.R. *Helv. Chim. Acta* **1981**, *64*, 1682–1687.
143. Johnson, W.S.; Wiedhaup, K.; Brady, S.F.; Olson, G.L. *J. Am. Chem. Soc.* **1968**, *90*, 5277–5279.
144. Johnson, W.S.; Wiedhaup, K.; Brady, S.F.; Olson, G.L. *J. Am. Chem. Soc.* **1974**, *96*, 3979–3984.
145. Johnson, W.S.; Telfer, S.J.; Cheng, S.; Schubert, U. *J. Am. Chem. Soc.* **1987**, *109*, 2517–2518.
146. Johnson, W.S.; Chen, Y.–Q.; Kellogg, M.S. *J. Am. Chem. Soc.* **1983**, *105*, 6653–6656.

147. Schmid, R.; Huesmann, P.L.; Johnson, W.S. *J. Am. Chem. Soc.* **1980**, *102*, 5122–5123.
148. Guay, D.; Johnson, W.S.; Schubert, U. *J. Org. Chem.* **1989**, *54*, 4731–4732.
149. Johnson, W.S.; Fletcher, V.R.; Chenera, B.; Bartlett, W.R.; Tham, F.S.; Kullnig, R.K. *J. Am. Chem. Soc.* **1993**, *115*, 497–504.
150. Johnson, W.S.; Buchanan, R.A.; Bartlett, W.R.; Tham, F.S.; Kullnig, R.K. *J. Am. Chem. Soc.* **1993**, *115*, 504–515.
151. Johnson, W.S.; Plummer, M.S.; Reddy, S.P.; Bartlett, W.R. *J. Am. Chem. Soc.* **1993**, *115*, 515–521.

5. Sakurai Reaction

Lewis Acid-Promoted Allylsilanes and Allylstannanes 1,4-Addition to Unsaturated Ketones

I. Introduction and History

In 1977, Sakurai and Hosomi reported that α,β-enones, when associated with $TiCl_4$ in CH_2Cl_2, undergo Michael addition (1,4-addition) from allyltrimethylsilane.[1,2]

$TiCl_4$, CH_2Cl_2, 85% yield

It is however worth noting that, also in 1977, Ojima showed that 3-trimethylsilylcyclopentene reacts with conjugated enones to afford high yields of Michael adducts.[3]

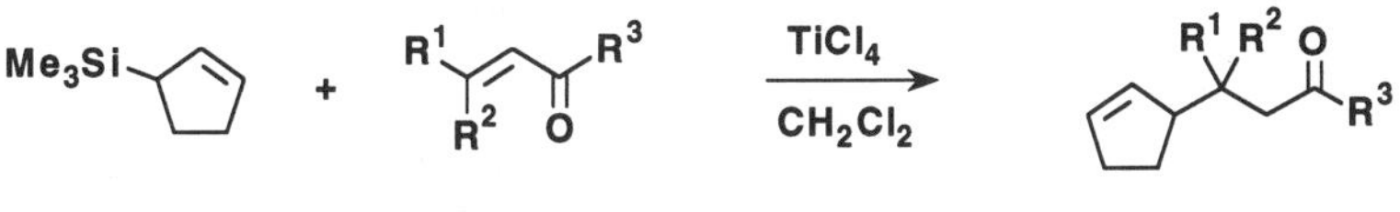

$TiCl_4$, CH_2Cl_2

These results were, and still are, particularly interesting considering that the addition of lithium diallylcuprates is severely substrate dependent.[4–6] The Sakurai reaction was hence very soon widely used.[7–11] It was particularly studied and developed in intramolecular reactions.

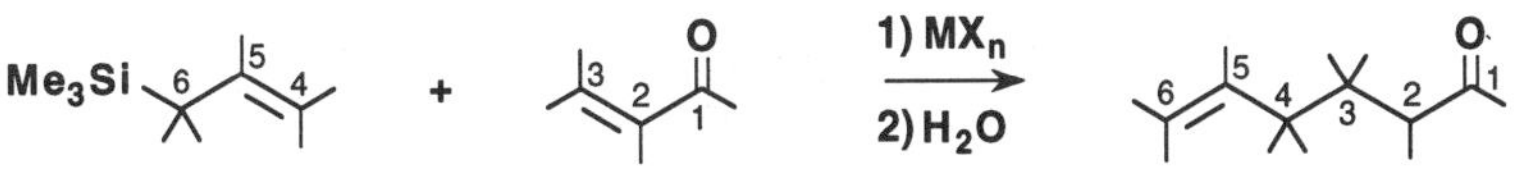

1) MX_n, 2) H_2O

The Sakurai reaction deals with the conjugate addition of allyltrialkylsilanes or allyltrialkylstannanes[12] to acyclic and cyclic[13] α,β-unsaturated ketones, alkyl alkylidenemalonates, alkyl alkylidenecyanoacetates, α,β-unsaturated acyl cyanides, α,β-unsaturated acylsilanes, α-nitro olefins or *N*-crotonyl oxazolidinones (*vide infra*). However α,β-unsaturated esters, amides and nitriles are not sufficiently reactive; nevertheless, Kemmitt reported the conjugate addition of allyltrimethylsilane to 1,1-dicarboxylate cyclopropane in the presence of $EtAlCl_2$.[14]

Conjugate allylation can also be performed on enones by using allylsilanes and fluoride catalysis.[15,16]

The allylsilanes used in the Sakurai reaction can be functionalized. Consequently, the allylic moiety thus obtained bears a functionality allowing further reactivity and cyclization in particular. Thus, Knapp proposed a [3+2] annulation by performing the addition of 2-chloromethyl-3-(trimethylsilyl) propene to cyclohexenone or to 2-phenylthiocyclopentenone. A basic treatment of the initially formed chloro ketone induces cyclization.[17]

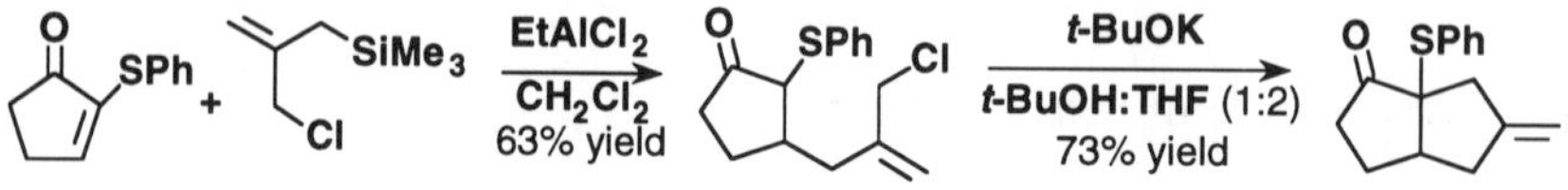

SPh + SiMe3 / Cl — $EtAlCl_2$, CH_2Cl_2, 63% yield → SPh, Cl — t-BuOK, t-BuOH:THF (1:2), 73% yield → O SPh

Similarly, Trost prepared bicyclo[3.3.0]octanols with *cis* and even *trans* ring junction, by adding 2-bromo-3-(trimethylsilyl)propene to acetyl cyclo pentene and by treating the adduct with lithium.[18]

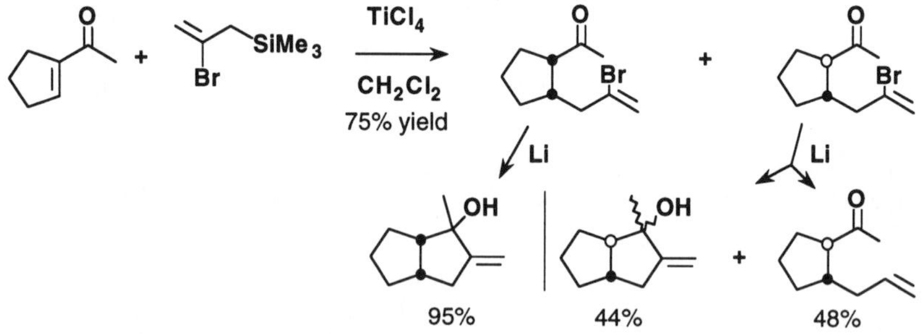

O + SiMe3 / Br — $TiCl_4$, CH_2Cl_2, 75% yield → Br + Br — Li → OH (95%); Li → OH (44%) + O (48%)

The addition of 1,3-bis(trimethylsilyl)propene allows the introduction of an allylsilane moiety in the β-position of α,β-unsaturated ketones.[19]

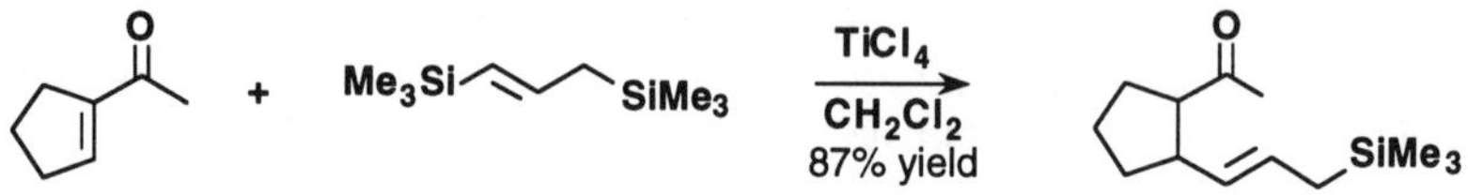

O + Me_3Si SiMe3 — $TiCl_4$, CH_2Cl_2, 87% yield → O, SiMe3

Danheiser established that the addition of allylsilanes to α,β-unsaturated acylsilanes takes place extremely quickly, even at –78°C. The acylsilanes thus obtained can easily be oxidized into the corresponding acids.[20]

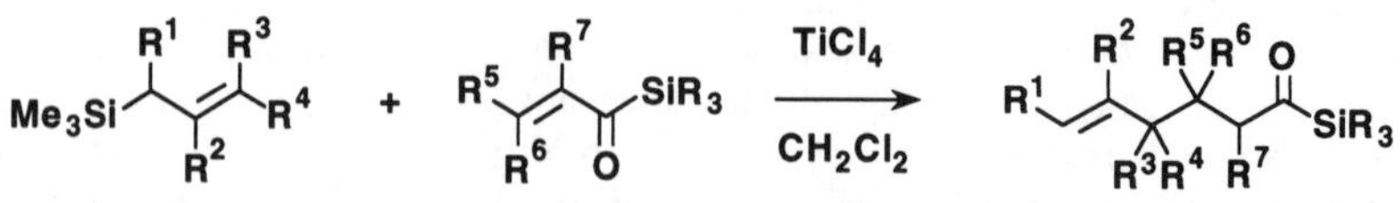

Me_3Si (R1, R3, R4, R2) + (R5, R7, SiR_3, R6, O) — $TiCl_4$, CH_2Cl_2 → R1, R2, R5R6, O, SiR_3, R3R4, R7

Fujita showed that the addition of allylsilanes to α-nitro olefins leads to unstable unsaturated nitronic acids which can be reduced, using $TiCl_3$, into γ,δ-unsaturated ketones.[21]

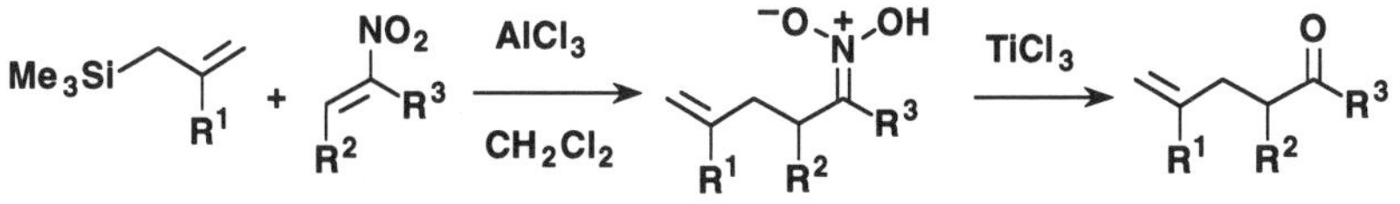

Allylation of *o*-benzoquinones or 1,2-naphthoquinones gives access, in fair to good yields, to naphthalen-1,2-diols.[22,23] These compounds proved to be useful intermediates in the synthesis of vitamin K derivatives.[24]

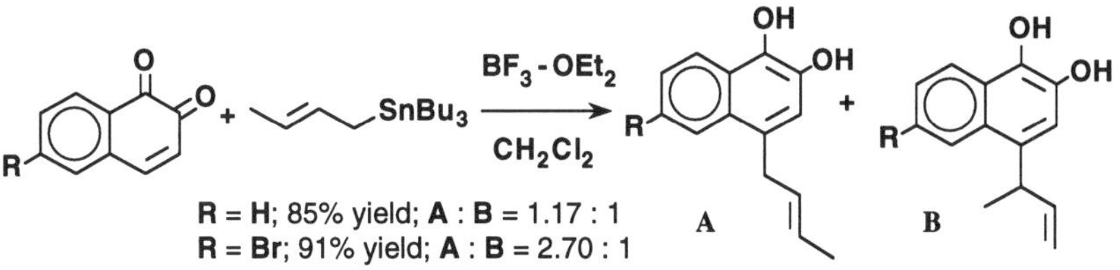

R = H; 85% yield; **A** : **B** = 1.17 : 1
R = Br; 91% yield; **A** : **B** = 2.70 : 1

Allylation have also been performed on substituted *p*-quinone.[25,26] Moreover, the use of pentadienyltins provides a route towards the synthesis of anthracyclinones skeletals. Thus, the tetracyclic system of these molecules is easily obtained by Lewis acid mediated tandem conjugate addition/Diels Alder reaction of 3-acryloyl quinone with pentadienyltins.[27,28] Various Lewis acids were used and overall yields from 30 to 70% were obtained in this sequence in which only the (*E*)-dienic intermediate participates in the intramolecular Diels-Alder step.[29]

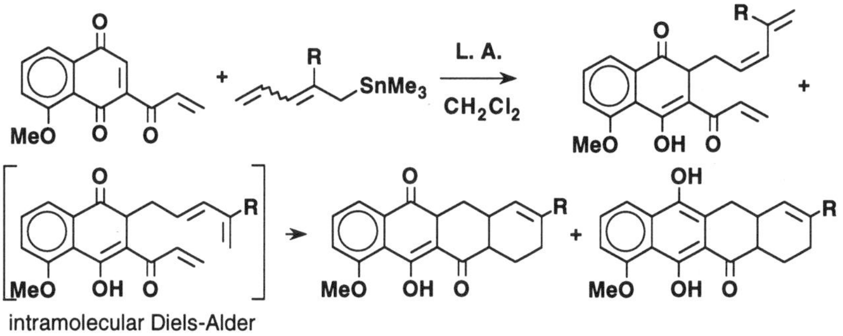

intramolecular Diels-Alder

The tetracyclic quinones are considered to be the precursors of various anthracyclinones such as (±)-11-deoxyanthracyclinone.

(±)-11-Deoxyanthracyclinone

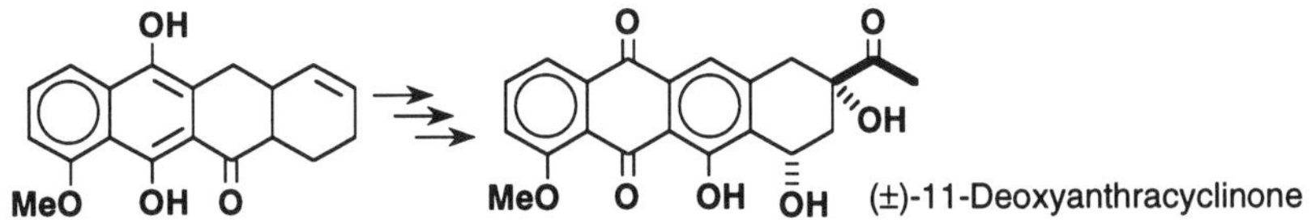

The Sakurai reaction generally requires the use of Lewis acids such as $TiCl_4$, $SnCl_4$, $AlCl_3$, or BF_3-Et_2O and is usually performed in CH_2Cl_2. A protodesilylation of the allylsilane moiety can occasionally occur; in such an eventuality, it proved better to use $EtAlCl_2$ which is both a Lewis acid and can serve as a "proton sponge", and thereby inhibits competitive protodesilylation. Mukaiyama showed that catalytic amounts of trityl perchlorates allow conjugate addition to proceed with good yields.[30] On the other hand, allyltributylstannanes can be used instead of allylsilanes with various Lewis acids and particularly with trialkylsilyl triflates.[31]

II. Mechanism

Sakurai did not propose a mechanism but, from well-established results on the reactivity of allylsilanes, the following one can be suggested:

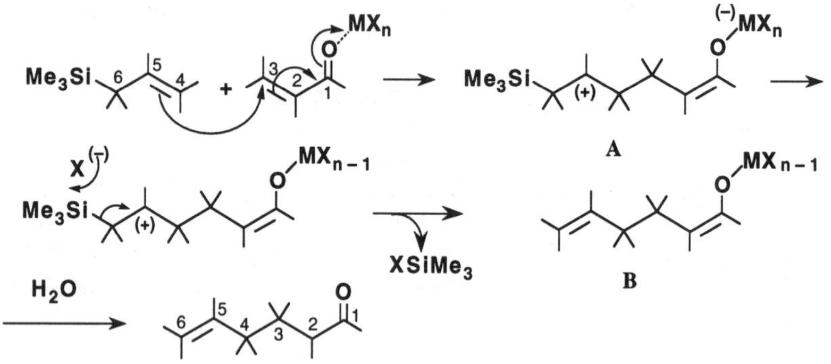

Denmark established that no metathesis was taking place, under the reaction conditions, between the allylsilane and $TiCl_4$ and was able to observe enolate **B** (M=Ti; X=Cl) by NMR (see Chapter 1, Section III.2).[32] The existence of β-silyl cation **A** is supported by the occurrence of a minor derivative resulting from an intramolecular *C*-alkylation of the titanium enolate **A**. The addition of allyltrimethylsilane to acetylcyclohexene illustrates that possibility.

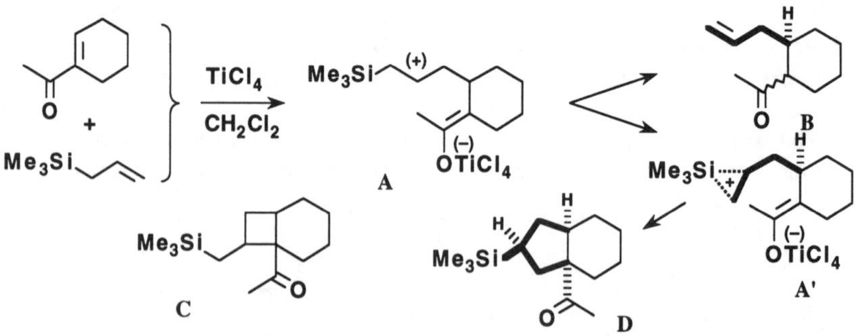

In 1979, Santelli proposed a cyclobutanic structure, analogous to **C** for these derivatives,[33] and several authors then confirmed that type of structure.[34–37] House even observed that a temperature increase favors the formation of these products.[38]

However in 1990, Knölker showed, using X-ray diffraction, that the correct structure of these derivatives was a trimethylsilycyclopentane such as **D**.[39] Starting from intermediate **A**, a 1,2-migration of the trimethylsilyl group would lead to **A'** which would then undergo cyclization (see Section IV).

Previous results must then be revised. Thus, the product, isolated by Danishefsky and Kahn[35] when adding allyltrimethylsilane to enedione **A** and initially thought to be a cyclobutanic derivative, must be the triquinane derivative **C**; the stereochemistry being assigned from analogous recent results.

H O O A + Me_3Si $TiCl_4$ CH_2Cl_2 H O O B (12–31%) H O O $SiMe_3$ C

This [3+2] cyclopentane annulation was then extensively investigated and proved to be particularly interesting because of its remarkable diastereo selectivity that will be discussed in Section IV.[40]

A similar result was reported by Danheiser as early as 1981. When adding (trimethylsilyl)allenes to conjugated enones, he observed the formation of acetyl cyclopentene derivatives (see Section IV). A bridged ion **A** (or an equilibrium between localized ions) was postulated as the intermediate of the reaction.[41]

Me_3Si R^1 R^3 R^2 + R^7 R^6 O R^4 R^5 $TiCl_4$ CH_2Cl_2 [Me_3Si + R^1 R^6 R^7 $OTiCl_n$ R^5 R^3 R^2 R^4] A → Me_3Si R^1 R^6 R^7 O R^5 R^2 R^3 R^4

On some occasions, an intramolecular *O*-alkylation between the β-silyl cation and the enolate can occur. Thus, the addition of allylsilanes to α,β-ethylenic acyl cyanides leads, along with the expected adducts, to dihydropyranes **B,** which result from the cyclization of intermediates **A**.[42,43]

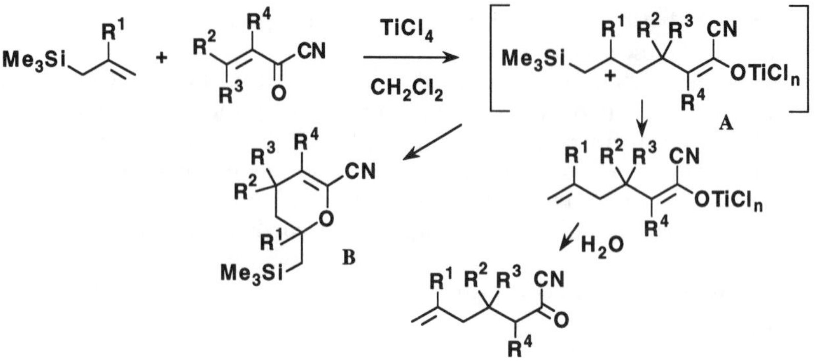

One can think that the cyclization regioselectivity is due to modifications of electronic factors resulting from the introduction of the cyano group. However, *O*-alkylation can also be favored by steric factors as in the following example.[44,45] It is worth noting that the same reaction, performed with allyltrimethylstannane, affords the expected allylation adduct.

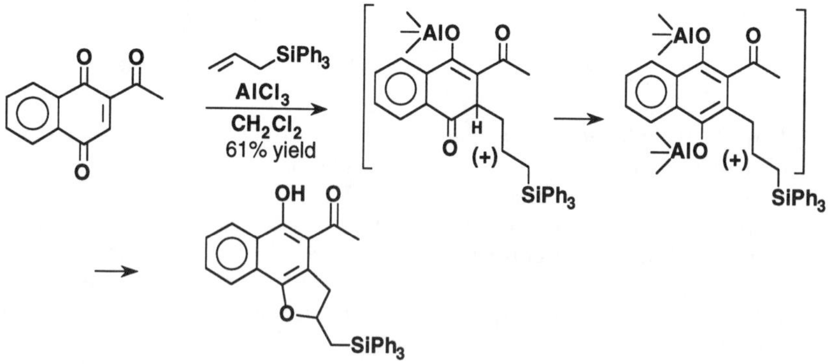

In this case, the second aluminum enolate is generated after the addition of the allylsilane. One can consider that the β-silyl cation is particularly persistent since it is not trapped by the kinetic enolate, resulting from the conjugated addition, but by the second one.

The addition of the pentadienyltrimethylsilane on mesityl oxide leads to the expected dienic ketone **A** but also to ketone **C** which may result from the cyclization of stabilized cation **B** (or from a Diels-Alder reaction).[46]

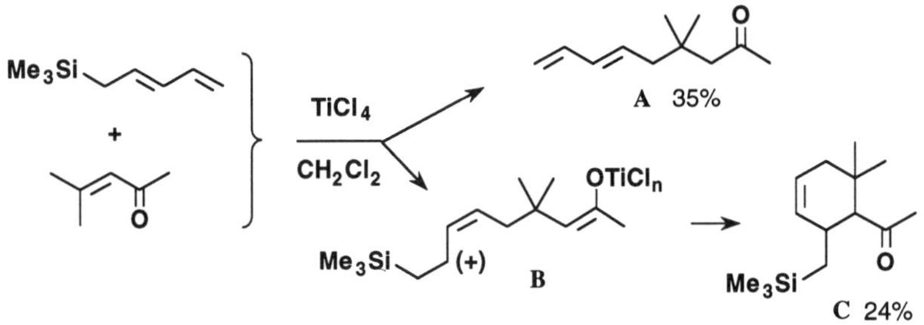

The enolates resulting from the standard addition of allylsilanes to enones can be used for addition reactions towards electrophiles such as aldehydes[47] or halides such as chloromethyl phenyl sulfide.[48]

III. Reaction Stereochemistry[49]

III.1 Diastereofacial selectivity

In his original paper, Sakurai showed that the addition of allylsilane to octalone led to the *cis* ring junction product, which therefore resulted from an axial attack. This result is supported by work from Heathcock in particular, as shown in the following example.[50,51]

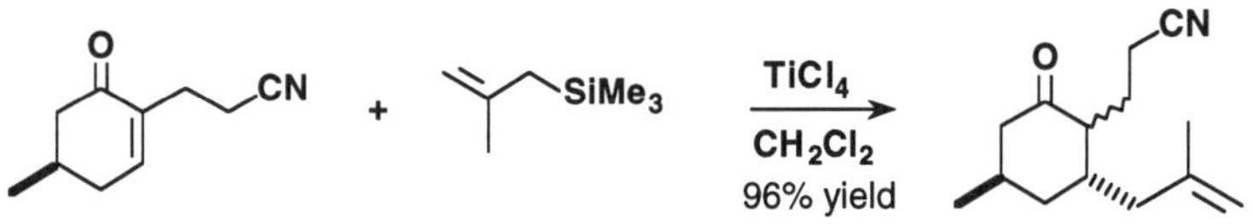

A similar result has been observed during the addition of allyltrimethylsilane to the following substituted cyclohexenone. The coordination of both carbonyl groups with the Lewis acid results in the formation of a chelate in which the *Re** face of the π system is the more accessible one.[52]

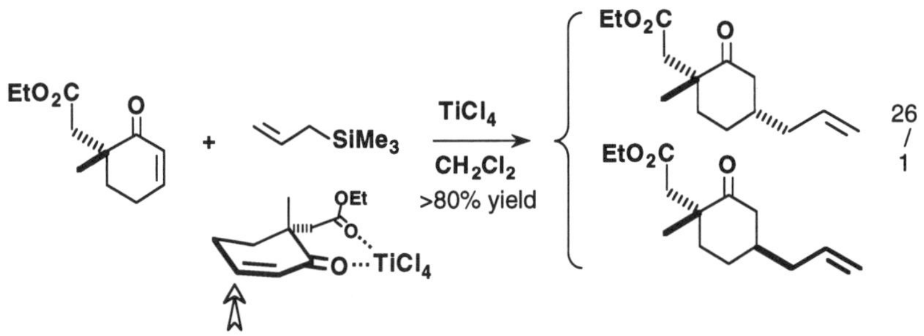

A good exocyclic diastereoselectivity was also observed by Yoshikoshi while adding allyltrimethylsilane or methallyltrimethylsilane to *trans* 3-

ethylidenenopinone, which is a key intermediate in the synthesis of (+)-nootkatone; the face opposed to the isopropylidene bridge being the more accessible one.[53]

R = H 89% yield; 4 : 1
R = Me 81% yield; 3.16 : 1

A very interesting example of asymmetric induction is given by Schultz who showed that the addition of allyltrimethylsilane to the chiral 2-amidocyclohexenone proceeds with high diastereoselectivity.[54] A chelated complex **A**, involving six atoms, was postulated. Ketones **B** (R = alkyl) are then easily obtained from the initial addition product.

Diastereofacial selectivity was also observed when adding allyltrimethyl silane to 1-acetyl-4-*tert*-butyl cyclopentene[55] and to conjugate dienone (1,6-addition).[56]

III.2 1,2-Diastereoselectivity in intermolecular reactions

Several reports in the literature deal with the diastereoselectivity resulting from the use of (*Z*)- or (*E*)-crotyltrimethylsilane or (*Z*)- or (*E*)-crotyltri-*n*-butylstannane. The first result was reported by Yamamoto who studied the addition of (*E*)-crotyltri-*n*-butylstannane to ethyl ethylidenemalonate.[57] The *anti* adduct is the major one with both $SnCl_4$ and $AlCl_3$ (52% yield with a *anti*:*syn* ratio of 9:1 with $SnCl_4$ and 94% yield with a *anti*:*syn* ratio of 4:1 with $AlCl_3$).[58]

Bu_3Sn + CO_2Et CO_2Et $\xrightarrow[CH_2Cl_2]{SnCl_4}$ 52% yield

CO_2Et CO_2Et *anti* 9 / 1

CO_2Et CO_2Et *syn*

Yamamoto proposed an antiperiplanar transition state in which the steric interaction, between substituents of the two bonding carbon atoms, is the major factor; the two methyl groups are in antiperiplanar position during the formation of the major isomer (*anti*), while they are synperiplanar during the formation of the *syn* isomer:

EtO_2C CO_2Et H H Bu_3Sn *ul*-approach (*R**) (*R**) CO_2Et CO_2Et *anti*

EtO_2C CO_2Et H H $SnBu_3$ *lk*-approach (*S**) (*R**) CO_2Et CO_2Et *syn*

A particular case is illustrated by the addition of allyltrimethylsilane to α,β-ethylenic carbonyl compounds bearing a γ stereogenic center. The first example was reported by Heathcock who added allyltrimethylsilane to 5-phenylpent-3-en-2-one.[59,60] The major product is the *anti* isomer, as predicted by the Felkin model (molecular mechanics calculations on the transition state structures involved in the conjugated addition of nucleophiles to γ-substituted crotonates have also been performed[61]).

Ph O + $SiMe_3$ $\xrightarrow[CH_2Cl_2]{TiCl_4}$ 74% yield Ph O + Ph O 4 : 1

The addition of allyltrimethylsilane to a *trans* γ-alkoxy-α,β-unsaturated ketone takes place with the same stereochemistry as previously; however, with the *cis* isomer, chelation induces the opposed diastereoselectivity.

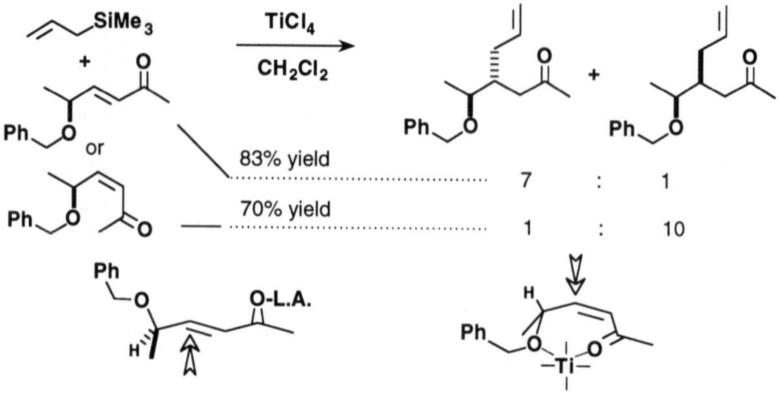

The $TiCl_4$-promoted addition of allyltrimethylsilane to 4-*tert*-butyl dimethylsilyloxy cyclohex-2-enone proceeds with high diastereoselectivity. Thus, the expected 1,4-adduct was obtained in 74% yield as a single *cis* diastereomer.[62]

Yamamoto added allyltrimethylsilane and allyltri-*n*-butylstannane on an ethyl alkylidenemalonate derivative, with excellent stereoselectivity.[57] The major adduct is the *anti* isomer resulting from an attack opposed to the phenyl group when this one is eclipsed with the π orbitals of the double bond (Felkin-Anh transition state model[63]).

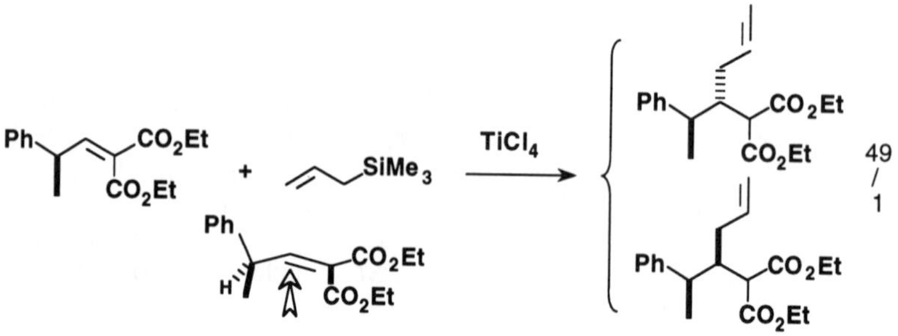

A series of studies was devoted to the addition of crotylsilanes to cycloalkenones. The first result, reported by Tokoroyama, is in connection with the synthesis of juvabione by addition of *(Z)*- or *(E)*- crotyltrimethyl silane to cyclohex-2-enone.[64]

Starting from (*E*)-crotyltrimethylsilane, the major product is the *unlike* (or *erythro*) diastereomer.

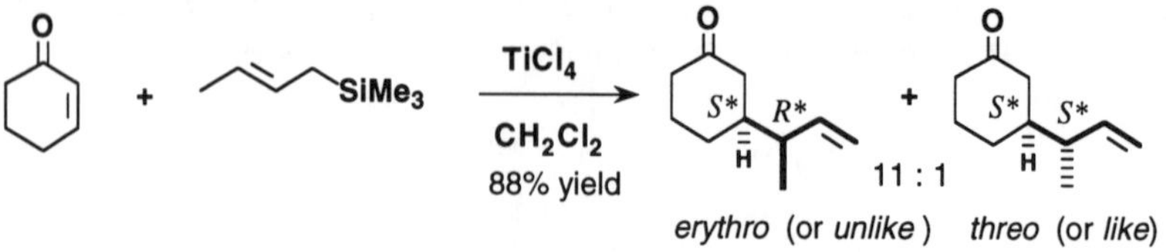

On the contrary, starting from (Z)-crotyltrimethylsilane, the *like* (or *erythro*) adduct is obtained as the major diastereomer, although with a much lower selectivity.

SiMe$_3$ — $TiCl_4$, CH_2Cl_2, 80% yield — S* S* + S* R* 3 : 1

threo (or *like*) *erythro* (or *unlike*)

The diastereoselectivity depends on the substituents borne by the silicon atom. When using (*E*)-crotylsilanes, the best *unlike*-diastereoselectivity is obtained with the diphenylmethylsilyl group (15.6:1), whereas when (Z)-crotylsilanes are used, the *like*-diastereoselectivity is highly enhanced by the ethoxydimethylsilyl group (11.2:1).

In order to account for the stereochemistry observed with (*E*)-crotylsilanes, synclinal transition state **A** was preferred to the open chain one (antiperiplanar) **A"**.

A, A', A" (*E*)-crotylsilane → *erythro* (or *unlike*) major — *ul*-approach

B, B', B" (*E*)-crotylsilane → *threo* (or *like*) minor — *lk*-approach

For (Z)-crotylsilanes, it is more difficult to find a transition state that would account for the results. The antiperiplanar model **C"**, that could lead to the major isomer, seems less favorable than **D"**. The best transition state **C'** would involve a synperiplanar approach.

From these results, the existence of several transition states and therefore several reaction paths seems probable.

C, C', C'' ((*Z*)-crotylsilane) → *threo* (or *like*), major, *lk*-approach

D, D', D'' ((*Z*)-crotylsilane) → *erythro* (or *unlike*), minor, *ul*-approach

Tokoroyama also studied the diastereoselectivity of the addition to cyclopentenones in order to prepare the neonepetalactone.[65] As previously, he observed an *erythro*-selectivity with (*E*)-crotylsilanes (though lower than with cyclohexenone) and a *threo*-selectivity with (*Z*)-crotylsilanes.

$TiCl_4$, CH_2Cl_2

		erythro		*threo*
R = H	(*Z*)-crotyl-$SiMe_3$	1	:	2
	(*E*)-crotyl-$SiMe_2Ph$	5.25	:	1
R = Me	(*Z*)-crotyl-$SiMe_3$	1	:	11.5
	(*E*)-crotyl-$SiMe_3$	5.25	:	1

The authors proposed a synclinal transition state such as **A** or **B** to rationalize their results. Nevertheless, an antiperiplanar transition **A''** or **B''** could also account for the observed *erythro*-selectivity.

A ≡ A, A'' → *erythro*, major, *ul*-approach

B ≡ B, B'' → *threo*, minor, *lk*-approach

Again, the high *threo*-selectivity, observed when starting from (*Z*)-crotyl trimethylsilane, is difficult to explain by comparing **C** and **D** transition states on one side and **C"** and **D"** on the other. However, synclinal transition state **C**, if favored by stereoelectronic factors, can account for the predominance of the *threo* isomer.

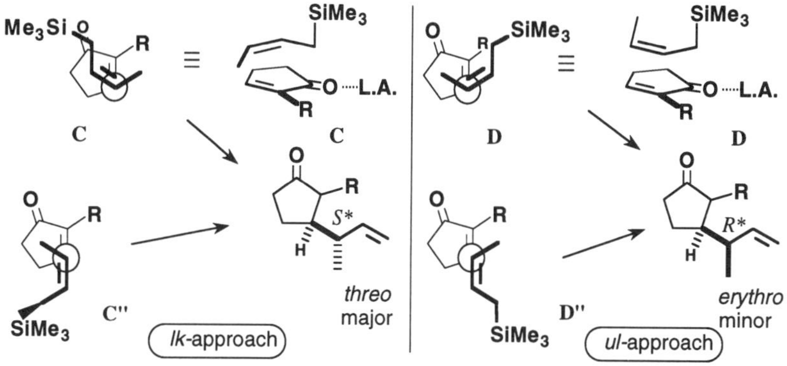

In the same paper, Tokoroyama also studied the stereochemistry of the $TiCl_4$ promoted addition of two different (*E*)- and (*Z*)-crotylsilanes to 2-ethoxycarbonylcyclopent-2-enone. He observed an enhancement of the diastereoselectivity when the Lewis acid was used in connection with HMPA.

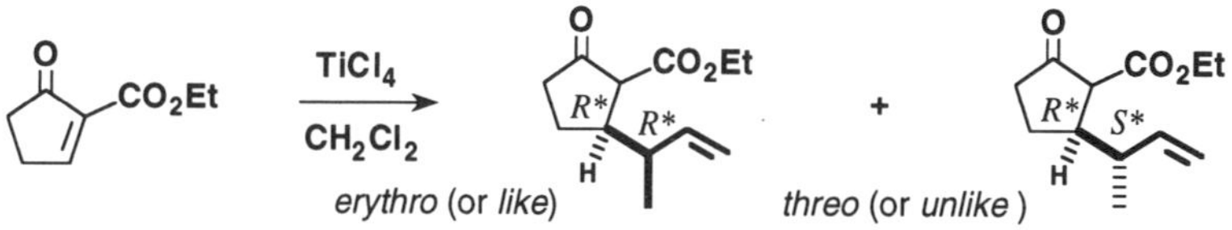

Crotylsilane		*erythro* (or *like*)		*threo* (or *unlike*)
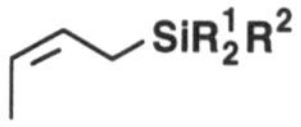 (*Z*)-$SiR^1_2R^2$	$R^1 = R^2 = Me$:	1	:	3
	$R^1 = Ph; R^2 = Me$:	1	:	2.57
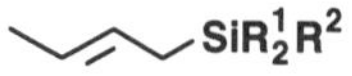 (*E*)-$SiR^1_2R^2$	$R^1 = R^2 = Me$:	3.76	:	1
	$R^1 = R^2 = Me$:	6.14	:	1[a]
	$R^1 = Ph; R^2 = Me$:	13.29	:	1[a]

[a]HMPA (2.8 equivalents) was added

Tokoroyama then reported on the addition of (*E*)- and (*Z*)-crotylsilanes to cyclopentenones bearing a chiral auxiliary in position 2. The *syn*-selectivity with (*E*)-crotylsilane and the *anti*-selectivity with (*Z*)-crotylsilane are again observed (**A** + **C** / **B** + **D**). In addition, the reaction shows a remarkable diastereofacial selectivity (**A** + **B** / **C** + **D**) (i.e., (*S*)/(*R*)-configuration in position 3), particularly when using $TiCl_4$ as a Lewis acid.[66]

L. A.
CH_2Cl_2

A *erythro* B *threo* C *erythro* D *threo*

R = —S(O)Tol	$TiCl_4$	(E)-crotyl-SiMePh$_2$	A + C : B + D = 9 : 1; A + B : C + D = 99 : 1
	BF_3-OEt_2	ditto	A + C : B + D = 12 : 1; A + B : C + D = 1 : 4
	$TiCl_4$	(Z)-crotyl-SiMePh$_2$	A + C : B + D = 1 : 14; A + B : C + D = 99 : 1
	BF_3-OEt_2	ditto	A + C : B + D = 1 : 1.5; A + B : C + D = 1 : 2
R = (8-phenylmenthyl)O_2C–	$TiCl_4$	(E)-crotyl-SiMePh$_2$	A + C : B + D = 20 : 1; A + B : C + D = 100 : 1
	$TiCl_4$	(Z)-crotyl-SiMePh$_2$	A + C : B + D = 1 : 1.5; A + B : C + D = 100 : 1

IV. [3+2] Annulation Reaction

The [3+2] adduct, once believed to be a cyclobutanic derivative, can become the major product of the reaction especially if allyltrialkylsilanes, with substituted alkyl groups, are used instead of allyltrimethylsilane.[40] Thus, Danheiser obtained the *trans*-1-acetyl-3-tri(isopropyl)silylcyclopentane with 77% yield from allyltriisopropylsilane and methylvinylketone.[67]

i-Pr_3Si (allyl) + methylvinylketone $\xrightarrow[CH_2Cl_2]{TiCl_4}$ (77% yield) *i*-Pr_3Si (trans) + *i*-Pr_3Si (cis), 99 : 1

The remarkable stereoselectivity of this annulation is even better than that of the 1,4-addition. One can note that the trialkylsilyl and the carbonyl groups are in 1,3-*trans* position. This stereoselectivity can be explained by the existence of a "compact" transition state **A** which would lead to the formation of the bridged ion **B** which, upon cyclization, would yield the cyclopentane derivative. This proposed sequence is illustrated with the addition of allyltrimethylsilane to acetylcyclohexene.

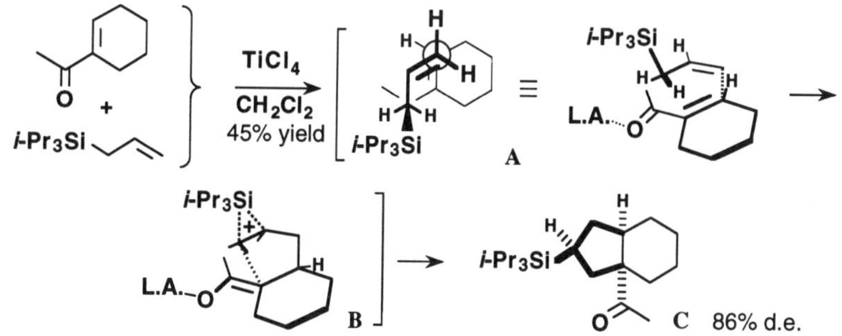

Danheiser also reported a very interesting result dealing with the addition of trialkylbutenylsilane to methyl vinyl ketone; in both cases studied, ketone **A** is obtained as the major diastereomer.

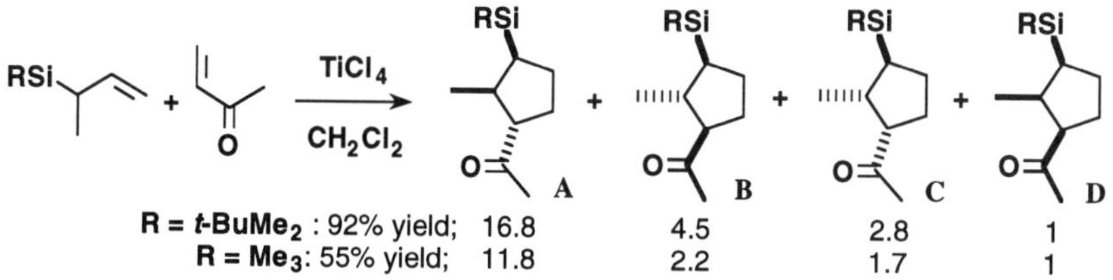

	A	B	C	D
R = *t*-BuMe$_2$: 92% yield;	16.8	4.5	2.8	1
R = Me$_3$: 55% yield;	11.8	2.2	1.7	1

One can note that the yield of the [3+2] annulation is enhanced when the silicon atom bears bulky groups. In fact, the kinetics of the departure of the trialkylsilyl group SiR_3 depends on the size of the alkyl group R. When the approach towards the silicon atom is difficult, the half-life of the β-silyl cation or of the bridged cation **A2** is extended, thus enhancing the cyclization possibilities. The "compact" transition state **A1** built from the *s-trans* conformation of the enone accounts for these results, but the butenylsilane is in a less favorable conformation.

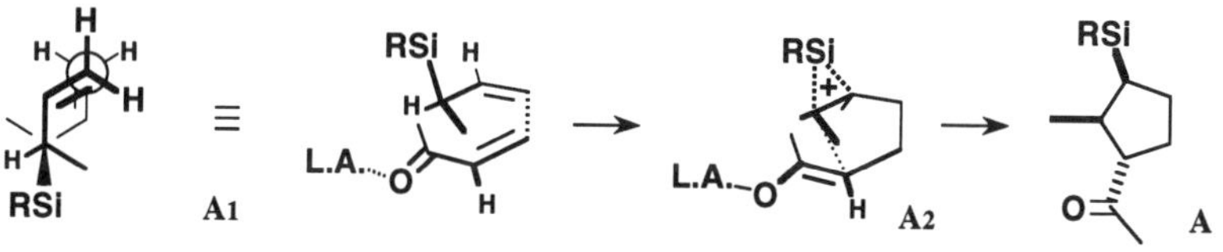

The second predominant product **B** results from a transition state **B1** in which both the methylvinyl ketone and the butenylsilane are in favorable conformations. The fact that **B1** does not lead to the major product might be caused by a greater distance between the two bonding atoms.

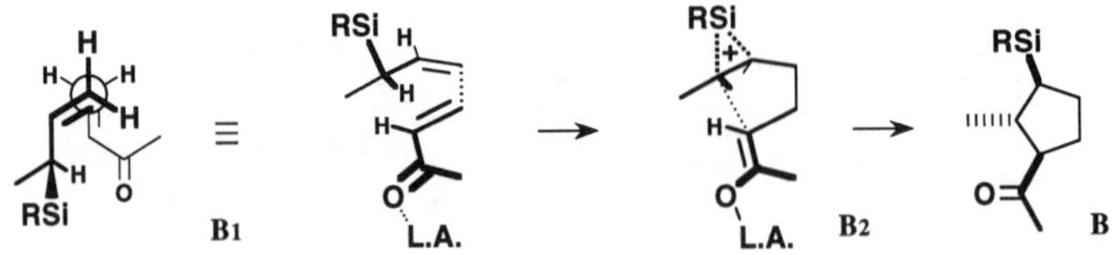

Transition state **A1** can also explain the stereochemistry of the acetylcyclopentane, obtained with a very high selectivity by addition of allyltriisopropylsilane to pentenone.[68]

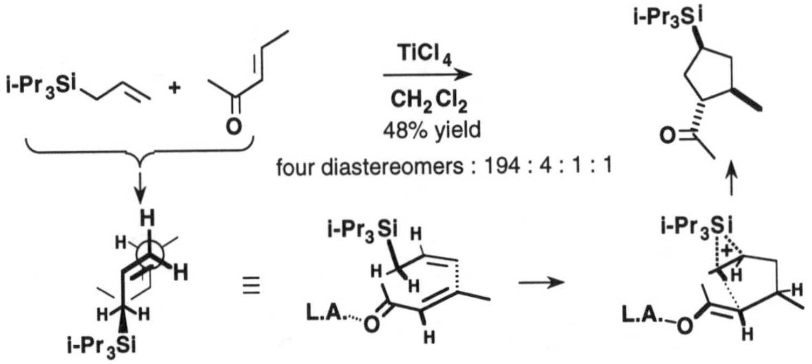

By using a chiral *N*-crotonyloxazolidinone **A**, Snider obtained an optically active cyclopentane **B**.[69] It is worth noting that if the cyclopentanylation is particularly stereoselective, this is not the case of the Sakurai reaction which leads to compounds **C** and **D** in a 2:1 ratio.

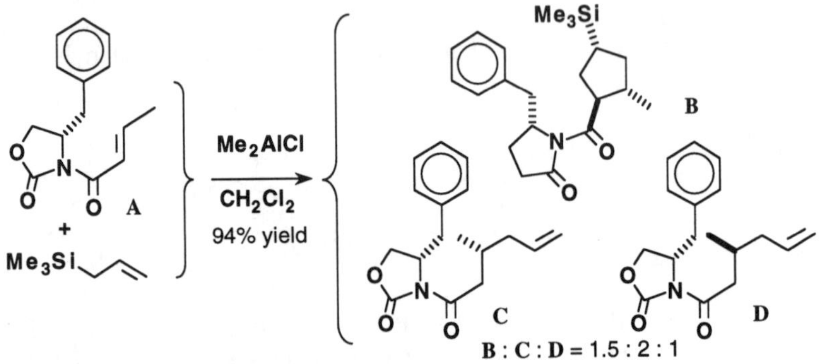

The exclusive formation of a single diastereomer **B** can be accounted for if the reaction proceeds through a "compact" transition state, in which the crotonyl moiety adopts an *s-cis* conformation.

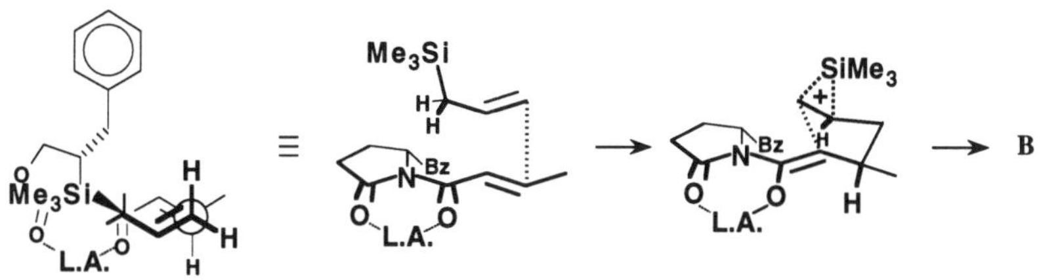

In the same way, Panek added an optically active crotylsilane to methacroleine. A “compact” transition state accounts for the absolute configuration of the four asymmetric centers borne by the cyclopentane.[70]

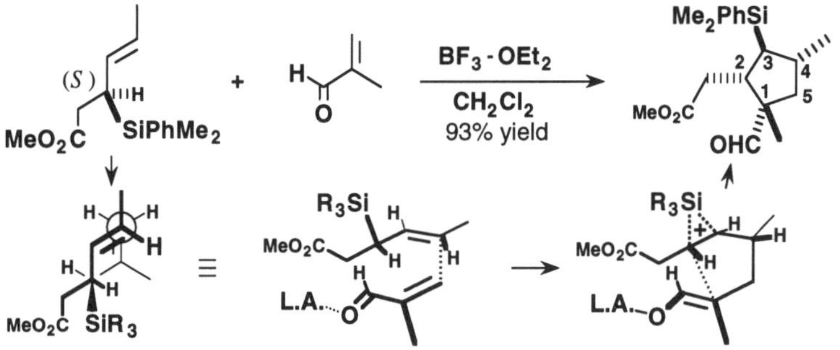

[3+2] Annulation was also reported while adding 3-(trimethylsilyl)but-1-ene to a 4-silyloxy-1-benzopyrylium salt.[71,72] However, allylstannanes led to the expected corresponding allylchromones.

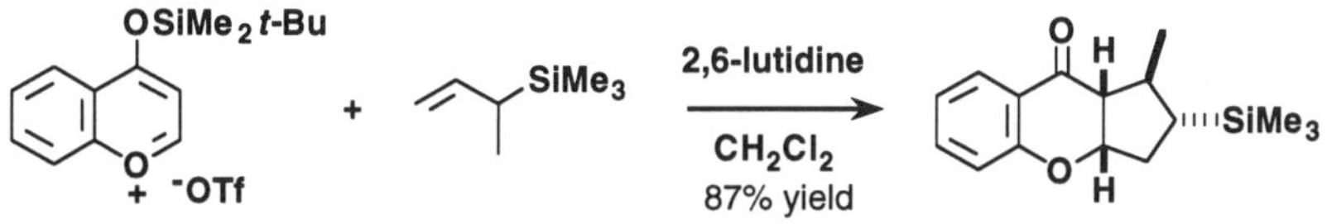

The interest in this new annulation reaction was even intensified by Knölker who obtained bicyclic derivatives by addition of allyltriisopropyl silane to acetylcycloalkenes.[73,74]

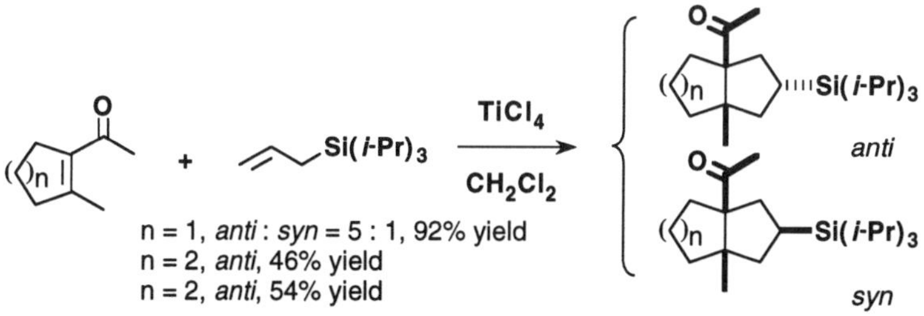

A “compact” transition state of chair type undergoes severe steric interactions, but can be stabilized by two factors:

1. A charge interaction. The attraction between opposed charges, formed in the transition state, can stabilize the "compact" conformation. Methylene chloride, because of its low dielectric constant, does not provide conditions propitious to charge separation. The yield increase in cyclopentanic derivatives, when using trialkylallylsilanes with branched alkyl groups, would result from a stabilization of the bridged ion.

2. An orbital interaction. The main interaction is between the allylsilane HOMO and the complexed enone LUMO (the indicated MO coefficients of the following scheme were calculated by us with the semi-empirical AM1 method).[75] Not surprisingly, it appears that the main bonding interaction takes place between the terminal carbon atom of the allylsilane and the β carbon atom of the enone. Moreover, the "compact" approach benefits from a stabilizing secondary interaction between the first carbon atom of the allyl moiety and the carbon atom of the carbonyl group. Finally, the chair-like geometry vs. the boat-like geometry, prevents an antibonding interaction between the two center carbon atoms.

A "compact" structure for the transition state has often been postulated for electrophilic cyclizations. In 1898, Stephan reported that linalol **A** rearranges into isomeric open-chain geraniol **E** (R = H) and nerol **F** (R = H) or cyclizes into α-terpineol **C** when treated in acidic media. Stephan also observed that optically active (–)-(*R*)-linalol **A** led to (+)-(*R*)-α-terpineol **C** even though the transformation was carried out under conditions that led to racemization of linalol[76] (cited by Vogel).[77] Similarly, hydrolysis of the linalol paranitrobenzoate **B** gives (+)-(*R*)-α-terpineol **C** and the corresponding terpineol paranitrobenzoate **D** (90% stereoselectivity).[78,79]

These results and others can result from the formation of intermediate **G** in a concerted process or from the formation of ion-pair intermediate **H** in which the π system stabilizes the developing allylic carbonium ion independently of steric considerations.[80] The fact that the attack of water should be concerted with the cyclization is supported by the formation of terpinol as the only cyclic product. Limonene is not formed although it is the major product of the hydrolysis of dinitrobenzoate terpineyl **D**; this difference would therefore result from the fact that **D** undergoes hydrolysis through terpineyl cations.

G or **H** display a "compact" transition state similar to the one postulated for the [3+2] annulation.

OR → C or D
ROH G
H → ROH

Other examples of transition states, in which the 3,3-interaction is governed by the frontier orbitals independently of steric considerations, are reported in the literature. The excellent diastereoselectivity observed during the addition of enamines to nitroolefins led Seebach to propose a "compact" transition state of chair conformation.[81] Enders and Steglich used this model for the addition of enamines to acyliminoacetates.[82] However, a boat-type transition state, stabilized by orbital interactions, was proposed by Santelli to account for the stereochemistry of the addition of allylic Grignard reagents to enones. The modification of the molecular orbital coefficients would be responsible for this reversal of regioselectivity, 1,2-addition instead of 1,4-addition.[83,84]

V. Reaction Involving Allenyl- and Propargylsilanes

V.1 Addition of allenylsilanes

The addition of allenylsilanes to α,β-unsaturated acyl cyanides leads to δ,ε-acetylenic acyl cyanides. These are obtained as a mixture of products with or without retention of the trimethylsilyl group (**C** and **D** respectively).[43,85] On two occasions, a pyrane derivative **B** appeared as a minor product; it could result from the intramolecular *O*-alkylation of titanium enolate **A** with the vinylic cation. Acyl cyanides **C** and **D** are then easily hydrolyzed to the corresponding acids.

Me_3Si (allenylsilane, R, R, H) + R^1, R^2, R^3, CN, O $\xrightarrow[CH_2Cl_2]{TiCl_4}$ Me_3Si (+) R R R^3 CN R^1R^2 $OTiCl_n$ H **A**

B: R^1 R^2 R^3 CN, R, R, O, Me_3Si, H

C: Me_3Si R R R^3 (OH) CN R^1R^2 O

D: R R R^3 (OH) CN R^1R^2 O

As already mentioned, Danheiser reported, in 1981, the formation of silylcyclopentene when adding allenyltrimethylsilanes to conjugated enones in the presence of $TiCl_4$.[86,87] The reaction, which is of general use, has a very good stereoselectivity, illustrated by the addition to cyclohexenone. A transition state similar to the one proposed for the [3+2] annulation of allylsilanes accounts for the stereochemistry of the major isomer.

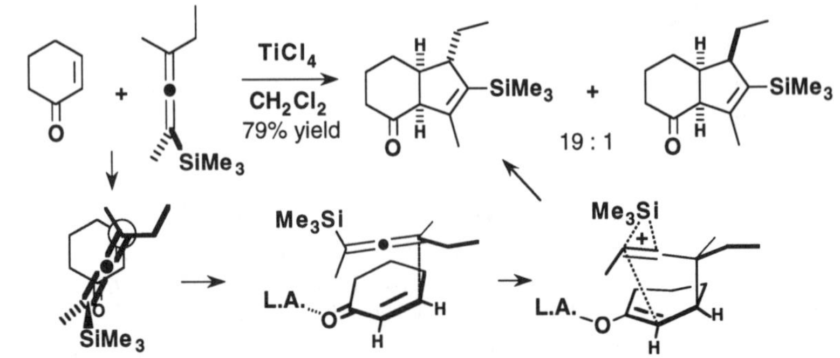

The addition of allenyltrimethylsilanes to α,β-unsaturated acylsilanes leads to the formation of silylcyclopentenyl acylsilanes **A**, but with 2-alkyl substituted α,β-unsaturated acylsilanes (R^6 = alkyl), a novel rearrangement occurs which leads to silylcyclohexenones **C**.[88]

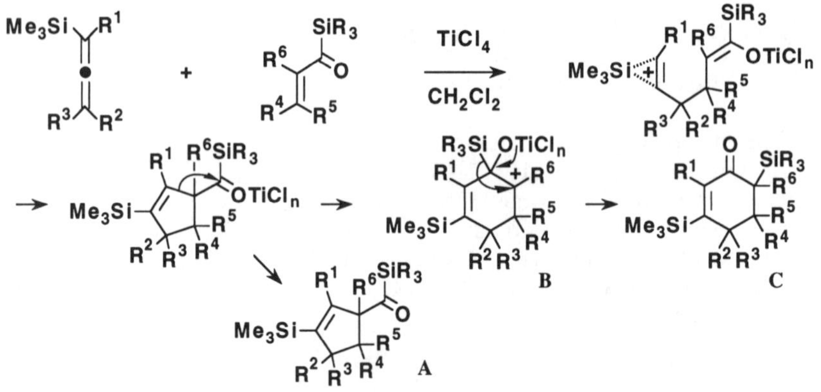

V.2 Addition of propargylsilanes

The $TiCl_4$ promoted addition of propargyltrimethylsilane to α,β-unsaturated acyl cyanides leads to γ-allenic acyl cyanides.[43]

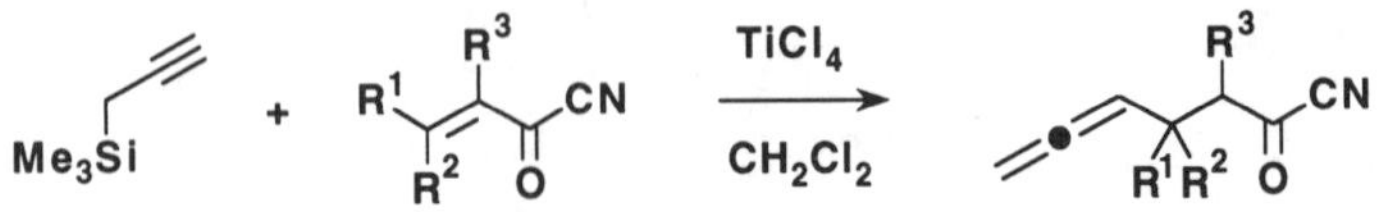

Danheiser showed that propargylsilanes also undergo [3+2] annulation; however the reaction between (trimethylsilyl)butyne and methylvinylketone is not stereoselective.[89]

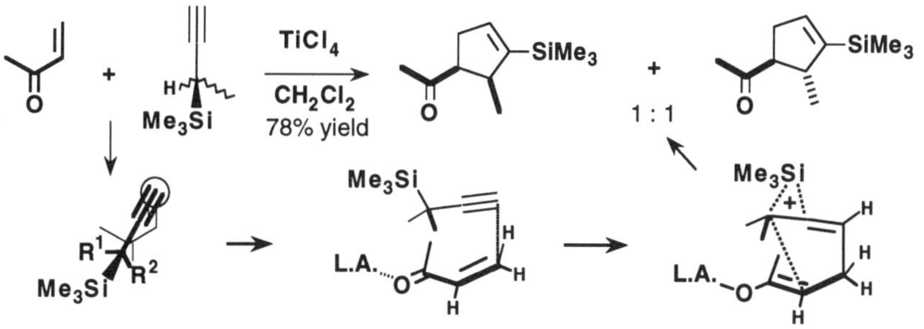

TiCl4, CH2Cl2, 78% yield; 1 : 1

VI. Intramolecular Reaction

VI.1 Addition of allylsilane or allylstannane moieties

The intramolecular Sakurai reaction rapidly became a widely used reaction. Thus, it enables the formation of cycles with, in general, good stereochemical control. Therefore, in 1988, Schinzer published a review on that reaction,[90] reported for the first time by Wilson.[91]

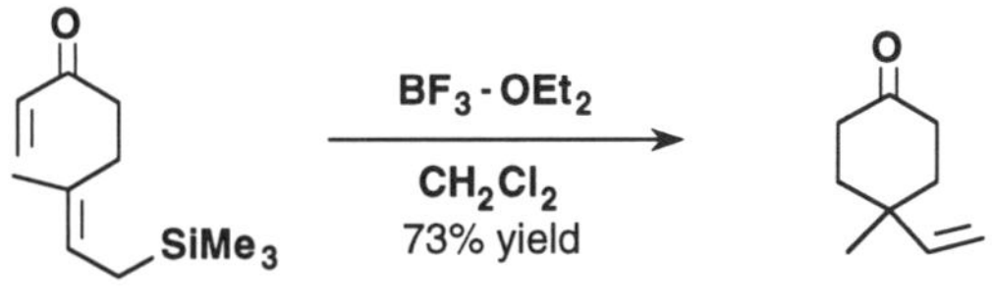

$BF_3 \cdot OEt_2$, CH_2Cl_2, 73% yield

Since 1983, Majetich has undertaken the synthesis of a number of natural products using, as the key step of his syntheses, the intramolecular Sakurai reaction.[92] However, at the beginning of his work, he had to use the fluoride anion to induce the cyclization because the Lewis acids would lead to protodesilylation.[93]

Schinzer then undertook a systematic study of the intramolecular 1,4-addition of allyltrimethylsilanes (and allyltri-*n*-butylstannanes) and propargyl trimethylsilanes to conjugated enones.[94–97] He showed that a spiro annulation can take place provided that a suitably substituted cyclic enone and $EtAlCl_2$, which is both a Lewis acid and a Bronsted base, are used. However BF_3 and $SnCl_4$ still induce only protodesilylation on the same substrate. The spiroannulation reaction leads to two diastereomers, with ratios varying according to the experimental conditions.[96,97]

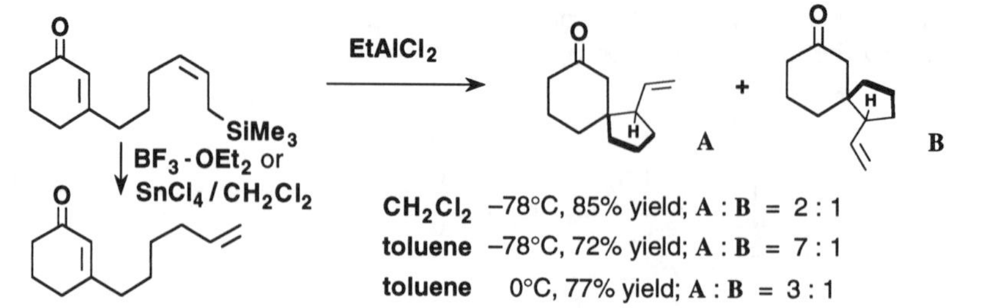

EtAlCl$_2$

SiMe$_3$

BF_3-OEt_2 or $SnCl_4$ / CH_2Cl_2

A + B

CH_2Cl_2 −78°C, 85% yield; **A** : **B** = 2 : 1
toluene −78°C, 72% yield; **A** : **B** = 7 : 1
toluene 0°C, 77% yield; **A** : **B** = 3 : 1

The formation of ketone **A**, which is always the major isomer, can be accounted for by the existence of a synclinal transition state.

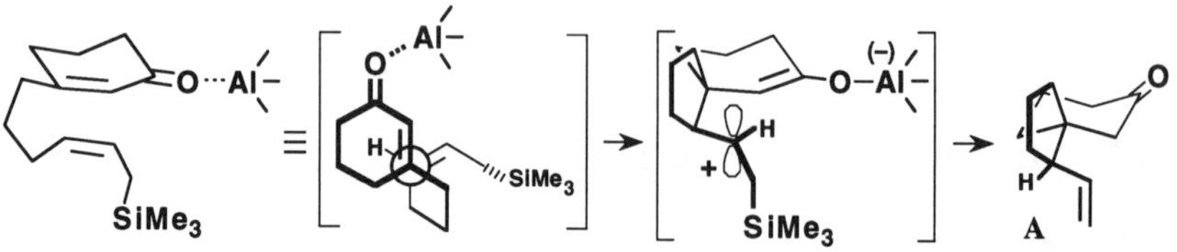

On the contrary, the formation of the other diastereomer, ketone **B**, involves a reaction path through an antiperiplanar transition state. It is worth noting that the synclinal transition state is favored by the use of a less polar solvent such as toluene.

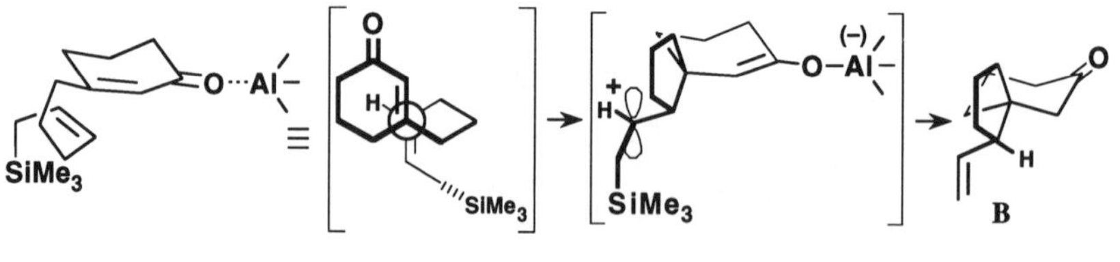

The stereochemistry of the vinyl hydrindenones, resulting from the addition of the allyl moiety to the conjugated system, provides useful information about the transition state of the reaction.[95,97]

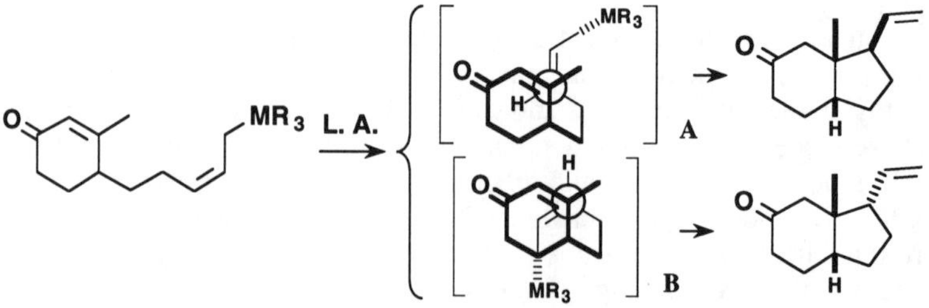

MR$_3$ L. A.

MR_3 = $SiMe_3$; L. A. = EtAlCl$_2$; toluene; 77% yield; **A** : **B** = 1 : 3
MR_3 = $SnBu_3$; L. A. = EtAlCl$_2$; toluene; 54% yield; **A** : **B** = 2 : 1
MR_3 = $SnBu_3$; L. A. = $TiCl_4$; CH_2Cl_2 ; 60% yield; **A** : **B** = 15 : 1

Both transition states **A** and **B** are synclinal. Transition state **B**, the more "compact", is the more probable when the reaction is performed with an allyltrimethylsilyl enone in toluene. The stereochemistry inversion observed with allylstannyl enones could result from an isomerization of the allylic moiety or from a transmetallation.

Among a large number of examples,[98] Schinzer observed the 1,6 intramolecular addition of allylsilane moieties to dienic enones, providing therefore an access to 7- or 8-membered ring bicyclic systems. In the following case, the reaction leads remarkably to a single regio- and stereoisomer, the structure of which is although not completely specified.

O SiMe$_3$ $EtAlCl_2$ / CH_2Cl_2 60% yield O

single regio- and stereoisomer

When the 1,6-addition leads to a cyclooctane, Schinzer established that the choice of the Lewis acid governs the regioselectivity. Thus, in the following example, $EtAlCl_2$ induces the formation of a mixture of regioisomers (1,4- and 1,6-addition) while $TiCl_4$ leads exclusively to a derivative of the bicyclo[6.4.0]dodecanone (1,6-addition), although the formation of a six-membered ring is kinetically preferred to that of an eight-membered ring by a factor of about 10^4.[99]

O SiMe$_3$

$EtAlCl_2$ / CH_2Cl_2 40% yield O + O H 1 : 1

4:1 mixture of diastereomers

$TiCl_4$ / CH_2Cl_2 45% yield O

single regio- and stereoisomer

Whereas all enones studied by Schinzer display a linear allylsilane moiety (*n*-alkenyl), Majetich has studied since 1985 the cyclization of 2-substituted allyltrimethylsilyl enones (iso-alkenylsilyl enones); he has thus obtained bicyclic ketones, bearing an exocyclic methylene.[100] When induced by the fluoride ion, the reaction has in general a different regioselectivity; in the following case, the allylsilane moiety adds to the enone in 1,2-position leading to an alcohol with bicyclo[3.2.1]octane skeletal, when treated with TBAF, while a 1,6-addition occurs in the presence of $EtAlCl_2$.

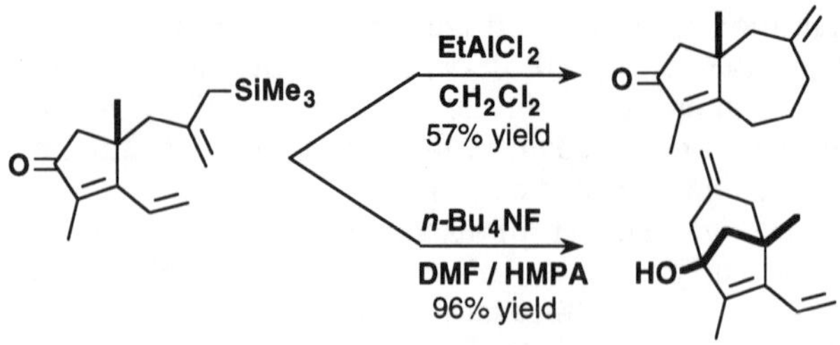
SiMe3
O
EtAlCl2
CH2Cl2
57% yield
O
n-Bu4NF
DMF / HMPA
96% yield
HO

On the other hand, with the corresponding cyclohexenone, Majetich observed also a 1,6-addition with $EtAlCl_2$ but a 1,4-addition with the fluoride anion. Similar studies have also been reported.[101]

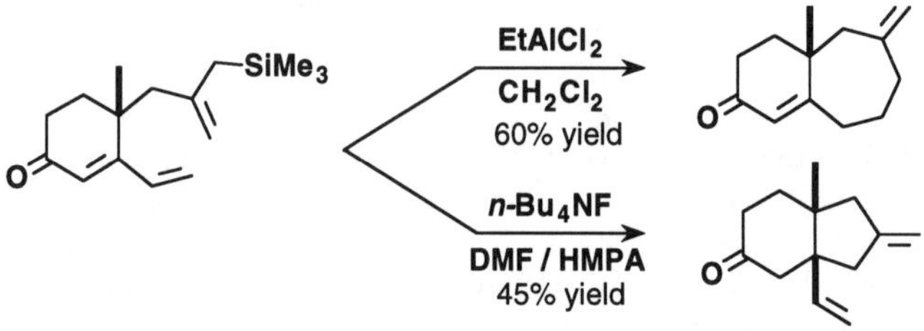
SiMe3
O
EtAlCl2
CH2Cl2
60% yield
O
n-Bu4NF
DMF / HMPA
45% yield
O

The cyclization promoted by intramolecular 1,6-addition of a linear allylsilane moiety provides an access to unsaturated bicyclic ketones bearing a vinylic moiety.[102] Both the high regio- and stereoselectivity of the reaction promoted by $EtAlCl_2$ are to be noted.

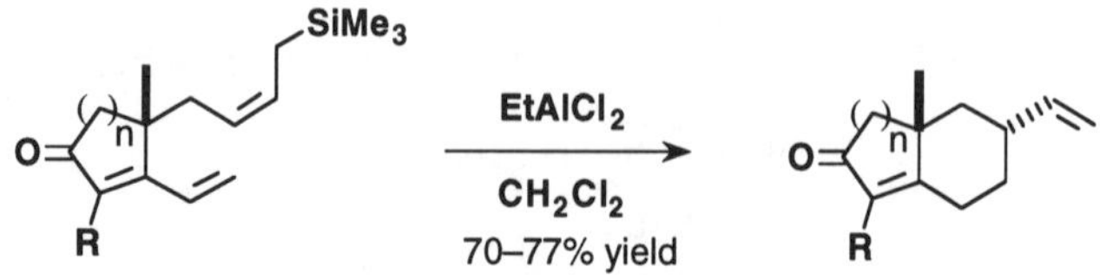
SiMe3
O
n
R
EtAlCl2
CH2Cl2
70–77% yield
O
n
R

n = 1 or 2, **R** = **H** or **Me**

With an isoalkenyl moiety on the side chain, it is possible to prepare methylenecycloheptenes by 1,6-addition.[103]

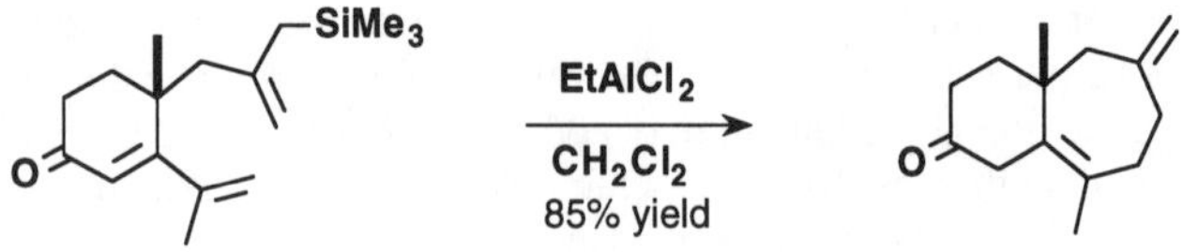
SiMe3
O
EtAlCl2
CH2Cl2
85% yield
O

The stereochemistry of the cyclization of 4-substituted cyclohexenones into hydrindanones is catalyst dependent.[104]

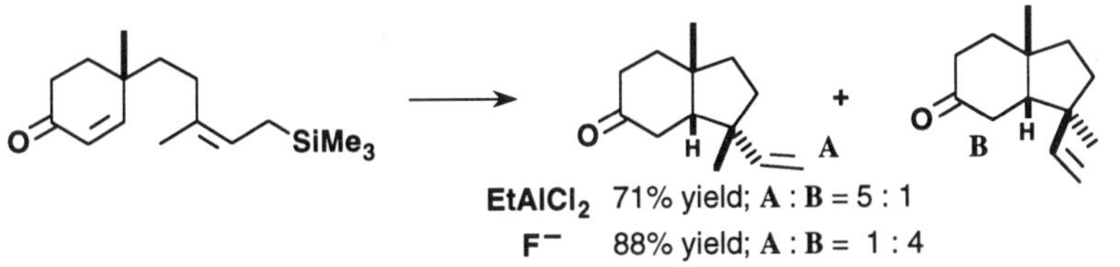

$EtAlCl_2$ 71% yield; **A** : **B** = 5 : 1

F^- 88% yield; **A** : **B** = 1 : 4

With a Lewis acid, the major diastereomeric ketone **A** is obtained through a "compact" (synclinal) transition state.

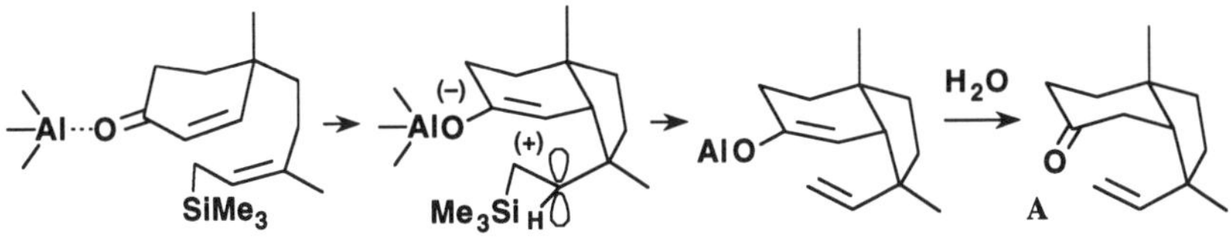

By contrast, minor isomer **B** (major when using fluoride ion) results from an antiperiplanar transition state.

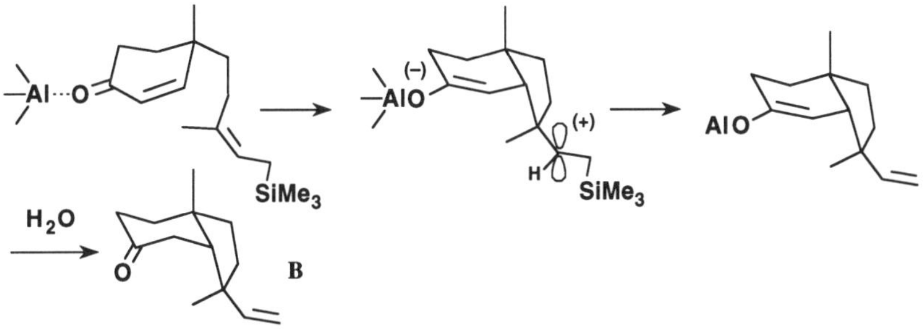

It seems reasonable to think that the stabilization of the synclinal transition state is due to the interaction between the positive charge and the enolate and that that stabilization is great enough to counterbalance the steric interactions in a positive way.

The cyclization of adequately substituted dienones to octalones provides a useful path to (±)-nootkatone.[105] Thus, the desired dienone is prepared from cyclohexan-1,3-dione through a sequence widely used by both Schinzer and Majetich.

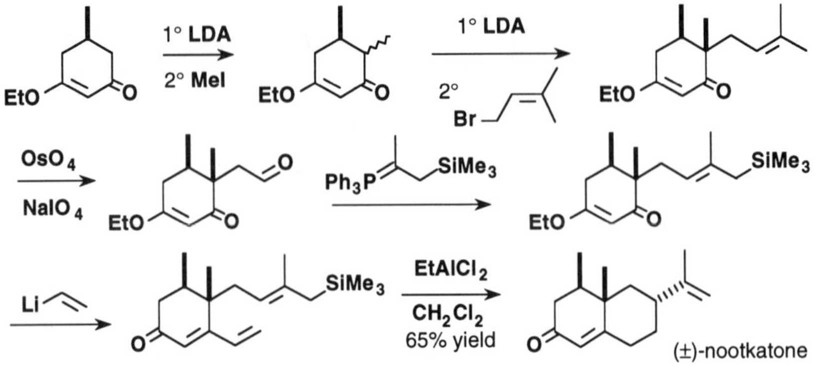

The best catalyst for the cyclization step is $EtAlCl_2$. The use of stronger Lewis acids such as $TiCl_4$ or BF_3-Et_2O generates low yields of nootkatone. Milder Lewis acids such as Et_2AlCl or Me_3Al have failed to promote cyclization.[106]

The ability of 4-isoalkenyl-3-vinylcyclohex-2-enones to undergo cyclization into bicyclo[5.4.0]undecane derivatives has been used to prepare three perforane derivatives.[107,108]

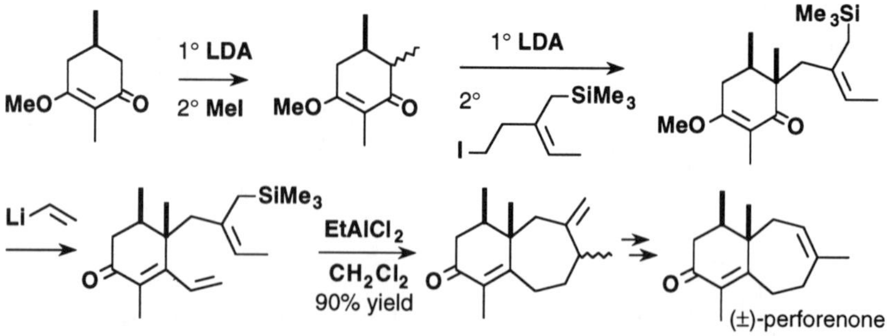

The intramolecular Sakurai reaction allows the preparation of triquinanes from cyclopentenone **A**. In order to synthesize hirsutene, Majetich prepared cyclopentenone **A**, which however has the wrong absolute configuration at carbon 4 of the enone. Fortunately, treatment of **A** with a Lewis acid promotes both epimerization and the intramolecular Sakurai reaction; the transition state being of synclinal geometry as in **C**. With $TiCl_4$, **D** is obtained with 73% yield, but the other Lewis acids used lead to **E** which results from an isomerization of the exocyclic double bond [BF_3-Et_2O (94%), $SnCl_4$ (77%), $EtAlCl_2$ (32%)].[109] These results are discussed in a review from Majetich.[110]

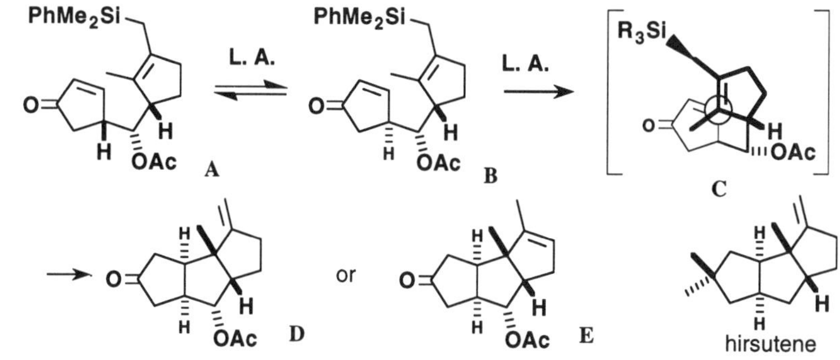

The cyclization of 2-substituted-3-vinylcyclohex-2-enones provides, through 1,4-addition, an access to octalones.[111] Again, $EtAlCl_2$ proved to be more efficient than $TiCl_4$.

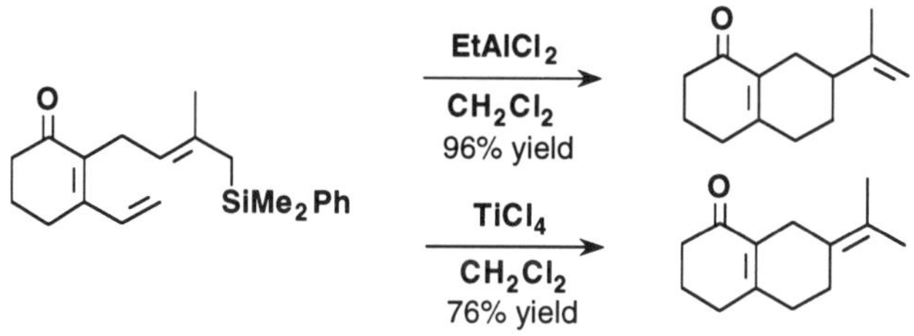

During the synthesis of 14-deoxyisoamijiol, Majetich showed that the reaction stereochemistry depends upon the relative stereochemistry of the cyclohexenyldimethylphenylsilane since the reaction always proceeds through an S_E2' mechanism with respect to the allylsilane moiety.[112,113]

From the following cyclopentenone derivative **A**, in which the two asymmetric centers are (*R**,*R**), 1,6-addition leads to the normally expected tricyclic compound **B**.

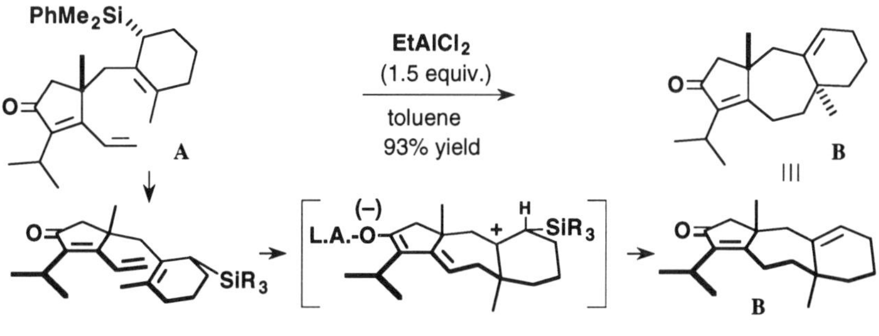

However, with the **C** epimer with structure (*R**,*S**), the reaction leads to enone **D** which results from a [3+2] annulation. This striking reactivity difference can be explained by the fact that the S_E2' mechanism (with respect to the allylsilane moiety) requires a very specific conformation of the transition state, which can be met from the **A** stereomer through a "compact"

synclinal transition state. This conformation cannot be reached from the **C** isomer for steric reasons.

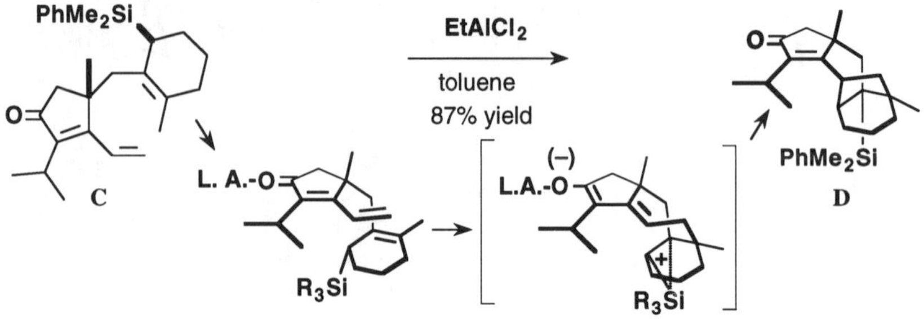

Majetich then proposed a use of the silyl enone resulting from the [3+2] annulation. He noted that upon treatment with $EtAlCl_2$ at moderate temperature the cycloadduct **D** undergoes a fragmentation reaction which leads to the addition product **E**, epimer of **B**, normally expected, with an excellent yield.

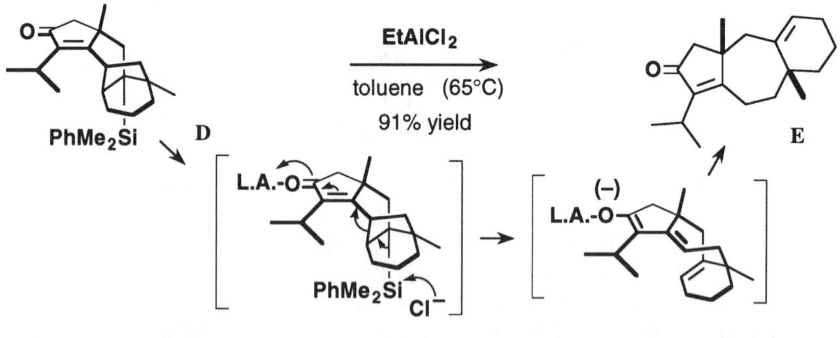

Tricyclic intermediate **B** proved to be a key intermediate in the synthesis of (±)-14-deoxyisoamijiol.

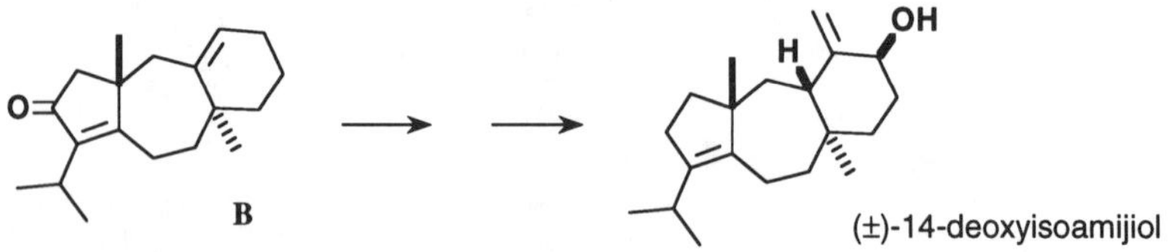

Works from Majetich showed that starting from the same precursor, various products can be obtained depending on the catalyst used. Thus, the following dienone leads either to an octalone derivative with $EtAlCl_2$ or to a bicyclo[6.4.0]dodecane derivative with the fluoride anion.

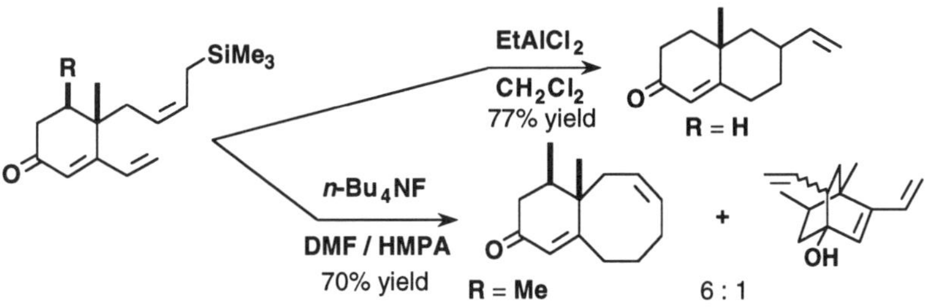

The latter adduct was used during the syntheses of (±)-neolemnanyl acetate and of (±)-neolemnane.[114]

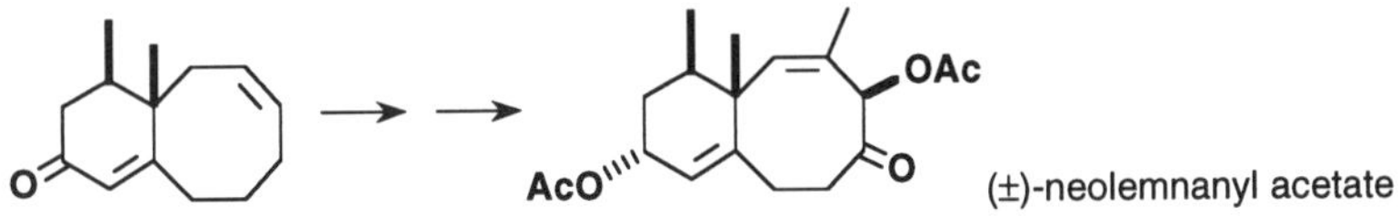

The cyclization of 2,3-disubstituted cyclopentenones leads to the guaianolide skeletal, providing that two equivalents or more of Lewis acid are involved. [115]

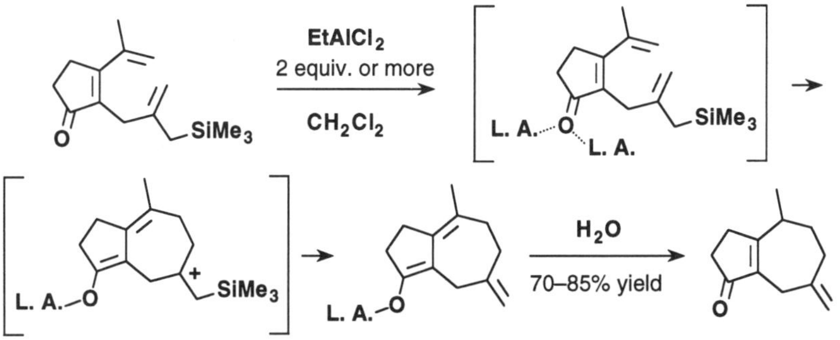

From the previous hexahydroazulenone, a particularly regioselective ene reaction allowed the functionalization of the seven-membered ring and provides a useful intermediate in the synthesis of graveolide and aromaticin.

In contrast, various attempts to promote the cyclization of the cyclopentenone lacking the isopropenyl group in the 3-position failed, only protodesilylation being observed.

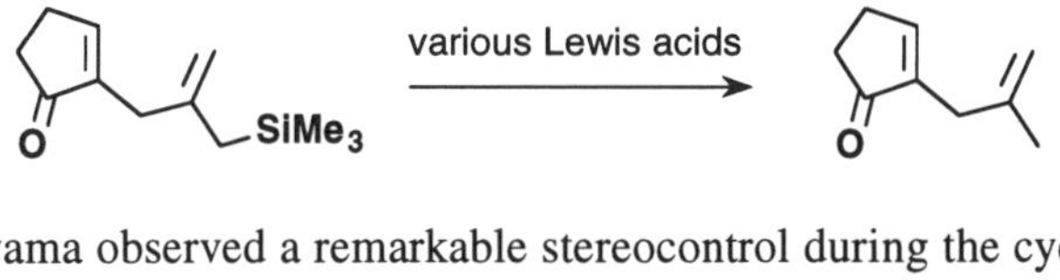

Tokoroyama observed a remarkable stereocontrol during the cyclization of the following cyclohexenone in which three stereogenic centers are controlled at the same time. The α-proton being epimerizable in basic medium, the *trans* decalone can be obtained as the sole product from the initial *cis:trans* mixture.[116]

TiCl$_4$, CH$_2$Cl$_2$, 80% yield; cis : trans = 1 : 3; MeONa / MeOH

The stereoselectivity is even more exceptional considering that the allylsilane moiety is a 1:1 mixture of (Z) and (E) forms. The *unlike* antiperiplanar transition state **A** is considerably more favorable than the *like* transition state **B** in which a strong interaction exists between the methyl group and the allylsilane moiety.

Not long ago, Tietze used a new and highly stereoselective sequential transformation: the tandem intramolecular Sakurai-carbonyl-ene reaction. First, the intramolecular addition of the allylsilane moiety to the conjugated aldehyde leads to a δ-ethylenic aldehyde which undergoes an intramolecular ene reaction. Second, the catalyst for both reactions is trimethylsilyl trifluoromethyl sulfonate. Third, although the starting aldehyde is a mixture of stereoisomers, a unique tricyclic compound is isolated with 52% yield.[117] The use of EtAlCl$_2$ leads to a complex mixture of stereoisomers.

TMS-OTf, CH$_2$Cl$_2$, 52% yield

intramolecular ene reaction

The cyclization of compounds **A** and **B** leads to the expected 1,4-adducts **C** and **D**. The stereochemistry of the products is highly dependent on the geometry of the starting material: *cis* keto ester **C** is predominant, regardless of the Lewis acid used, when the reaction is performed on (Z) precursor **A**, and *trans* keto ester **D** is predominant when the reaction is carried out from (E) precursor **B**.[118]

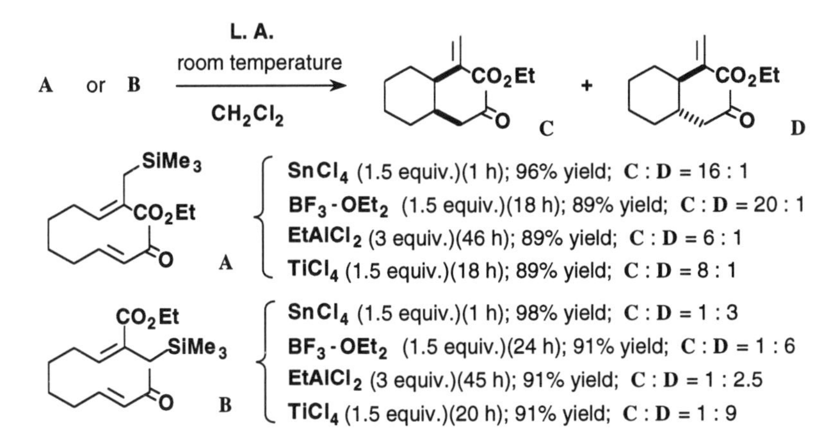

For each cyclization, four transition states can exist. From **A**, the formation of keto ester **C** (the major product) can proceed either through synclinal transition states **C1** or **C2.** It is, however, difficult to explain why **C1** should be more probable than **D1**, which would lead to keto ester **D**.

C1 (*unlike*) C2 (*unlike*) D1 (*like*) D2 (*like*)

From **B**, both transition states **D'1** (antiperiplanar) and **D'2** (synclinal) lead to keto ester **D** (the major product). However, if transition state **D'1** was the more probable one, it should also have been the case for **D1** during the cyclization of **A**.

C'1 (*unlike*) C'2 (*unlike*) D'1 (*like*) D'2 (*like*)

In conclusion, it appears that transition states **C2** and **D'2** are the most likely to explain the formation of **C** and **D**. Both these transition states are "compact" and have a short distance between the charged atoms.

VI.2 Addition of propargylylsilane moieties

Propargylsilanes are very good nucleophiles that enable the preparation of bicyclic compounds bearing a vinylidene moiety.[90,94,95,98] When the side chain is in β position of the α,β-unsaturated ketones, one obtains an allenyl spiro ketone with a moderate yield.[98]

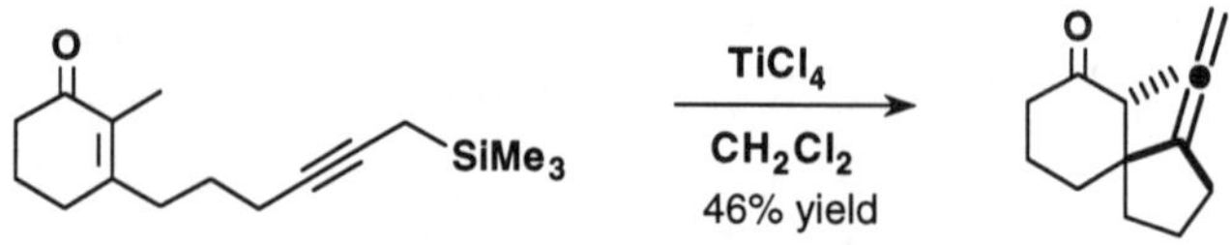

SiMe3 — $TiCl_4$, CH_2Cl_2, 46% yield

On the contrary, if the side chain is in γ position, allenic hydrindenones can be obtained with excellent yields, upon treatment of the suitable precursor with $TiCl_4$.[98]

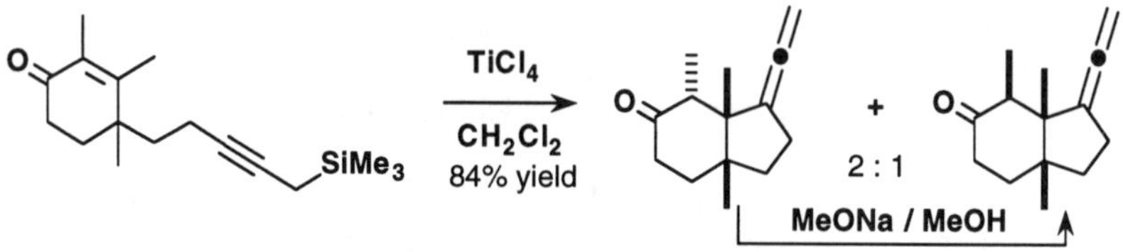

SiMe3 — $TiCl_4$, CH_2Cl_2, 84% yield — 2 : 1 — MeONa / MeOH

With conjugated dienic ketones, the propargyl moiety adds both in 1,4- and 1,6-position but with a clear preference for the former.[98]

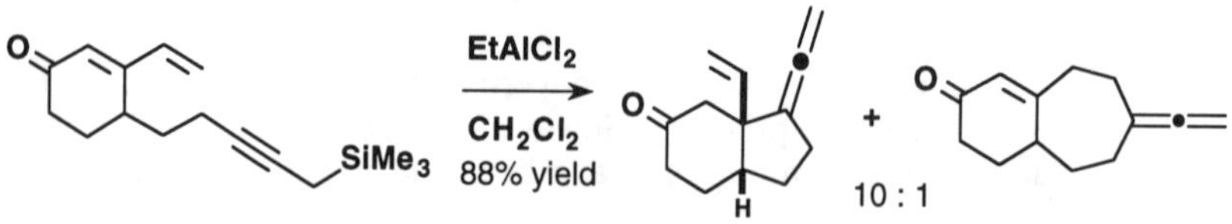

SiMe3 — $EtAlCl_2$, CH_2Cl_2, 88% yield — H — 10 : 1

With a longer side chain, no cyclization occurs, despite the fact that 1,4-addition would have led to a decalone.[98]

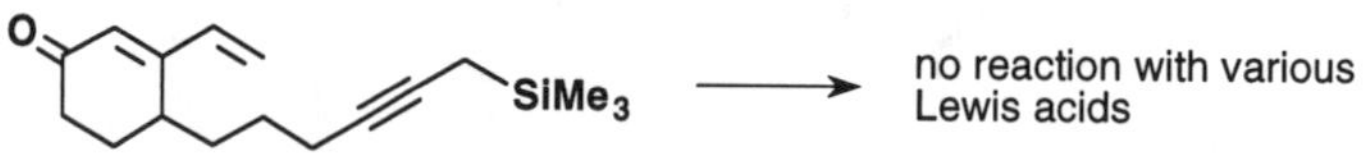

SiMe3 — no reaction with various Lewis acids

VII. Conclusion

Although it has only been discovered less than twenty years ago, the Sakurai reaction, and more generally the Lewis acid promoted 1,4-addition of allylsilanes or allylstannanes to α,β-unsaturated carbonyl compounds, already has a long history.

Unlike the ene reaction or the Diels-Alder reaction, the Sakurai reaction has been a Lewis acid catalyzed reaction since its discovery. The reaction can be performed with allenyl and propargyl silanes too and also in an intramolecular way. The intramolecular 1,4–addition of allylsilane or propargylsilane moieties has been particularly studied since it enables the synthesis of polycyclic compounds. The correct choice of the Lewis acid can in some cases steer the stereoselectivity of the reaction in the desired direction or prevent protodesilylation which is a usual drawback of the reaction.

Finally, the [3+2] annulation, initially only a side reaction, is nowadays an established reaction which exhibits remarkable diastereoselectivity. Along with the 1,4–addition, inter– and intramolecular, these reactions have found numerous applications in synthesis, including the total synthesis of natural products.

References

1. Hosomi, A.; Sakurai, H. *J. Am. Chem. Soc.* **1977**, *99*, 1673–1675.
2. Sakurai, H.; Hosomi, A.; Hayashi, J. *Org. Synth.* **1984**, *62*, 86–94.
3. Ojima, I.; Kumagai, M.; Miyazawa, Y. *Tetrahedron Lett.* **1977**, 1385–1388.
4. Posner, G.H. *Org. React.* **1972**, *19*, 1–113.
5. Majetich, G.; Casares, A.M.; Chapman, D.; Behnke, M. *Tetrahedron Lett.* **1983**, *24*, 1909–1912.
6. Majetich, G.; Casares, A.; Chapman, D.; Behnke, M. *J. Org. Chem.* **1986**, *51*, 1745-1753.
7. Hosomi, A.; Kobayashi, H.; Sakurai, H. *Tetrahedron Lett.* **1980**, *21*, 955–958.
8. Sakurai, H. *Pure Appl. Chem.* **1982**, *54*, 1–22.
9. Fleming, I.; Paterson, I. *Synthesis* **1979**, 446–448.
10. Yamamoto, Y.; Sasaki, N. *The Stereochemistry of the Sakurai Reaction,* in *Chemical Bonds – Better Ways to Make Them and Break Them*, Bernal, I. Ed., Elsevier, Amsterdam, **1989**, pp. 363–441.
11. Yamamoto, Y.; Asao, N. *Chem. Rev.* **1993**, *93*, 2207–2293.
12. Hosomi, A.; Iguchi, H.; Endo, M.; Sakurai, H. *Chem. Lett.* **1979**, 977–980.
13. Leu, H.–R.; Schori, H.; Keese, R. *Chimia* **1985**, *35*, 12–14.
14. Bambal, R.; Kemmitt, R.D.W. *J. Chem. Soc., Chem. Commun.* **1988**, 734–735.
15. Hosomi, A.; Shirahata, A.; Sakurai, H. *Tetrahedron Lett.* **1978**, 3043–3046.
16. Ricci, A.; Fiorenza, M.; Grifagni, M. A.; Bartolini, G.; Seconi, G. *Tetrahedron Lett.* **1982**, *23*, 5079–5082.
17. Knapp, S.; O'Connor, U.; Mobilio, D. *Tetrahedron Lett.* **1980**, *21*, 4557–4560.
18. Trost, B. M.; Coppola, B. P. *J. Am. Chem. Soc.* **1982**, *104*, 6879–6881.
19. House, H. O.; Gaa, P. C.; Lee, J. H. C.; VanDerveer, D. *J. Org. Chem.* **1983**, *48*, 1670–1678.
20. Danheiser, R. L.; Fink, D. M. *Tetrahedron Lett.* **1985**, *26*, 2509–2512.
21. Ochiai, M.; Arimoto, M.; Fujita, E. *Tetrahedron Lett.* **1981**, *22*, 1115–1118.
22. Maruyama, K.; Takuwa, A.; Naruta, Y.; Satao, K.; Soga, O. *Chem. Lett.* **1981**, 47–50.
23. Takuwa, A.; Soga, O.; Mishima, T.; Maruyama, K. *J. Org. Chem.* **1987**, *52*, 1261–1265.
24. Naruta, Y. *J. Org. Chem.* **1980**, *45*, 4097–4104.
25. Naruta, Y.; Maruyama, K. *Chem. Lett.* **1979**, 881–884.

26. Naruta, Y.; Uno, H.; Maruyama, K. *Tetrahedron. Lett.* **1981**, *22*, 5221–5224.
27. Naruta, Y.; Kashiwagi, M.; Nishigaichi, Y.; Uno, H.; Maruyama, K. *Chem. Lett.* **1983**, 1687bis–1690.
28. Naruta, Y.; Nishigaichi, Y.; Maruyama, K. *Chem. Lett.* **1986**, 1703–1706.
29. Naruta, Y.; Nishigaichi, Y.; Maruyama, K. *J. Org. Chem.* **1988**, *53*, 1192–1199.
30. Hayashi, M.; Mukaiyama, T. *Chem. Lett.* **1987**, 289–292.
31. Kim, S.; Lee, J.M. *Synth. Commun.* **1991**, *21*, 25–29.
32. Denmark, S. E.; Almstead, N. G. *Tetrahedron* **1992**, *48*, 5565–5578.
33. Pardo, R.; Zahra, J.P.; Santelli, M. *Tetrahedron Lett.* **1979**, 4557–4560.
34. Hosomi, A.; Kobayashi, H.; Sakurai, H. *Tetrahedron Lett.* **1980**, *21*, 955–958.
35. Danishefsky, S.; Kahn, M. *Tetrahedron Lett.* **1981**, *22*, 485–488.
36. Nickisch, K.; Laurent, H. *Tetrahedron Lett.* **1988**, *29*, 1533–1536.
37. Majetich, G.; Defauw, J.; Ringold, C. *J. Org. Chem.* **1988**, *53*, 50–68.
38. House, H.O.; Gaa, P.C.; VanDerveer, D. *J. Org. Chem.* **1983**, *48*, 1661–1670.
39. Knölker, H.–J; Jones, P.G.; Pannek, J.–B. *Synlett* **1990**, 429–430.
40. Knölker, H.–J.; Foitzik, N.; Graf, R.; Pannek, J.–B., Jones, P.G. *Tetrahedron* **1993**, *49*, 9955–9972.
41. Danheiser, R.L.; Carini, D.J.; Basak, A. *J. Am. Chem. Soc.* **1981**, *103*, 1604–1607.
42. Jellal, A.; Santelli, M. *Tetrahedron Lett.* **1980**, *21*, 4487–4490.
43. Santelli, M.; El Abed, D.; Jellal, A. *J. Org. Chem.* **1986**, *51*, 1199–1206.
44. Naruta, Y.; Uno, H.; Maruyama, K. *Tetrahedron Lett.* **1981**, *22*, 5221–5224.
45. Uno, H. *J. Org. Chem.* **1986**, *51*, 350–358.
46. Seyferth, D.; Pornet, J.; Weinstein, R.M. *Organometallics* **1982**, *1*, 1651.
47. Hosomi, A.; Hashimoto, H.; Kobayashi, H.; Sakurai, H. *Chem. Lett.* **1979**, 245–248.
48. Neunert, D.; Klein, H.; Welzel, P. *Tetrahedron* **1989**, *45*, 661–672.
49. For a review, see Yamamoto, Y.; Sasaki, N. *Stereochem. Organomet. Inorg. Compounds*, **1989**, *3*, 363–441. *Chem. Abstr.* **1990**, *113*, 151509a.
50. Heathcock, C.H.; Kleinman, E.F.; Binkley, E.S. *J. Am. Chem. Soc.* **1982**, *104*, 1054–1068.
51. Blumenkopf, T.A.; Heathcock, C.H. *J. Am. Chem. Soc.* **1983**, *105*, 2354–2358.
52. Mobilio, D.; De Lange, B. *Tetrahedron Lett.* **1987**, *28*, 1483–1486.
53. Yanami, T.; Miyashita, M.; Yoshikoshi, A. *J. Org. Chem.* **1980**, *45*, 607–612.
54. Schultz, A.G.; Lee, H. *Tetrahedron Lett.* **1992**, *33*, 4397–4400.
55. House, H.O.; Yau, C.C.; Vanderveer, D. *J. Org. Chem.* **1979**, *44*, 3031–3036.
56. Kirk, D.N.; Miller, B.W. *J. Chem. Res. (S)* **1988**, *9*, 278–279.
57. Yamamoto, Y. *Acc. Chem. Res.* **1987**, *20*, 243–249.
58. Yamamoto, Y.; Nishii, S. *J. Org. Chem.* **1988**, *53*, 3597–3603.
59. Heathcock, C.H.; Kiyooka, S–i.; Blumenkopf, T.A. *J. Org. Chem.* **1984**, *49*, 4214–4223.
60. Heathcock, C.H.; Kiyooka, S–i.; Blumenkopf, T.A. *J. Org. Chem.* **1986**, *51*, 3252.
61. Bernardi, A.; Capelli, A.M.; Gennari, C.; Scolastico, C. *Tetrahedron: Asymmetry* **1990**, *1*, 21–32.

62. Jeroncic, L.O.; Cabal, M.-P., Danishefsky, S.J.; Shulte, G.M. *J. Org. Chem.* **1991**, *56*, 387–395.
63. Anh, N.T.; Eisenstein, O. *Nouv. J. Chim.* **1977**, *1*, 61–70. .
64. Tokoroyama, T.; Pan, L.-R. *Tetrahedron Lett.* **1989**, *30*, 197–200.
65. Pan, L.-R; Tokoroyama, T. *Chem. Lett.* **1990**, 1999–2002.
66. Pan, L.-R.; Tokoroyama, T. *Tetrahedron Lett.* **1992**, *33*, 1469–1472.
67. Danheiser, R.L.; Dixon, B.R.; Gleason, R.W. *J. Org. Chem.* **1992**, *57*, 6094–6097.
68. Danheiser, R.L.; Takahashi, T.; Bertok, B.; Dixon, B.R. *Tetrahedron Lett.* **1993**, *34*, 3845–3848.
69. Snider, B.B.; Zhang, Q. *J. Org. Chem.* **1991**, *56*, 4908–4913.
70. Panek, J.S.; Jain, N.F. *J. Org. Chem.* **1993**, *58*, 2345–2348.
71. Ohkata, K.; Ishimaru, K.; Lee, Y.-G.; Akiba, K.-Y. *Chem. Lett.* **1990**, 1725–1728.
72. Lee, Y.-G.; Ishimaru, K.; Iwasaki, H.; Ohkata, K.; Akiba, K.-Y. *J. Org. Chem.* **1991**, *56*, 2058–2066.
73. Knölker, H.-J.; Foitzik, N.; Goesmann, H.; Graf, R. *Angew. Chem., Int. Ed. Engl.* **1993**, *32*, 1081–1083; *Angew. Chem.* **1993**, *105*, 1104.
74. Knölker, H. -J.; Graf, R. *Tetrahedron Lett.* **1993**, *34*, 4765–4768.
75. Dewar, M.J.S.; Zoebisch, E.G.; Healy, E.F.; Steward, J.P. *J. Am. Chem. Soc.* **1985**, *107*, 3902–3909.
76. Stephan, K. *J. Prakt. Chem.* **1898**, *58*, 109.
77. Vogel, P. *Carbocation Chemistry*, Elsevier, Amsterdam, **1985**, p. 470.
78. Winstein, S.; Valkanas, G.; Wilcox, Jr., C. F. *J. Am. Chem. Soc.* **1972**, *94*, 2286–2290.
79. Astin, K. B.; Whiting, M. C. *J. Chem. Soc., Perkin Trans. 2* **1976**, 1160–1165.
80. Godfredsen, S.; Obrechit, J.P.; Arigoni, D. *Chimia* **1977**, *31*, 62.
81. Seebach, D.; Golinski, J. *Helv. Chim. Acta* **1981**, *64*, 1413–1423.
82. Kober, R.; Papadopoulos, K.; Miltz, W.; Enders, D.; Steglich, W. *Tetrahedron* **1985**, *41*, 1693–1701.
83. El Idrissi, M.; Santelli, M. *J. Org. Chem.* **1988**, *53*, 1010–1016.
84. Zair, T.; Santelli-Rouvier, C.; Santelli, M. *J. Org. Chem.* **1993**, *58* , 2686–2693.
85. Jellal, A.; Santelli, M. *Tetrahedron Lett.* **1980**, 4487.
86. Danheiser, R.L.; Carini, D.J.; Basak, A. *J. Am. Chem. Soc.* **1981**, *103*, 1604–1606.
87. Danheiser, R.L.; Carini, D.J.; Fink, D. M.; Basak, A. *Tetrahedron* **1983**, *39*, 935–947.
88. Danheiser, R.L.; Fink, D. M. *Tetrahedron Lett.* **1985**, *26*, 2513–2516.
89. Danheiser, R.L.; Dixon, B. R.; Gleason, R. W. *J. Org. Chem.* **1992**, *57*, 6094–6097.
90. Schinzer, D. *Synthesis* **1988**, 263–273.
91. Wilson, S. R.; Price, M. F. *J. Am. Chem. Soc.* **1982**, *104*, 1124–1126.
92. Majetich, G. in *Organic Synthesis: Theory and Application*, Hudlicky, T., Ed.; JAI Press Inc., Greenwich, CT, **1989**, Vol. 1, pp. 173–240.
93. Majetich, G.; Desmond, R.; Casares, A. M. *Tetrahedron Lett.* **1983**, *24*, 1913–1916.
94. Schinzer, D.; Steffen, J.; Solyom, S. *J. Chem. Soc., Chem. Commun.* **1986**, 829–830.
95. Schinzer, D.; Solyom, S.; Becker, M. *Tetrahedron Lett.* **1985**, *26*, 1831–1834.
96. Schinzer, D. *Angew. Chem., Int. Ed. Engl.* **1984**, *23*, 308–309.

97. Schinzer, D.; Allagiannis, C.; Wichmann, S. *Tetrahedron* **1988**, *44*, 3851–3868.
98. Schinzer, D.; Dettmer, G.; Ruppelt, M.; Solyom, S.; Steffen, J. *J. Org. Chem.* **1988**, *53*, 3823–3828.
99. Galli, C.; Illuminati, G.; Mandolini, L.; Tamborra, P. *J. Am. Chem. Soc.* **1977**, *99*, 2591–2597.
100. Majetich, G.; Hull, K.; Defauw, J.; Desmond, R. *Tetrahedron Lett.* **1985**, *26*, 2747–2750.
101. Majetich, G.; Desmond Jr., R. W.; Soria, J. J. *J. Org. Chem.* **1986**, *51*, 1753–1769.
102. Majetich, G.; Hull, K.; Desmond, R. *Tetrahedron Lett.* **1985**, *26*, 2751–2754.
103. Majetich, G.; Hull, K.; Defauw, J.; Shawe, T. *Tetrahedron Lett.* **1985**, *26*, 2755–2758.
104. Majetich, G.; Defauw, J.; Hull, K.; Shawe, T. *Tetrahedron Lett.* **1985**, *26*, 4711–4714.
105. Majetich, G.; Behnke, M.; Hull, K. *J. Org. Chem.* **1985**, *50*, 3615–3618.
106. Majetich, G.; Hull, K. *Tetrahedron* **1987**, *43*, 5621–5635.
107. Majetich, G.; Ringold, C. *Heterocycles* **1987**, *25*, 271–275.
108. Majetich, G.; Defauw, J.; Ringold, C. *J. Org. Chem.* **1988**, *53*, 50–68.
109. Majetich, G.; Defauw, J. *Tetrahedron* **1988**, *44*, 3833–3849.
110. Majetich, G.; Hull, K.; Lowery, D.; Ringold, C.; Defauw, J. in *Selectivities in Lewis Acid–Promoted Reactions*, Schinzer, D., Ed.; Kluwer Academic Publishers, Dordrecht, Holland, **1989**, pp. 169–188.
111. Majetich, G.; Hull, K.; Casares, A.M.; Khetani, V. *J. Org. Chem.* **1991**, *56*, 3958–3973.
112. Majetich, G.; Song, J.–S.; Ringold, C.; Nemeth, G. A. *Tetrahedron Lett.* **1990**, *31*, 2239–2242.
113. Majetich, G.; Song, J.–S.; Ringold, C.; Nemeth, G. A.; Newton, M. G. *J. Org. Chem.* **1991**, *56*, 3973–3988.
114. Majetich, G.; Lowery, D.; Khetani, V.; Song, J.–S.; Hull, K.; Ringold, C. *J. Org. Chem.* **1991**, *56*, 3988–4001.
115. Majetich, G.; Song, J.–S.; Leigh, A. J.; Condon, S. M. *J. Org. Chem.* **1993**, *58*, 1030–1037.
116. Tokoroyama, T.; Tsukamoto, M.; Iio, H. *Tetrahedron Lett.* **1984**, *25*, 5067–5070.
117. Tietze, L. F.; Rischer, M. *Angew. Chem., Int. Ed. Engl.* **1992**, *31*, 1221–1222.
118. Kuroda, C.; Ohnishi, Y.; Satoh, J. Y. *Tetrahedon Lett.* **1993**, *34*, 2613–2616.

6. Lewis Acid-Promoted Diels-Alder Reaction

I. Introduction and History

The Diels-Alder reaction has long been recognized as one of the most important reactions in the field of organic synthesis. Due to the concerted and secondary orbital controlled pathway, usually high and predictable stereoselectivity can be realized, making this reaction particularly useful in the stereoselective synthesis of various useful synthetic intermediates.

In 1906, Albrecht observed the formation of 1:1 and 2:1 adducts by heating cyclopenta-1,3-diene and *p*-benzoquinone.[1] Diels and Alder determined in 1928 the structure of these adducts and demonstrated the generality and elucidated the basic regiochemical and stereochemical principles of this reaction which now bears their names.[2]

O O O
+ Δ +
O O O

This focuses on the carbocyclization and on most recent developments in its asymmetric version. However, a section (Section V) is devoted to hetero Diels-Alder reactions with special attention paid to the cycloaddition between activated dienes and aldehydes. The Diels-Alder reaction has been extensively reviewed.[3–13]

I.1 The thermal reaction

A most attractive feature of the Diels-Alder reaction is the simultaneous, regioselective formation of two bonds leading to the creation of up to four chiral centers at the binding sites with largely predictable relative stereochemistry. In particular, endo adducts are formed preferentially in reactions conducted under kinetic control. This endo rule was originally formulated for additions of cyclic dienes and dienophiles.[14]

Qualitative molecular orbital theory accounts for the observed products. In the most utilized class of Diels-Alder reactions, the highest occupied molecular orbital (HOMO) of the diene interacts with the lowest unoccupied molecular orbital (LUMO) of the dienophile (frontier molecular orbital (FMO)). The reactivity increases as the energy difference between interacting FMOs decreases and the atoms with the larger coefficients are bonded preferentially in the transition state.[15–19] Woodward and Hoffmann explained the endo stereopreference by a stabilization of the transition state in which secondary orbital overlap can occur between the π system of the diene and a π system in conjugation with the dienophilic double bond (secondary orbital overlap (SOO)).[20]

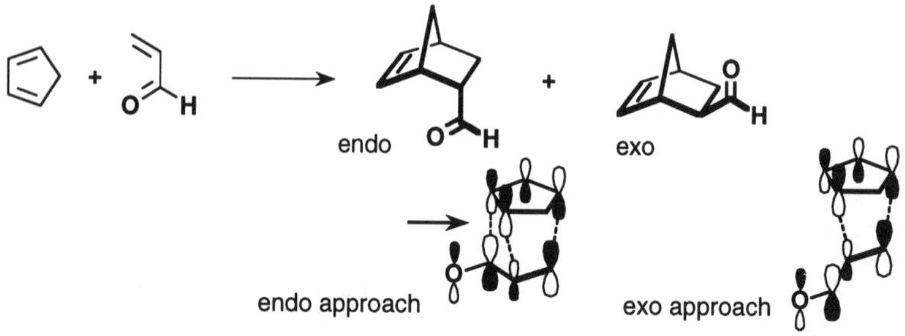

endo exo

endo approach exo approach

Although the observed stereopreference for endo adducts can be explained by SOO, several striking examples show that other factors must also influence it. For instance, SOO cannot explain the endo preference in the kinetic addition of cyclopentene to cyclopentadiene since no secondary interactive sites are present.[21]

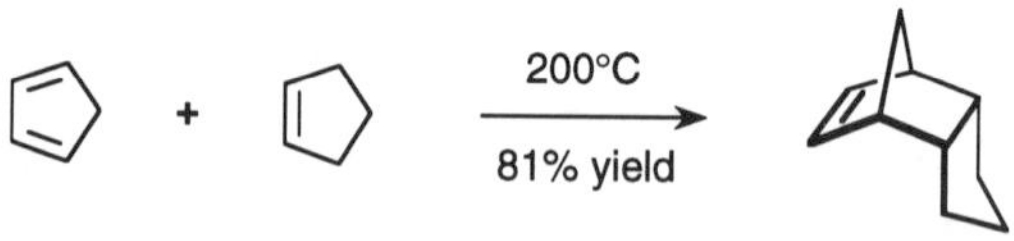

200°C

81% yield

Numerous theoretical studies on the thermal Diels-Alder reaction are reported.[22–59]

I.2 The Lewis acid-promoted reaction

The increase of the Diels-Alder reaction rate in acid media has been known for many years,[60,61] and the catalytic effect of Lewis acids has been demonstrated by Yates and Eaton in 1960.[62] Moreover, in most Diels-Alder reactions the important increment of the reaction rate is accompanied by an increase of selectivity.[63–70] Thus, Diels-Alder reactions that are catalyzed with a Lewis acid are known to exhibit remarkable differences in reactivity, stereoselectivity and regioselectivity from their thermal counterparts.

For example, dramatically different results were obtained by Trost with juglone **A** as dienophile. While the thermal reaction with acetoxybutadiene gave **B** and **C** in a 3:1 ratio, the addition of 5 mol% of BF_3-Et_2O enhanced this ratio to such an extent that essentially only **B** was obtained at room temperature.[71]

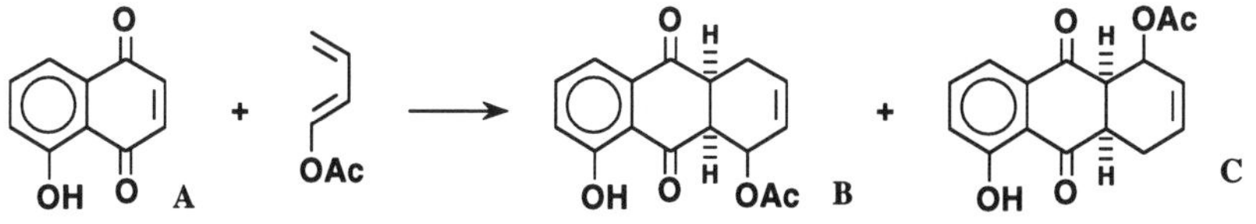

thermal; **B** : **C** = 3 : 1; BF_3-OEt_2 (5 mol%); **B** : **C** = 20 : 1

It was proposed early that the catalytic action is due to the formation of a complex between the Lewis acid and the dienophile.[72] However, several theoretical works have been devoted to the catalytic effect on the selectivity of the Diels-Alder reaction.[73–79] When complexed to the dienophile, which is the most frequent case, the Lewis acid induces a lowering of the dienophile LUMO and therefore diminishes the energy gap with the diene HOMO; this effect being responsible for the rate enhancement of the reaction.

According to Houk and Valenta, in the plane-nonsymmetrical following diene, the higher stereoselectivity of the Lewis acid catalyzed reactions is postulated to arise from an "earlier" rather "tighter" transition state, which has flatter addends and closer approach of the out-of-plane substituents.[80,81]

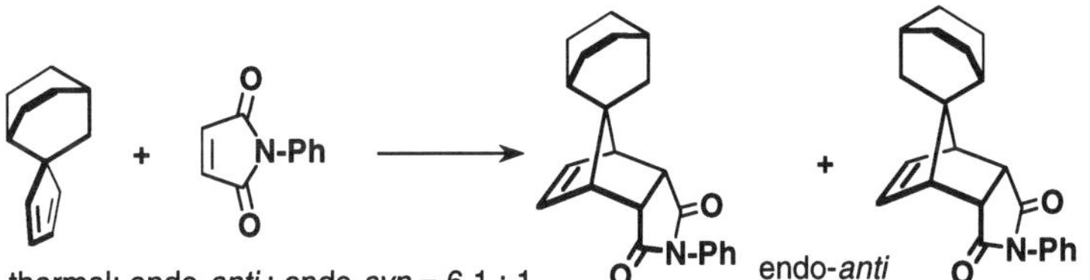

thermal; endo-*anti* : endo-*syn* = 6.1 : 1

Lewis acid catalyzed reaction; endo-*anti* : endo-*syn* = 13.3 : 1

Along with the most common Lewis acids, less usual ones, including $LiClO_4$,[82–84] have also been involved in the Diels-Alder reaction.[85–93]

I.3 Stereochemical issues

As already mentioned, a large part of the synthetic value of the Diels-Alder reaction comes from its high regio-, stereo- and even enantioselectivity.

The regioselectivity of the Lewis acid version of the reaction can be explained in a number of cases, as for the thermal version, by secondary orbital overlap. Moreover, it is frequently increased.

The stereoselectivity of the catalyzed reaction can often be anticipated relying on the Alder endo-rule (again, the endo/exo ratio frequently increases with the use of Lewis acids) and the *cis*-rule ("the relative stereochemistry of the substituent groups in the diene and dienophile is maintained in the cycloadduct").[12] However, one must be aware that the endo-rule is far from being universal and applies only to the kinetic products of the reaction. Particularly, not all intramolecular reactions are endo selective.

As for enantioselectivity in asymmetric reactions, it depends on the facial selectivity of the addition. As in most cases, the Lewis acid (which can be the source of chirality) is complexed to a functionalized vinyl derivative (the other main source of chirality), the outcome of the reaction will depend on whether the Lewis acid is *syn* or *anti* to the double bond and (therefore) on whether the conjugated unsaturated system adopts an *s-cis* or *s-trans* conformation as shown with the following Lewis acid complexes (see also Chapter 1, Section III).

L. A. O R — *anti s-cis*; L. A. O R — *anti s-trans*; L. A. O R — *syn s-cis*; L. A. O R — *syn s-trans*

R = H, OR, NRR'

II. Intermolecular Reaction. Regio- and Diastereofacial Selectivities

II.1 Cycloaddition of buta-1,3-diene and methylbuta-1,3-dienes with various dienophiles

Cycloalkenones proved to be particularly versatile dienophiles and have therefore often been used in studies related to the stereoselectivities of the Diels-Alder reaction. Particularly, the stereo-outcome of the reaction between various substituted cyclohexenones and classical dienes has been examined systematically. Moreover, the substituted octalones obtained are synthetically useful.

II.1.a Buta-1,3-diene

Addition to acetoxycyclohexenone **A** gave rise to a mixture of endo-*syn* and endo-*anti* adducts.[94]

+ OAc A —$AlCl_3$, toluene, 60% yield→ B (H, OAc) + C (H, OAc) 1.3 : 1

The lack of reactivity of 3-methylcyclohex-2-enone **D** precluded the preparation of octalone **E** but the conversion of **B** and **C** overcame this obstacle.

+ D —$AlCl_3$ ✗→ E (H, O) ⇇ B + C, 36% overall yield

Danishefsky reported a greater stereoselectivity with 4-silyloxy cycloalkenones.[95] The major products resulted from an endo-*anti* transition state.

$AlCl_3$, toluene, 76% yield

Si-*t*-$BuMe_2$ — Si-*t*-$BuMe_2$ 16.6 : 1 Si-*t*-$BuMe_2$

The reaction with 5-methylcyclohex-2-enone took place exclusively from the *anti* position. The subsequent epimerization proved to be highly dependent upon the reaction conditions.[96]

L. A., solvent

A **B**

L. A. : BF_3-OEt_2; solvent : benzene; **A** : **B** = 1 : 5
L. A. : $SnCl_4$; solvent : acetonitrile; **A** : **B** = 1 : 1.5

Oppolzer[96] then used octalone **A** as a key intermediate in the synthesis of (±)-luciduline.

A → (±)-luciduline

II.1.b (*E*)-Piperylene

The reaction of (*E*)-piperylene with cycloalkenones leads to bicyclic ketones with an "ortho" structure. Due to the presence of a methyl group, (*E*)-piperylene enables the determination of the endo–exo diastereoselectivity of the reaction. With 2-cyclohexenone, epimerization occurred during the reaction, but very high endo selectivity was observed[97] (for the reaction of 2-methylcyclohex-2-enone, see ref.98).

$AlCl_3$ (0.25 equiv.), toluene, 80% yield

A endo >32 : <1 exo **B**

The cycloaddition with 2-methylcyclohept-2-enone led exclusively, in good yield, to the endo adduct.[99]

$AlCl_3$, toluene, 85% yield

Endo Diels-Alder adducts **B** and **C** were also predominantly obtained from 2,4-dimethylcyclohex-2-enone **A**. However, diastereofacial selectivity was poor.[100]

$AlCl_3$ (0.25 equiv.), toluene, 95% yield

B : **C** : **D** : **E** = 5.7 : 4.2 : 1 : 1.5
endo : exo = 4 : 1
anti : *syn* = 1.17 : 1

endo-*anti* → **B**; endo-*syn* → **C**; exo-*anti* → **D**; exo-*syn* → **E**

In contrast, very high diastereofacial endo-*anti* selectivity was observed with 4-*tert*-butylcyclohex-2-enone (**G** resulting from epimerization of **F**).[101,102]

$AlCl_3$ (0.5 equiv.), toluene, 83% yield

endo-*anti* **F** : endo-*anti* **G** = 99 : 1

Likewise, important endo-*anti* selectivity occurred with 5-alkylcyclohexenones (**B** resulting from epimerization of **A**).[103]

$AlCl_3$ (0.5 equiv.), toluene

A endo-*anti* + **B** endo-*anti* + **C** exo-*syn*

R = Me; 80% yield; **A** : **B** : **C** = 22.7:1.2:1
R = *i*-Pr; 84% yield; **A** : **B** : **C** = 200:1:0

The cycloaddition with octalone **A** proceeded also exclusively in an endo fashion and with high level of *anti* diastereofacial selectivity leading to **B**.[104]

AlCl3 (0.25 equiv.) toluene 90% yield

B : C : D = 45 : 4 : 1

In the case of methyl *trans*-4-oxobutenoate **A**, a bifunctional dienophile, the exclusive formation of **B** indicates that the complexed formyl group determines the orientation (regioselectivity) and the more favorable endo state (stereoselectivity) as shown in transition state **C**.[105]

$SnCl_4$, CH_2Cl_2, 95% yield

An interesting "guided-by-catalysis" regioselectivity was observed by Reusch with quinone **A**. While thermal addition produced a 1:1 mixture of adducts **B** and **C**, the $SnCl_4$ catalyzed addition mainly gave adduct **C** resulting from a chelation controlled reaction, whereas, BF_3-OEt_2, a monodentate Lewis acid, induced the predominant formation of **B**.[106]

thermal (100°C), 80% yield, **B** : **C** = 1 : 1 | $SnCl_4$, (−16°C), >85% yield, **B** : **C** = 1 : 20
BF_3-OEt_2, (−16°C), >85% yield, **B** : **C** = 4 : 1

Kanematsu also observed a high regioselectivity during the cycloaddition with benzofuranquinone **A** when using a bidentate Lewis acid such as $TiCl_2$(O-*i*-Pr)$_2$.[107,108]

$TiCl_2$(O-*i*-Pr)$_2$, CH_2Cl_2, 2° air, 66% yield

B 20 : 1 **C**

Liu studied the facial selectivity of the addition of dienes to various 4,4-disubstituted-cyclohexa-2,5-dienones. Depending on the substitution of the diene and the dienophile, a reversal of selectivity was observed.[109]

OHC O $(CH_2)_n$ CO_2Et **A** + $ZnCl_2$ CH_2Cl_2 → CO_2Me O H $(CH_2)_n$ CO_2Et **B** + CO_2Me O H $(CH_2)_n$ CO_2Et **C**

n = 0, 95% yield; **B** : **C** = 1 : 0. n = 2, 90% yield; **B** : **C** = 1 : 1.2

II.1.c Isoprene

The reaction of isoprene with cycloalkenones leads predominantly to bicyclic ketones with a "para" structure. Thus, the formation of endo adduct **A** occurred with high regioselectivity from isoprene and 2-methylcyclopent-2-enone[99] (see also ref.110).

+ O $AlCl_3$ toluene 79% yield → O H **A** 32 : 1 **B** O H

The regioselectivity even increased in the case of 1,1-difunctionalized alkene **A** to reach a ratio greater than 99:1.[111]

+ Ar O O O **A** Ar = *p*-nitrophenyl BF_3 - OEt_2 CH_2Cl_2 83% yield → O O O Ar **B** >99 : <1 **C** O O O Ar

Isoprene reacts with high regioselectivity with hex-2-enone to yield the "para" adduct (under BF_3-OEt_2 catalysis)[112] but a loss of regioselectivity was however observed when the reaction was carried out in the presence of a cyclohexenone bearing a gem-dimethyl group.[113,114]

+ O **A** $AlCl_3$ (0.25 equiv.) toluene 95% yield → O H **B** 1.4 : 1 **C** O H

Similarly, a mixture of "para" and "meta" products were formed from dienone **A**; the regioselectivity being moreover dependent on the nature of the Lewis acid.[115]

MeO$_2$C, O, A; L. A. (0.5 equiv.), Et_2O; MeO$_2$C O, H, B; MeO$_2$C O, H, C

$SnCl_4$; 99% yield; **B** : **C** = 4.6 : 1; $AlBr_3$; 64% yield; **B** : **C** = 1.3 : 1
BF_3-Et_2O ; 73% yield; **B** : **C** = 1 : 2.3

In the case of 4-isopropylcyclohex-2-enone, adducts were obtained with very high regioselectivity. However, the primary Diels-Alder product **C**, resulting from an endo-*anti* transition state, underwent epimerization to yield enone **A**.[116]

$AlCl_3$ (0.25 equiv.), toluene, 75% yield; H O, H, endo-*anti* **A**; H O, H, endo-*syn* **B**; H O, H, endo-*anti* **C**

A : **B** : **C** = 180 : 19 : 1

The same diastereofacial selectivity was observed with (–)-carvone and led to endo-*anti* adduct **A** but with a much lower ratio.[117]

$AlCl_3$ (0.2 equiv.), toluene, 80% yield; O, H, endo-*anti* **A** 9.5 : 1 **B** O, H, endo-*syn*

The high regio- and diastereofacial selectivity of the cycloaddition was employed to prepare, from functionalized enone **A**, key precursor **B** of (–)-khusimone, a component of the vetiver oil.[118]

O, CO_2Me, **A**; $SnCl_4$, CH_2Cl_2, 41% yield; H O, H, CO_2Me, **B**; H, O, (–)-khusimone

While with pyperilene and a methoxy paraquinone, the reaction was oriented by the carbonyl α to the methoxy group,[106] an apparent activation of the carbonyl β to the methoxy substituent was observed in the reaction of quinone **D**.[119]

thermal (100°C), **E** : **F** = 1 : 1; **$SnCl_4$** ; 98% yield, **E** : **F** = 1.9 : 1
$TiCl_4$; 98% yield, **E** : **F** = 1.9 : 1

Isoprene was also shown to react with an anthracene-1,4-dione derivative to produce in good yield the corresponding tetracyclic compound.[120]

Finally, an interesting spiroannulation was observed with dienone **A**. Thus, **B** was obtained with both high regio- ("para" compounds) and facial selectivity.[121]

$AlCl_3$, $CHCl_3$, 60–70% yield; **B** 19 : 1 **C**

II.2 Cycloaddition of substituted 1,3-dienes with various dienophiles

An interesting rearrangement occurring in the Lewis acid mediated cycloaddition of 2,3-dimethylbutadiene with methacrolein was established by Baldwin. Ketone **A** resulting from the thermal reaction was similarly rearranged in the presence of $SnCl_4$.[122]

150°C, 65% yield; $SnCl_4$ (0.14 equiv.), toluene, 58% yield; $SnCl_4$ (0.1 equiv.), toluene, 66% yield

The following mechanism was proposed to account for this rearrangement.

Silyl dienes have also been used on many occasions with variable success. Thus, 1-trimethylsilylbutadiene has reacted with citraconic anhydride without regioselectivity.[123]

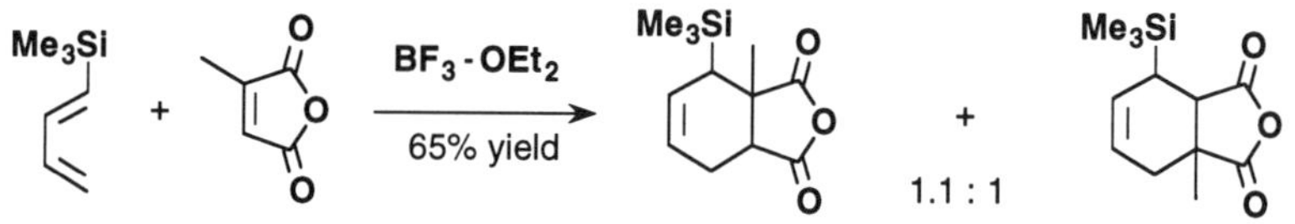

However, Hosomi and Sakurai reported the formation of almost pure "para" adducts from 2-trimethylsilylmethylbuta-1,3-diene **A** and various α,β-unsaturated carbonyl compounds or esters such as methyl acrylate.[124] The reactivity of more functionalized isoprenylsilane was also reported by the same authors.[125]

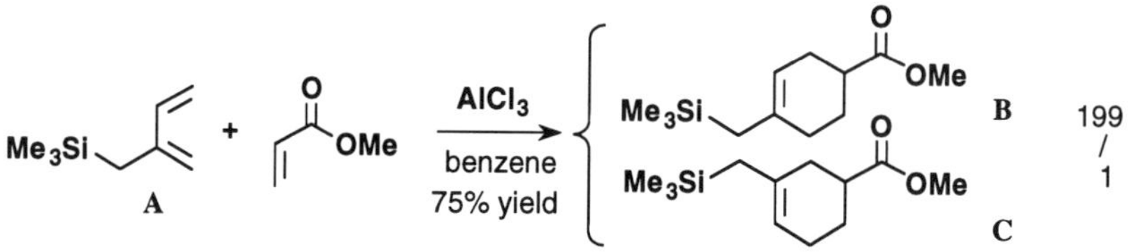

Bisabolene and cadinane derivatives were even prepared in good yield from adduct **E** obtained, as a single regioisomer, from isoprenylsilane **A** and dienone **D**.

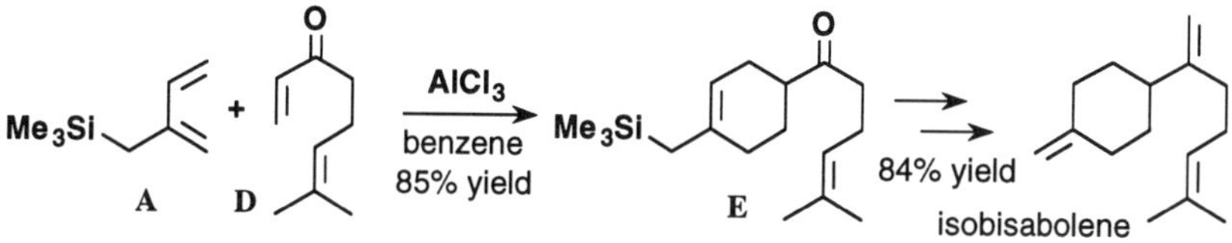

On the contrary, with **A** high "ortho" regioselectivity occurred.[111]

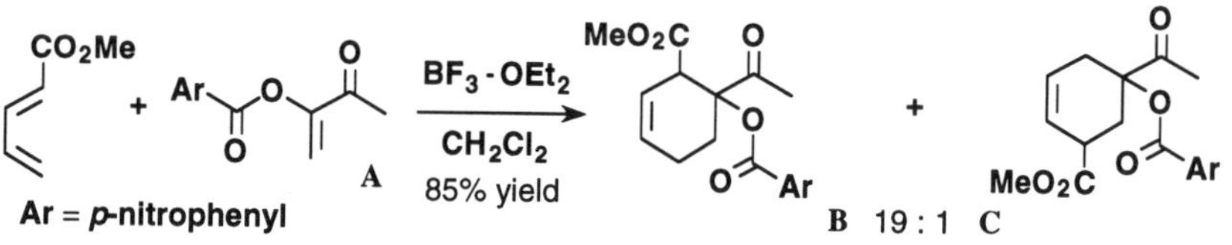

An inventive activation of 2,2-dimethoxyethyl acrylate **A**, due to its transformation upon Lewis acid action into the reactive cationic species **B** was proposed by Saigo; it enabled the formation of endo adducts, either with excellent selectivity or even exclusively as for **C**.[126]

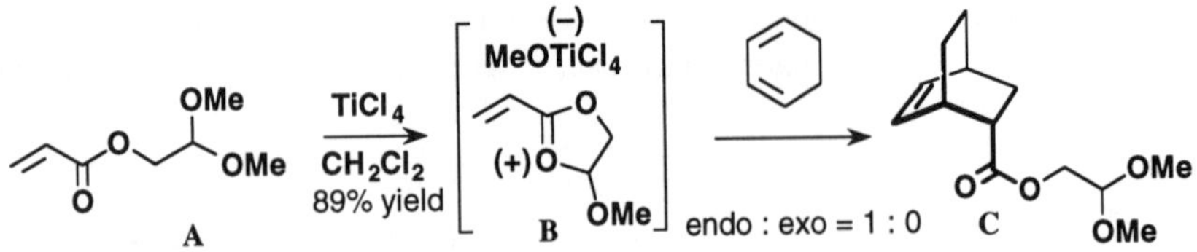

Versatile octalones were prepared by Liu by $SnCl_4$ catalyzed cycloaddition of *trans* 2-diethylphosphoryloxypenta-1,3-diene with a variety of α,β-unsaturated ketones.[127–129]

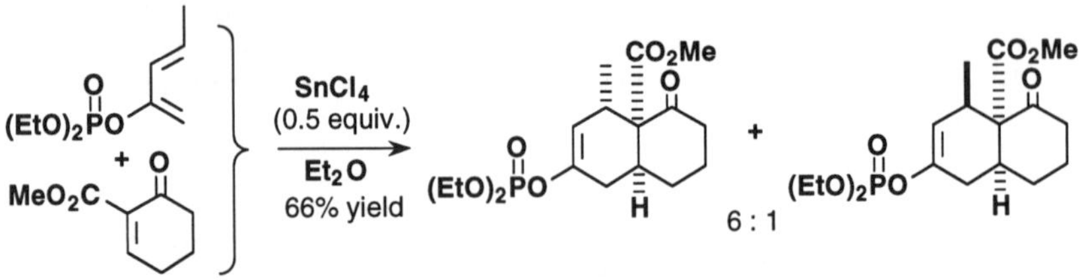

Ketone **C**, a key intermediate in the synthesis of (±)-quassin, was obtained almost pure by Grieco through a cycloaddition between cyclohexenone **A** and functionalized diene **B**.[130,131]

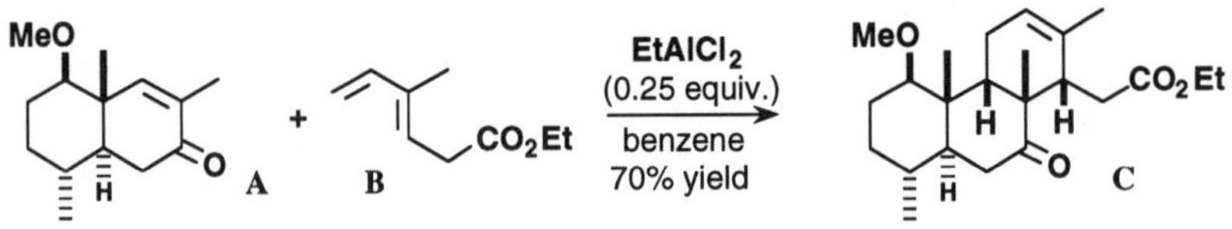

High-pressure Diels-Alder reactions are also accelerated by mild Lewis acids. In the following reaction between diene **A** and quinone **B**, the use of $ZnBr_2$ also induced both better regio- and stereoselectivities; thus, dihydroquinone **C** was formed in 92% yield as a single isomer.[132]

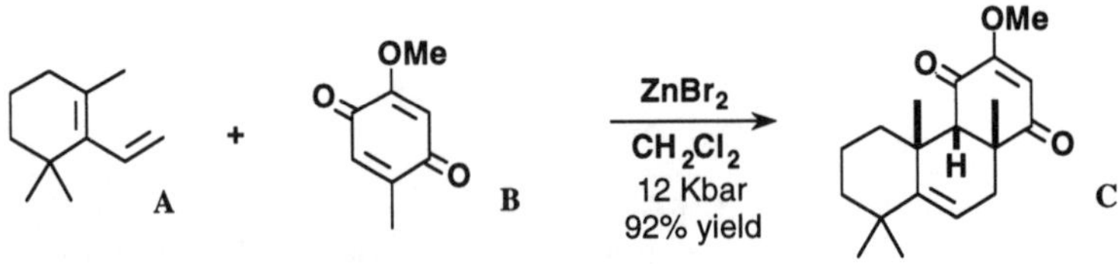

The use of this strategy enabled Engler to formally synthesize, through adducts **E** and **F**, antitumor diterpenes taxodione and royleanone from diene **A** and quinone **D**.[132]

A + D → (ZnBr$_2$, CH$_2$Cl$_2$, 12 Kbar, 55–74 % yield) E + F (E : F 4–5 : 1)

E + F → → → (±)-royleanone or (±)-taxodione

As part of his systematical study of the behavior of cyclohexenone in Diels-Alder reactions, Wenkert showed that (+)-car-3-en-2-one reacted, under Yb(fod)$_3$ catalysis, with 1-methoxybuta-1,3-diene to selectively give the *anti* adduct.[133]

(0.5 equiv.), toluene, 64% yield

Trost developed a general approach, directed toward anthracycline antitumor compounds, based on the Diels-Alder reaction between 1-oxy and 2-oxy butadiene and juglone **A**.[71]

thermal, **B** : **C** = 3 : 1
Lewis acid catalysis, **BF$_3$-OEt$_2$** (0.04 equiv), **B** : **C** = >20 : 1

Dione **B** was then transformed into enone **D** and a second Diels-Alder reaction with 2-oxy-3-thiobutadiene **E** and only 0.06 equivalent of BF$_3$-OEt$_2$ enabled the formation, with very high diastereofacial selectivity (addition *syn* to the ring junction hydrogen atoms), of **F**, an intermediate in the synthesis of deoxypillaromycinone.[134]

Finally, siloxydienes have also been used successfully towards α,β-unsaturated carbonyl compounds.[135]

II.3 Cycloaddition of cyclopentadiene with various dienophiles

In the course of the study of the reaction of acrolein with cyclopentadiene, Yamamoto observed enhanced endo selectivity upon use of Lewis acids (DAD: dimethylaluminum-2,6-di-*tert*-butyl-4-methylphenoxide).[136]

thermal; endo : exo = 3 : 1
L. A. : **Me_3Al**; 57% yield; endo : exo = 24 : 1
L. A. : **DAD**; 56% yield; endo : exo = 1 : 24

In contrast, the endo rule does not prevail for the reaction between cyclopentadiene and methacrolein since the α-methyl group of the dienophile possesses an appreciable exo orienting ability. Thus, a marked "α-methyl effect" leading to the predominant formation of the exo adduct was observed with various ethylenic dienophiles.

thermal; endo : exo = 1 : 5
L. A. : **Me_3Al**; 57% yield; endo : exo = 1 : 15
L. A. : **DAD**; 56% yield; endo : exo = 1 : 24

The stereooutcome of the cycloaddition between mesityl oxide and cyclopentadiene is also Lewis acid dependent. While $ZnCl_2$ and $AlCl_3$ gave rise to endo adducts, $TiCl_4$ favored the formation of exo epimers[137] (see also the study reported by Corcoran on the geometrical aspects of the activation of enones by $TiCl_4$[138]).

L. A., CH_2Cl_2; endo + exo

L. A. = $ZnCl_2$; endo : exo = 4 : 1

L. A. = $AlCl_3$; endo : exo = 3.35 : 1; L. A. = $TiCl_4$; endo : exo = 1 : 5.25

Good endo selectivity was reported by Buono in the $AlCl_3$ catalyzed reaction of vinyl phosphonate **A** with cyclopentadiene even with the sterically demanding dienophile **A3**.[139]

A; $AlCl_3$, CH_2Cl_2; **B**, **C**

R = P(O)(OEt)$_2$ **1**; R = P(O)Ph$_2$ **2**; R = P(O)(Ph)(2-MeO-C$_6$H$_4$) **3**

A1; 85% yield; **B1** : **C1** = 5.7 : 1

A2; 82% yield; **B2** : **C2** = 4.6 : 1

A3; 98% yield; **B3** (mixt. of diasteromers) : **C3** (mixt. of diastereomers) = 8.1 : 1

Wenkert and co-workers also undertook a systematic study of the reaction between cyclopentadiene and various cyclohexenones. Good endo selectivity was reported for the reaction with cyclohexenone.[140]

$AlCl_3$ (0.25 equiv.), toluene, 80% GLC yield; endo 8.1 : 1 exo

The presence of a *gem*-dimethyl group at carbon atom 4 even enhanced the endo selectivity. In contrast, when the *gem*-dimethyl group was located on the carbon atom 5, the endo selectivity decreased.

$AlCl_3$ (0.25 equiv.), toluene, 92% GLC yield; endo 19 : 1 exo

$AlCl_3$ (0.25 equiv.), toluene, 99% GLC yield; endo 3.8 : 1 exo

Again the "α-methyl effect" operates for reactions performed from 2-methylated cyclohexenones.

$AlCl_3$ (0.25 equiv.), toluene, 63% GLC yield; endo 1 : 1.4 exo

$AlCl_3$ (0.25 equiv.), toluene, 57% GLC yield; endo 1.5 : 1 exo

$AlCl_3$ (0.25 equiv.), toluene, 99% GLC yield; endo 3.8 : 1 exo

A remarkable endo and *anti* diastereoselectivity occurred for the reaction of pyranoid enolone **A** with cyclopentadiene. Under thermal or high pressure conditions, the same substrates led to a mixture of diastereomers.[141]

$TiCl_4$, CH_2Cl_2, 82% yield; endo : exo = 81 : 1

A very interesting Lewis acid dependent reaction occurred between dimethyl thionofumarate **A** and cyclopentadiene. Among the various Lewis acids used, hard Lewis acids (except BF_3-OEt_2) promoted the predominant formation of the endo-ester diastereomer, whereas soft Lewis acids mainly gave the endo-thionoester. In the thermal reaction, the ester and thionoester have a comparable endo-directing ability, the latter being slightly superior.[142]

95% yield; **B** endo-thionoester; **C** exo-thionoester

thermal (−54°C); **B** : **C** = 2.3 : 1
BF_3-OEt_2; **B** : **C** = 15.7 : 1
$Cu(OTf)_2$; **B** : **C** = 8.1 : 1
$TiCl_4$; **B** : **C** = 1.1 : 1
$Eu(fod)_3$; **B** : **C** = 1 : 1
Et_2AlCl; **B** : **C** = 1 : 3.8
BCl_3; **B** : **C** = 1 : 24

These results suggested that the endo selectivity was controlled by complexation site of the Lewis acid.

The endo rule was operating in the case of the following trimethylsilyl cyclopentadiene. Adducts obtained were used as bicyclic precursors of verrucarol.[143]

II.4 Cycloaddition of allenic dienophiles with various dienes

Allenic derivatives, including allenyl cations,[144] can also be used as dienophiles. Thus, Hoffmann established that while the thermal cycloaddition between allenic esters and cyclopentadiene is poorly endo selective, the addition of $AlCl_3$ resulted in a significant increase of the proportion of endo isomers.[145]

thermal (benzene, reflux), 90% yield; endo : exo = 1.8 : 1
$AlCl_3$, (benzene, 25°C), 95% yield, endo : exo = 6.1 : 1

As could be expected, the thermal Diels-Alder reaction of α-methylated allenic esters violates the Alder endo rule ("α-methyl effect"). However, the addition of $AlCl_3$ restored the endo rule (similar results were obtained by Gandhi[146]).

thermal (benzene, reflux), 80% yield; endo : exo = 1 : 1.5
$AlCl_3$, (benzene, 25°C), 95% yield, endo : exo = 3.2 : 1

With allenic esters, bearing an α-sulfone group, Mattay showed that the endo directing effect was due to the ester moiety.[147]

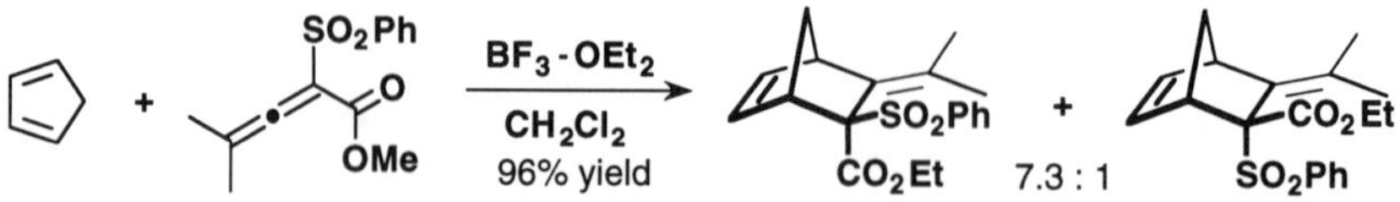

The cycloaddition of γ-substituted α-allenic ketones afforded β-alkylidene ketones with both high endo selectivity and high control of the stereochemistry of the ethylenic double bond.[148]

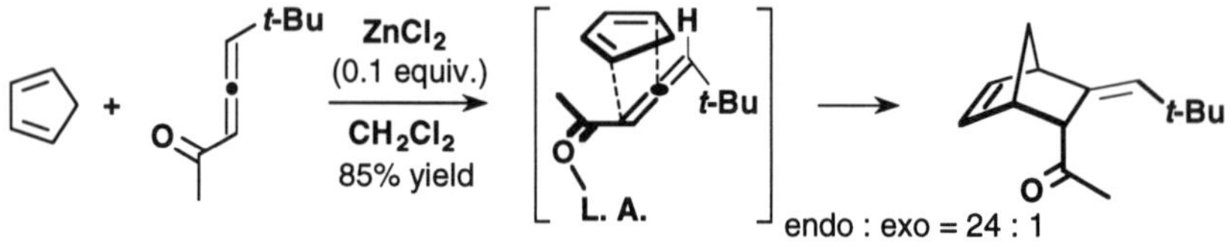

Gras also established that the Diels-Alder addition of α-ethenylidene cyclanones with dienes generated spiro dienic ketones with high stereocontrol.[149]

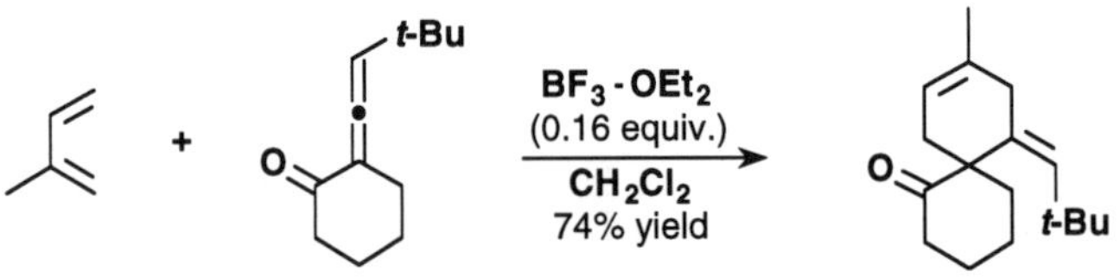

The cycloaddition between allenic lactone **B** and dienoxysilane **A** produced mainly the endo spiro adduct, exo methylene decalone enol ether **C** while the thermal reaction gave predominantly the exo product **D**. Spirolactone **C** is a precursor of plaunols B and C.[150]

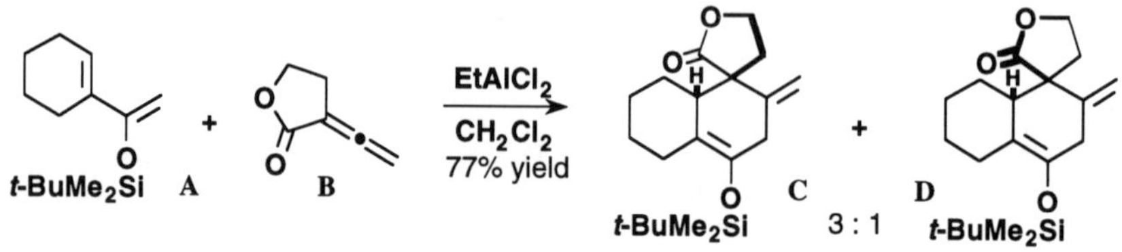

III. Intramolecular Reaction

In the intramolecular Diels-Alder reaction, of the two rings formed, only one results from the [4+2] cycloaddition. Though four cases are theoretically possible, the most important is Type I, leading to "ortho" products (fused ring systems), and possibly Type II leading to "meta" products (bridged ring systems) (for reviews on the intramolecular reaction, both thermal and Lewis acid catalyzed, see refs.151–153).

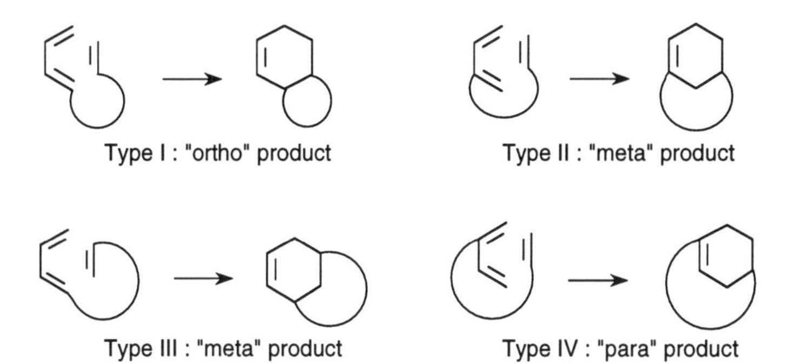

III.1 Cyclization of Type I

III.1.a Cyclization of nona-1,3,8-triene derivatives

The cyclization of nona-1,3,8-triene derivatives provides a versatile route toward bicyclo[4.3.0] derivatives. Thus, Roush showed that the cyclization of methyl dienoates **A** and **D** afforded mixtures of cycloadducts, among which the *trans*-fused products **B** predominated.[154–156]

R = CO_2Me **A** → CO_2Me **B** + CO_2Me **C**

thermal (150°C), 72% yield, **B** : **C** = 2.6 : 1
$AlCl_3$, (0.1 equiv., 23°C), 75–84% yield, **B** : **C** = 1 : 0

CO_2Me **A** → CO_2Me **E** + CO_2Me **F**

thermal (180°C), 75% yield, **E** : **F** = 2 : 1
$EtAlCl_2$, (0.9 equiv., 23°C), 59% yield, **E** : **F** = 1.7 : 1

Transition states **G** and **H** can account for the formation of *trans*-fused products **B** and **E**.

MeO_2C **G** → **B** | CO_2Me **H** → **E**

The intramolecular cycloaddition of cyclopentenone **A** occurred according to the endo rule. Thus, the very strained tricyclic ketone **B** was obtained in fair yield.[157]

A → B: $EtAlCl_2$, CH_2Cl_2, 55% yield

III.1.b Cyclization of deca-1,3,9-triene derivatives

In that case, the reaction leads to bicyclo[4.4.0] derivatives, a class of compounds of particular usefulness. Indeed, a number of results deal with the synthesis of octahydronaphthalene derivatives which are prepared with a very good control of the stereochemistry. Thus, Marshall reported the cyclization of ester **A** or aldehyde **C** gave rise to bicyclic compounds **B** and **D** with the simultaneous creation of four contiguous chiral centers.[158]

A or C → B or D: $EtAlCl_2$, CH_2Cl_2

R = CO_2*i*-Pr; **B** : 60% yield.

R = CHO; **D** : 64% yield.

In the course of studies directed towards the synthesis of kijanolide and related natural products, the cyclization of aldehyde **A** gave rise to bicyclic aldehyde **B** in excellent yield.

A (R = CHO) → B: Et_2AlCl, CH_2Cl_2, 93% yield

α-OMOM : β-OMOM = 1 : 1.22

Endo transition state **C** is relevant with the stereochemistry of aldehyde **B**.

Various substituted *trans*-fused bicyclo[4.4.0]decenes can be prepared by this way[159] (see also ref.160).

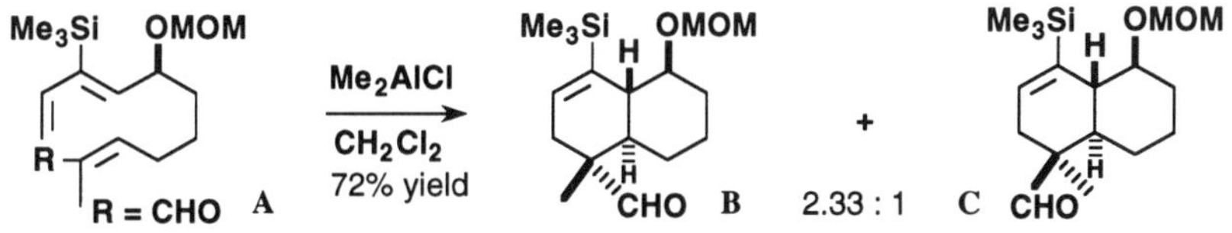

Similarly, trienal **D** underwent Diels-Alder cyclization to yield **E**, a synthetic precursor of chlorothricolide and kijanolide.[161,162]

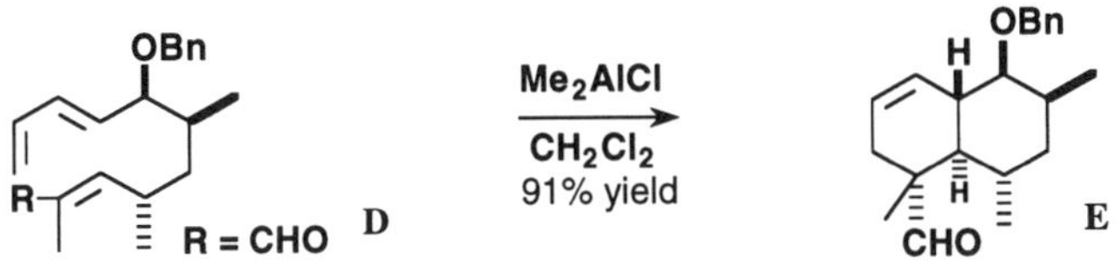

Isomeric alkylidene octalones **B** and **C** could even be prepared by intramolecular Diels-Alder reaction of very reactive enone siloxyvinylallene **A**.[163]

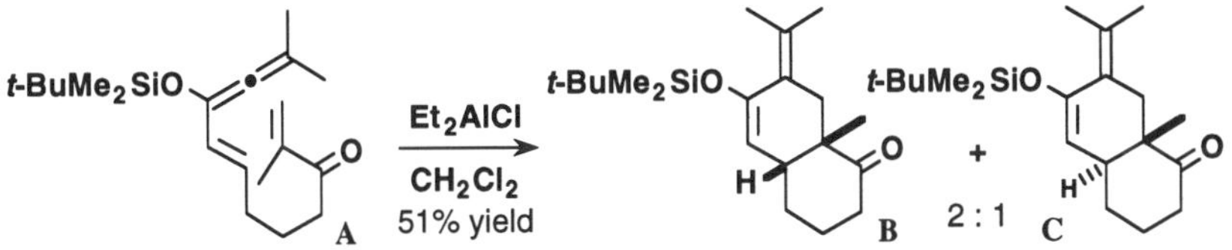

The intramolecular Diels-Alder reaction of a furan ring connected to a dienophile by a four carbon chain is known to have an equilibrium which lies towards the starting material. $MeAlCl_2$ proved to be the best catalyst to overcome this unfavorable equilibrium.[164–166] Rogers and Keay then applied that reaction to the synthesis of (±)-1,4-epoxycadinane.[167]

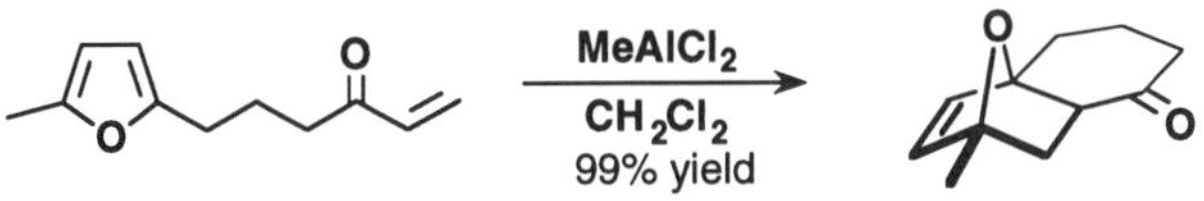

III.1.c Cyclization of 1,3,ω-triene derivatives

The Lewis acid mediated cyclization of a vinyl ketone group tethered to a diene moiety provides a useful means for the preparation of bicyclic[n.4.0] derivatives with *cis* ring junction, n being comprised between 5 and 9.[168]

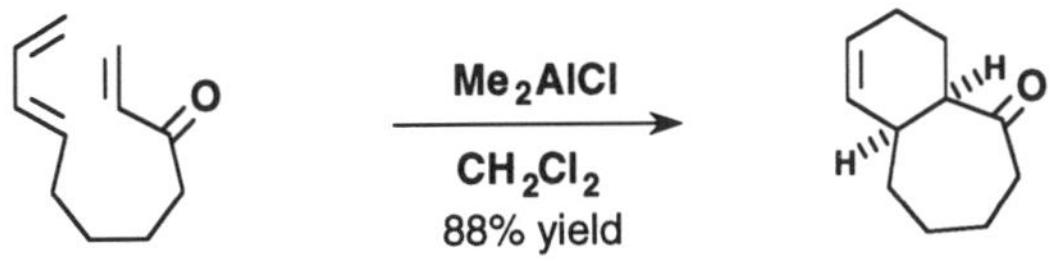

In connection with the synthesis of taxane diterpenes such as taxol and cephalomannine, Sakan studied the construction of the basic carbon framework of these molecules by intramolecular Diels-Alder reaction.[169,170]

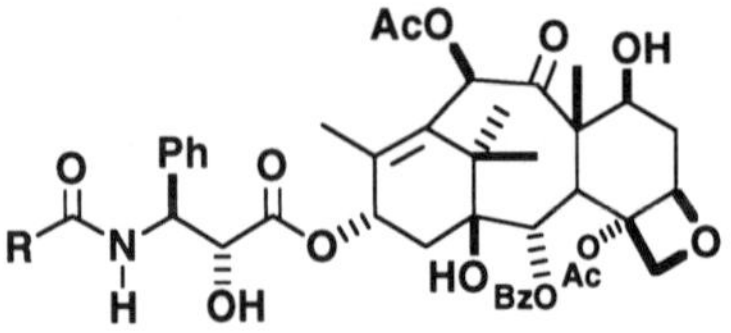

R = Ph : taxol
R = C(Me)CHMe : cephalomannine

The thermal and Lewis acid catalyzed reactions of **A** proceeded in high yield to give *trans*- and *cis*-fused products. Uncatalyzed and catalyzed reactions occurred with reverse stereochemistry.

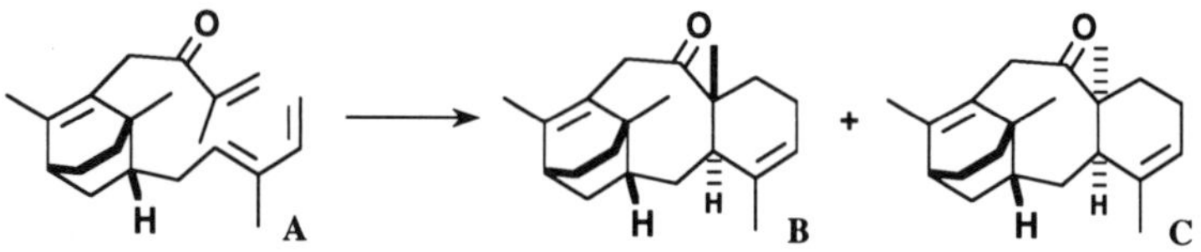

thermal (160°C; 5 equiv. **B(OMe)** $_3$; 70% yield; **B** : **C** = 4 : 1
Me_2AlCl; 90% yield; **B** : **C** = 1 : 19

III.2 Cyclization of Type II

The taxane ring system could also result from a Type II intramolecular reaction performed on 3-alkylidene-1,9-decadienes. The two following examples illustrate this strategy.[171–173]

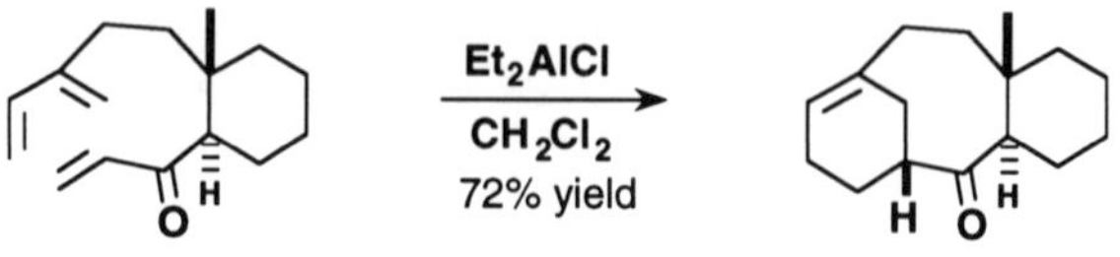

Et_2AlCl, CH_2Cl_2, 72% yield

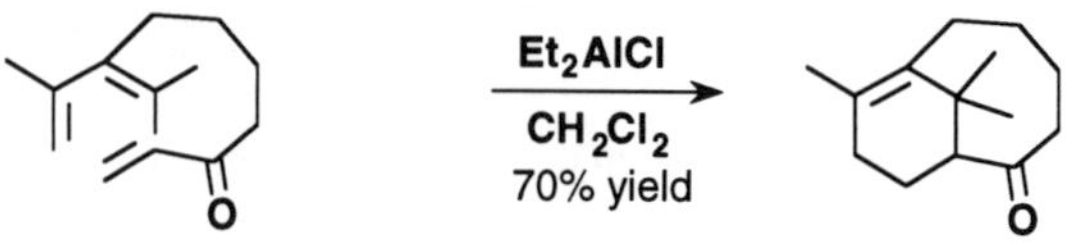

Et_2AlCl, CH_2Cl_2, 70% yield

A rather unique opportunity to probe the relationship between transition state conformation and the product outcome, in a Lewis acid catalyzed intramolecular Diels-Alder reaction, was given by the cyclization of **A**. At –70°C interconversion between the conformational isomers **A**-exo and **A**-endo is fast. Interconversion of the conformational isomers of cycloadduct **B**, however, is slow. The ratio **B**-endo/**B**-exo represents the kinetically controlled rate of conformational isomer formation (k_{endo} / k_{exo} = 70 at –70°C).[174]

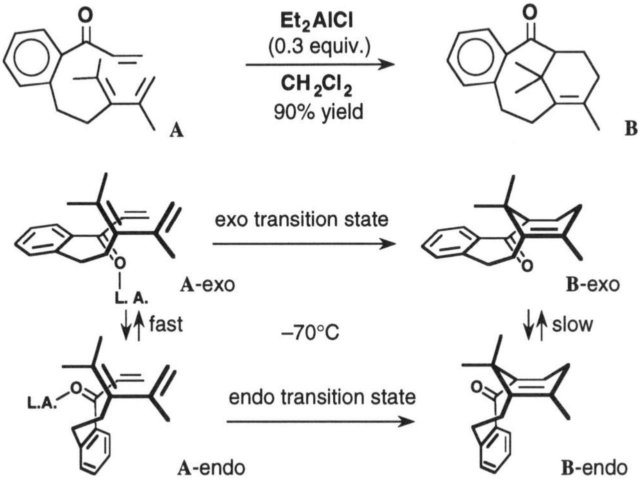

Et2AlCl (0.3 equiv.) CH2Cl2 90% yield

A B

exo transition state

A-exo B-exo

L. A.

fast slow

−70°C

endo transition state

L.A.

A-endo B-endo

IV. Asymmetric Reaction

Due to the importance of the Diels-Alder reaction, much interest has been invested in the development of the enantioselectivity of the reaction. Optically active cyclohexene derivatives can be obtained either by using homochiral dienophiles (or dienes) as starting materials or by running the reaction under homochiral Lewis acid catalysts (for reviews, see ref.175–184). The first approach was initiated in 1949 by Korolev,[185] who observed that optically active cycloadducts were obtained from menthyl fumarate and butadiene, and later improved by Walborsky who showed that the introduction of Lewis acids greatly enhanced the enantioselectivity of the reaction.[186,187]

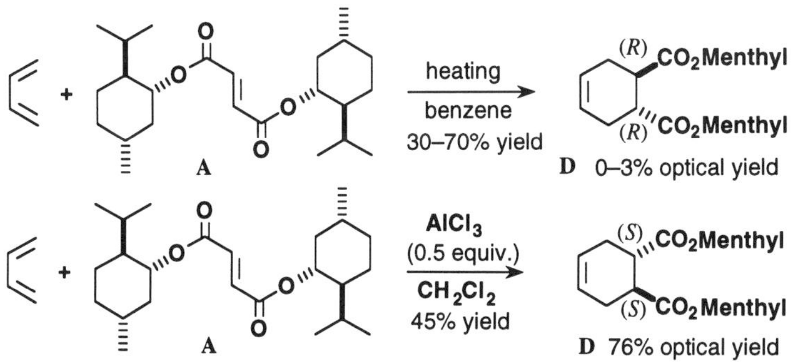

heating benzene 30–70% yield

A

(R) CO2Menthyl (R) CO2Menthyl

D 0–3% optical yield

AlCl3 (0.5 equiv.) CH2Cl2 45% yield

A

(S) CO2Menthyl (S) CO2Menthyl

D 76% optical yield

The use of chiral Lewis acids is much more recent; thus, the first report, due to Guseinov, was only published in 1976.[188] However, the enantiomeric excess reported at that time was not very promising and really encouraging results were only published three years later by Koga who reported the formation of cyclohexene derivatives with up to 72% e.e. when using chiral alkoxyaluminum dichlorides as Lewis acids.[189]

IV.1 Chiral dienophiles

After Korolev's pioneering work and Walborsky's encouraging results, considerable efforts were made to improve the enantioselective version of the reaction by using chiral dienophiles. Among these, acrylate and substituted acrylate derivatives are of particular importance.

IV.1.a Chiral acrylate and alkylacrylate derivatives

The cycloaddition between acrylates of chiral alcohols and cyclopentadiene has been the subject of numerous studies; four diastereomers, two endo and two exo, can be obtained.

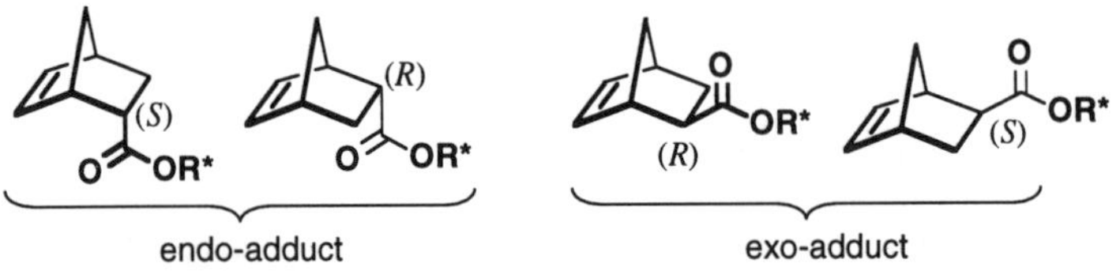

(*S*) O OR* (*R*) O OR* O OR* (*R*) O OR* (*S*)

endo-adduct exo-adduct

Pioneer works deal with the cycloaddition between menthyl acrylate and cyclopentadiene.[190,191] In 1975, Corey proposed to use 8-phenylmenthol which proved to have, compared to menthol, a dramatically superior chiral directing ability.[192]

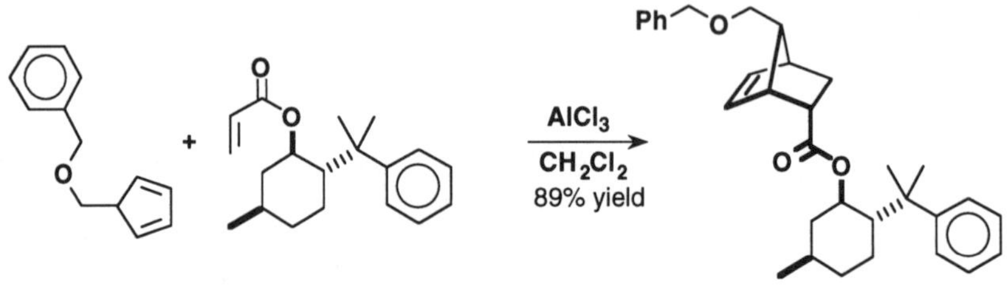

$AlCl_3$, CH_2Cl_2, 89% yield

Rather early studies are also due to Oppolzer who examined the reaction between cyclopentadiene and various acrylates derived from chiral *sec*-alcohols.[193] He also reinvestigated the relative efficiency of acrylates derived from menthol and 8-phenyl menthol as a function of the nature of the Lewis acid involved in the reaction. The best results were obtained from acrylate **A** in the presence of $TiCl_4$ in toluene[194] (recently, a perhydronaphthalene based chiral auxiliary was also found to give good results[195]).

Ph
O
TiCl$_4$
toluene
86% yield
endo : exo = 7.33 : 1
A
L. A.
C
Ph
B
93% d.e.

Acrylates derived from isoborneol were found later to give, under $TiCl_2(O\text{-}i\text{-}Pr)_2$ catalysis, even better results.[196,197]

$TiCl_2(O\text{-}i\text{-}Pr)_2$
CH_2Cl_2
95% yield
endo : exo = 24 : 1
(S)
97% d.e.

$TiCl_2(O\text{-}i\text{-}Pr)_2$
CH_2Cl_2
96% yield
endo : exo = 24 : 1
(R)
>99.2% d.e.

$TiCl_2(O\text{-}i\text{-}Pr)_2$
CH_2Cl_2
89% yield
endo : exo = 11.5 : 1
C
(S)
D 94.3% d.e.

Chiral sulfonamido-isobornyl acrylates can also be used to induce enantioselectivity.[198–201] These derivatives, due to Oppolzer, proved to be very effective.

$i\text{-}Pr_2NSO_2$
$TiCl_2(O\text{-}i\text{-}Pr)_2$
CH_2Cl_2
83% yield
100% endo
(R)
$i\text{-}Pr_2NSO_2$
99% d.e.

The influence of polar groups borne by the chiral auxiliary was particularly studied by Helmchen who used the acrylate derived from (*S*)-ethyl lactate. He established that the absolute configuration of the endo adduct depended upon the nature of the Lewis acid involved.[202]

L. A. : $EtAlCl_2$; A : B = 1 : 3.5
L. A. : $TiCl_4$; A : B = 13.3 : 1

The predominant formation of adduct **A** when using $TiCl_4$ results from the formation of chelate **C** whose existence and structure was established by X-ray analysis (first crystal structure of a Lewis acid complex of a chiral dienophile, see Chapter 1, Section III.3).[203]

Diastereofacial selectivity was also very good during the $TiCl_4$-catalyzed reaction between the acrylate of D-pantolactone and cyclopentadiene, isoprene or butadiene[204] (see also ref.205).

$TiCl_4$, CH_2Cl_2 / hexane, 81% yield; A 32 : 1 B; endo : exo = >20 : 1

Gras showed that acrylates derived from threitol derivatives could, despite their relative flexibility as compared for example to camphor derivatives, induce very high diastereofacial selection. The formation of a Lewis acid-acrylate complex, of semirigid conformation, was postulated to account for these results.[206] Related results on various other sugar residue based chiral acrylates were also reported.[207,208]

$EtAlCl_2$, CH_2Cl_2, 88% yield; >99% d.e.

Another very convincing example of diastereofacial differentiation due to the formation of a chelate between a Lewis acid and a conformationally rigid acyclic chiral auxiliary was provided by Yamamoto[209] (see also ref.210).

O O O O Ph A — $TiCl_4$ (0.5 equiv.), CH_2Cl_2, 96% yield — [$TiCl_4$ … C] → *t*-Bu *t*-Bu OBz B, 82% d.e.

N-Acryloyl derivatives were also used successfully.[211] Thus, Cativiela used derivatives of L-prolines, L-phenylalanine, L-alanine and *N*-methyl-L-alanine esters. With $TiCl_4$, the structure of major adduct **A** is relevant with the formation of complex **C** in which the Lewis acid is *syn* to the double bond and the acryloyl moiety adopts an *s-trans* conformation.[212]

H Ph N (*S*) O CO_2Me — $TiCl_4$ (0.75 equiv.), CH_2Cl_2, 77% yield → (*R*) O NH (*S*) Ph CO_2Me + (*S*) O NH (*S*) Ph CO_2Me

endo : exo = 20.4 : 1 **A** 4.6 : 1 **B**

[H-----O N OMe O H Ph L. A.] **C**

Cyclopentadiene and *N*-enoyl sultam **A** led, in the presence of $TiCl_4$, to the exclusive formation of endo adduct **B** with 93% d.e.[213] X-ray structure of complex **C** provided support for the explanation proposed for this remarkable result.[214] This approach was then used in an enantioselective synthesis of loganin, a preeminent member of the iridoid family[215] (see also ref.216).

N O SO_2 **A** — $TiCl_4$, CH_2Cl_2, 98% yield → [N Ti-O S O O **C**] → O N SO_2 **B** 93% e.e. endo : exo = 99 : 1

An asymmetric bornan sultam directed intramolecular reaction was used by Oppolzer as its key-step en route to the synthesis of (–)-pulo'upone.[217]

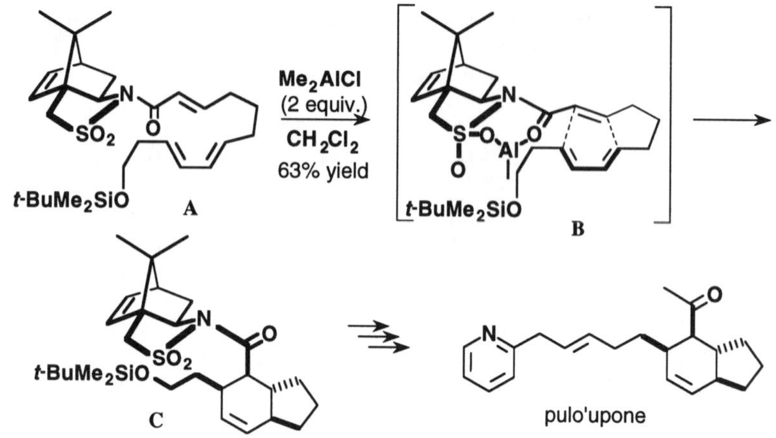

Moreover, Kocienski showed that large scales of both (*R*)- and (*S*)-cyclohex-3-ene carboxylic acids, which are key starting materials in the synthesis of many natural products, could be prepared via an $EtAlCl_2$ catalyzed asymmetric reaction between butadiene and *N*-acryloylbornanesultams; moreover, it was established that the polymerization which complicates the reaction can be suppressed by adding 3 mol% of radical inhibitor galvinoxyl.[218]

Asymmetric Diels-Alder reactions were also performed from chiral α,β-unsaturated *N*-acyloxazolidinones. Evans established that these compounds are both highly reactive and highly distereoselective in the presence of dialkylaluminum chlorides.[219]

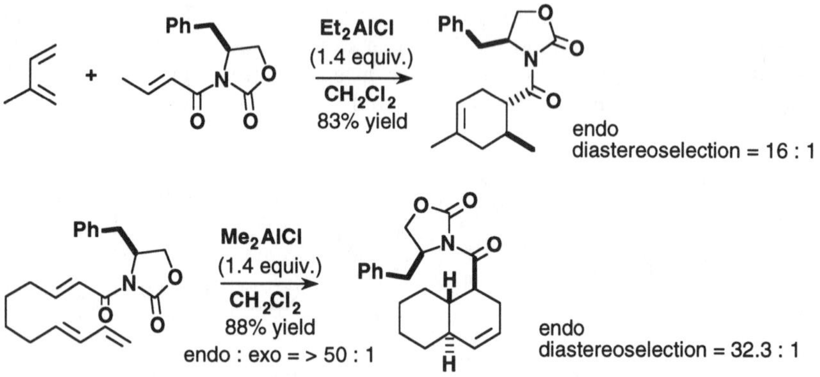

Cationic complex **E** was proposed to account for the rate enhancement and increased selectivity observed when adding more than one Lewis acid equivalent.

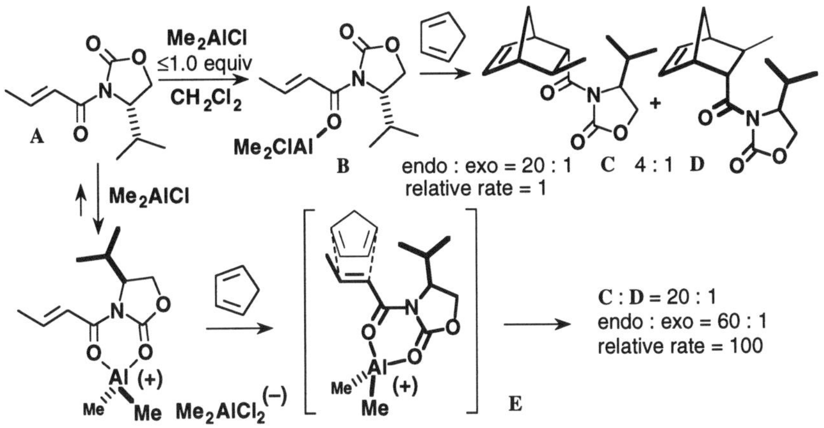

Tanaka used chiral tricyclic oxazolidinones to induce the predominant formation of a single endo adduct. Both enantiomers of the endo carboxylic acid could even be obtained, after removal of the chiral auxiliary, depending on whether the reaction had been carried out on the endo or the exo tricyclic starting material.[220]

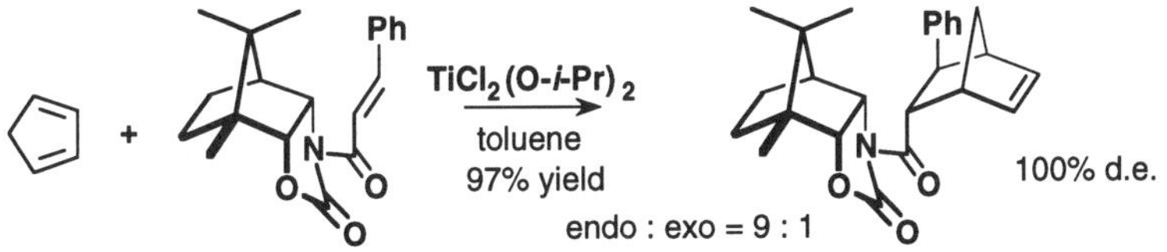

Finally, silica and alumina modified by Lewis acids,[221] and clay[222] were used to promote asymmetric cycloaddition with optically active acrylates.

IV.1.b Functionalized vinyl derivatives

Among the other chiral dienophiles commonly used, the most important are functionalized acrylates, *N*-acryloyl derivatives, vinyl sulfones, and for historical reasons, fumarates. Indeed, after Walborsky's early works,[186,187] Yamamoto reinvestigated the reaction by changing $AlCl_3$ for homogeneous organoaluminium reagents. The reaction of dimenthyl fumarate **A** with cyclopentadiene gave rise to adduct **B** with remarkably high diastereoselectivity.[223]

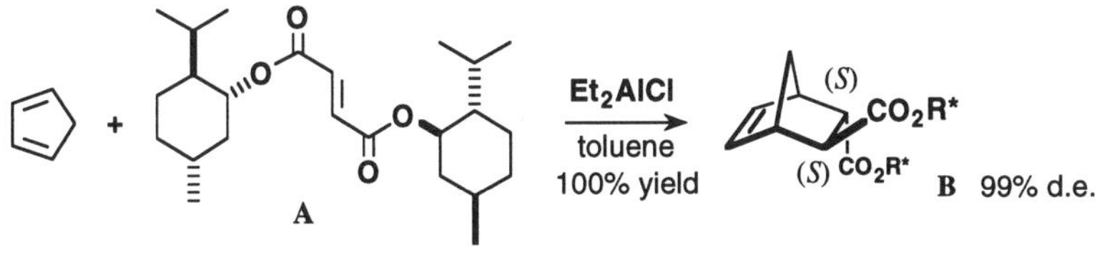

A cooperative blocking effect occurring during the Diels-Alder process was invoked to account for such a result. Thus, the two likely transition states **C** and **D** account for the observed diastereoselectivity.

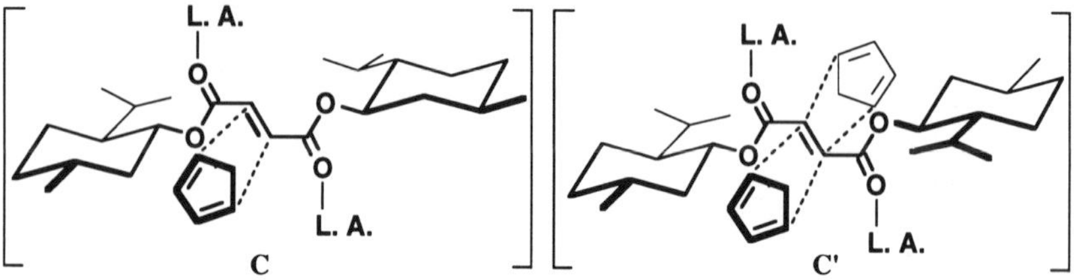

Similarly, the reaction with isoprene leads to the corresponding cyclohexene with good d.e.

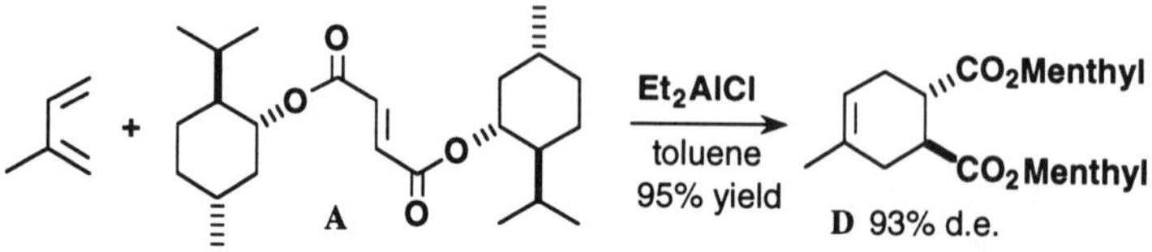

The Lewis acid catalyzed and thermal reactions of (*S*)-ethyl lactate fumarate **A** with cyclopentadiene led to the highly diastereoselective formation of adducts **B** and **C** with inverse stereoselectivity for both variants[224] (similar results were also reported by Charlton[225]).

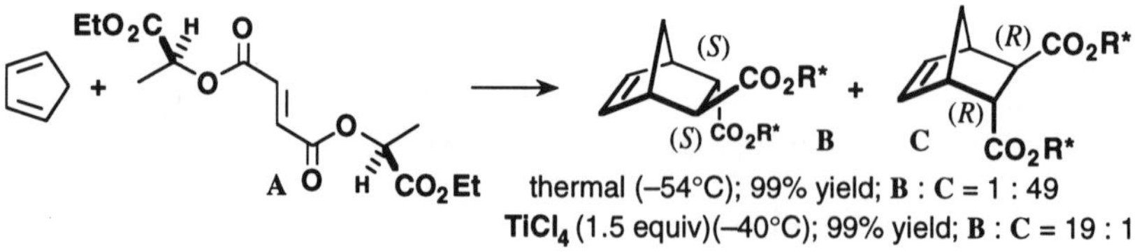

The Diels-Alder reaction with unsymmetrical chiral fumarates was also studied;[226] it provides an efficient route to 1,2-dicarboxylatecyclohex-4-ene derivatives.[227]

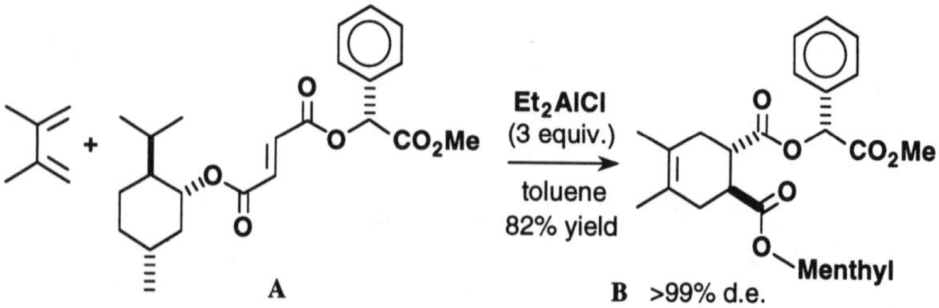

The $TiCl_4$-catalyzed reaction between the 2-(trifluoromethyl)propenoate ester of D-pantolactone and cyclopentadiene or isoprene gave cycloadducts with very high d.e.[228]

The reaction between cyclopentadiene and chiral dienophile **A** occurred with total diastereofacial selectivity and a preference for adduct **B** in which the ester group is in endo position.[229]

Cativiela studied the reaction between several homochiral derivatives of (*E*)-2-cyanocinnamic acid and cyclopentadiene. (*S*)-Ethyl lactate and (*R*)-pantolactone acrylate derivatives led, in the presence of $TiCl_4$, to virtually enantiomerically pure endo (ester) cycloadducts.[230,231]

L. A. = $TiCl_4$; 99% yield; **B** : **C** : **D** = 49 : 1 : 6.5 + **D** (diastereomeric mixture)
L. A. = $AlCl_2Et$; 60% yield; **B** : **C** : **D** = 1.5 : 1.2 : 1

A : **B** = 99 : 1; **A** + **B** : **C** + **D** = 85 : 15

Chiral α,β-unsaturated sulfoxides can also be used as dienophiles; however the presence of an additional electron-withdrawing group, such as an ester group, in the α or β position is needed (these compounds can therefore be regarded as functionalized acrylates). A number of examples exist in the literature; for instance, Koizumi proposed a new route to (–)-neplanocin A starting with the cycloaddition between cyclopentadiene and sulfoxide **A**.[232,233]

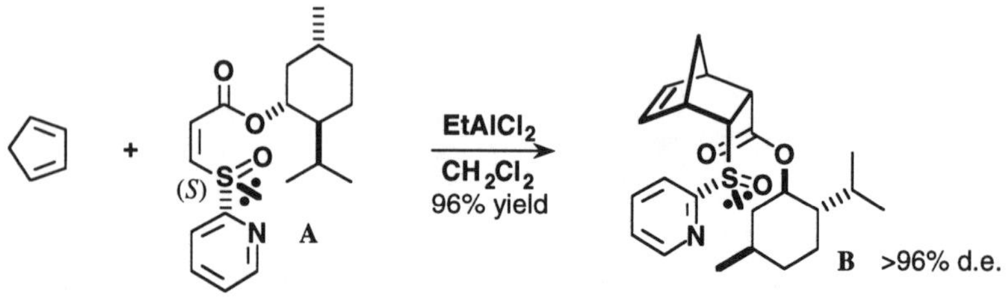

Carretero established that in the presence of mild Lewis acids, such as $Eu(fod)_3$, unsaturated sulfoxide **A** and cyclopentadiene underwent a cycloaddition to yield **B** in fair yield with 84% d.e.[234]

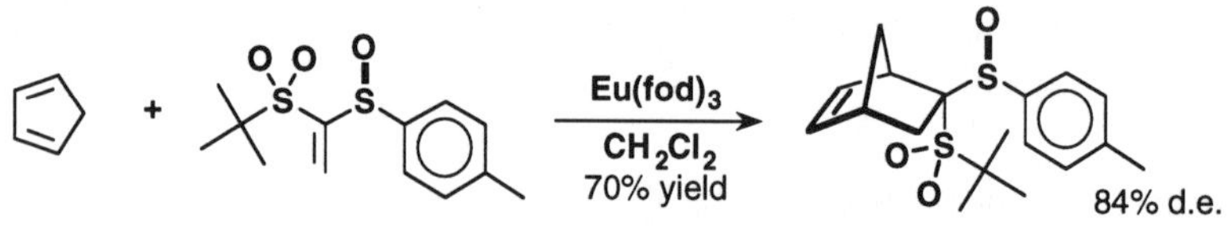

The quasi exclusive formation of an endo adduct was also reported by Kagan who reacted sulfonium salt **A** with cyclopentadiene to obtain sulfoxide **B** with over 99% d.e.[235]

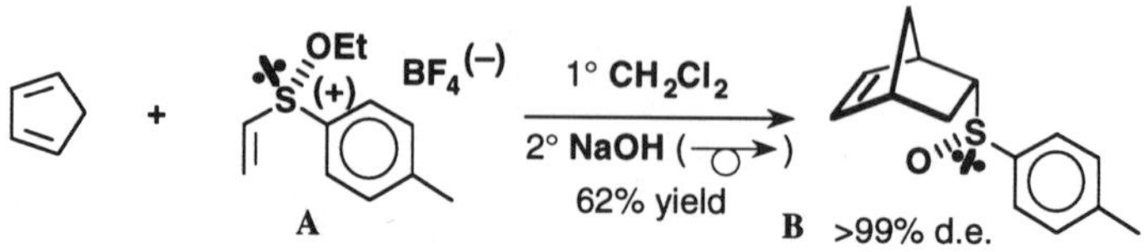

Katagiri proposed using dimenthyl acetoxy methylenemalonate **A** as chiral dienophile; cycloadducts **B** and **C** were used in an enantioselective approach toward carboxylic analogues of *C*-nucleoside;[236] the effects of high-pressure and of the nature of the Lewis acid were also studied.[237]

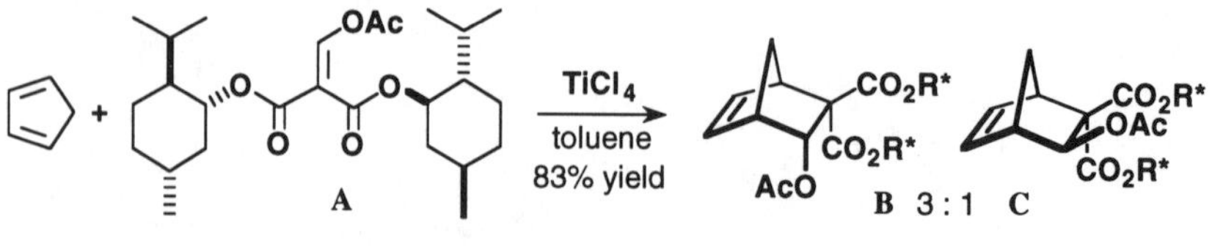

B + C → → AcO, CO2R*, CO2R* → HO, CO2R*, CO2R* → → 92% d.e.

exo approach OAc O O O O TiCl4 endo approach

Chiral allenic esters also give very interesting results as dienophiles. Thus, Oppolzer proposed an enantioselective synthesis of (–)-β-santalene starting with a remarkably high diastereoselective cycloaddition between a camphor derived allenic ester and cyclopentadiene.[238,239]

$TiCl_2(O\text{-}i\text{-}Pr)_2$, CH_2Cl_2, 98% yield, endo : exo = 49 : 1, (S), 99% d.e.

(S) (R) (R) (–)-β-santalene

Although examples are rather rare, chiral α-enones have also been used successfully as dienophiles. Thus, hydroxy enone **A** led, upon various Lewis acids catalysis, to cycloadduct with high stereoselectivity.[240]

$BF_3\text{-}OEt_2$ (0.25 equiv.), CH_2Cl_2, 100% yield, endo : exo = 35 : 1; **A**; **B** 35 : 1 **C**

$ZnCl_2$, toluene, 83% yield; **A**; **D** >98% d.e.

IV.2 Chiral dienes

The use of homochiral dienes to induce enantioselectivity in thermal Diels-Alder reaction has not been, by a long way, as developed as the use of dienophiles. With Lewis acids, examples are even scarcer. Trost reported the reaction between juglone **A** and chiral diene **B** in the presence of boron triacetate leading with over 97% d.e. to cycloadduct **D**[241,242] (see also Bloch's work[243]).

A + B → [C] → D: $B(OAc)_3$ (1.6 equiv.), $CHCl_3$, 98% yield; > 97% d.e.

The diastereofacial reactivity of (*E*)-3-(*tert*-butyldimethylsilyloxy)-1-(D-glucopyranosyloxy)buta-1,3-dienes was studied by Stoodley.[244]

Grieco observed a remarkable diastereofacial selectivity for the reaction of diene **A** with maleic anhydride in 5.0 M $LiClO_4$-Et_2O.[245]

thermal (55°C), toluene; 86% yield; **B** : **C** = 7.3 : 1
5.0 M $LiClO_4$ - Et_2O; 99% yield; **B** : **C** = 15.6 : 1

Isobenzofurandione **B**, upon exposure to trifluoroacetic acid, generated a tetrahydro isoindolone nucleus that is present in a variety of cytochalasans.

B —TFA→ … → cytochalasan (proxiphomin)

IV.3 Chiral Lewis acids

Although the use of chiral dienophiles (or dienes) provides a versatile entry to optically active cyclohexene derivatives, this approach requires the introduction, removal and moreover the use of at least a stoichiometric amount of the chiral auxiliary. In principle, all these drawbacks should be overcome by using homochiral Lewis acids in catalytic amounts.

After his first encouraging report,[189] Koga studied the influence of the structure of the chiral source on the absolute configuration of the major product formed. The best result was obtained with chiral catalyst **A** which enabled the formation of **C**, through transition state **B**, with 61% e.e.[246]

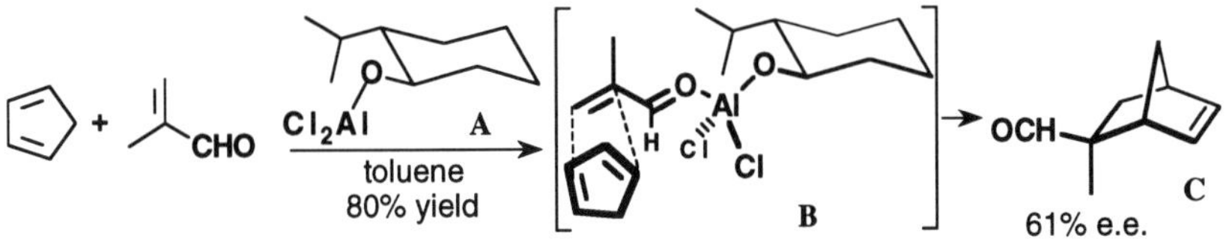

Early studies on the use of chiral Lewis acids are also due to Seebach who prepared a titanium centered Lewis acid derived from binaphthol and thus obtained an endo adduct with 50% e.e.[247]

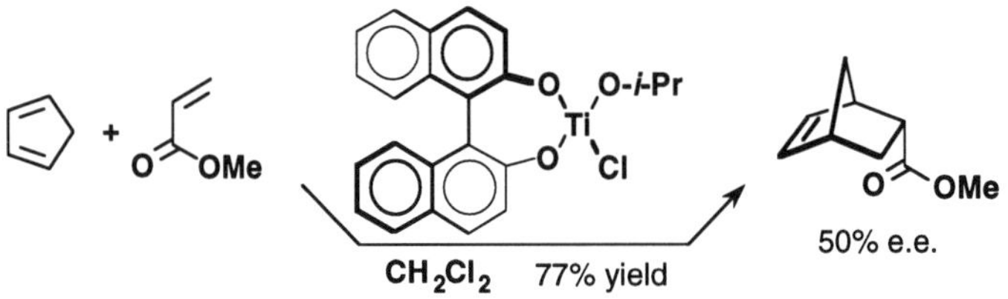

Dimethyl fumarate **A** and 2,3-dimethylbutadiene led, under the action of a titanium catalyst prepared by Oh from (*R,R*)-hydrobenzoin, to the cyclohexene derivative **B** in 92% e.e.[248]

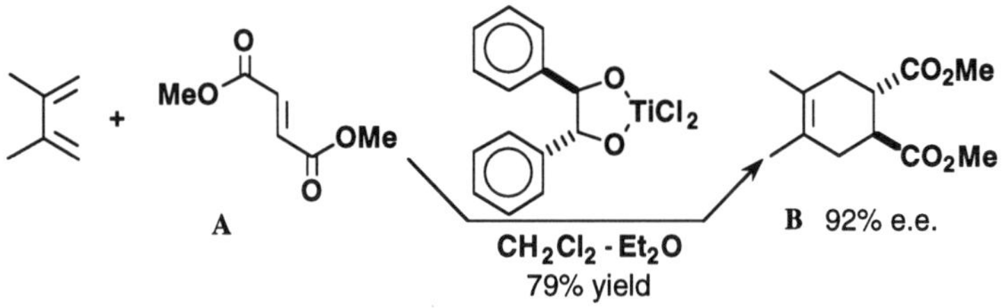

As for Narasaka, he obtained high e.e. by combining the use of dienophiles, such as **A**, prepared from α,β-unsaturated acids and 1,3-oxazolidin-2-one and chiral alkoxytitanium[249] (for related results, see also refs.250–253). Interesting results were also obtained with a chiral bis-sulfoxide Fe (III) complex.[254]

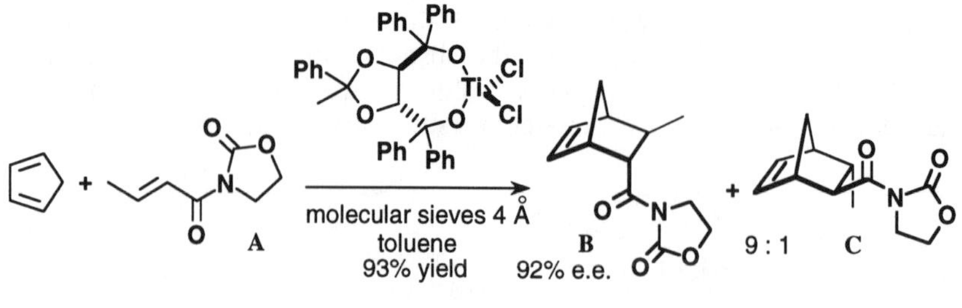

These results, which also received intramolecular applications,[255] were later improved by using various dienophiles and mixed solvents. Thus, **B** was prepared with 94% e.e. from dienophile **A** and the same titanium chiral catalyst in toluene-petroleum ether.[256]

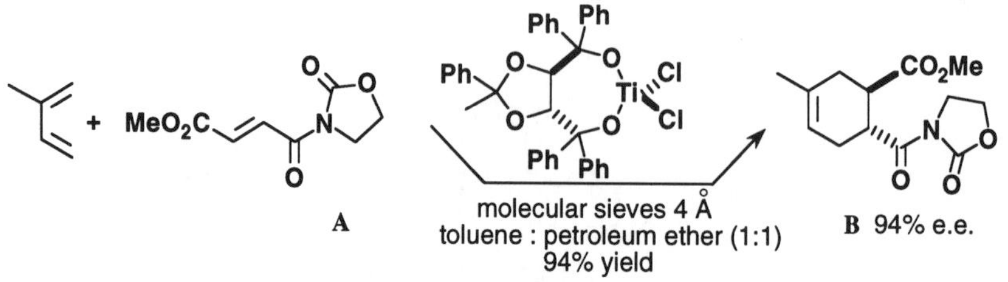

Even better results were reported with 3-(3-borylpropenoyl) oxazolidinones as dienophiles.[257]

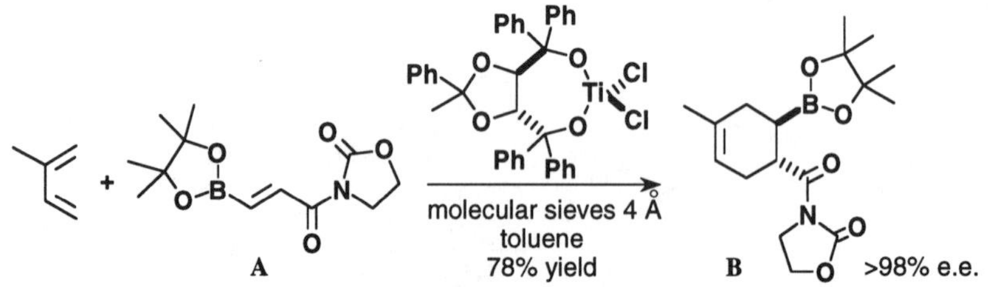

Boron centered chiral Lewis acids have also been used successfully. Thus, in 1986 Kelly and Yamamoto independently reported the preparation of optically active polycyclic derivatives obtained under boron chiral catalysts. Kelly prepared **B**, a precursor to (–)-bostrycin, with over 90% e.e. from quinone **A** in the presence of a catalyst derived from BH_3.[258]

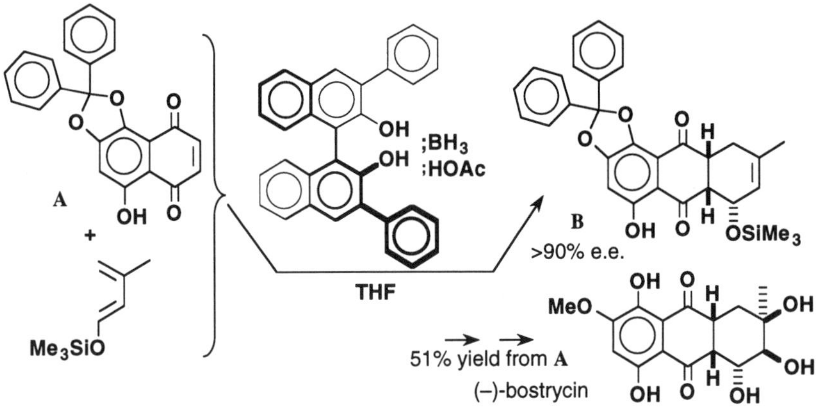

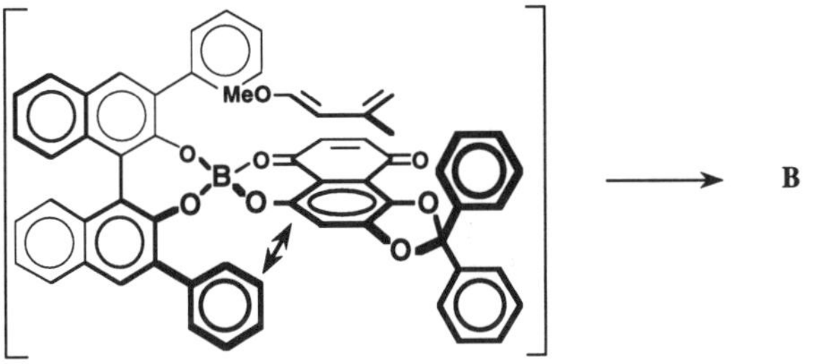

Similarly Yamamoto reacted juglone **A** with 1-trialkylsiloxy-1,3-butadiene, in the presence of a chiral catalyst prepared from trimethyl borate, to afford anthracyclin intermediate **B**.[259]

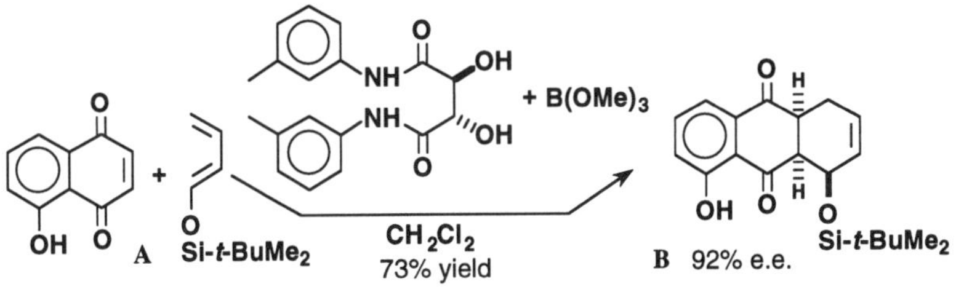

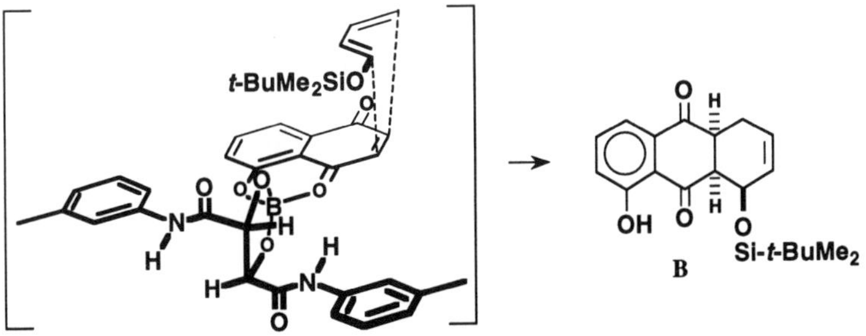

Over the years, many chemists got involved in this area of research[260–264](for reviews, see refs.175–184). Particularly, specific efforts were made to obtain high e.e. with only catalytic amounts of chiral Lewis acids. The following most recent examples illustrate the achievement reached in the field. Evans used bis(imino)-copper (II) complex **A** in 9 mol% to promote the asymmetric reaction between various α,β-unsaturated imides and cyclopentadiene.[265]

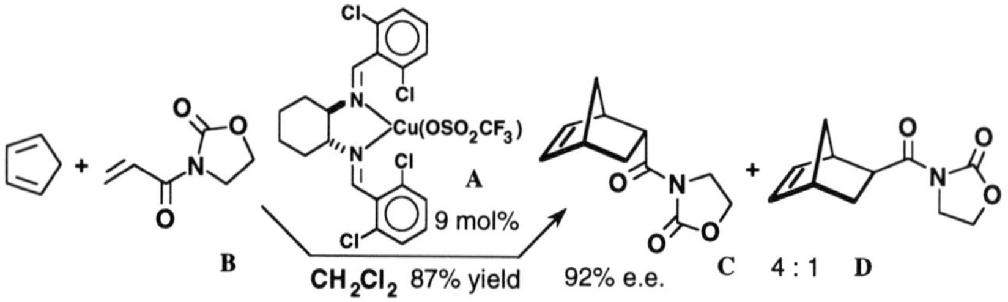

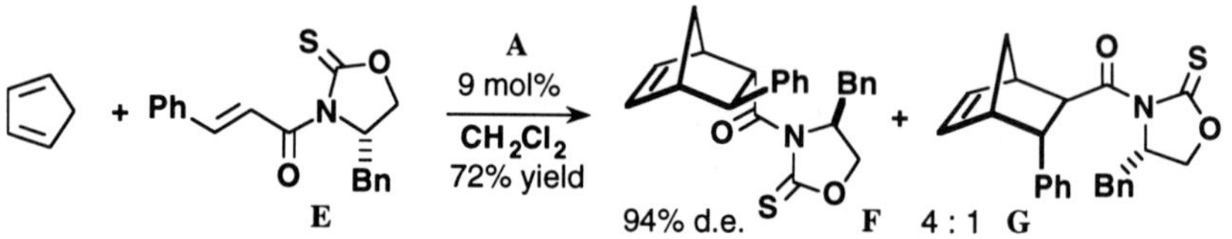

CH_2Cl_2 87% yield 92% e.e. **B** **A** 9 mol% **C** 4 : 1 **D**

A 9 mol% CH_2Cl_2 72% yield **E** 94% d.e. **F** 4 : 1 **G**

Yamamoto designed chiral titanium helical Lewis acids and was able to use them in only 10 mol% to react acrolein with cyclopentadiene, the endo adduct being obtained with 92% e.e.[266]

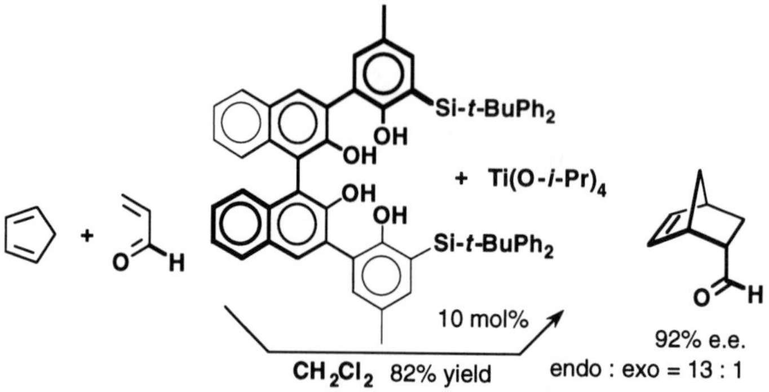

+ $Ti(O\text{-}i\text{-}Pr)_4$ 10 mol% CH_2Cl_2 82% yield 92% e.e. endo : exo = 13 : 1

Taking advantage of their remarkable results in the asymmetric ene reaction (Chapter 2, Section V), Mikami and Nakai used their (*R*)-binaphthol titanium catalyst in 10 mol% to induce the reaction between butadienyl carbamate **A** and methacrolein. **B** was obtained virtually as a single diastereomer with 86% e.e.[267]

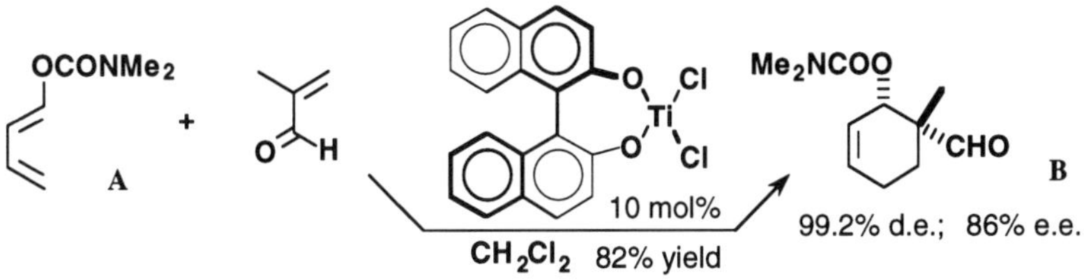

Fe(III)[268] and Al[269] C_2-symmetric chiral Lewis acids were also proposed by Corey who, for example, obtained endo adduct **B** with 95% e.e. by using only 10 mol% of catalyst **A**.

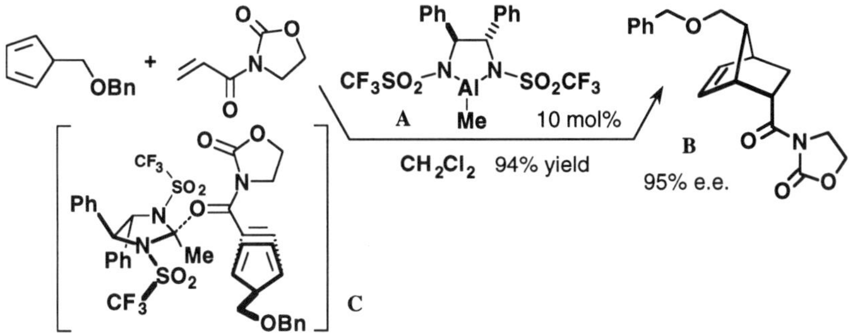

A great achievement was due to Wulff who reported the quasi quantitative preparation of exo adduct **B**, obtained in 98% e.e. with only 0.5 mol% of catalyst **A** involved in the reaction.[270]

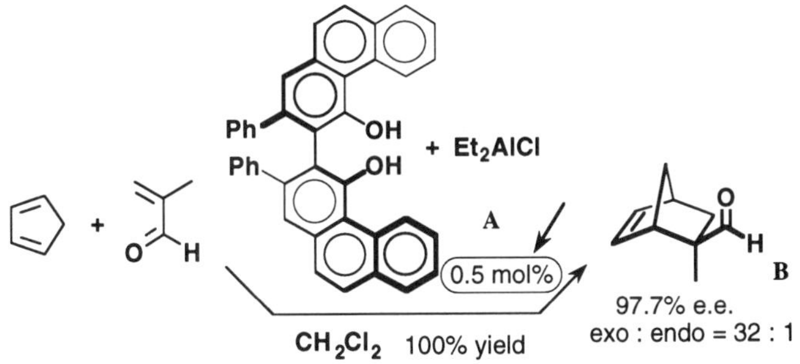

Examples of non-C_2-symmetric chiral Lewis acids involved in asymmetric reactions are also numerous. Thus, Yamamoto used his chiral(acyloxy)borane complex to prepare, from methacrolein and cyclopentadiene, the expected endo adduct with up to 96% e.e.[271,272]

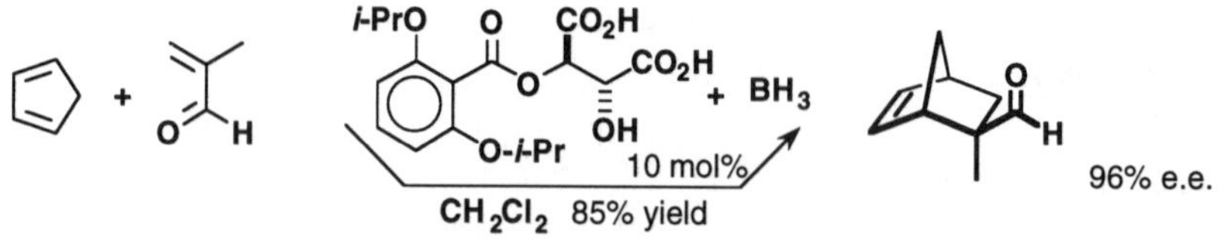

Corey reported the enantioselective reaction of α-bromo α,β-enals with cyclopentadienes in the presence of titanium[273] and boron[274] complexes. In both cases, the exo adduct was obtained with high e.e. with only catalytic amounts of Lewis acids.

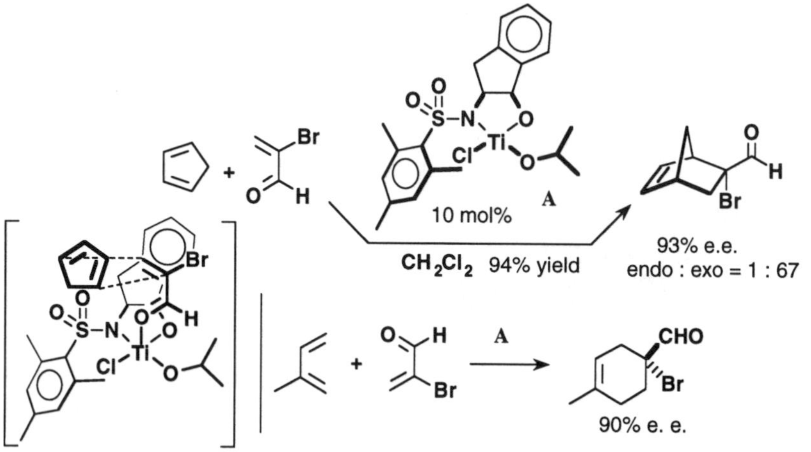

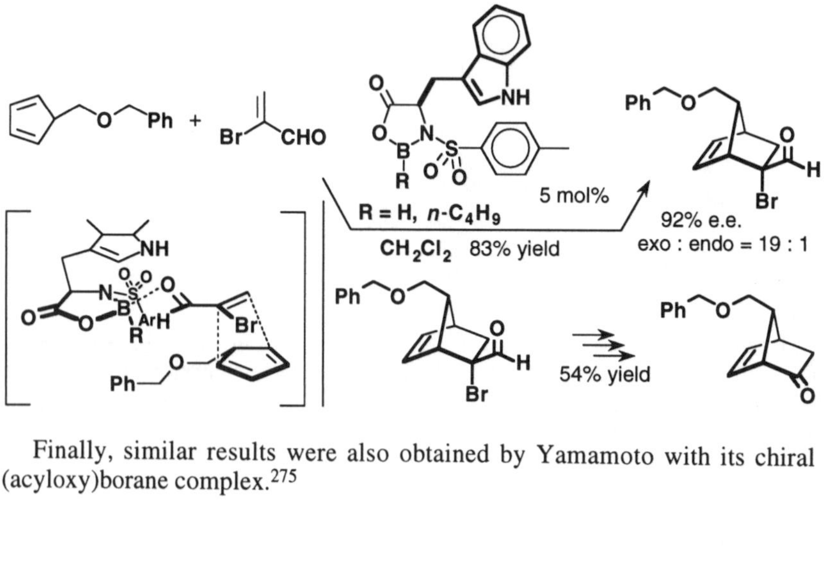

Finally, similar results were also obtained by Yamamoto with its chiral (acyloxy)borane complex.[275]

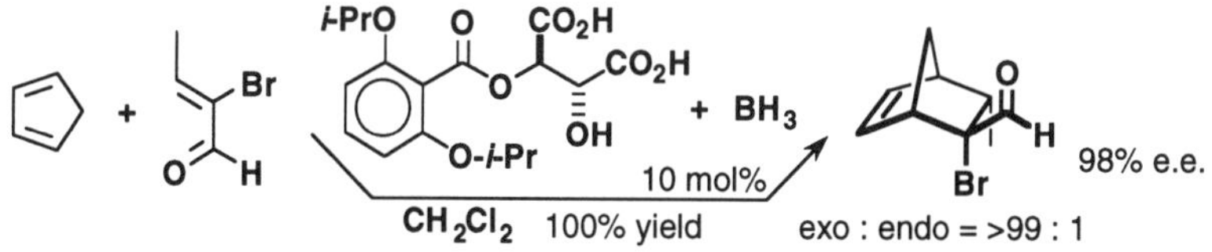

V. Hetero Diels-Alder Reaction

Although less studied than the all-carbon version of the reaction, the hetero Diels-Alder reaction has attracted over the last two decades a greater attention. Indeed, it provides a versatile regio- and stereoselective approach toward heterocyclic compounds from heterodienes[276–278] or hetero dienophiles.[279,280] More recently, the asymmetric hetero Diels-Alder reaction has also been studied.[281]

The use of Lewis acid catalysis in these reactions is overwhelmingly associated with heterodienophiles. Particularly, the cyclocondensation of activated dienes with aldehydes, developed by Danishefsky, is of particular importance.[282,283]

By comparison, heterodienes are not frequently associated with Lewis acids; however, recent examples are promising.

V.1 Lewis acid-mediated cyclocondensation of activated dienes with aldehydes

Danishefsky demonstrated that activated dienes, such as siloxydiene **A** (commonly referred to as Danishefsky's diene), react with a wide spectrum of aldehydes to afford 5,6-dihydro-γ-pyrones **C**. Pyrones **C** can be envisioned as arising from precursors **B**, the formal cycloadducts between **A** and aldehydes, or from **D**, the Mukaiyama-like adducts.[284]

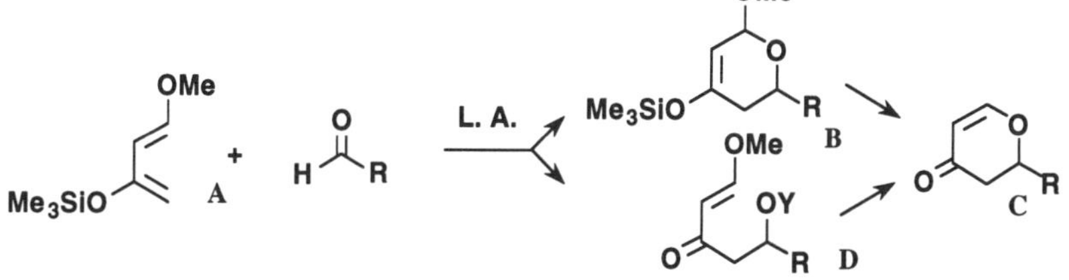

In order to determine whether precursors **B** arise from a [4+2] cycloaddition process or whether it was a consequence of a Mukaiyama-like aldol reaction followed by cyclization, Danishefsky studied the reaction of siloxydiene **A** with cinnamaldehyde. Quenching the reaction after 5 min at –78°C, with aqueous $NaHCO_3$, afforded Mukaiyama-like adducts **B** (40%) and cycloadducts **D** (38%) whereas **D** was the only product found after 4 h at the same temperature. It appears therefore that two distinct pathways are operative, a cycloaddition reaction giving rise to **D** via **C** and the Mukaiyama-like aldolization leading to **B** which then could undergo conversion into **D**.[285]

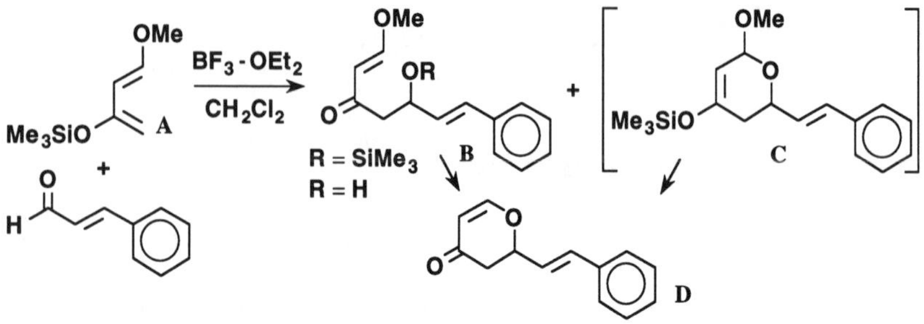

Moreover, the cyclocondensation between siloxydiene **A**, bearing a methyl group on C-2, and benzaldehyde, a remarkable reversal of product stereochemistry could be achieved by changing the catalyst/solvent couple. When the reaction was catalyzed by $ZnCl_2$ in THF and quenched with $NaHCO_3$ without aqueous workup conditions, enol ether **D** was isolated in addition to pyrone **B**. This was the first example of the isolation of an enol ether such as **D** in this cyclocondensation process. **D** could then be transformed into **B** by acidic treatment. Thus, when the reaction was carried out in THF with $ZnCl_2$ as catalyst, virtually complete *cis* stereoselectivity was observed. With BF_3-Et_2O, the reaction led to a diastereomeric mixture of pyrones **B** and **C**.[286,287]

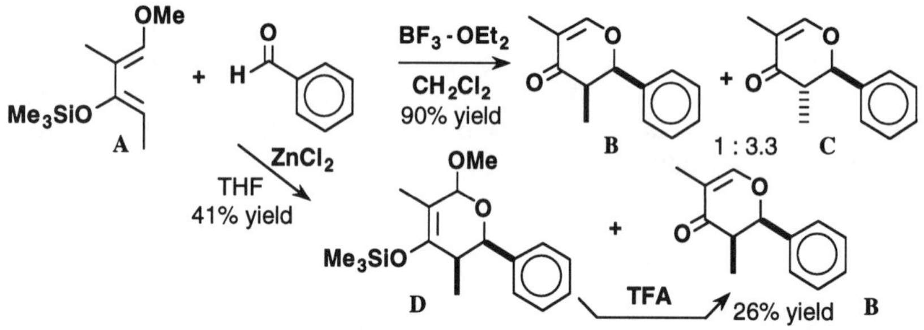

In such cycloaddition reactions, the relative configuration of C_6–R^{1*} and C_3–R^{2*} depends upon the diastereofacial selectivity of the diene. Similarly, the C_2–R^{3*} relation depends upon the diastereofacial selectivity of the dienophile alone. However the relative configuration of C_2–C_3 depends upon both the diene and dienophile diastereofacial selectivities.

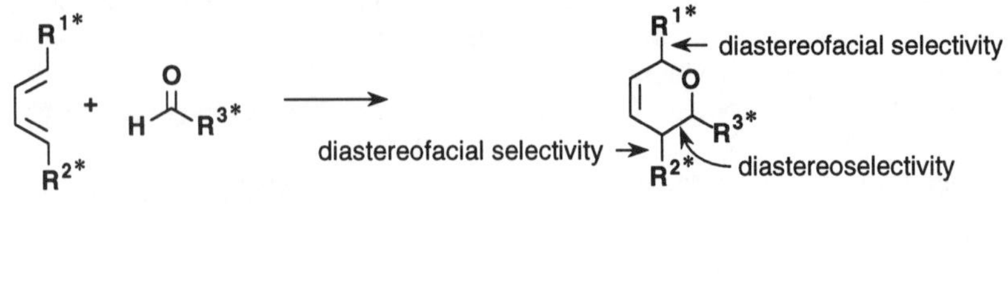

The cyclocondensation between siloxydienes and aldehydes can be catalyzed with a wide variety of Lewis acids[288] including BF_3-Et_2O,[289] $ZnCl_2$[290] and soluble lanthanide complexes.[291–293]

V.1.a Cyclocondensation with achiral aldehydes

Though most reactions reported in this section involve activated dienes bearing alkoxy and/or siloxy groups, it is worth noting that Hosomi and Sakurai established that 2-trimethylsilylmethylbuta-1,3-diene could also react with aldehydes to yield dihydropyranes in useful yield.[125]

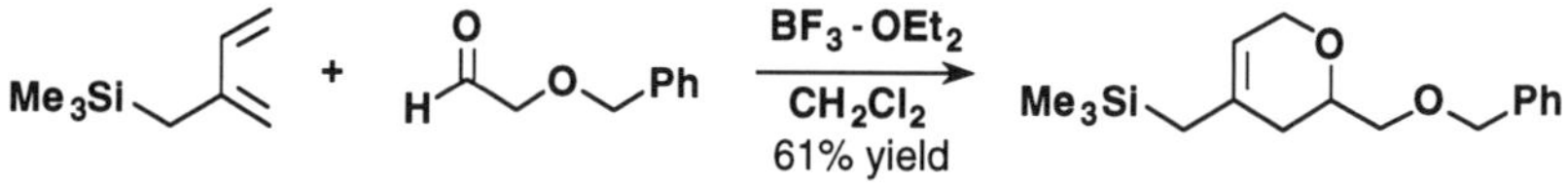

The *cis* relationship present in pyrone **B**, resulting from the cyclocondensation between siloxydiene **A** and acetaldehyde, enabled the synthesis of (±)-fucose and (±)-daunosamide.[294]

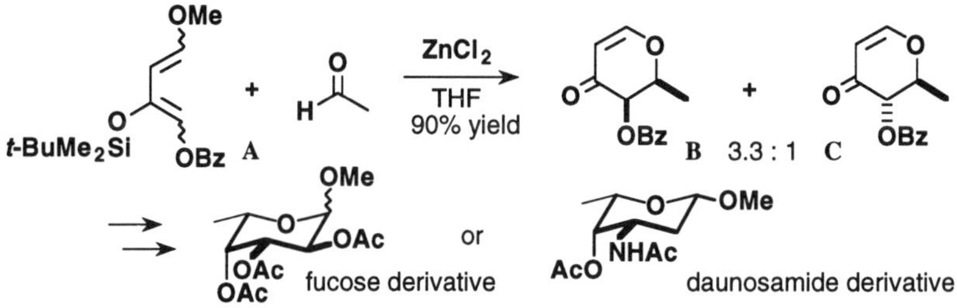

The cyclocondensation between siloxydiene **A** and benzyloxy acetaldehyde, followed by treatment with trifluoroacetic acid, gave rise to pyrone **B**, a precursor of (±)-talose derivative.[284]

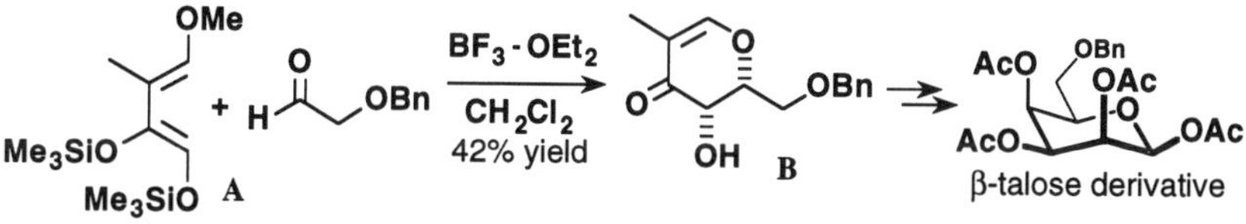

Silylation of β-diketone **A** affords a mixture of (*E,Z*)-bis-siloxydiene **B**. The BF_3-OEt_2 promoted reaction of **B** with benzaldehyde afforded mainly *cis* pyrone **C**.[295] The origin of this dramatic effect of the substitution state at C-2 of dienes on the topography of the cyclocondensation reaction is not clear.

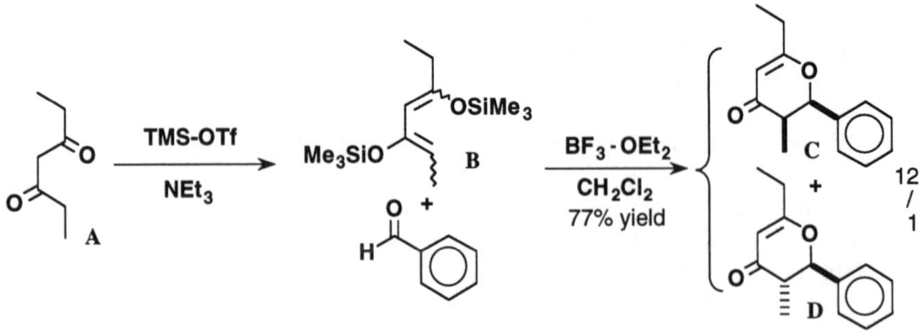

Several examples dealing with lanthanide reagents as Lewis acids are present in the literature.[291–293] Thus, $Eu(fod)_3$ catalyzes the reaction between siloxydiene **A** and hexanal to yield pyrone **B**,[295] and between **A** and cinnamaldehyde to yield (±)-kawain[296] (see also ref.297).

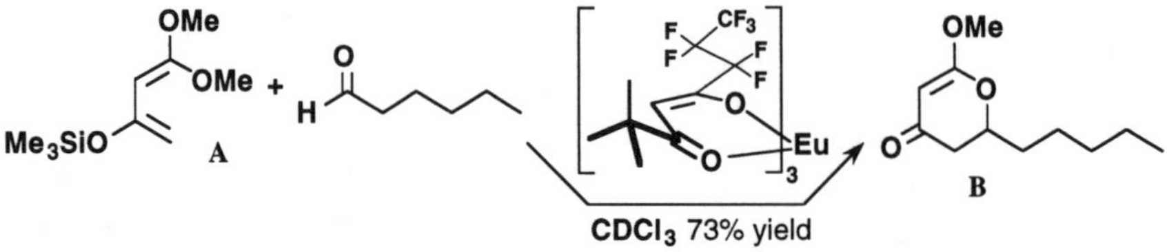

Moreover, the cyclocondensation of furfural with siloxydiene **A**, followed by treatment with TFA, afforded pyrone **B**, a precursor of the galacto derivative **C**.[298]

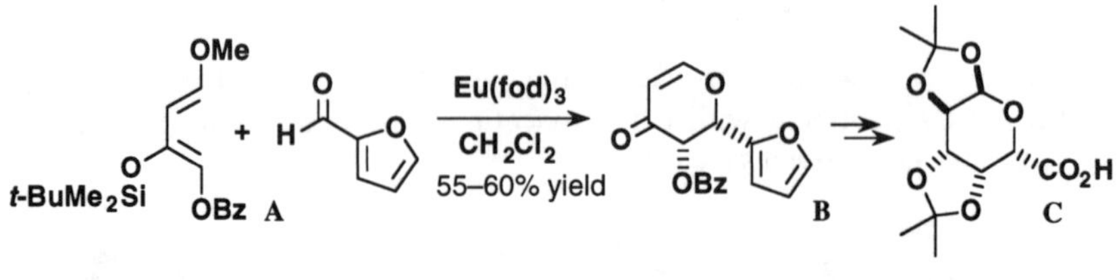

Although chiral (+)-$Eu(hfc)_3$ showed only modest enantiofacial selectivity with achiral dienes, its combination with chiral diene **A** exhibited striking interactivities. Thus, diastereofacial excesses greater than 95% could be obtained in some instances.[299,300] These findings were applied to the synthesis of optically pure L-glucose. Jankowski also used (+)-$Eu(hfc)_3$ and prepared some dihydropyran carbohydrates with up to 64% e.e.[301]

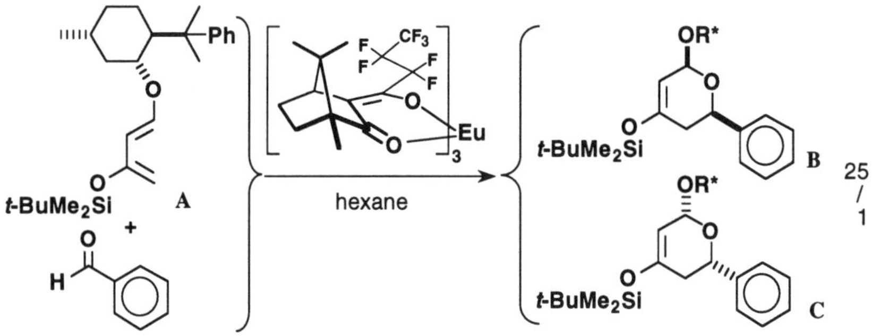

$VO(hfc)_2$ catalyzed the cycloaddition of siloxydiene **A** with benzaldehyde, with both high yield and important enantiofacial selectivity, to yield *cis* pyrone **B**.[302] Better enantioselectivity was reached by Yamamoto with a chiral organoaluminum reagent used in only 10 mol%.[303]

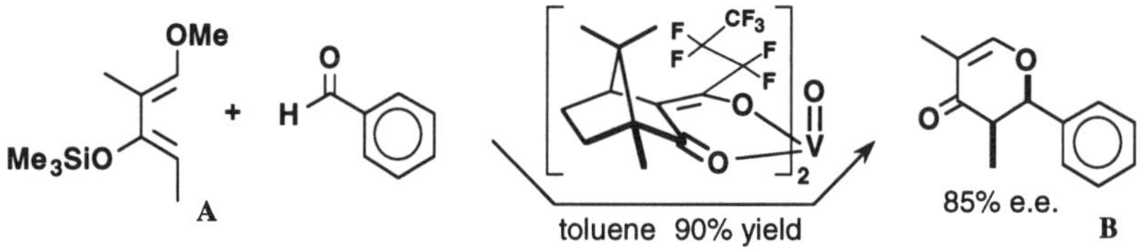

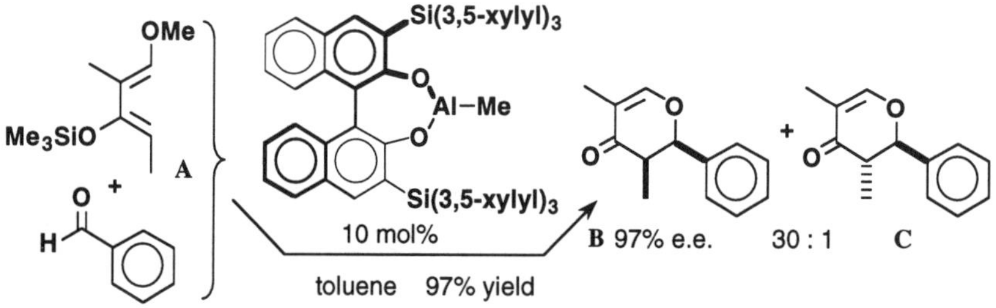

Finally, Nakai and Mikami prepared dihydropyran dicarboxylates **B** and **C** with high e.e. from methoxy butadiene **A** and methyl glyoxylate in the presence of a (*R*)-BINOL-titanium catalyst (10 mol%).[304]

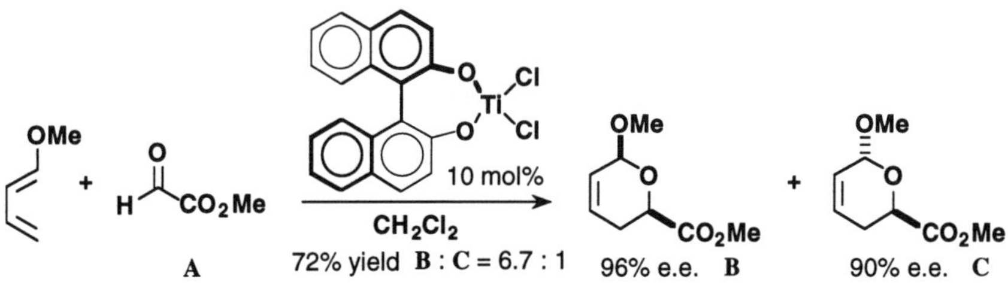

V.1.b Cyclocondensation with chiral aldehydes

The stereochemistry of the BF_3-OEt_2 mediated cycloaddition of siloxydiene **A** with hydratropaldehyde **B** was affected by solvent changes.[305]

BF3 · OEt2; 2° TFA / THF; 95 % yield
solv. : CH_2Cl_2; **C** : **D** = 2 : 1
solv. : toluene; **C** : **D** = 1 : 10

A short synthesis of the Prelog-Djerassi lactone from readily available pyrone **C** was reported by Danishefsky.[306]

DIBAL; TsOH / i-PrOH; H2 / Pd; O3
Prelog-Djerassi lactone

Aldehyde **A**, derived from the Prelog-Djerassi lactone, underwent a cycloaddition reaction ($ZnCl_2$/THF) with the branched siloxydiene **B** to yield pyrone **C**, a precursor of the C1–C9 fragment of 6a-deoxy-erythronolide.[286,307]

Prelog-Djerassi lactone; ZnCl2 / THF; 43% yield
carbons 1-9 of
6a-deoxy-erythronolide B aglycone

Siloxy furyldiene **A** reacted with (*S*)-2-(phenylseleno)propioaldehyde **B** with good selectivity to yield pyrones **C** and **D**. Pyrone **C** was the starting point of a synthetic route to *N*-acetylneuraminic acid developed by Danishefsky.[308]

OMe + H O Ph-Se — BF_3-OEt_2, CH_2Cl_2, 76% yield →
Me_3Si O O-Bz **A** **B**
O O Bz-O Se-Ph **C** + O O O Bz-O Se-Ph **D**
4.85 : 1
C ⟶ HO CO_2H O HO OH HO AcNH OH
N-acetylneuraminic acid

Moderate to high diastereoselectivities were observed for the cycloaddition of siloxydiene **A** (known as Brassard's diene[309,310]) with α-alkoxy aldehydes such as **B**. Results observed by Midland under $Eu(hfc)_3$ and $MgBr_2$ catalysis indicated a possible "chelation-control" pathway.[311,312]

OMe $OSiMe_3$ MeO **A** + H O OBn **B** — L. A., CH_2Cl_2 →
MeO O O OBn **C** *anti*
MeO O O OBn **D** *syn*

L. A. = $Eu(hfc)_3$; 70% yield; **C** : **D** = 9 : 1
L. A. = Et_2AlCl; 65% yield; **C** : **D**= 1 : 7.3
L. A. = $MgBr_2$; 50% yield; **C** : **D** = 9 : 1

Similar results were observed with α-dialkylamino aldehydes.[313]

OMe $OSiMe_3$ MeO + H O NBn_2 — $Eu(fod)_3$ (0.05 equiv), CH_2Cl_2, 87% yield →
MeO O O NBn_2 *anti*
MeO O O NBn_2 *syn*
anti >99 / 1 *syn*

The chelation-controlled cyclocondensation of siloxydienes with α-alkoxy aldehydes was applied to a synthesis of the mouse androgen.[314] Thus, pyrone **C**, obtained as a single diastereomer from diene **A** and chiral aldehyde **B**, is a precursor in the synthesis of the mouse androgen.

OSiMe$_3$ + H OBn; Me$_3$SiO; MgBr$_2$; THF; 76–80% yield; O; OBn; H; A; B; C; mouse androgen

The cyclocondensation of siloxydiene **A** with functionalized aldehydes also identified a new strategy for the synthesis of polypropionate segments of various natural products. Thus, pyrone **B** could be transformed into aldehyde **C**, a precursor of rifamycin S.[315]

OMe; Me$_3$SiO; A; +; H; OBn; TiCl$_4$; CH$_2$Cl$_2$; 80% yield; B; OBn; H; C; OMe; OSiMe$_3$; Cl; Ti; O; OBn; H; OMe; Me$_3$SiO; OBn; B

Reiterative diene-aldehyde cyclocondensation of **C** with siloxydiene **E** gave rise to pyrone **F** with a β-methyl at C-5.

OMe; Me$_3$SiO; E; +; H; OBn; C; BF$_3$·OEt$_2$; CH$_2$Cl$_2$; OBn; F

In contrast, cyclocondensation of the same aldehyde **C** with methyl siloxydiene **A** led to pyrone **D** bearing an α-methyl at C-5.

OMe; Me$_3$SiO; A; +; H; OBn; C; BF$_3$·OEt$_2$; CH$_2$Cl$_2$; OBn; D

The key step of the total synthesis of zincophorin was the cyclocondensation of siloxydiene **A** and aldehyde **B**. Pyrone **C** was obtained according to an exo chelation control transition state.[316,317]

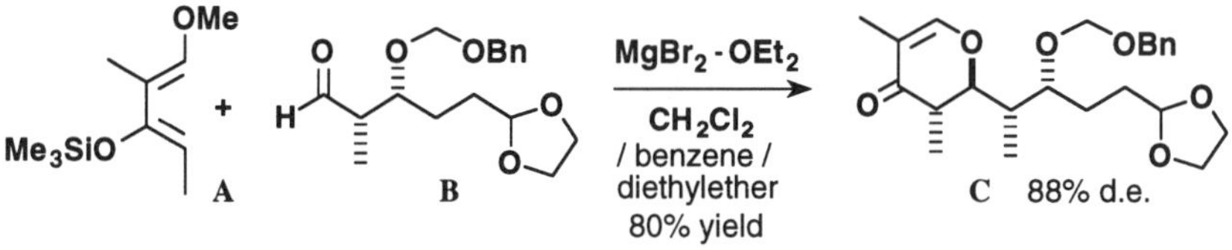

V.2 Reaction with other heterodienophiles

The cyclocondensation of activated dienes with imine dienophiles was also studied and proved to be successful. Thus the reaction between siloxydiene **A** and the parent Δ^1-pyrroline **B** provided, through the obtention of **C**, the basis for a rapid synthesis of (±)-ipalbidine.[318]

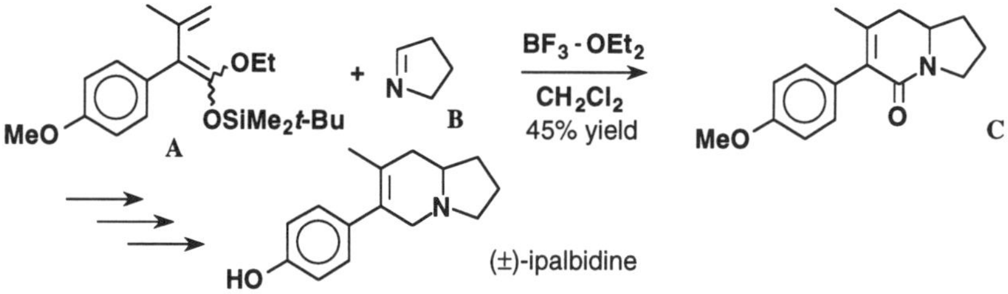

Asymmetric hetero Diels-Alder reactions were also performed from the same kind of substrates.[281] For instance, Waldmann used amino acid esters as chiral auxiliaries with Brassard's diene with success.[319]

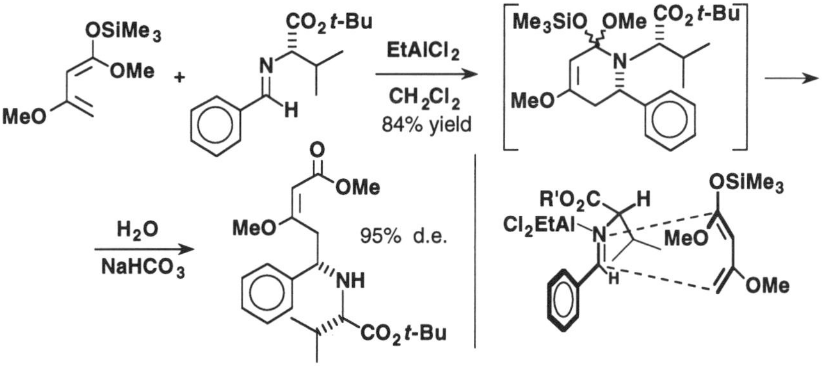

Yamamoto used Danishefsky's diene, chiral imine **A** and a large variety of achiral Lewis acids; best results were obtained with $B(OPh)_3$.[320] With (*R*)-binaphthol B(OPh), imine **D** led to the cycloadduct **E** with 85% e.e.[321] (see also ref.322).

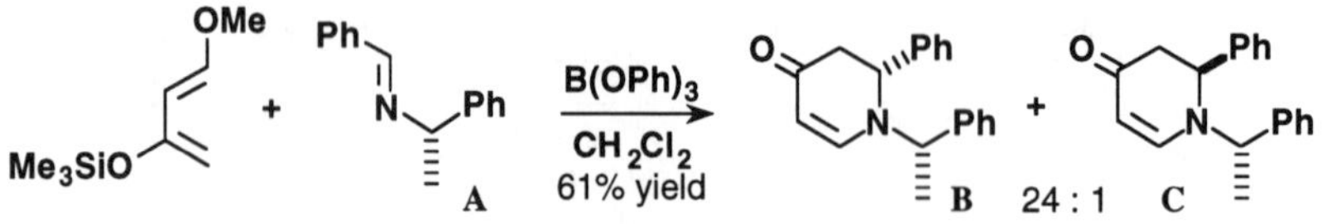

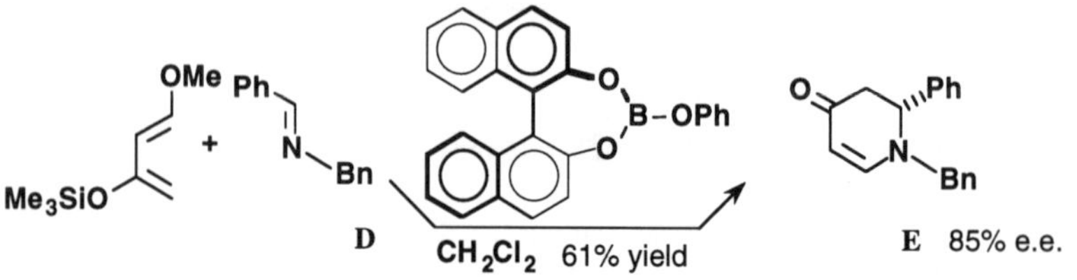

Finally, *N*-sulphinyl carbamates were also used as dienophiles by Whitesell,[323] with dimethyl butadiene, and Weinreb,[324] with cyclohexadiene.

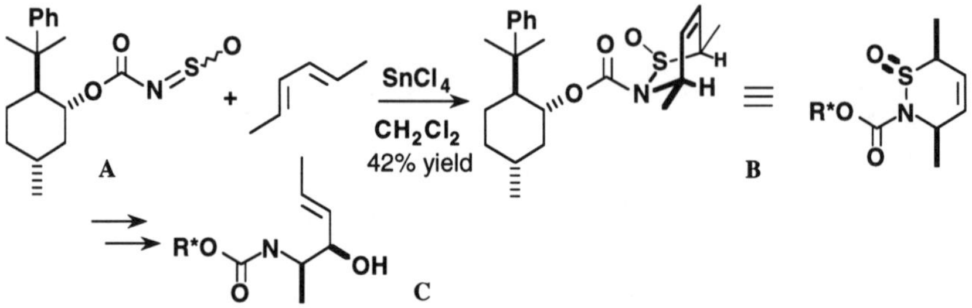

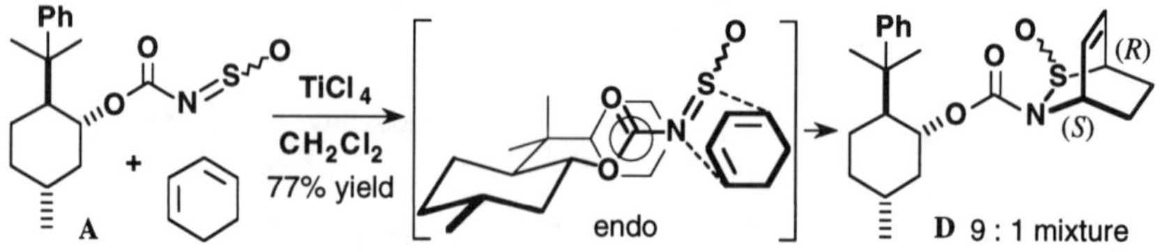

V.3 Reaction involving heterodienes

α,β-Unsaturated carbonyl compounds were found to behave as reactive heterodiene in Diels-Alder reaction with inverse electron demand.[325] Thus, Chapleur showed that, in the presence of Eu(fod)$_3$, α-enone **A** and (*E*)-1-ethoxypropene led to cycloadduct **B** with good yield.[326] Tietze described an intramolecular high pressure reaction catalyzed by a chiral Lewis acid[327] and Hiroi demonstrated a reaction involving a chiral diene bearing an optically active sulfinyl methyl group.[328]

Ghosez provided a remarkable example of stereochemical variations due to the nature of the Lewis acid involved in the reaction. Indeed, whereas, the reaction between heterodiene **A** and *N,N*-dimethyl acrylamide leads predominantly to endo adduct **B** under *t*-BuMe$_2$SiOTf catalysis, only exo adduct **C** is obtained with Eu(fod)$_3$.[329]

L. A. : **$CF_3SO_2OSiMe_2$*t*-Bu** ; 98% yield; **B** : **C** = 4.44 : 1
L. A. : **Eu(fod)$_3$** ; C : 95% yield

Recently, Laschat proposed a new intramolecular aza-Diels-Alder reaction of *N*-aryl imines, such as **A**, with unactivated olefin tethered to the 2-azadiene moiety. Octahydroacridines, such as **B**, were obtained in very good yield and, in most cases, with complete *trans* stereoselectivity.[330]

In 1986, Denmark showed that nitroalkenes could undergo Lewis acid promoted inter-[331] and intramolecular[332] [4+2] cycloaddition with olefins. Later, he also established that 2-nitrostyrenes could also be used in both of these reactions.[333]

More recently, this nitroalkene [4+2] cycloaddition was proposed as a general and stereoselective route to the synthesis of 3,3- and 3,4-disubstituted pyrrolidines such as **C**.[334]

MAD (3 equiv.), toluene, 80% yield; **A** → **B** anomer ratio 2.3 : 1

H_2, PtO_2, MeOH, 78% yield → **C**

MAD =

VI. Conclusion

The Diels-Alder reaction has been used by chemists worldwide for several decades. Yet, the introduction of Lewis acids, leading to better regio- and stereoselectivities, in the sixties and the development of its asymmetric version, using homochiral dienes or dienophiles, in the seventies and eighties has even increased its usefulness and established it as part of the standard synthetic repertoire.

A better understanding of the mechanism of the reaction and particularly of the role of the Lewis acid has recently led to the design of homochiral Lewis acids which, not only can be used in catalytic amount, but in most cases lead to virtually optically pure compounds. So far, the best results are obtained in the intermolecular all-carbon reaction but recent reports in intramolecular and heteronuclear reactions are promising and will certainly lead to exciting improvements in the near future.

References

1. Albrecht, W. *Leibigs Ann. Chem.* **1906**, *348*, 31.
2. Diels, O.; Alder, K. *Leibigs Ann. Chem.* **1928**, *460*, 98–122.
3. Kloetzel, M.C. *Org. React.* **1948**, *4*, 1–59.
4. Holmes, H.L. *Org. React.* **1948**, *4*, 60–173.
5. Wollweber, H. *Diels–Alder Reaktion*, Verlag, G. T.: Stuttgart, **1972**.
6. Oppolzer, W. *Angew. Chem., Int. Ed. Engl.* **1977**, *16*, 10–23.
7. Sauer, J.; Sustmann, R. *Angew. Chem., Int. Ed. Engl.* **1980**, *19*, 779–807.
8. Brieger, G.; Bennett, J.N. *Chem. Rev.* **1980**, *80*, 63–97.
9. Petrzilka, M.; Grayson J.I. *Synthesis*, **1981**, 753–786.
10. Ciganek, E. *Org. React.* **1984**, *32*, 1–374.
11. Fringuelli, F.; Taticchi, A. *Dienes in the Diels–Alder Reaction*, J. Wiley & Sons: New York, **1990**.
12. Carruthers, W. *Cycloaddition Reactions in Organic Synthesis*, Pergamon, Oxford, **1990**, Chap. 1, pp. 1–90 and Chap. 2, pp. 91–139.

13. Oppolzer, W. in *Comprehensive Organic Synthesis*, Trost, B.M.; Fleming, I., Eds; Pergamon, Oxford, **1991**, Vol. 5, Chap. 4.1, pp. 315–399.
14. Alder, K.; Stein, G. *Angew. Chem.* **1937**, *50*, 510–519.
15. Hoffmann, R.; Woodward, R.B. *J. Am. Chem. Soc.* **1965**, *87*, 4388–4389.
16. Fleming, I. *Frontier Orbitals and Organic Chemical Reactions*, Wiley, New York, **1976**.
17. Houk, K.N. *Acc. Chem. Res.* **1975**, *8*, 361–369.
18. Dewar, M.J.S. *The Molecular Orbital Theory of Organic Chemistry*, McGraw–Hill: New York, **1969**.
19. Fukui, K. *Theory of Orientation and Stereoselection*, Springer–Verlag: Berlin, **1975**.
20. Woodward, R.B.; Hoffmann, R. *Angew. Chem., Int. Ed. Engl.* **1969**, *8*, 781–853.
21. Fox, M.A.; Cardona, R.; Kiwiet, N.J. *J. Org. Chem.* **1987**, *52*, 1469–1474.
22. Fukui, K. *Acc. Chem. Res.* **1971**, *4*, 57–64.
23. Houk, K.N. *J. Am. Chem. Soc.* **1973**, *95*, 4092–4094.
24. McIver, Jr., J.W. *Acc. Chem. Res.* **1974**, *7*, 72–77.
25. Burke, L.A.; Leroy, G.; Sana, M. *Theor. Chim. Acta* **1975**, *40*, 313–321.
26. Townshend, R.E.; Ramunni, G.; Segal, G.; Hehre, W.J.; Salem, L. *J. Am. Chem. Soc.* **1976**, *98*, 2190–2198.
27. Basilevsky, M.V.; Shamov, A.G.; Tikhomirov, V.A. *J. Am. Chem. Soc.* **1977**, *99*, 1369–1372.
28. Eisenstein, O.; Lefour, J.M.; Anh, N.T.; Hudson, R.F. *Tetrahedron* **1977**, *33*, 523–531.
29. Burke, L.A.; Leroy, G. *Theor. Chim. Acta* **1977**, *44*, 219–221.
30. Fleming, I.; Michael, J.P.; Overman, L.E.; Taylor, G.F. *Tetrahedron Lett.* **1978**, 1313–1314.
31. Dewar, M.J.S.; Olivella, S.; Rzepa, H.S. *J. Am. Chem. Soc.* **1978**, *100*, 5650–5659.
32. Pancir, J. *J. Am. Chem. Soc.* **1982**, *104*, 7424–7430.
33. Ortega, M.; Oliva, A.; Lluch, J.M.; Bertran, J. *Chem. Phys. Lett.* **1983**, *102*, 317.
34. Brown, F.K.; Houk, K.N. *Tetrahedron Lett.* **1984**, *25*, 4609–4612.
35. Brown, F.K.; Houk, K.N. *Tetrahedron Lett.* **1985**, *26*, 229–230.
36. Burke, L.A. *Theor. Chim. Acta* **1985**, *68*, 101–105.
37. Bernardi, F.; Bottini, A.; Robb, M.A.; Field, M.J.; Hillier, I.H.; Guest, M.F. *J. Chem. Soc., Chem. Commun.* **1985**, 1051–1052.
38. Houk, K.N.; Lin, Y.–T.; Brown, F.K. *J. Am. Chem. Soc.* **1986**, *108*, 554–556.
39. Kahn, S.D.; Pau, C.F.; Overman, L.E.; Hehre, W.J. *J. Am. Chem. Soc.* **1986**, *108*, 7381–7396.
40. Branchadell, V.; Oliva, A.; Bertran, J. *J. Mol. Struct. (Theochem)*, **1986**, *136*, 25–33.
41. Dewar, M.J.S.; Olivella, S.; Stewart, J.J.P. *J. Am. Chem. Soc.* **1986**, *108*, 5771–5779.
42. Guner, O.F.; Ottenbrite, R.M.; Shillady, D.D.; Alston, P.V. *J. Org. Chem.* **1987**, *52*, 391–394.
43. Kahn, S.D.; Hehre, W.J. *J. Am. Chem. Soc.* **1987**, *109*, 663–666.
44. Gajewski, J.J.; Peterson, K.B.; Kagel, J.R. *J. Am. Chem. Soc.* **1987**, *109*, 5545–5546.
45. Hancock, R.A.; Wood, Jr., B.F. *J. Chem. Soc., Chem. Commun.* **1988**, 351–353.

46. Bernardi, F.; Bottoni, A.; Field, M.J.; Guest, M.F.; Hillier, I.H.; Robb, M.A.; Venturini, A. *J. Am. Chem. Soc.* **1988**, *110*, 3050–3055.
47. Sustmann, R.; Daute, P.; Sauer, R.; Sicking, W. *Tetrahedron Lett.* **1988**, *29*, 4699–4702.
48. Bach, R.D.; McDouall, J.J.W.; Schlegel, H.B.; Wolber, G.J. *J. Org. Chem.* **1989**, *54*, 2931–2935.
49. Loncharich, R.J.; Brown, F.K.; Houk, K.N. *J. Org. Chem.* **1989**, *54*, 1129–1134.
50. Houk, K.N.; Loncharich, R.J.; Blake, J.F.; Jorgensen, W.L. *J. Am. Chem. Soc.* **1989**, *111*, 9172–9176.
51. Birney, D.M.; Houk, K.N. *J. Am. Chem. Soc.* **1990**, *112*, 4127–4133.
52. Ishida, M.; Beniya, Y.; Inagaki, S.; Kato, S. *J. Am. Chem. Soc.* **1990**, *112*, 8980–8982.
53. Branchadell, V.; Sodupe, M.; Ortuno, R.M.; Oliva, A.; Gomez–Pardo, D.; Guingant, A.; d'Angelo, J. *J. Org. Chem.* **1991**, *56*, 4135–4141.
54. McCarrick, M.A.; Wu, Y.–D.; Houk, K.N. *J. Am. Chem. Soc.* **1992**, *114*, 1499–1500.
55. Jorgensen, W.L.; Lim, D.; Blake, J.F. *J. Am. Chem. Soc.* **1993**, *115*, 2936–2942.
56. Brown, F.K.; Singh, U.C.; Kollman, P.A.; Raimondi, L.; Houk, K.N.; Bock, C.W. *J. Org. Chem.* **1992**, *57*, 4862–4869.
57. Yamabe, S.; Kawajiri, S.; Minato, T.; Machiguchi, T. *J. Org. Chem.* **1993**, *58*, 1122–1127.
58. McCarrick, M.A.; Wu, Y.–D.; Houk, K.N. *J. Org. Chem.* **1993**, *58*, 3330–3343.
59. Ruiz–Lopez, M.F.; Assfeld, X.; Garcia, J.I.; Mayoral, J.A.; Salvatella, L. *J. Am. Chem. Soc.* **1993**, *115*, 8780–8787.
60. Wassermann, A. *J. Chem. Soc.* **1942**, 618–621.
61. Rubin, W.; Steiner, H.; Wassermann, A. *J. Chem. Soc.* **1949**, 3046–3057.
62. Yates, P.; Eaton, P. *J. Am. Chem. Soc.* **1960**, *82*, 4436–4437.
63. Sauer, J.; Kredel, J. *Tetrahedron Lett.* **1966**, 731–736.
64. Inukai, T.; Kojima, T. *J. Org. Chem.* **1966**, *31*, 1121–1123.
65. Inukai, T.; Kojima, T. *J. Org. Chem.* **1966**, *31*, 2032–2033.
66. Williamson, K.L.; Li Hsu, Y.–F. *J. Am. Chem. Soc.* **1970**, *92*, 7385–7389.
67. Inukai, T.; Kojima, T. *J. Org. Chem.* **1971**, *36*, 924–928.
68. Kakushima, M.; Espinosa, J.; Valenta, Z. *Can. J. Chem.* **1976**, *54*, 3304–3306.
69. Lindsay Smith, J.R.; Norman, R.O.C.; Stillings, M.R. *Tetrahedron*, **1978**, *34*, 1381–1383.
70. Cohen, T.; Kosarych, Z. *J. Org. Chem.* **1982**, *47*, 4005–4008.
71. Trost, B.M.; Ippen, J.; Vladuchick, W.C. *J. Am. Chem. Soc.* **1977**, *99*, 8116–8118.
72. Lutz, E.F.; Bailey, G.M. *J. Am. Chem. Soc.* **1964**, *86*, 3899–3901.
73. Anh, N.T.; Seyden–Penne, J. *Tetrahedron*, **1973**, *29*, 3259–3265.
74. Houk, K.N.; Strozier, R.W. *J. Am. Chem. Soc.* **1973**, *95*, 4094–4096.
75. Alston, P.V.; Ottenbrite, R.M. *J. Org. Chem.* **1975**, *40*, 1111–1116.
76. Imamura, A.; Hirano, T. *J. Am. Chem. Soc.* **1975**, *97*, 4192–4198.
77. Epiotis, N.D.; Shaik, S. *J. Am. Chem. Soc.* **1978**, *100*, 1–17.
78. Branchadell, V.; Oliva, A.; Bertran, J. *Chem. Phys. Lett.* **1983**, *97*, 378–380.
79. Branchadell, V.; Oliva, A.; Bertran, J. *Chem. Phys. Lett.* **1985**, *113*, 197–201.

80. Burnell, D.J.; Goodbrand, H.B.; Kaiser, S.M.; Valenta, Z. *Can. J. Chem.* **1984**, *62*, 2398–2400.
81. Brown, F.K.; Houk, K.N.; Burnell, D.J.; Valenta, Z. *J. Org. Chem.* **1987**, *52*, 3050–3059.
82. Jenner, G. *High Pressure Research* **1993**, *11*, 257–262.
83. Reetz, M.T.; Gansäuer, A. *Tetrahedron* **1993**, *49*, 6025–6030.
84. Braun, R.; Sauer, J. *Chem. Ber.* **1986**, *119*, 1269–1274.
85. Bonnensen, P.V.; Puckett, C.L.; Honeychuck, R.V.; Hersh, W.H. *J. Am. Chem. Soc.* **1989**, *111*, 6070–6081.
86. Faller, J.W.; Smart, C.J. *Tetrahedron Lett.* **1989**, *30*, 1189–1192.
87. Kelly, T.R.; Maity, S.K.; Meghani, P.; Chandrakumar, N.S. *Tetrahedron Lett.* **1989**, *30*, 1357–1360.
88. Saha, A.K.; Hossain, M.M. *Tetrahedron Lett.* **1993**, *34*, 3833–3836.
89. Eklund, L.; Axelsson, A.K.; Nordahl, A.; Carlson, R. *Acta Chem. Scand.* **1993**, *47*, 581–591.
90. Pagni, R.M.; Kabalka, G.W.; Hondrogiannis, G.; Bains, S.; Anosike, P.; Kurt, R. *Tetrahedron* **1993**, *49*, 6743–6756.
91. Cativiela, C.; Fraile, J.M.; Garcia, J.I.; Mayoral, J.A.; Pires, E.; Royo, A.J.; Figueras, F.; de Ménorval, L.C. *Tetrahedron* **1993**, *49*, 4073–4084.
92. Kobayashi, S.; Hachiya, I.; Araki, M.; Ishitani, H. *Tetrahedron Lett.* **1993**, *34*, 3755–3758.
93. Veselovsky, V.V.; Gybin, A.S.; Lozanova, A.V.; Moiseenkov, A.M.; Smit, W.A.; Caple, R. *Tetrahedron Lett.* **1988**, *29*, 175–178.
94. Angell, E.C.; Fringuelli, F.; Pizzo, F.; Minuti, L.; Taticchi, A.; Wenkert, E. *J. Org. Chem.* **1989**, *54*, 1217–1218.
95. Jeroncic, L.O.; Cabal, M.–P.; Danishefsky, S.J.; Shulte, G.M. *J. Org. Chem.* **1991**, *56*, 387–395.
96. Oppolzer, W.; Petrzilka, M. *J. Am. Chem. Soc.* **1976**, *98*, 6722–6723.
97. Fringuelli, F.; Pizzo, F.; Taticchi, A.; Halls, T.D.J.; Wenkert, E. *J. Org. Chem.* **1982**, *47*, 5056–5065.
98. Nagakura, I.; Ogata, H.; Ueno, M.; Kitahara, Y. *Bull. Chem. Soc. Jpn* **1975**, *48*, 2995–2996.
99. Fringuelli, F.; Pizzo, F.; Taticchi, A.; Wenkert, E. *J. Org. Chem.* **1983**, *48*, 2802–2808.
100. Angell, E.C.; Fringuelli, F.; Pizzo, F.; Porter, B.; Taticchi, A.; Wenkert, E. *J. Org. Chem.* **1986**, *51*, 2642–2649.
101. Angell, E.C.; Fringuelli, F.; Halls, T.D.J.; Pizzo, F.; Porter, B.; Taticchi, A.; Touris, A.P.; Wenkert, E. *J. Org. Chem.* **1985**, *50*, 4691–4696.
102. Angell, E.C.; Fringuelli, F.; Minuti, L.; Pizzo, F.; Porter, B.; Taticchi, A.; Wenkert, E. *J. Org. Chem.* **1986**, *51*, 2649–2652.
103. Angell, E.C.; Fringuelli, F.; Minuti, L.; Pizzo, F.; Porter, B.; Taticchi, A.; Wenkert, E. *J. Org. Chem.* **1985**, *50*, 4686–4690.
104. Angell, E.C.; Fringuelli, F.; Pizzo, F.; Taticchi, A.; Wenkert, E. *J. Org. Chem.* **1988**, *53*, 1424–1426.
105. Kakushima, M.; Espinosa, J.; Valenta, Z. *Can. J. Chem.* **1976**, *54*, 3304–3306.
106. Tou, J.S.; Reusch, W. *J. Org. Chem.* **1980**, *45*, 5012–5014.
107. Aso, M.; Ojida, A.; Yang, G.; Cha, O.–J.; Osawa, E.; Kanematsu, K. *J. Org. Chem.* **1993**, *58*, 3960–3968.
108. Aso, M.; Ojida, A.; Yang, G.; Kanematsu, K. *Heterocycles* **1993**, *35*, 33–36.
109. Liu, H.–J.; Han, Y. *Tetrahedron Lett.* **1993**, *34*, 423–426.

110. Minuti, L.; Scheeren, H.W.; Selvaggi, R.; Taticchi, A. *Synth. Commun.* **1992**, *22*, 2965–2969.
111. Aguilar, R.; Reyes, A.; Tamariz, J.; Birbaum, J.L. *Tetrahedron Lett.* **1987**, *28*, 865–868.
112. Hauser, F.M.; Prasanna, S. *Tetrahedron* **1984**, *40*, 4711–4717.
113. Fringuelli, F.; Pizzo, F.; Taticchi, A.; Wenkert, E. *Synth. Commun.* **1986**, *16*, 245–249.
114. Angell, E.C.; Fringuelli, F.; Minuti, L.; Pizzo, F.; Taticchi, A.; Wenkert, E. *J. Org. Chem.* **1986**, *51*, 5177–5182.
115. Liu, H.–J.; Browne, E.N.C. *Can. J. Chem.* **1987**, *65*, 1262–1278.
116. Fringuelli, F.; Pizzo, F.; Taticchi, A.; Ferreira, V.F.; Michelotti, E.L.; Porter, B.; Wenkert, E. *J. Org. Chem.* **1985**, *50*, 890–891.
117. Angell, E.C.; Fringuelli, F.; Pizzo, F.; Porter, B.; Taticchi, A.; Wenkert, E. *J. Org. Chem.* **1985**, *50*, 4696–4698.
118. Sakurai, K.; Kitahara, T.; Mori, K. *Tetrahedron* **1988**, *44*, 6581–6588.
119. Hendrickson, J.B.; Haestier, A.M.; Stieglitz, S.G.; Foxman, B.M. *New J. Chem.* **1990**, *14*, 689–693.
120. Gupta, R.C.; Jackson, D.A.; Stoodley, R.J. *J. Chem. Soc., Chem. Commun.* **1982**, 929–930.
121. McCurry Jr., P.M.; Singh, R.K. *J. Chem. Soc., Chem. Commun.* **1976**, 59–60.
122. Baldwin, J.E.; Lusch, M.J. *J. Org. Chem.* **1979**, *44*, 1923–1927.
123. Fleming, I.; Percival, A. *J. Chem. Soc., Chem. Commun.* **1976**, 681.
124. Hosomi, A.; Iguchi, H.; Sasaki, J.–i. Sakurai, H. *Tetrahedron Lett.* **1982**, *23*, 551–554.
125. Hosomi, A.; Sakata, Y.; Sakurai, H. *Tetrahedron Lett.* **1985**, *26*, 5175–5178.
126. Hashimoto, Y.; Nagashima, T.; Kobayashi, K.; Hasegawa, M.; Saigo, K. *Tetrahedron* **1993**, *49*, 6349–6358.
127. Liu, H.–J.; Ngoi, T.K. *Synth. Commun.* **1982**, *12*, 715–722.
128. Liu, H.–J.; Feng, W.M. *Synth. Commun.* **1986**, *16*, 1485–1492.
129. Liu, H.–J.; Feng, W.M. *Synth. Commun.* **1987**, *17*, 1777–1786.
130. Grieco, P.A.; Vidari, G.; Ferrino, S.; Haltiwanger, R.C. *Tetrahedron Lett.* **1980**, *21*, 1619–1622.
131. Vidari, G.; Ferrino, S.; Grieco, P.A. *J. Am. Chem. Soc.* **1984**, *106*, 3539–3548.
132. Engler, T.A.; Sampath, U.; Naganathan, S.; Vander Velde, D.; Takusagawa, F.; Yohannes, D. *J. Org. Chem.* **1989**, *54*, 5712–5727.
133. Minuti, L.; Radics, L.; Taticchi, A.; Venturini, L.; Wenkert, E. *J. Org. Chem.* **1990**, *55*, 4261–4265.
134. Trost, B.M.; Caldwell, C.G.; Murayama, E.; Heissler, D. *J. Org. Chem.* **1983**, *48*, 3252–3265.
135. Devine, P.N.; Oh, T. *J. Org. Chem.* **1991**, *56*, 1955–1958.
136. Maruoka, K.; Nonoshita, K.; Yamamoto, H. *Synth. Commun.* **1988**, *18*, 1453–1459.
137. Bachner, J.; Huber, U.; Buchbauer, G. *Monatsh. Chem.* **1981**, *112*, 679–682.
138. Corcoran, R.C.; Ma, J. *J. Am. Chem. Soc.* **1991**, *113*, 8973–8975.
139. Maffei, M.; Buono, G. *New J. Chem.* **1988**, *12*, 923–930.
140. Angell, E.C.; Fringuelli, F.; Guo, M.; Minuti, L.; Taticchi, A.; Wenkert, E. *J. Org. Chem.* **1988**, *53*, 4325–4328.

141. Dauben, W.G.; Kowalczyk, B.A.; Lichtenthaler, F.W. *J. Org. Chem.* **1990**, *55*, 2391–2398.
142. Braddock, D.C.; Brown, J.M.; Guiry, P.J. *J. Chem. Soc., Chem. Commun.* **1993**, 1244–1246.
143. Roush, W.R.; D'Ambra, T.E. *J. Org. Chem.* **1981**, *46*, 5045–5047.
144. Mayr, H.; Halberstadt-Kausch, I.K. *Chem. Ber.* **1982**, *115*, 3479–3515
145. Ismail, Z.M.; Hoffmann, H.M.R. *J. Org. Chem.* **1981**, *46*, 3549–3550.
146. Gandhi, R.P.; Ishar, M.P.S.; Wali, A. *J. Chem. Soc., Chem. Commun.* **1988**, 1074–1075.
147. Conrads, M.; Mattay, J. *Chem. Ber.* **1991**, *124*, 867–873.
148. Gras, J.L. *J. Chem. Res. (S),* **1982**, 300–301, *(M)*, **1982**, 3032–3047.
149. Gras, J.L.; Guerin, A. *Tetrahedron Lett.* **1985**, *26*, 1781–1784.
150. Jung. M.E.; Zimmerman, C.N.; Lowen, G.T.; Khan, S.I. *Tetrahedron Lett.* **1993**, *34*, 4453–4456.
151. Taber, D.F. *Intramolecular Diels–Alder and Alder Ene Reactions*, Springer–Verlag, Berlin, **1984**.
152. Carruthers, W. *Cycloaddition Reactions in Organic Synthesis*, Pergamon, Oxford, **1990**, Chap. 3, pp. 140-208.
153. Roush, W.R. in *Comprehensive Organic Synthesis*, Trost, B.M.; Fleming, I., Eds; Pergamon, Oxford, **1991**, Vol. 5, Chap. 4.4, pp. 513–550.
154. Roush, W.R.; Gillis, H.R. *J. Org. Chem.* **1980**, *45*, 4267–4268.
155. Roush, W.R.; Gillis, H.R.; Ko, A.I. *J. Am. Chem. Soc.* **1982**, *104*, 2269–2283.
156. Roush, W.R.; Gillis, H.R. *J. Org. Chem.* **1982**, *47*, 4825–4829.
157. Haufe, R.; Jansen, M.; Tobias, K.M.; Winterfeldt, E.; Wray, V. *Chem Ber.* **1987**, *120*, 2007–2013.
158. Marshall, J.A.; Audia, J.E.; Grote, J. *J. Org. Chem.* **1984**, *49*, 5277–5279.
159. Marshall, J.A.; Shearer, B.G.; Crooks, S.L. *J. Org. Chem.* **1987**, *52*, 1236–1245.
160. Marshall, J.A.; Audia, J.E.; Grote, J.; Shearer, B.G. *Tetrahedron* **1986**, *42*, 2893–2902.
161. Marshall, J.A.; Audia, J.E.; Grote, J. *J. Org. Chem.* **1986**, *51*, 1155–1157.
162. Marshall, J.A.; Audia, J.E.; Shearer, B.G. *J. Org. Chem.* **1986**, *51*, 1730–1735.
163. Reich, H.J.; Eisenhart, E.K. *J. Org. Chem.* **1984**, *49*, 5282–5283.
164. Keay, B.A. *J. Chem. Soc., Chem. Commun.* **1987**, 419–421.
165. Rogers, C.; Keay, B.A. *Synlett* **1991**, 353–355.
166. Rogers, C.; Keay, B.A. *Can. J. Chem.* **1992**, *70*, 2929–2947.
167. Rogers, C.; Keay, B.A. *Can. J. Chem.* **1993**, *71*, 611–622.
168. Smith, D.A.; Sakan, K.; Houk, K.N. *Tetrahedron Lett.* **1986**, *27*, 4877–4880.
169. Sakan, K.; Craven, B.M. *J. Am. Chem. Soc.* **1983**, *105*, 3732–3734.
170. Sakan, K.; Smith, D.A.; Babirad, S.A.; Fronczek, F.R.; Houk, K.N. *J. Org. Chem.* **1991**, *56*, 2311–2317.
171. Brown, P.A.; Jenkins, P.R.; Fawcett, J.; Russell, D.R. *J. Chem. Soc., Chem. Commun.* **1984**, 253–255.
172. Brown, P.A.; Jenkins, P.R. *J. Chem. Soc., Perkin Trans. 1* **1986**, 1303–1309.
173. Shea, K.J.; Gilman, J.W. *Tetrahedron Lett.* **1983**, *24*, 657–660.
174. Shea, K.J.; Gilman, J.W. *J. Am. Chem. Soc.* **1985**, *107*, 4791–4792.
175. Helmchen, G.; Karge, R.; Weetman, J. in *Modern Synthetic Methods*, Scheffold, R.; Ed.; Springer–Verlag, Berlin–Heidelberg, **1986**, Vol. 4, pp. 262–306.

176. Krohn, K. *Nachr. Chem. Tech. Lab.* **1987**, *35*, 836–841.
177. Narasaka, K. *Synthesis* **1991**, 1–11.
178. Maruoka, K.; Yamamoto, H. in *Catalytic Asymmetric Synthesis*, Ojima, I., Ed.; VCH Verlag, Weinheim, **1993**, Chap. 9, pp. 413–440.
179. Krohn, K. in *Organic Synthesis Highlights*, VCH Publishers, Weinheim, New York, **1991**, pp. 54–65.
180. Oppolzer, W. *Tetrahedron* **1987**, *43*, 1969–2004.
181. Oppolzer, W. *Angew. Chem., Int. Ed. Engl.* **1984**, *23*, 876–889.
182. Kagan, H.B.; Riant, O. *Chem. Rev.* **1992**, *92*, 1007–1019.
183. Wurzinger, H. *Kontakte* **1984**, *2*, 3–19.
184. Oh, T.; Reilly, M. *Org. Prep. Proc. Int.* **1994**, *26*, 129–158.
185. Korolev, A.; Mur, V. *Dokl. Akad. Nauk S.S.S.R.* **1949**, *59*, 251–253.
186. Walborsky, H.M.; Barash, L.; Davis, T.C. *J. Org. Chem.* **1961**, *26*, 4778–4779.
187. Walborsky, H.M.; Barash, L.; Davis, T.C. *Tetrahedron* **1963**, *19*, 2333–2351.
188. Guseinov, M.M.; Akhmedov, I.M.; Mamedov, E.C. *Azerb. Khim. Zhur.* **1976**, 46. *Chem. Abs.* **1976**, *85*, 176925z.
189. Hashimoto, S.–i.; Komeshima, N.; Koga, K. *J. Chem. Soc., Chem. Commun.* **1979**, 437–438.
190. Farmer, R.F.; Hamer, J. *J. Org. Chem.* **1966**, *31*, 2418–2419.
191. Sauer, J.; Kredel, J. *Tetrahedron Lett.* **1966**, 731–736.
192. Corey, E.J.; Ensley, H.E. *J. Am. Chem. Soc.* **1975**, *97*, 6908–6909.
193. Oppolzer, W.; Kurth, M.; Reichlin, D.; Chapuis, C.; Mohnhaupt, M.; Moffatt, F. *Helv. Chim. Acta* **1981**, *64*, 2802–2807.
194. Oppolzer, W.; Kurth, M.; Reichlin, D.; Moffatt, F. *Tetrahedron Lett.* **1981**, *22*, 2545–2548.
195. Hamon, D.P.G.; Holman, J.W.; Massy–Westropp, R.A. *Tetrahedron* **1993**, *49*, 9593–9604.
196. Oppolzer, W.; Chapuis, C.; Dao, G.M.; Reichlin, D.; Godel, T. *Tetrahedron Lett.* **1982**, *23*, 4781–4784.
197. Oppolzer, W.; Chapuis, C. *Tetrahedron Lett.* **1984**, *25*, 5383–5386.
198. Oppolzer, W.; Chapuis, C.; Kelly, M.J. *Helv. Chim. Acta* **1983**, *66*, 2358–2361.
199. Oppolzer, W.; Chapuis, C.; Bernardinelli, G. *Tetrahedron Lett.* **1984**, *25*, 5885–5888.
200. Bauer, T.; Chapuis, C.; Kozak, J.; Jurczak, J. *Helv. Chim. Acta* **1989**, *72*, 482–486.
201. Oppolzer, W.; Kelly, M.J.; Bernardinelli, G. *Tetrahedron Lett.* **1984**, *25*, 5889–5892.
202. Poll, T.; Helmchen, G.; Bauer, B. *Tetrahedron Lett.* **1984**, *25*, 2191–2194.
203. Poll, T.; Metter, J.O.; Helmchen, G. *Angew. Chem., Int. Ed. Engl.* **1985**, *24*, 112–114.
204. Poll, T.; Sobczak, A.; Hartmann, H.; Helmchen, G. *Tetrahedron Lett.* **1985**, *26*, 3095–3098.
205. Maruoka, K.; Oishi, M.; Yamamoto, H. *Synlett* **1993**, 683–685.
206. Gras, J.–L.; Pellissier, H. *Tetrahedron Lett.* **1991**, *32*, 7043–7046.
207. Gras, J.–L.; Poncet, A.; Nouguier, R. *Tetrahedron Lett.* **1992**, *33*, 3323–3326.
208. Nouguier, R.; Gras, J.–L.; Giraud, B.; Virgili, A. *Tetrahedron* **1992**, *48*, 6245–6252.

209. Suzuki, I.; Kin, H.; Yamamoto, H. *J. Am. Chem. Soc.* **1993**, *115*, 10139–10146.
210. Maruoka, K.; Shioara, K.; Oishi, M.; Saito, S.; Yamamoto, H. *Synlett* **1993**, 421–422.
211. Bueno, M.P.; Cativiela, C.; Mayoral, J.A.; Avenoza, A.; Charro, P.; Roy, M.A.; Andres, J.M. *Can. J. Chem.* **1988**, *66*, 2826–2829.
212. Bueno, M.P.; Cativiela, C.A.; Mayoral, J.A.; Avenoza, A. *J. Org. Chem.*, **1991**, *56*, 6551–6555.
213. Oppolzer, W.; Chapuis, C.; Bernardinelli, G. *Helv. Chim. Acta* **1984**, *67*, 1397–1401.
214. Oppolzer, W.; Rodriguez, I.; Blagg, J.; Bernardinelli, G. *Helv. Chim. Acta* **1989**, *72*, 123–130.
215. Vandewalle, M.; Van der Eycken, J.; Oppolzer, W.; Vullioud, C. *Tetrahedron* **1986**, *42*, 4035–4043.
216. Pindur, U.; Lutz, G.; Fischer, G.; Schollmeyer, D.; Massa, W.; Schröder, L. *Tetrahedron* **1993**, *49*, 2863–2872.
217. Oppolzer, W.; Dupuis, D.; Poli, G.; Raynham, T.M.; Bernardinelli, G. *Tetrahedron Lett.* **1988**, *46*, 5885–5888.
218. Thom, C.; Kocienski, P.; Jarowicki, K. *Synthesis* **1993**, 475–477.
219. Evans, D.A.; Chapman, K.T.; Bisaha, J. *J. Am. Chem. Soc.* **1988**, *110*, 1238–1256.
220. Tanaka, K.; Uno, H.; Osuga, H.; Suzuki, H. *Tetrahedron: Asymmetry* **1993**, *4*, 629–632.
221. Cativiela, C.; Figueras, F.; Garcia, J.I.; Mayoral, J.A.; Pires, E.; Royo, A.J. *Tetrahedron: Asymmetry* **1993**, *4*, 621–624.
222. Cativiela, C.; Figueras, F.; Fraile, J.M.; Garcia, J.I.; Mayoral, J.A. *Tetrahedron: Asymmetry* **1993**, *4*, 223–228.
223. Furuta, K.; Iwanaga, K.; Yamamoto, H. *Tetrahedron Lett.* **1986**, *27*, 4507–4510.
224. Hartmann, H.; Hady, A.F.A.; Sartor, K.; Weetman, J.; Helmchen, G. *Angew. Chem., Int. Ed. Engl.* **1987**, *26*, 1143–1145.
225. Charlton, J.L.; Maddaford, S.; Koh, K.; Boulet, S.; Saunders, M.H. *Tetrahedron: Asymmetry* **1993**, *4*, 645–648.
226. Helmchen, G.; Schmierer, R. *Angew. Chem., Int. Ed. Engl.* **1981**, *20*, 205–207.
227. Furuta, K.; Hayashi, S.; Miwa, Y.; Yamamoto, H. *Tetrahedron Lett.* **1987**, *28*, 5841–5844.
228. Hanzawa, Y.; Suzuki, M.; Kobayashi, Y.; Taguchi, T.; Iitaka, Y. *J. Org. Chem.* **1991**, *56*, 1718–1725.
229. Cativiela, C.; Lopez, P.; Mayoral, J.A. *Tetrahedron: Asymmetry* **1991**, 2, 449–456.
230. Avenoza, A.; Cativieda, C.; Mayoral, J.A.; Peregrina, J.M.; Sinou, D. *Tetrahedron: Asymm.* **1990**, *1*, 765–768.
231. Cativiela, C.; Mayoral, J.A.; Avenoza, A.; Peregrina, J.M.; Lahoz, F.J.; Gimeno, S. *J. Org. Chem.* **1992**, *57*, 4664–4669.
232. Arai, Y.; Hayashi, Y.; Yamamoto, M.; Takayama, H.; Koizumi, T. *Chem. Lett.* **1987**, 185–186.
233. Arai, Y.; Takadoi, M.; Koizumi, T. *Chem. Pharm. Bull.* **1988**, *36*, 4162–4166.
234. Lopez, R.; Carretero, J.C. *Tetrahedron: Asymmetry* **1991**, 2, 93–96.
235. Ronan, B.; Kagan, H.B. *Tetrahedron: Asymmetry* **1991**, 2, 75–90.
236. Katagiri, N.; Haneda, T.; Hayasaka, E.; Watanabe, N.; Kaneko, C. *J. Org. Chem.* **1988**, *53*, 226–227.

237. Katagiri, N.; Watanabe, N.; Kaneko, C. *Chem. Pharm. Bull.* **1990**, *38*, 69–72.
238. Oppolzer, W; Chapuis, C. *Tetrahedron Lett.* **1983**, *24*, 4665–4668.
239. Oppolzer, W.; Chapuis, C.; Dupuis, D.; Guo, M. *Helv. Chim. Acta* **1985**, *68*, 2100–2114.
240. Masamune, S.; Reed, III, L.A.; Davis, J.T.; Choy, N. *J. Org. Chem.* **1983**, *48*, 4441–4444.
241. Trost, B.M.; Godleski, S.A.; Genêt, J.–P. *J. Am. Chem. Soc.* **1978**, *100*, 3930–3931.
242. Trost, B.M.; O'Krongly, D.; Belletire, J.L. *J. Am. Chem. Soc.* **1980**, *102*, 7595–7596.
243. Bloch, R.; Chaptal–Gradoz, N. *Tetrahedron Lett.* **1992**, *33*, 6147–6150.
244. Beagley, B.; Larsen, D.S.; Pritchard, R.G.; Stoodley, R.J. *J. Chem. Soc., Perkin Trans. 1* **1990**, 3113–3126.
245. Grieco, P.A.; Beck, J.P. *Tetrahedron Lett.* **1993**, *34*, 7367–7370.
246. Takemura, H.; Komeshima, N.; Takahashi, I.; Hashimoto, S.–i.; Ikota, N.; Tomioka, K.; Koga, K. *Tetrahedron Lett.* **1987**, *28*, 5687–5690.
247. Seebach, D.; Beck, A.K.; Imwinkelried, R.; Roggo, S.; Wonnacott, A. *Helv. Chim. Acta* **1987**, *70*, 954–974.
248. Devine, P.N.; Oh, T. *J. Org. Chem.* **1992**, *57*, 396–399.
249. Narasaka, K.; Inoue, M.; Okada, N. *Chem. Lett.* **1986**, 1109–1112.
250. Narasaka, K.; Inoue, M.; Yamada, T. *Chem. Lett.* **1986**, 1967–1968.
251. Narasaka, K.; Inoue, M.; Yamada, T.; Sugimori, J.; Iwasawa, N. *Chem. Lett.* **1987**, 2409–2412.
252. Narasaka, K.; Tanaka, H.; Kanai, F. *Bull. Chem. Soc. Jpn* **1991**, *64*, 387–391.
253. Corey, E.J.; Matsumara, Y. *Tetrahedron Lett.* **1991**, *32*, 6289–6292.
254. Khiar, N.; Fernandez, I.; Alaudia, F. *Tetrahedron Lett.* **1993**, *34*, 123–126.
255. Iwasawa, N.; Sugimori, J.; Kawase, Y.; Narasaka, K. *Chem. Lett.* **1989**, 1947–1950.
256. Narasaka, K.; Iwasawa, N.; Inoue, M.; Yamada, T.; Nakashima, M.; Sugimori, J. *J. Am. Chem. Soc.* **1989**, *111*, 5340–5345.
257. Narasaka, K.; Yamamoto, I. *Tetrahedron* **1992**, *48*, 5743–5754.
258. Kelly, T.R.; Whiting, A.; Chandrakumar, N.S. *J. Am. Chem. Soc.* **1986**, *108*, 3510–3512.
259. Maruoka, K.; Sakurai, M.; Fijiwara, J.; Yamamoto, H. *Tetrahedron Lett.* **1986**, *27*, 4895–4898.
260. Ketter, A.; Glahsl, G.; Herrmann, R. *J. Chem. Res. (S)* **1990**, 278–279.
261. Ketter, A.; Herrmann, R. *Z. Naturforsch.* **1990**, *45b*, 1684–1688.
262. Sartor, D.; Saffrich, J.; Helmchen, G.; Richards, C.J.; Lambert, H. *Tetrahedron: Asymmetry* **1991**, *2*, 639–642.
263. Sartor, D.; Saffrich, J.; Helmchen, G. *Synlett* **1990**, 197–198.
264. Takasu, M.; Yamamoto, H. *Synlett* **1990**, 194–196.
265. Evans, D.A.; Lectka, T.; Miller, S.J. *Tetrahedron Lett.* **1993**, *34*, 7027–7030.
266. Maruoka, K.; Murase, N.; Yamamoto, H. *J. Org. Chem.*, **1993**, *58*, 2938–2939.
267. Mikami, K.; Terada, M.; Motoyama, Y.; Nakai, T. *Tetrahedron: Asymmetry* **1991**, *2*, 643–646.
268. Corey, E.J.; Imai, N.; Zhang, H.–Y. *J. Am. Chem. Soc.* **1991**, *113*, 728–729.

269. Corey, E.J.; Imwinkelried, R.; Pikul, S.; Xiang, Y.B. *J. Am. Chem. Soc.* **1989**, *111*, 5493–5495.
270. Bao, J.; Wulff, W.D.; Rheingold, A.L. *J. Am. Chem. Soc.* **1993**, *115*, 3814–3815.
271. Furuta, K.; Shimizu, S.; Miwa, Y. Yamamoto, H. *J. Org. Chem.* **1989**, *54*, 1481–1483.
272. Ishihara, K.; Gao, Q.; Yamamoto, H. *J. Am. Chem. Soc.* **1993**, *115*, 10412–10413.
273. Corey, E.J.; Roper, T.D.; Ishihara, K.; Sarakinos, G. *Tetrahedron Lett.* **1993**, *34*, 8399–8402.
274. Corey, E.J.; Loh, T.–P. *J. Am. Chem. Soc.* **1991**, *113*, 8966–8967.
275. Ishihara, K.; Gao, Q.; Yamamoto, H. *J. Org. Chem.* **1993**, *58*, 6917–6919.
276. Boger, D.L. *Chem. Rev.* **1986**, *86*, 781–793.
277. Boger, D.L. *Tetrahedron* **1983**, *39*, 2869–2939
278. Boger, D.L. in *Comprehensive Organic Synthesis*, Trost, B.M.; Fleming, I., Eds; Pergamon, Oxford, **1991**, Vol. 5, Chap. 4.3, pp. 451–512.
279. Weinreb, S.M.; Staib, R.R. *Tetrahedron* **1982**, *38*, 3087–3128.
280. Weinreb, S.M. in *Comprehensive Organic Synthesis*, Trost, B.M.; Fleming, I., Eds; Pergamon, Oxford, **1991**, Vol. 5, Chap. 4.2, pp. 401–449.
281. Waldmann, H. *Synthesis* **1994**, 535–551.
282. Danishefsky, S. *Acc. Chem. Res.* **1981**, *14*, 400–406.
283. Bednarski, M.D.; Lyssikatos, J.P. in *Comprehensive Organic Synthesis*, Trost, B.M.; Fleming, I., Eds; Pergamon, Oxford, **1991**, Vol. 2, Chap. 2.1, pp. 661–706.
284. Danishefsky, S.; Kerwin Jr., J.F.; Kobayashi, S. *J. Am. Chem. Soc.* **1982**, *104*, 358–360.
285. Larson, E.R.; Danishefsky, S. *Tetrahedron Lett.* **1982**, *23*, 1975–1978.
286. Danishefsky, S.J.; Larson, E.; Askin, D.; Kato, N. *J. Am. Chem. Soc.* **1985**, *107*, 1246–1255.
287. Larson, E.R.; Danishefsky, S. *J. Am. Chem. Soc.* **1982**, *104*, 6458–6460.
288. Danishefsky, S.; Bednarski, M. *Tetrahedron Lett.* **1985**, *26*, 3411–3412.
289. Danishefsky, S.J.; Larson, E.; Springer, J.P. *J. Am. Chem. Soc.* **1985**, *107*, 1274–1280.
290. Danishefsky, S.; Webb II, R.B. *J. Org. Chem.* **1984**, *49*, 1955–1958.
291. Bednarski, M.; Danishefsky, S. *J. Am. Chem. Soc.* **1983**, *105*, 3716–3717.
292. Danishefsky, S.; Bednarski, M. *Tetrahedron Lett.* **1984**, *25*, 721–724.
293. Danishefsky, S.; Bednarski, M. *Tetrahedron Lett.* **1985**, *26*, 2507–2508.
294. Danishefsky, S.J.; Maring, C.J. *J. Am. Chem. Soc.* **1985**, *107*, 1269–1274.
295. Danishefsky, S.; Harvey, D.F.; Quallich, G.; Uang, B.J. *J. Org. Chem.* **1984**, *49*, 392–393.
296. Castellino, S.; Sims, J.J. *Tetrahedron Lett.* **1984**, *25*, 4059–4062.
297. Castellino, S.; Sims, J.J. *Tetrahedron Lett.* **1984**, *25*, 2307–2310.
298. Danishefsky, S.; Maring, C. *J. Am. Chem. Soc.* **1985**, *107*, 7762–7764.
299. Bednarski, M.; Danishefsky, S. *J. Am. Chem. Soc.* **1983**, *105*, 6968–6969.
300. Bednarski, M.; Danishefsky, S. *J. Am. Chem. Soc.* **1986**, *108*, 7060–7067.
301. Quimpère, M.; Jankowski, K. *J. Chem. Soc., Chem. Commun.* **1987**, 676–677.
302. Togni, A. *Organometallics* **1990**, *9*, 3106–3113.

303. Maruoka, K.; Itoh, T.; Shirasaka, J.; Yamamoto, H. *J. Am. Chem. Soc.* **1988**, *110*, 310–312.
304. Terada, M.; Mikami, K.; Nakai, T. *Tetrahedron Lett.* **1991**, *32*, 935–938.
305. Danishefsky, S.; Chao, K.–H.; Schulte, G. *J. Org. Chem.* **1985**, *50*, 4650–4652.
306. Danishefsky, S.; Kato, N.; Askin, D.; Kerwin Jr., J.F. *J. Am. Chem. Soc.* **1982**, *104*, 360–362.
307. Danishefsky, S.; Larson, E.R.; Askin , D. *J. Am. Chem. Soc.* **1982**, *104*, 6457–6458.
308. Danishefsky, S.J.; DeNinno, M.P.; Chen, S.–h. *J. Am. Chem. Soc.* **1988**, *110*, 3929–3940.
309. Savard, J.; Brassard, P. *Tetrahedron Lett.* **1979**, 4911–4914.
310. Savard, J.; Brassard, P. *Tetrahedron* **1984**, *40*, 3455–3464.
311. Midland, M.M.; Graham, R.S. *J. Am. Chem. Soc.* **1984**, *106*, 4294–4296.
312. Midland, M.M.; Koops, R.W. *J. Org. Chem.* **1990**, *55*, 5058–5065.
313. Midland, M.M.; Afonso, M.M. *J. Am. Chem. Soc.* **1989**, *111*, 4368–4371.
314. Danishefsky, S.J.; Pearson, W.H.; Harvey, D.F.; Maring, C.J.; Springer, J.P. *J. Am. Chem. Soc.* **1985**, *107*, 1256–1268.
315. Danishefsky, S.J.; Myles, D.C.; Harvey, D.F. *J. Am. Chem. Soc.* **1987**, *109*, 862–867.
316. Danishefsky, S.J.; Selnick, H.G.; Zelle, R.E.; DeNinno, M.P. *J. Am. Chem. Soc.* **1988**, *110*, 4368–4378.
317. Danishefsky, S.J.; Pearson, W.H.; Harvey, D.F. *J. Am. Chem. Soc.* **1984**, *106*, 2456–2458.
318. Danishefsky, S.; Vogel, C. *J. Org. Chem.* **1986**, *51*, 3915–3916.
319. Waldmann, H.; Braun, M.; Dräger, M. *Angew. Chem., Int. Ed. Engl.* **1990**, *29*, 1468–1471.
320. Hattori, K.; Yamamoto, H. *Synlett* **1993**, 129–130.
321. Hattori, K.; Yamamoto, H. *J. Org. Chem.* **1992**, *57*, 3264–3265.
322. Akiba, K.–Y.; Motoshima, T.; Ishimaru, K.; Yabuta, K.; Hirota, H.; Yamamoto, Y. *Synlett* **1993**, 657–659.
323. Whitesell, J.K.; James, D.; Carpenter, J.F. *J. Chem. Soc., Chem. Commun.* **1985**, 1449–1450.
324. Remiszewski, S.W.; Yang, J.; Weinreb, S.M. *Tetrahedron Lett.* **1986**, *27*, 1853–1856.
325. Bakker, C.G.; Scheeren, J.W.; Nivard, R.J.F. *Recl. Trav. Chim. Pays–Bas* **1981**, *100*, 13–20.
326. Chapleur, Y.; Euvrard, M.–N. *J. Chem. Soc., Chem. Commun.* **1987**, 884–885.
327. Tietze, L.F.; Ott, C.; Gerke, K.; Bubach, M. *Angew. Chem., Int. Ed. Engl.* **1993**, *32*, 1485–1486.
328. Hiroi, K.; Umemura, M.; Tomikawa, Y. *Heterocycles* **1993**, *35*, 73–76.
329. Lamy–Schelkens, H.; Gioni, D.; Ghosez, L. *Tetrahedron Lett.* **1989**, *30*, 5887–5890.
330. Laschat, S.; Lauterwein, J. *J. Org. Chem.* **1993**, *58*, 2856–2861.
331. Denmark, S.E.; Cramer, C.J.; Sternber, J.A. *Helv. Chim. Acta* **1986**, *69*, 1971–1989.
332. Denmark, S.E.; Dappen, M.S.; Cramer, C.J. *J. Am. Chem. Soc.* **1986**, *108*, 1306–1307.
333. Denmark, S.E.; Kesler, B.S.; Moon, Y.–C. *J. Org. Chem.* **1992**, *57*, 4912–4924.
334. Denmark, S.E.; Marcin L.R. *J. Org. Chem.* **1993**, *58*, 3857–3868.

General Index

Ab initio
calculation, 4
Acetaldehyde
monothioacetals, 201
Acetals, pentan-2,4-diol, 188, 195, 206
Acetals, butan-1,3-diol, 210
Acetals, butan-2,3-diol, 213, 222
Acetals, 2,3-dimethyl-
butan-2,3-diol, 194
Acrylates, 11, 22, 57, 277–278, 283, 301
homochiral, 290–293
Acyl iminium ions, 220
Acylsilanes, 232, 250
Africanol, 173
Aklavinone, 212
Alanine derivatives, 293
Alder endo-rule, 269
(Alkenylimino)malonates, 73
Alkoxyallyltri-*n*-butyltin, 113, 116
homochiral, 119, 122
Alkoxyallyltrimethylsilanes, 219
Alkynyltrimethylsilanes, 205
Allenyl dienophiles, 283
homochiral, 299
Allenylsilanes, 249
homochiral, 100, 153
Allenylstannanes, 110, 154
α-Allokainic acid, 67
Allylbenzene, 26
Allylsilanes
homochiral, 98, 102, 118, 128, 161, 247
functionalized, 165, 168, 174, 200
Allylsilane ene-ketals, 221
Allyltins, trihalogenated, 146
Allylstannanes, 91
homochiral, 127
functionalized, 156, 166, 171
Aminoaldehydes, 50, 141, 313
acetals, 219
β-Amyrin, 225
Androgen, mouse, 313
Anthracyclinone, 216
homochiral, 303
Annulation reaction
[3+2], 232, 234, 244, 250, 257
[4+2], 317
Aromaticin, 259
1-Arylcyclopentenes, 28
Azimic acid, 201

Benzaldehyde dimethyl acetals, 203
Benzene sulfinyl chloride, 60
1-Benzopyrylium salts, 247
Benzyloxyaldehydes, 51, 53, 155
BINOL, 60, 77–78, 129, 301, 305, 311, 315
1,3-Bis(tributylstannyl)-2-methylenepropane,
117
1,3-Bis(trimethylsilyl)-2-methylenepropane,
198
1,3-Bis(trimethylsilyl)hexa-
2,4-diene, 111
1,3-Bis(trimethylsilyl)octa-2,6-diene,110,
197
Bornan sultam, 293
Bostrycin, 302
Brassard's diene, 313, 315
Brevicomin, 117
Butan-1,3-diol, 210
But-2-ene, 24, 44
But-3-yn-2-one, 57
homochiral
Butan-2,3-diol, 211
Bases,
Lewis, 2, 15
Biomimetic
cyclization,221
Bonding
σ-, 2
π-, 2

ε-Cadinene, 78
Calcitriol lactone, 206
Carbamates, 204
Carbohydrate derivative, 143
"Carbon-Ferrier" displacement, 218
β-Carotene, 203
Cephalomannine, 288
Chelation control, 134, 145, 147, 150, 237,
240, 313
Chiral orthoesters, 215
Chloro(phenylthio)acetate, 59
Chlorothricolide, 287
Cieplak model, 131
Cinnamic acid derivatives, 297
Cinnamyltrimethylsilane, 143
Cinnamyltriphenyltin, 107
"*Cis*-effect", ene reaction, 31
cis-rule, 269
Citronellal, 65, 77

"Compact" transitions state, 245, 255, 258, 261
Cornforth's dipolar trans. stat., 138
Cram's rule, 131, 145, 208
Crotonic acid derivatives, 293– 295, 302
N-Crotonyloxazolidone, 246
Crotyltri-*n*-butyltin, 91, 97, 104, 106, 122, 147, 200, 233, 239
Crotyl-*n*-butylchlorotins, 111
Crotyltitanium compounds, 97, 101
Crotyltrimethylsilane, 92, 196, 218, 240
Cyanotrimethylsilane, 210
α-Cyanovinylic sulfoxide, 65
Cyclobutanones, 204, 220
Cyclocitral dimethyl acetal, 203
Cyclohexene, 36
Cyclohexenyltrimethyltin, 109
Cyclooctene, 36
Cyclopenta-1,3-diene, 280, 301, 304
Cyclopentenylmethyltrimethyl-silane,140
Cyclopentenyltrimethylsilane,93, 100, 186, 232
Cycloocta-1,5-diene, 32
Cytochalasan, 300
Carbonyl
  Lewis acid complexes, 2

DAD, 280
Danishefsky's diene, 307
Daunosamide derivative, 309
Deca-1,3,9-triene derivatives, 286
11-Deoxyanthracyclinone, 233
11-Deoxyanthracyclinone, 213
6a-Deoxyerythronolide B, 312
14-Deoxyisoamijiol, 257
Deoxypillaromycinone, 279
2,3-Di-*O*-benzyl-glyceraldehydes, 54
Diels-Alder reaction, 267
1,3-Dienes, chiral, 299
Dienophiles, chiral, 289, 299
2-Diethylphosphoryloxy-penta-1,3-diene278
2,3-Dimethylbuta-1,3-diene, 276
2,3-Dimethylbut-2-ene, 26

Ene reaction, 21
Enoxysilanes, 202
Enyne trimethylsilane, 204
Epoxy alcohols, 67
Epoxy-1,4-cadinane, 287
Epoxycyclohexanones, 75
1-Ethoxy-3-trimethylsilyl-propyne, 118
Ethylidenemalonate, 238
1-Ethylidene-2-methylcyclo-pentane,38
α,β-Ethylenic acyl cyanides, 235, 249
Ethylidenemalonate, 238
2-Ethylthiobut-1-ene, 48

Felkin-Anh model, 131, 137, 151, 159, 161, 207, 239
Formaldehyde, 27, 37
3-Formyl-Δ^2-isoxazolines, 49, 140
Fucose derivatives, 309
Fumarate, 301
  homochiral, 289, 295

Geranyltri-*n*-butyltin, 107
Glycopyranoside derivatives, 215
Glycal allylation, 217
Glyoxylates, 40, 51, 80
Graveolide, 259
Glycofuranoside derivatives, 214

α-Haloaldehydes, 51
Hemibrevetoxin, 173
Heptadienylsilane, 101
Hept-5-enal derivatives, 62
Hetero Diels-Alder reaction, 307
Heterodienes, 315
Hexa-1,5-dienes, 37

α-Imino esters, 59
Inhoffen-Lythgoe diol, 224
Ipalbidine, 315
Ipsdienol, 80, 82
Iron tricarbonyl complexes, 153
Isobisabolene, 277
Isoprene, 274, 302
Isoprenyltrimethylsilane, 186, 199
Isovaleraldehyde
  ene reaction, 45
Isotope
  effect, 26, 195

Kawain, 310
α-Keto esters, 93, 104
Khusimone, 275
Ketone silyl ethers, 83
Kijanolide derivatives, 286
Kumausallene, 145

γ-Lactols, 49
Lewis G. N., 1
Lewis acids
  carbonyl comp. complexes, 2
  σ-bonding, 2, 13
  π-bonding, 2, 13
    theoretical studies, 4
    NMR studies, 8
    X-ray studies, 14
  properties, 2
  hard and soft, 282

homochiral, 77, 80, 128, 300, 310, 316

Lewis bases, 2, 15
Limonene, 22, 45
α-Lipoic acid, 210
Longiborneol derivatives, 221
Luciduline, 271
Lanthanides, shift reagents, 46, 279, 282, 298, 310, 311, 313, 317

Magelanine derivatives, 75
Maleic anhydride, 21, 24
Menthene, 22, 24
Menthone ketal, 209
Mesoxalate, 26, 56
1-Methoxycyclohexene, 47
2-Methoxypropene, 47
2-Methoxy oxazolidine, 200
α-Methyl effect, 280
2-Methylbut-2-ene, 31, 34
2-Methylcrotyltri-*n*-butyltin, 148
1-Methylcyclohexene, 38
Methylenemalonate deriv., 298
Methylhepta-2,6-diene, 28
3-Methylpent-2-ene, 36, 38, 46
3-Methylpent-3-en-2-yltri-*n*-butyltin, 108
Mevinolin analog, 211
Monothioacetals, allylation, 218

Neolemnanyl derivatives, 259
Neuraminic acid derivative, 313
Nigericin, 217
Nitroalkenes, 317
NMR
complexes, 8
Nona-1,3,8-triene derivatives, 285
Nonlinear effect, 81
Nootkatone, 255
Norbornene, 31

Octenal derivatives
ene reaction, 63
ω-Olefinic trifluoromethyl-ketones, 68
Oct-1-yne, 40
Orbital interactions, 5, 7, 31, 248, 267
Oxazolidinone derivatives, 13, 294, 302, 304
Oxepanes, 74, 172
Oxocenes, 73
Oxyallyltrimethylsilanes, 114

D-Pantolactone, 58, 292, 296
Pentadienyltins, 109, 116
Pentadienylsilane homochiral, 101
Pentadienyltrimethylsilanes, 110, 187, 236
Pentan-2,4-dione mono ketal, 198
Pent-2-ene, 42
Pent-3-en-2-yltri-*n*-butyltin, 107, 151
Pent-3-en-2-yltrimethylsilane, 152
Phenylalanine derivatives, 293
8-Phenylmenthyl esters, 66, 152, 169, 290, 316
Phenylthioacetals, 197
β-Pinene, 11, 22, 29, 33
(*E*)-Piperylene, 271
Polyolefin cyclization, 222
Polyols, 123
Prelog-Djerassi lactone, 150, 312
Proline derivatives, 293
Propargylsilanes, 204, 223, 250, 262
Propiolates, 56
Pulegol, 21
Pumiliotoxin, 72
Pulo'upone, 293
Pyrrolizinone, 205

Quassin, 278

Ribofuranose derivatives, 215
Rifamycin S, 314

Sakurai reaction, 231
β-Santalene, 299
S_E2' mechanism, 97, 257
Semi-empirical studies, 4
Siloxyvinylallene derivatives, 287
β-Silyl cation, 42
Steroidal aldehydes, 51, 54
Steroidal olefins, 41, 47, 49, 57, 225
Steroid side chain, 133, 207
Stork-Eschenmoser postulate, 185
Stratine derivatives, 212
N-Sulfinylcarbamates, 59, 316
Sulfoxides, homochiral, 298

Taxane ring system, 288
Thermal
Diels-Alder reaction, 267
ene reaction, 23
Thionofumarate, 282
Threitol derivatives, 292
α-Tri-*n*-butylstannylacetal, 209
Trichodermol, 69
1,3,ω-Triene derivatives, 287
Trimethylsilylcyclopentadiene, 283
Trimethylsilylmethyl-buta-1,3-diene,277, 309

Verrucarin derivatives, 130, 153
Veticadinol, 66
Vinylcyclopropane, 29

Vinylcyclopropyloxysilanes, 204
Vinyl phosphonates, 281
Vitamine A, 203
Vitamine K, 233

Wacker-type oxidation, 106

X-ray studies, 14, 292

Yamamoto's trans. state, 97, 133

Zimmerman-Traxler trans. state, 95
Zincophorin, 218, 314

Lewis Acids Index

$AgSbF_6$, 13
$AlBr_3$, 10, 105
$AlCl_3$, 2, 9, 22, 31, 34, 37, 56, 91–92, 102, 105, 124, 136, 140, 162, 186, 190, 197, 203, 233, 236, 270, 272, 274, 276–277, 281, 283, 285, 289
$AlCl_2$(OMenthyl), 301

$Ba(ClO4)_2$, 13
$B(OAc)_3$, 300
BBr_3, 6
BCl_3, 6, 10, 190, 282
Boron chiral complex, 306
BF_3, 2, 5, 9, 14, 22, 39, 66, 71–72, 74, 76, 93, 97, 102, 104, 106–107, 110, 113, 115, 117, 121, 124, 127, 132, 135, 137, 143, 145, 151, 153, 155, 158, 163, 166, 169, 171, 173, 175, 187, 189, 193, 200, 213, 215, 217, 221, 233, 244, 247, 251, 261, 269, 271, 273, 277, 279, 282, 284, 299, 308–309, 312–313
Boron-H (BINOL), 303
Boron-H-Tartaric acid derivative, 306–307
Boron-OMe(diol chiral), 303
$B(OPh)_3$, 316

Cu-bis-imino complex, 304
$Cu(OTf)_2$, 282

DAD, 280

$EtAlCl_2$, 10, 27, 50, 56, 58, 68, 76, 164, 174, 205, 232, 252, 254, 257, 261, 278, 284, 285, 286, 292, 297, 315
Et_2AlCl, 65, 67, 69, 102, 114, 166, 282, 286, 288, 294–296, 313
Et_2AlI, 47
$Eu(fod)_3$, 282, 298, 310, 317
$Eu(hfc)_3$, 311, 313

$FeCl_3$, 66, 102

$Hg(OTf)_2$, 29

$LiClO_4$, 13, 269, 300

MAD, 318
MAPH, 38
MABR, 70

Me_3Al, 10, 40, 280
Me_2AlCl, 10, 27, 29, 37, 39–40, 45, 47-48, 51, 57, 62, 70, 72, 246, 287, 288, 294–295
$MeAlCl_2$, 61–62, 141, 287
MeAlCl(OTf), 44
MeAlBINOL, 80
Me_2Al-ethylenediamine, 305
Me_2AlOTf, 44
$MeAlOTf_2$, 44, 51
$MeTiCl_3$, 12
$MgBr_2$, 15, 51, 110, 124, 137, 147, 149, 154, 156, 313–315
$Mg(ClO_4)_2$, 13

$NaClO_4$, 13

Rh(I)-complex, 64

$SbCl_5$, 13
$SnBr_4$, 123
$SnCl_4$, 2, 9, 12, 26, 28, 40, 42, 44, 50–51, 54, 59, 61, 65, 67, 69, 73–74, 93, 102, 104, 122, 124, 132, 135, 140, 149, 152, 162, 164, 169, 173, 175, 185, 190, 192, 196, 200, 205, 212, 219, 222, 224, 239, 261, 273, 275, 278, 316

Tartaric acid, borane derivatives, 128–129
$TiBr_2$BINOL, 81
$TiCl_4$, 2, 6, 9, 12, 15, 34, 43, 59, 68, 76, 92, 94, 97–98, 107, 109, 113, 115, 118, 124, 128, 132, 134, 136, 141, 145, 149, 163, 165, 168, 171, 175, 186, 188, 191, 194, 197, 200, 202, 204, 207, 209, 217, 219, 221, 223, 231, 232, 234, 237, 239, 244, 249, 252, 260, 282, 291–292, 295–296, 298, 314, 316
$TiCl_3$(0R), 13
$TiCl_3$(O*i*-Pr), 51, 219
$TiCl_2(Oi\text{-}Pr)_2$, 50, 53, 78, 193, 200, 203, 206, 209, 225, 274, 291, 299
$TiCl_2$BINOL, 79–80, 82, 129–130, 305, 311
TiCl(O*i*-OPr)BINOL, 301
$TiCl(Oi\text{-}OPr)_2$BINOL, 130–131
$TiCl_2$(diol chiral), 302
$TiCl_2$(hydrobenzoin), 301
$Ti(Oi\text{-}OPr)_4$, 67
Ti-chiral tetrahydroxy-binaphthol, 304
Ti-chiral complex, 306

TMS-OTf, 73, 114, 125, 167, 189, 190–191, 198, 200–201, 203–204, 214, 216, 219–220, 260
TBDMS-OTf, 317

$VOCl_2(OR)$, 10
$VO(hfc)_3$, 311
$VO(OET)_3$, 67

$Yb(fod)_3$, 46, 279

ZnBINOL, 77
$ZnBr_2$, 2, 64, 149, 214–215, 218, 278$ZnCl_2$, 60, 124, 144, 205, 274, 281, 284, 299, 308–309, 312
ZnI_2, 57, 64
$ZrCl_4$, 103